Reviews in Fluorescence 2004

Reviews in Fluorescence 2004

Edited by

CHRIS D. GEDDES
Institute of Fluorescence
University of Maryland Biotechnology Institute
Baltimore, Maryland

and

JOSEPH R. LAKOWICZ
Center for Fluorescence Spectroscopy
University of Maryland
Baltimore, Maryland

Kluwer Academic/Plenum Publishers
New York, Boston, Dordrecht, London, Moscow

ISBN: 0-306-48460-9 (hardback)
ISBN: 0-306-48672-5 (eBook)

233 Spring Street, New York, N.Y. 10013

http://www.wkap.nl/

10 9 8 7 6 5 4 3 2 1

A C.I.P. record for this book is available from the Library of Congress

Printed in the United States of America

PREFACE

The early 1990's saw the beginning of a rapid growth phase for fluorescence spectroscopy. Instrumentation became more capable and user friendly, and fluorophore chemistry more versatile. The power of fluorescence spectroscopy would soon become a major tool for the forthcoming revolutions in structural biology and biotechnology.

During the past 13 years fluorescence has been significantly transformed from a methodology practiced in a few specialized laboratories, to one widely practiced in many laboratories encompassing a vast spectrum of scientific disciplines, such as gene expression, flow cytometry and diagnostics, to name but just a very few. Joseph R. Lakowicz foresaw many of these exciting changes in 1991, and responded by founding the *Journal of Fluorescence*. Since that time the *Journal of Fluorescence* has substantially grown and under the leadership of Chris D. Geddes is the dominant repository for peer reviewed original fluorescence research articles in the world. In addition, the number of workers employing fluorescence methodology has also substantially grown. Subsequently the new *Who's Who in Fluorescence Annual Volume*, launched last year, now connects 350 workers from no fewer than 35 countries, disseminating both specialty and contact details.

Continuing to respond to the ever changing face of fluorescence we are pleased to announce the launch of the new *Annual Reviews in Fluorescence*. This new volume addresses the requirement for detailed fluorescence review articles, both reflecting and archiving the yearly progress in fluorescence. In this first volume we have invited notable scientists from around the world to progress their work in fluorescence with applications including Molecular Thermometers, Saccharide Sensors, Semiconductor Quantum Dot Nanoassemblies and the application of luminescence to Information Processing, to name a few. We hope you find this volume a useful resource and we look forward to receiving any suggestions you may have in the future.

Finally we would like to thank Caroleann Aitken and Kadir Aslan for typesetting, and Mary Y. Rosenfeld for administrative support.

Chris D. Geddes
Joseph R. Lakowicz

CONTENTS

Reviews in Fluorescence 2004

ARENEDICARBOXIMIDES AS VERSATILE BUILDING BLOCKS FOR FLUORESCENT PHOTOINDUCED ELECTRON TRANSFER SACCHARIDE SENSORS

Michael D. Heagy*

1. INTRODUCTION

Within the last decade a considerable amount of effort has been directed towards the detection of saccharides by fluorescent chemosensors.[1-3] Such studies have shown that the response which signals an interaction between carbohydrate and receptor is frequently communicated by changes in fluorescence intensity either through chelation enhanced-quenching (CHEQ) or chelation-enhanced fluorescence (CHEF).[4,5] While significant advances continue in the areas of chemosensors for saccharides, invariably, one or more of the requisite conditions necessary for biologists to measure these analytes goes unmet. For carbohydrate measurements, conditions such as neutral pH as well as selectivity in an aqueous testing environment are essential. In addition to these physiological requirements, for effective photoinduced electron transfer (PET), the signaling properties of the chemosensor must also meet certain criteria. Three critical prerequisites of the fluorescent sensor that must be satisfied for carbohydrate recognition have been outlined by Shinkai: strong fluorescence intensity, large pH dependent change in I_{max}, and shift of the pH-I_{max} profile to lower pH region in the presence of saccharides.[6]

With few exceptions,[7-9] current designs in saccharide-sensors have relied exclusively upon a common anthracene or similar PAH-based fluorophore as the reporting unit, with synthetic modification of the phenylboronic acid or methylene spacer that binds these two groups together. The consequences of such unrelieved hydrophobicity often necessitate the addition of organic cosolvents to increase the solubility of the sugar sensor.[10] In order to overcome this limitation, we have been developing boronic acid saccharide sensors which utilize more polar reporting groups based upon arenecarboximide chromophores.[11,12]

---

*Michael D. Heagy, New Mexico Institute of Mining and Technology, Socorro, New Mexico USA, 87801. Department of Chemistry, Jones Hall Rm 259; Tel: 505.835.5417, Email: mheagy@nmt.edu

Derivatives of naphthalimide display useful photochemical properties and have been utilized as photoactivatable DNA-cleaving agents, fluorescent tags, and receptor antagonists.[13-19] Recently, *N*-phenylnaphthalimides have been found to exhibit dual fluorescence when appropriately substituted at both the *N*-phenyl ring and naphthalene π–system. These compounds display dual luminescence with two clearly resolved emission bands in the visible region from a locally excited state and a strongly red-shifted band emitted by an internal charge transfer state.[20] Such unique optical properties prompted us to explore *N*-phenyl derivatives of naphthalimide as fluorescent components for saccharide sensors. Here we present a rare case where a large family of saccharide sensors is synthesized from a single source reaction type using commercially available materials – arenedicarboxylic anhydrides and aminophenylboronic acids. This versatile class of chemosensors possess boronic acid receptors for saccharide complexation and several arenedicarboximide as well as tetracarboxomide fluorophores addressable at different wavelengths. The saccharide sensing properties of this family were examined from a fluorophore-spacer-receptor design where the spacer component has been characterized as a virtual C_0 spacer via *N*-phenyl to arenedicarboximide linkage.[21] The nature of their optical properties was found to be dependent on the position of the boronic acid group, the various substituents groups located on the fluorophore and internal conversion pathways between phenylboronic acid and naphthalene moiety.

2. ARENECARBOXIMIDES IN PET COMPONENTS

Arenecarboximides in either the di- or tetracarboximide form have been shown to be good PET acceptors.[22] Pyromellitic anhydride (**1**) shows an affinity for electrons with an E_{red} = -0.55 V which is only slightly lower than that of benzoquinone E_{red} = -0.51 V.[23] Whereas two reduction potentials have been observed for 1,4,5,8-napthalene tetracarboxylic diimide (**2**) at -0.385 and -0.695 V.[24] Naphthalene imides (**3** and **4**) have less electron accepting E_{red} values than their tetracarboxylic derivatives, such as the case of 1,8-naphthalenedicarboxylic imide, at -1.31 V.[25]

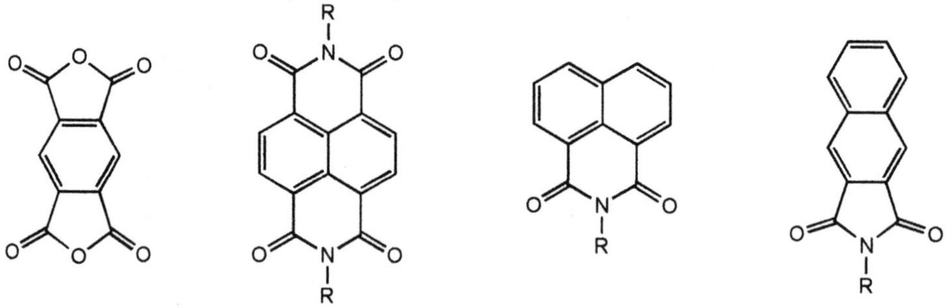

1 **2** **3** **4**

We begin this review with the 1,8-naphthalene dicarboximides as these systems represent the most simple monoboronic acid saccharide sensors. In particular, we probe their fluourescence features through a number of substituted versions at either the 3 or 4-position of the naphthalene ring. Next, we investigate the differences between positional isomers of this system from the 2,3-naphthalene dicarboximides as well as placement of the boronic acid group. Finally, we examine bis-boronic acid systems derived from tetracarboxylic anhydrides of **1** and **2** and compare the effects of benzene vs naphthalene platforms with respect to their photophysical properties and saccharide selectivity.

3. SUBSTITUENT EFFECTS ON MONOBORONIC ACID DERIVATIVES OF N-PHENYL-1,8-NAPHTHALENEDICARBOXIMIDES

A striking feature of the *N*-phenylnaphthalimides (compared to N-alkyl derivatives) is the presence of two emitting states, capable of yielding both short wavelength (SW) and long wavelength (LW) fluorescence.[20] Recent studies indicate that the geometry of the SW state is similar to that of the ground state whereas twisting of the phenyl group toward a coplanar conformation is thought to form the LW state. The identification of the first and second excited singlet states is supported by Hückel MO calculations adapted from Berces *et. al* and carried out for the parent *N*-phenylnaphthalimide.[20] The calculated electron distributions for the relevant molecular orbitals are given in Figure 1.

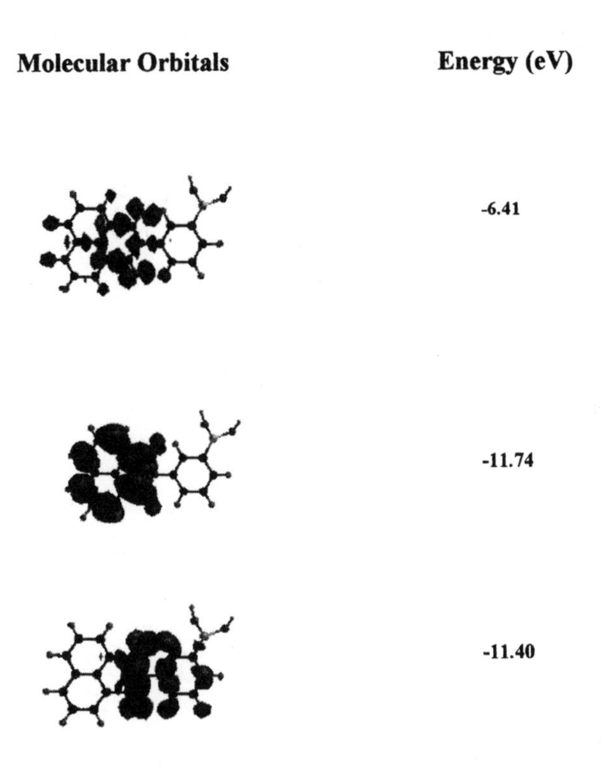

Figure 1. Relevant molecular orbitals of *N*-phenyl naphthalimide scaffold (adapted from reference 20).

Electron transfer from the HOMO to the LUMO occurs from the naphthalene moiety to the π^* orbitals of the carbonyl groups, in which the electrons of the aniline group do not participate. This excited state transition is expected to relax via radiative decay as SW emission. A comparison between the electron distributions in HOMO-1 and LUMO orbitals, however indicates that electron density shifts from the aniline moiety to the π^*

orbitals of the carbonyl group. This S_2 state reverses the direction of the dipole moment in relation to the ground state thus giving rise to charge-transfer character (ICT) states responsible for LW fluorescence. In addition, substituent groups with electron releasing and electron withdrawing properties on the naphthalene moiety have been shown to significantly affect the photophysical behavior of these fluorophores.

3.1. Synthesis

To explore these effects, we synthesized a series of sensors **5-10**, using a variety of different fluorophores derived from 1,8-naphthalenedicarboxylic anhydride. Sensors **5**, **6**, **10** were prepared in single step reactions from commercially available naphthalic anhydrides. 3-aminophenylboronic acid was selected for the synthesis of all six sensors because of its low cost and commercial availability. The other sensors were obtained by reduction (**6** to **7**), acetylation (**7** to **8**) and introduction of methoxy substitution of *N*-(4'-bromo-1', 8'-naphthaloyl)-3-aminophenyl boronic acid to give **9** (Figure 1). Sulfo- and amino derivatives of naphthalic anhydride were prepared to increase solubility in aqueous solvent systems. Nitro, methoxy and acetamido were prepared for their potential charge transfer properties.

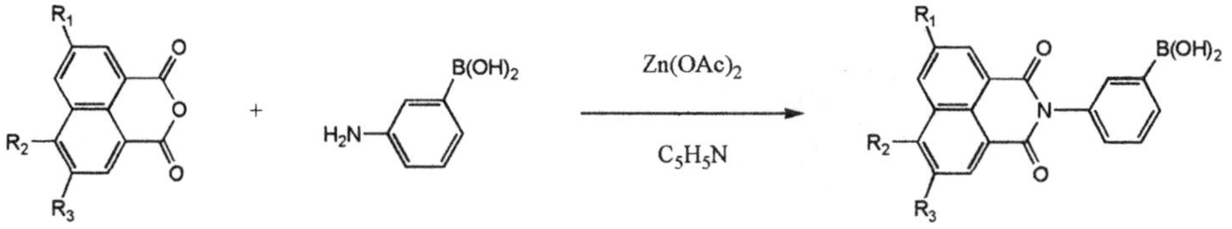

Scheme 1.

Table 1. Synthesis of 3-Phenylboronic acid-1,8-naphthalenedicarboimide derivatives

1,8 naphthalic anhydride			yield
5 R_1=H	R_2= H	R_3=H	50%
6 R_1=H	R_2= H	$R_3=NO_2$	67%
7 R_1=H	R_2= H	$R_3=NH_2$ (Reduced from **2**)	87%
8 R_1=H	R_2= H	$R_3=CH_3CONH$, (Acetylated from **3**)	91%
9 R_1=H	$R_2=OCH_3$	R_3=H,	40%
10 $R_1=SO_3K$	$R_2=NH_2$	$R_3=SO_3K$	91%

Table 2 summarizes the photophysical properties of the six compounds synthesized for this study. Sensors **6** and **7** display significant Stokes shifts and sensors **9** and **10** show relatively high quantum yields. In an effort to explain their optical features, we calculated the energy gap between HOMO to LUMO and HOMO-1 to LUMO for all six of the compounds synthesized in this study using the extended Hückel software from Chem 3D version 5.0.

Table 2. Photophysical properties of 3-Phenylboronic acid-1,8-naphthalimide and derivatives.

Entry	Sensor	λ_{ex}, nm	λ_{em}, nm	ϕ_F	ΔE HOMO-LUMO (eV)
5	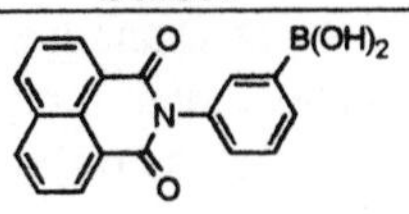 O, O, N, $B(OH)_2$	345	400	0.010	3.59
6	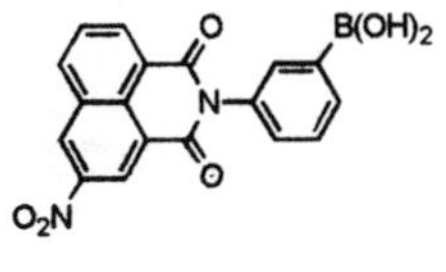 O, O, N, $B(OH)_2$, O_2N	337	430/550	0.006	3.68
7	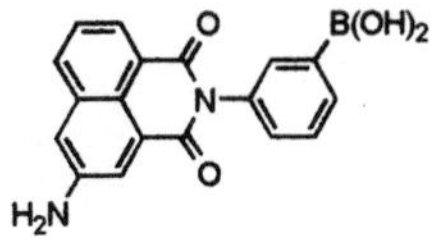 O, O, N, $B(OH)_2$, H_2N	347	581	0.017	3.57
8	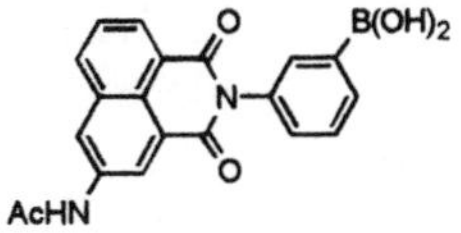 O, O, N, $B(OH)_2$, AcHN	349	407	0.014	3.55
9	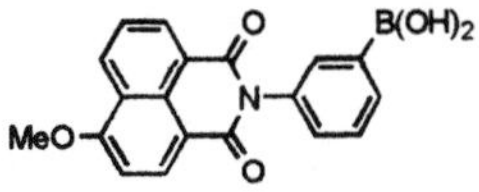 O, O, N, $B(OH)_2$, MeO	363	440	0.407	3.42
10	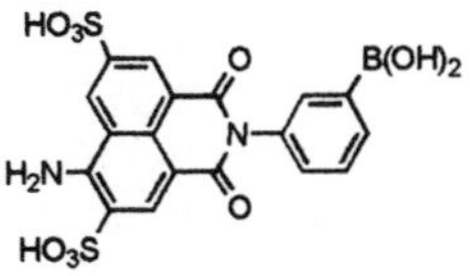 O, O, N, $B(OH)_2$, HO_3S, H_2N, HO_3S	429	534	0.165	2.89

As shown in table 2, sensors **6** and **7** display significant Stokes shifts and sensors **9** and **10** show relatively high quantum yields. Sensors **6** and **9** display the lowest and highest quantum yield, respectively. In addition sensor 2 displays two emission bands which are characteristic of *S*1 and *S*2 states. With the exception of entry **6**, the experimentally determined HOMO-LUMO energy gaps are in agreement with similarly substituted fluorophores.[26,27]

3.2. Photoelectrochemical Model

Photoinduced electron transfer has been widely used as a tool in the design of fluorescent sensors for saccharides. These fluorescent sensors are based on the boronate ester complex between carbohydrate and boronic acid receptor and typically display optical signals through changes in fluorescence intensity either through chelation enhanced-quenching (CHEQ), or chelation-enhanced fluorescence (CHEF). As with fluorescent chemosensors for ion detection, carbohydrate sensors usually consist of three parts; fluorophore, spacer and receptor. Fluorescent probes without spacers are far fewer in number and have been classified as orthogonal systems. Current examples include twisted biaryls where the π molecular orbitals of the fluorophore and receptor are separated due to steric hindrance between their σ frameworks.[21] Because orthogonality of their molecular orbitals is concomitant with geometric orthogonality, the signaling behavior of these systems has been interpreted as a PET process in a "fluor-spacer-receptor" assembly with a virtual C_0 spacer.

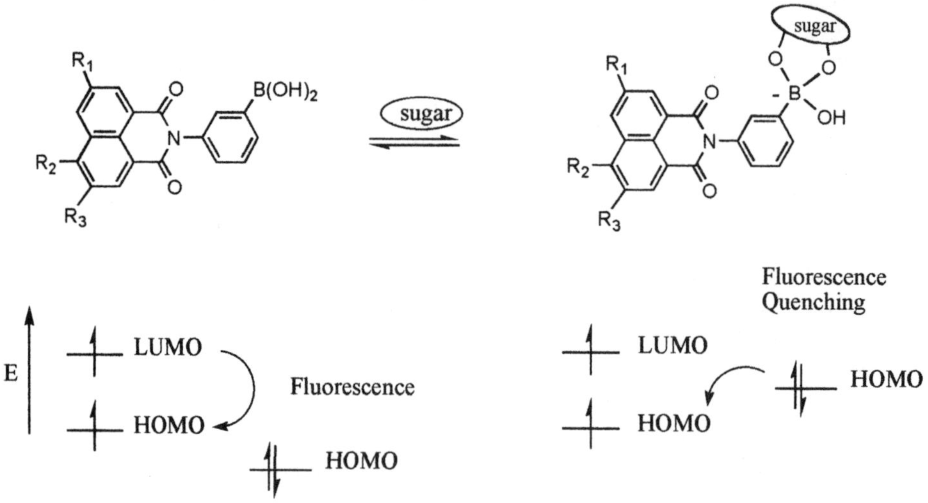

R1 B(OH)2 O N O R2 R3 sugar R1 O N O R2 R3 sugar O O - B OH

Fluorescence Quenching

E LUMO HOMO Fluorescence HOMO LUMO HOMO HOMO

Figure 2. Photochemical model used in describing PET fluorescence quenching and interaction between saccharide complexation.

Here, our model deviates from the photophysical model to allow for conformational changes associated with sugar binding and treats the phenylboronic acid MO as a separate π system from the naphthalene MO.[28] X-ray crystal studies find that the plane of the *N*-phenyl group makes a dihedral angel of 69.4° with the plane of the naphthalimide.[29] From this description, we have a C_0 spacer design that places the naphthalimide group as fluorophore and phenylboronic acid moiety (receptor) in a non-planar arrangement. Because saccharide complexation alters the oxidation state of phenylboronic acid, our model proposes a photoelectrochemical mechanism similar to one generally accepted for ion-responsive probes. In this case, the boronic acid receptor has an sp^2-hybridized boron atom with a trigonal planar geometry in the unbound form. Upon complexation with saccharides, a boronate anion forms which possesses an sp^3-hybridized boron with tetrahedral geometry. This saccharide complex alters the orbital energy of the HOMO for the phenyl boronic π-system via occupation of the next highest molecular orbital as shown in Figure 2. Fluorescence quenching upon saccharide binding occurs since the electron transfer process (PET) from phenyl group to naphthalic imide becomes energetically favorable. This photoelectrochemical model is expected to agree with factors set forth in the Weller equation, where the free energy of electron transfer is given by $\Delta G_{ET} = -E_{S.fluor} - E_{red.fluor} + E_{ox.receptor} - E_{ion\ pair}$.[30,31] Thus any decrease in the singlet energy component ($E_{S.fluor}$) is expected to display a reduced PET quenching response.

3.3. Saccharide Complexation Results

From the model described above, we predict chelation enhanced quenching (CHEQ) to be the predominant signaling pathway. For chemosensors **5** and **6**, this fluorescence quenching mechanism is supported by a pH-dependent change in I_{max} and a significant shift of the pH- profile to lower pH- I_{max} region in the presence of saccharides. The pH-fluorescence profiles of **5** and **6** obtained from buffered solution are shown in Figures 3a and 4a, respectively.

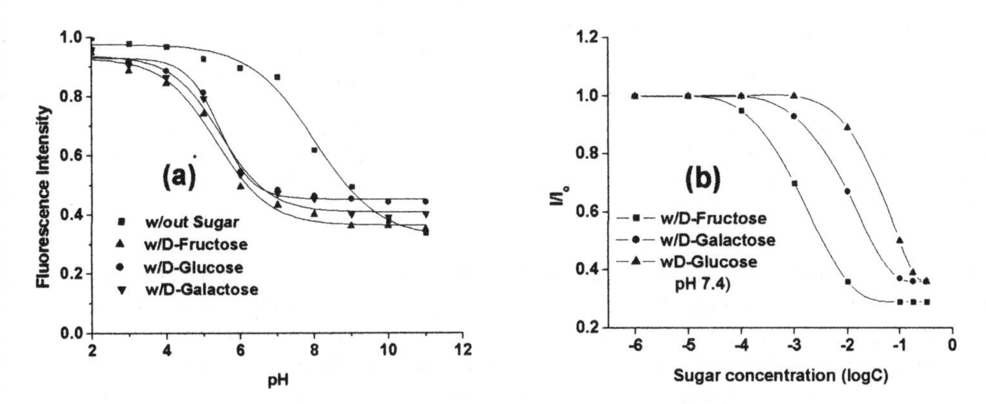

Figure 3. a) Fluorescence intensity versus pH profile for sensor **1** (3.0×10^{-5}M) measured in 1% DMSO (v/v) buffer solution, saccharides (0.05M) at 25°C. (b) Relative fluorescence as a function of saccharide concentration for sensor **1** measured in 1% DMSO (v/v) phosphate buffer (100mM), pH 7.4 at 25°C (λ_{ex} = 345nm, λ_{ex} = 400nm)

From figure 3a, sensor **5** displays one emission band at 400 nm and a pK value calculated to be 7.7. In the presence of fructose, ester formation between sensor **5** and fructose was observed as a function of pH to obtain a pK_a of hydroxyboronate nearly 2 pK_a units lower than unbound sensor. As shown in figure 4a, the higher degree of quenching in sensor **6** compared to **5** is attributed to the greater E_{red} values which are characteristic of nitroaromatic compounds. The observed effect from this substituent leads to enhanced PET quenching relative to the parent sensor **5**. Similar to sensor **5**, the intensity changes recorded at 430 nm for **6** correlate with a mono-acid titration curve and pK_a of 8.0. The acidity of the boronic acid group increases in the presence of glucose, giving a pK_a value of 6.6

Two unique features appear in both saccharide sensing systems relative to several other early saccharide probes at this time. Specifically, a minimum amount of organic cosolvent is required for these probes and secondly, both operate at relatively neutral pH. Such properties are attributed to the polarity of the naphthalimide platform as well as to the electron withdrawing nature of the imide functionality. In addition, sensor **6** shows two emission bands (430/550 nm) found in figure 5 from a single excitation wavelength (337 nm). While this probe has a significantly lower quantum yield, the two well resolved bands indicate that the dual fluorescence signal is readily detected and may find use in practical applications such as ratiometric detection.

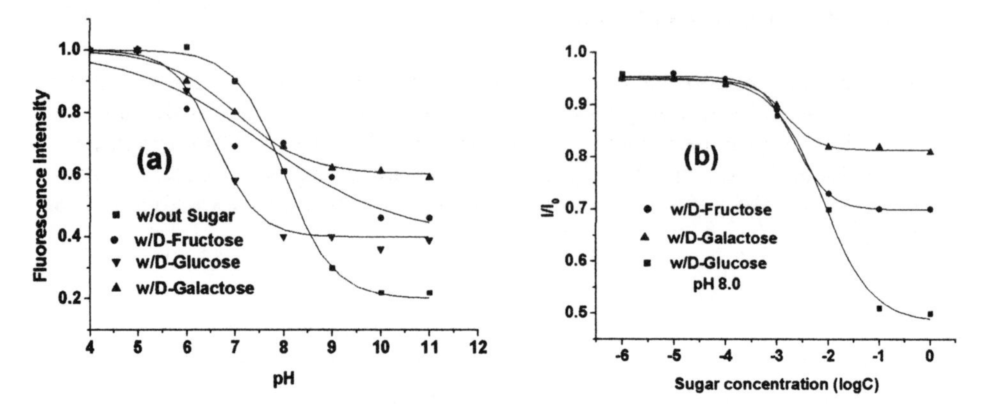

Figure 4. (a) Fluorescence intensity versus pH profile for sensor **2** (3.0×10^{-5}M) measured in 1% DMSO (v/v) buffer solution, saccharides (0.05M) at 25°C. (b) Relative fluorescence as a function of saccharide concentration for sensor **2** (3.0×10^{-5}M) measured in 1% DMSO (v/v) phosphate buffer (100mM), pH 7.4 8.0 at 25°C. (λ_{ex} = 337nm, λ_{ex} = 430nm and 550nm)

Next, we examined the selectivity of these sensors to common monosaccharides at neutral pH conditions. Figure 3b shows the relative fluorescence of **5** at 400 nm as a function of carbohydrate concentration. The decrease in fluorescence intensity (I in the presence of saccharide/ I_0 in the absence of saccharide) for this series is about 0.25. The selectivity of sensor **1** compares with other monoboronic acid probes and shows the greatest association constant with D-fructose. Based on a 1:1 complex (obtained from Job's plot), the dissociation constant was found to be 1 mM for fructose while a higher K_d of 250 mM was calculated for glucose.

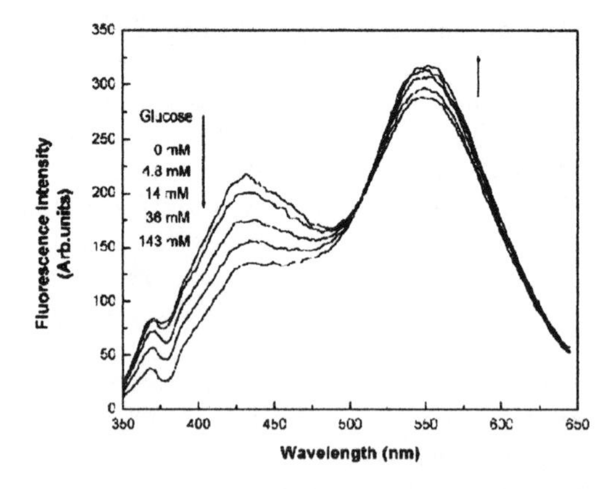

Figure 5. Fluorescence response for **6** (3.0×10^{-5}M) with glucose (0.05 M) measured in 1% DMSO (v/v) phosphate buffer (100mM), pH 8.0 at 25°C (λ_{ex} = 337nm).

As the graph demonstrates in figure 6, sensor **7** displays marked fluorescence quenching response in the lower pH region between pH 3 to 5. Given this sensor's dynamic response to galactose at low pH we titrated **7** with monosaccharides at pH 4. Here, the pH modulation of **7**'s ammonium group to amino is expected to alter its E_{red} values. Between pH of 3-5, the ammonium ion predominates in aqueous solution but as the ammonium group is converted to the free amine, less effective PET quenching occurs.

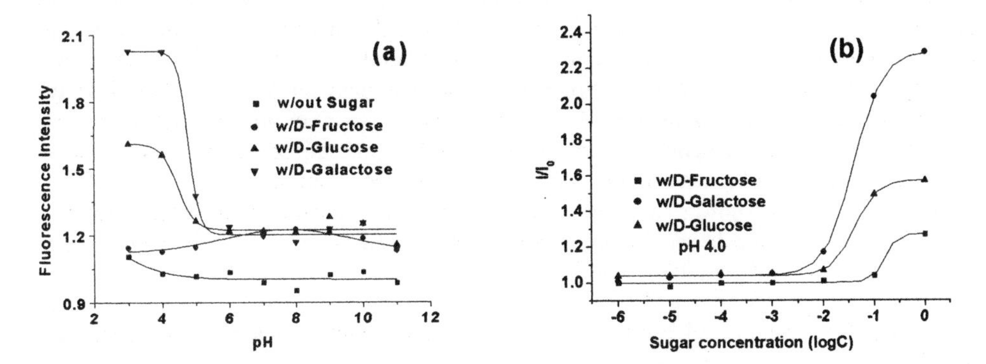

Figure 6: (a) Fluorescence intensity versus pH profile for **7** (3.0×10^{-5}M) measured in 1% DMSO (v/v) buffer solution, saccharides (0.05M) at 25°C. (b) Relative fluorescence as a function of saccharide concentration for sensor **7** (3.0×10^{-5}M) measured in 1% DMSO (v/v) phosphate buffer (100mM), pH 4.0 at 25°C. (λ_{ex} = 347nm, λ_{ex} = 581nm)

Sensor **10** shows the least effective PET quenching of the four sensors shown (figure 7). As evidenced from the HOMO-LUMO energy gaps in table 1, the 4-amino-3,6 -disulfo substituents raise the HOMO of the fluorophore which is expected to reduce PET from hydroxyboronate anion. A molecular modeling analysis which appears in the next section provides a possible rationale for the dynamic response to galactose relative to fructose and glucose.

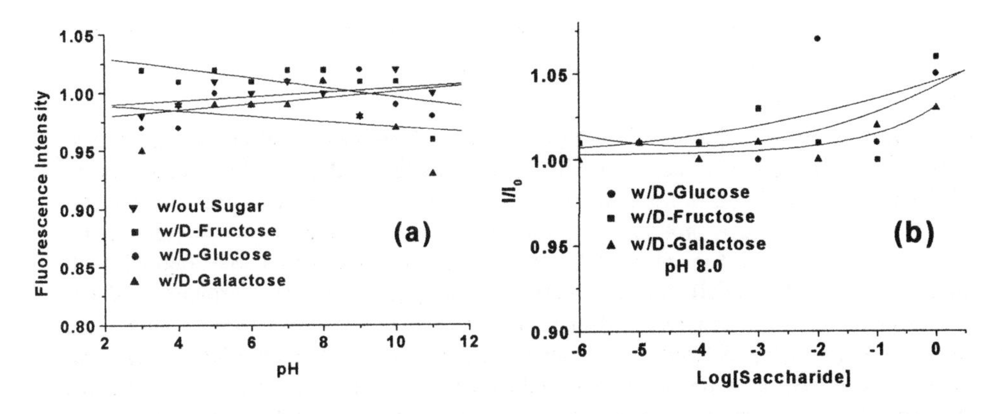

Figure 7: (a) Fluorescence intensity versus pH profile for **10** (3.0×10^{-5}M) measured in 1% DMSO (v/v) buffer solution, saccharides (0.05M) at 25°C. (b) Relative fluorescence as a function of saccharide concentration for sensor **6** (3.0×10^{-5}M) measured in 1% DMSO (v/v) phosphate buffer (100mM), pH 8.0 at 25°C. (λ_{ex} = 349nm, λ_{ex} = 407nm)

Although sensors **5-10** displayed unique optical properties such as **9** with good quantum yield ($\phi_F = 0.407$), **10** with moderate quantum yield ($\phi_F = 0.165$) and enhanced solubility in water, none of these sensors reported significant optical changes in the presence of saccharides. Results similar to those of **10** were also observed for sensors **8** and **9**. These results are consistent with a less negative ΔG_{ET} term from the Weller equation as electron donating substituent groups located on the naphthalene ring are expected to decrease the energy gap ($\Delta G_{S,fluor}$) between HOMO and LUMO. For the saccharide titrations with **3**, a fluorescence increase was observed only at high concentrations of sugar. While such low binding reflects the reduced affinity of trigonal boronic acid for sugars, a significantly enhanced fluorescence response for **7** with galactose is observed in the titration curve. To explain this unusual response, we compared MMX minimized geometries of these complexes using both tetrahedral and trigonal versions of boron.[32] Using the predominant form of boron (tetrahedral) and free amino substituent in basic solution, Table 2 shows the calculated torsional angles along the biaryl bond for the known ester bonds of furanoid and pyranoid geometries.

Table 3: MMX minimized N-imide/C-phenyl dihedral angles of **6** and **7**: saccharide esters[33]

Saccharide Complex	Probable di- or triols in ester bonds	Dihedral angle (± 0.5°) **6(-NO$_2$)** tetrahedral boron	**6(-NO$_2$)** trigonal boron	Dihedral angle (± 0.5°) **7(-NH$_2$)** tetrahedral boron	**7(-NH$_3^+$)** trigonal boron
D-Fructose	β-2,3,6-furanoid	61°	-	60°	-
	β-2,3-furanoid	58°	59°	58°	57°
	β-3,6-furanoid	57°	61°	60°	57°
D-Glucose	1,2-furanoid	58°	59°	59°	57°
	1,2-pyranoid	57°	61°	60°	56°
D-Galactose	3,4-furanoid	59°	62°	61°	55°
	4,6-pyranoid	56°	59°	60°	55°

As indicated in the second to last column for complexes of tetrahedral boron, minimized geometries display little variance (±2°) in dihedral angle for all possible boronate esters of 7. Under acidic conditions, the geometries were minimized with trigonal boron as the organizational node and a protonated amino substituent (last column). A comparison between dihedral angles for these complexes reveals a reduced torsional angle for all complexes with both possible galactose complexes giving the closest interplaner angle at 55°. Although this 5-6° angular difference between biaryl systems appears small, the fluorescence intensity of *N*-aryl carboximides has been shown to be particularly sensitive to dihedral angle. Because the sensor 7:galactose complex shows the largest fluorescence enhancement followed by glucose and fructose, these differences in torsional angles may account for the fluorescence titration data trends observed in Figure 7. This fluorescence intensity dependence on torsional angle agrees with findings for all six sensors where smaller dihedral angles were calculated for trigonal boron-containing probes relative to their tetrahedral boronate analogs. Finally, it should be mentioned that galactose did not give the smallest torsional angle when this conformational analysis was carried out with the other five probes.

An extension of this conformational analysis may account for the unusually large fluorescence change observed with sensor **6** and glucose. Table 3 also provides a comparative modeling study between nitro derivatives of **6** in the trigonal and tetrahedral state of phenylboronic acid. Although these angular differences are less pronounced than those found in the case of sensor **7**, glucose shows a 4° increase in dihedral angle in its conversion from trigonal to tetrahedral boron. Thus the increase in dihedral angle upon saccharide binding is expected to reduce fluorescence intensity. Again, the sensitivity of *N*-arylnaphthalimides to twist angle may be responsible for the dynamic response observed for glucose in pH 8.

3.4. Conclusion

In conclusion, a series of monoboronic acid fluorescent sensors were synthesized based on the *N*-phenyl naphthalimide fluorophore, in most cases with one step. These carbohydrate sensors exhibit interesting photochemical and photophysical properties which provided useful insights into the relatively limited area of C_0 design chemosensor. While many saccharide probes include a spacer group, such as Wulff's use of a benzylic amine spacer to detect in neutral conditions, sensor **5** gives a large fluorescence response at pH 7.4 with no spacer.[35] To the best of our knowledge, sensor **5** displays the largest CHEQ response for a monotopic receptor in aqueous solution. The large fluorescence response lies within detection requirements for fructose but remains less sensitive to physiological level of glucose. Whereas several bisboronic acid sensors have been developed to provide a greater fluorescence response of glucose over fructose, sensor **6** represents the first monoboronic acid sensor to display higher sensitivity for glucose over fructose.[10, 36-39] This anomalous chelation-enhanced quenching from **6** opposes the expected signal response involving monosaccharides. In this case, the fluorescence data show that the optical change which results from fructose complexation is weak relative to glucose. From the obtained K_d values, it appears that the binding affinity is not proportional to the observed optical change for this saccharide sensor. These findings suggest that other competing factors may be simultaneously operating, such as conformational dynamics between the phenylhydroxyboronate:saccharide complex and its influence on the excited state. Such geometrical changes are also attributed to sensor **7**'s unusual response which is specific to galactose. Probes **5-10** gave poor signal response upon titration with monosaccharides, however their substituent effects lend support to the chelation enhanced quenching model proposed for these systems. Based on this model a reduced PET effect is expected to be more pronounced in sensors **8** to **10**. In addition, the large Stoke shift and bright fluorescence that these compounds display may be useful in other applications such as saccharide labeling experiments.

4. POSITIONAL ISOMERS OF NAPHTHALENE DICARBOXIMIDES

4.1. Ortho-substituted monoboronic acid sensor

To observe the effects of positional isomers with these monoboronic acid sensors, **11** was prepared from 2-aminophenylboronic acid and 1,8-naphthalic anhydride. Sensor **11** possesses similar photophysical properties to isomer **5** with regard to its excitation and emission wavelengths. Any differences in its CHEQ response are therefore attributed to the placement of the boronic acid group.

Scheme 2

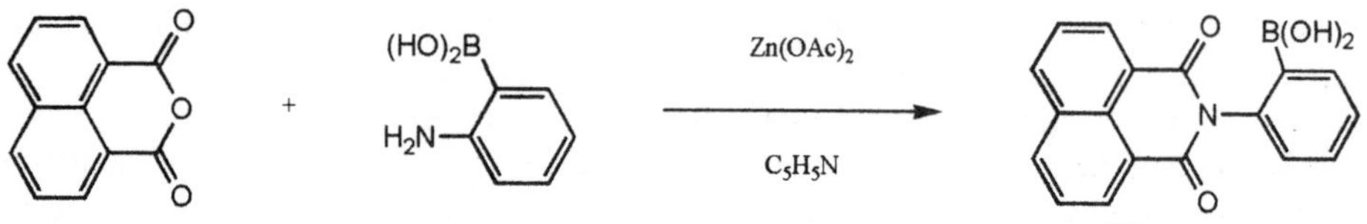

+ $(HO)_2B$, H_2N — $Zn(OAc)_2$, C_5H_5N → $B(OH)_2$ **11**

A pH vs fluorescence intensity profile (figure 8a) demonstrated that PET signal transduction mechanism is operating as fluorescence intensity becomes suppressed upon ionization to the hydroxyboronate anion. This sensor showed behavior similar to *meta*-phenylboronic acid based sensor giving the expected trend in saccharide selectivity with fructose>glucose>galactose (figure 8b). K_D values for fructose and glucose were calculated to be 1.6 and 57, respectively.[40] A key difference between sensor **11** and **5**, however, is evident in the degree of CHEQ that these sensors display. *Ortho*-substituted sensor **11** quenches fluorescence less effectively (I/I_o = 0.4) than *meta*-isomer **5** (I/I_o = 0.2).

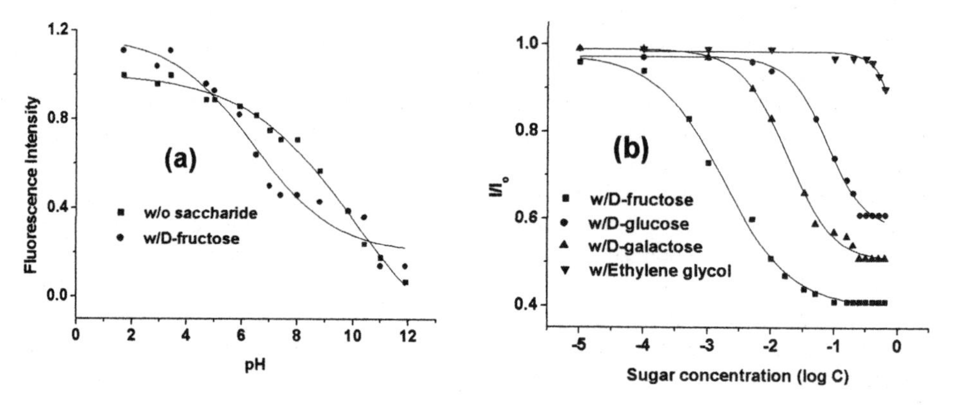

Figure 8. (a) Titration curve vs pH for **11** (6.0×10^{-3} M, λ_{exc} = 346 nm at 25 °C) (b) Titration curves against monosaccharides for **11**, measured in phosphate buffer (100 mM), pH 7.4 at 25 °C.

The energy difference of their respective HOMO-LUMO gaps is expected to have little to no dependence on the position of the boronic acid goup. Based on our photoelectrochemical model for PET fluorescence quenching, these findings point to a less effective conversion of sensor **11** to its hydroxyboronate ester relative to **5**. Molecular modeling studies indicate that increased steric congestion is likely to be responsible for this weaker binding.

4.2. 3-Phenylboronic acid-2,3-naphthalenedicarboximide

2,3-naphthalenedicarboxylic imide is yet another useful platform to probe the fluorescence properties of arenecarboximides. Given the shared symmetry between 1,8 and 2,3-napthalene imides, both are expected to possess similar electronic structures.

Scheme 3

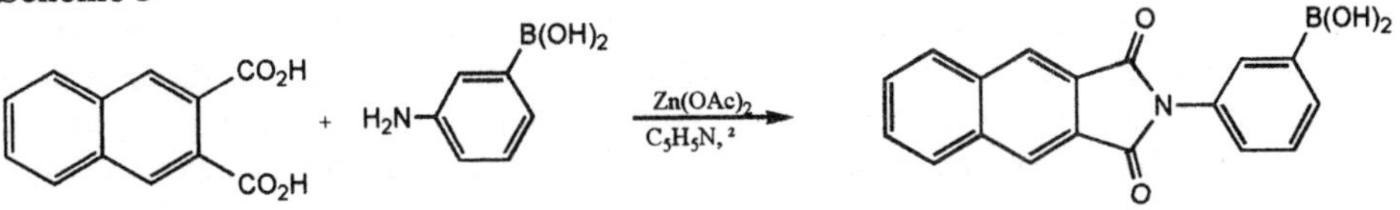

CO_2H CO_2H + H_2N $B(OH)_2$ $\xrightarrow[C_5H_5N, \Delta]{Zn(OAc)_2}$ O N O $B(OH)_2$

12

Indeed, Demeter *et al* [20] have shown that both systems are capable of dual fluorescence originating from S_1 and S_2 excited states. The synthesis of this isomeric monoboronic sensor was readily achieved from 2,3-naphthalene dicarboxylic acid and 3-aminophenylboronic acid (scheme 5).

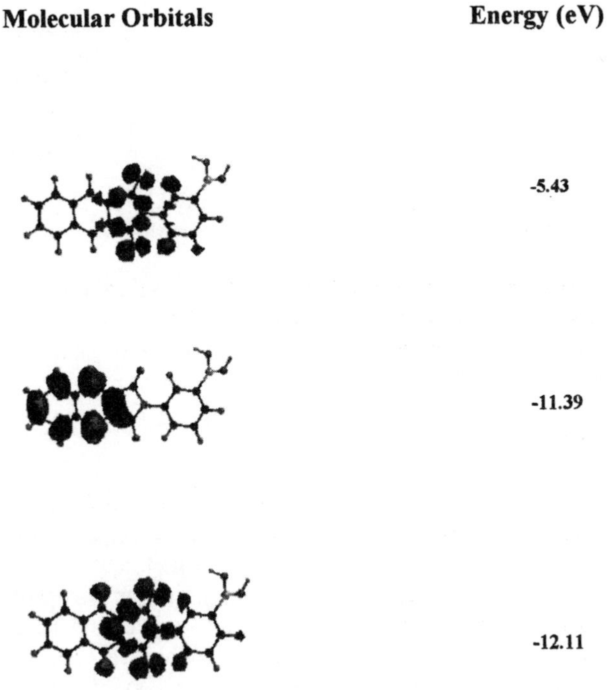

Molecular Orbitals	**Energy (eV)**
	-5.43
	-11.39
	-12.11

Figure 9. Relevant Hückel molecular orbitals for compound **12** based on extended Hückel calculations from Chem 3D v 5.0.

Molecular orbitals for the HOMO-1, HOMO and LUMO resemble those of the 1,8-naphthalene system whereby the S_2 state is characterized by an extensive shift in electron density relative to the ground state. Upon excitation at 365 nm in acetonitrile solution the fluorescence response from **12** shows a characteristic dual emission at 410 and 490 nm (spectrum not shown). However, in 5% DMSO:H_2O this probe gave a single emission band at 412 nm. The pH-fluorescence intensity graph (figure 10a) shows an exceptionally large dynamic range for galactose with $I/I_o \sim 0.4$ relative to glucose and fructose ($I/I_o \sim 0.6$). Comparisons from saccharide titrations at pH 7 (figure 10b), indicate an unusual trend for monoboronic acid:saccharide complexes with the strongest binding to galactose ($K_D = 0.87$) followed by fructose and glucose with K_D's of 6.8 and 12.6, respectively.

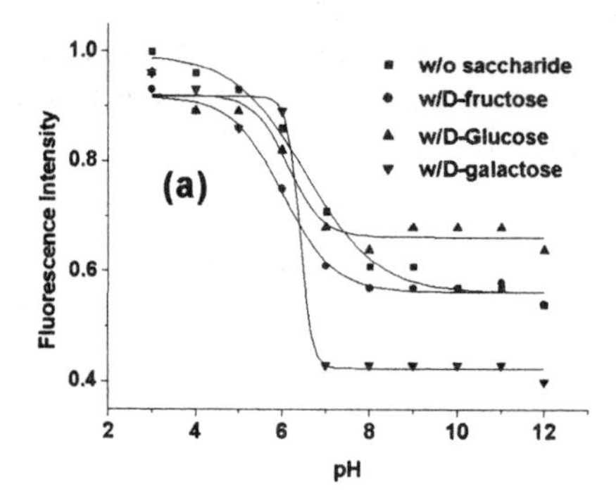

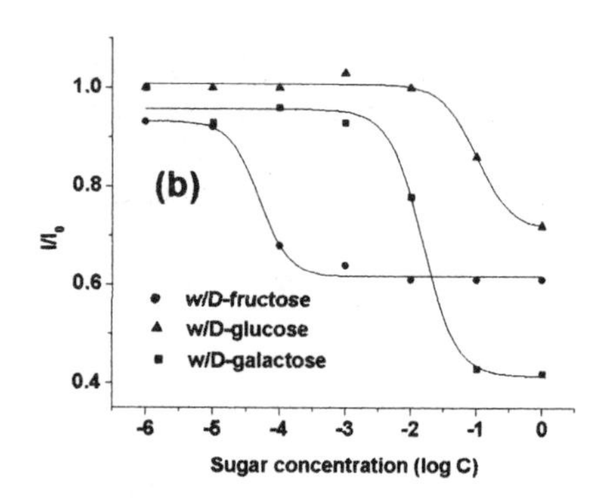

Figure 10. (a) pH vs fluorescence profile of **12** (5.0×10^{-5} M, λ_{exc} = 370 nm, at 25 °C). (b) with monosaccharides measured in phosphate buffer (100 mM) pH 7.4 at 25 °C.

As in the anomalous case of sensor **6**, a monoboronic acid probe demonstrates a greater optical sensitivity to a monosaccharide other than fructose. To the best of our knowledge, **6** represents the first monoboronic acid sensor to displays such optical sensitivity for galactose.

5. BIS-BORONIC ACID SENSORS

Our interest in designing bis-boronic sensors using the diimide platform stems from comparative studies with several well-studied saccharide sensors from Shinkai. The bis-boronic acid probes from this group have shown a preference for glucose by way of a good conformational fit between two phenylboronic groups appended to a naphthalene platform. In their work, a benzylamine-spacer provides a CHEF response with an appropriate distance for a bis-chelation effect. Having synthesized several versions of 1,8-naphthallic imides, the diimide system seemed a natural extension of our work to investigate.

5.1. Naphthalene 1,4,5,8-tetracarboxylic diimide

For bis-dentate boronic acid complexing of saccharides, naphthalic dianhydride provided the necessary platform to directly append two boronic acid groups. Sensor **13** was prepared in one step from 1,4,5,8-naphthalenetetracarboxylic anhydride and 2-aminophenylboronic acid[41] (Scheme 4).

Scheme 4

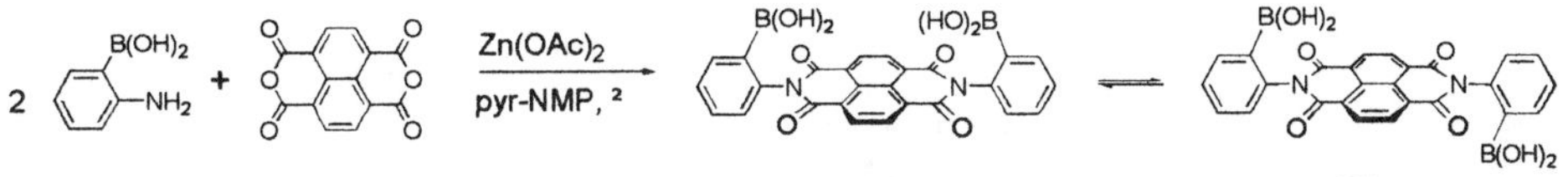

13a 13b

The presence of atropisomers due to restricted rotation about the two $C_{(aryl)}$-$N_{(imide)}$ bonds was confirmed by the use of variable temperature ^{1}H NMR. As the temperature is raised to 37 °C, these peaks coalesce to one doublet giving a rotational frequency of 16.7 s^{-1} with an energy barrier to rotation of 16.5 kcal/mol at 310 K.[42] Sensor **13** proved to be non-fluorescent within the detection limits of our fluorescence spectrometer. Relative to parent naphthalic anhydride fluorophore ϕ_F = 0.34, the *N*-phenylboronic acid 1,8-naphthalenedicarboximide quantum yield drops to 0.01. Addition of a second phenylboronic acid completely quenches the fluorescence beyond our detection. These findings agree with previous reports on the photophysics of *N*-phenylnaphthalimides where phenyl substitution has been shown to enhance internal conversion relative to *N*-alkylnaphthalimides.[20] UV-Vis absorption spectroscopy and NMR titration measurements as a function of saccharide concentration indicate minimal binding interactions occur with this sensor. ^{11}B NMR chemical shifts have provided more direct evidence of boron geometries for this molecule. Comparisons between a bis-*meta*-phenylboronic acid analog (δ = 2.8) and bis *ortho*-**13** (δ = 28.5) demonstrate a trigonal boron preference for **13**.[43] This sterically favorable boron geometry in **13** vs. tetrahedral may be responsible for its reduced saccharide affinity.

5.2 Benzene 1,2,4,5-tetracarboxylic diimide

An alternative platform used to append two phenylboronic acid groups and differing by one benzene ring is found in the pyromellitic anhydride fluorophore **1**. The synthesis of isomeric bisboronic sensors that vary in the position of the boronic acid group was readily achieved from 1,2,4,5 tetracarboxylic anhydride with either 2 or 3-aminophenylboronic acid giving isomers **14** and **15**, respectively.

14a 14b

15a 15b

Unlike the non-emissive naphthalene derived bisboronic probe, both compounds displayed quantum yields of about 0.25. Using a Job plot analysis, both sensors gave a 1:1 binding isotherm in the presence of monosaccharide (fructose).

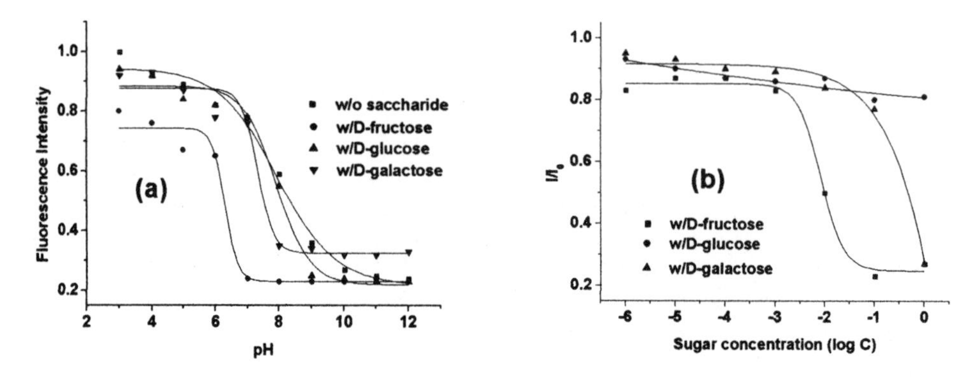

Figure 11. (a) Titration curves vs pH for **14** (6.0×10^{-5} M, λ_{exc} = 315 nm at 25 °C). (b) Titration curves against monosaccharides measured in phosphate buffer at pH 7.4.

Other measurements, however, indicate a marked difference between these isomeric probes. The response of sensor **14** to pH in the presence of saccharides is shown in figure 11a. Its behavior appears similar to monoboronic based probes as it displays the largest chelation response to fructose. Figure 11b gives the relative fluorescence quenching response to saccharide concentrations. Of interest is the very large CHEQ response (I/I_0 = 0.20) for this system in the millimolar region of fructose at neutral pH. Glucose and galactose give a negligible fluorescence quenching with glucose providing significant CHEQ only at very high concentrations. Dissociation constant comparisons favor fructose complexation (K_D = 3.4) followed by K_D of 18 for glucose.

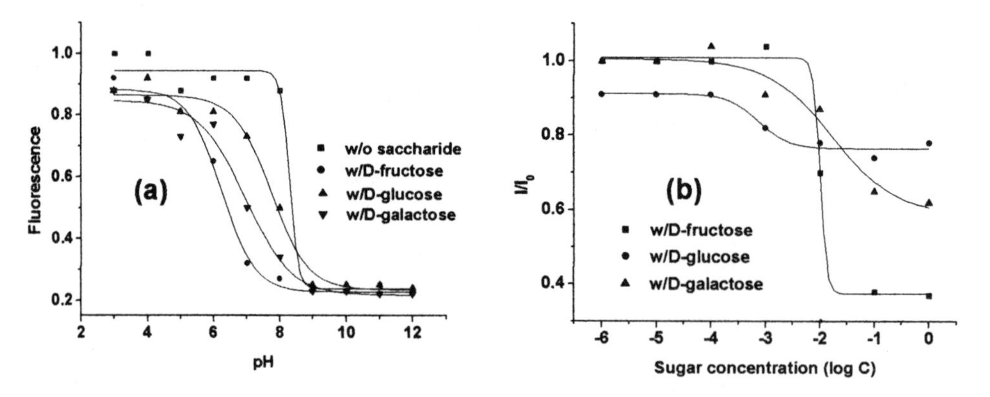

Figure 12 (a) Titration curves vs pH for **15** (6.0×10^{-5} M, λ_{exc} = 320 nm at 25 °C). (b) Titration curves against monosaccharides measured in phosphate buffer at pH 7.4.

With **15**, we find in figure 12b that fructose quenches our probe to the greatest extent (I/I_o = 0.4), but at considerably higher saccharide concentrations relative to glucose. Here CHEQ begins with glucose solutions of 100 μM thereby indicating a geometrical preference between this sugar and **15** (K_D = 1.4) relative to **14**. Although a large CHEQ response for glucose does not occur with **15**, the bi-dentate probe appears to show a greater affinity to glucose. Whereas fructose gives the strongest saccharide complex for probe **14**, a greater K_D value with **15** (K_D = 3.4) indicates that fructose is weakly complexed by the bis-*ortho* positional isomer. Both **14** and **15** exhibited the least affinity to galactose as their dissocation constant values (K_D = 35 and 34, respectively) were quite similar.

6. SUMMARY AND OUTLOOK

Herein three approaches were described by which arenecarboximides and arenedicarboximides have been employed as versatile fluorescenct platforms for saccharide sensing. By investigating the effects of substuent groups, placement of the boronic acid and appending two boronic acids per sensor a number of saccharide sensors were prepared with usually one step reactions. All three of these approaches have resulted in the discovery of chemosensors with various saccharide selectivities. Though by no means an exhaustive sampling of this modular synthetic approach, those sensors that were generated exhibited interesting photochemical and photophysical properties and provided useful insights into the relatively limited area of C_0 design chemosensor.

Among several monoboronic sensors we investigated, sensor **5** displayed a large change in CHEQ signal response under physiological condition. Sensor **6** exhibits dual emission with remarkable sensitivity for glucose relative to fructose and galactose through subtle changes in pH. At low pH sensor **7** signaled enhanced fluorescence (CHEF) in the presence of galactose. Probes **8-10** gave poor signal response upon titration with monosaccharides, yet their substituent effects lend support to the chelation enhanced quenching model proposed for these systems.

Our investigations into positional isomers of simple monoboronic acid probes gave additional evidence for the photoelectrochemical model involving CHEQ. A comparison between aromatic platforms of the 1,8- and 2,3-naphthalenecarboximides indicates a dramatic shift in saccharide selectivity from fructose to galactose, respectively. An additional comparison between chemosensors **6** and **12**, these studies have shown that the optical response arising from PET signaling can be quite different from the expected trends in binding constant selectivity. With bis-boronic acid systems it appears that our bis-*meta* boronic acid probe binds more selectively to fructose than any of the monoboronic acid generated in our lab. Whereas in the bis-*ortho*phenylboronic acid system, a slight preference due to geometrical constraints is observed for glucose relative to fructose or galactose.

Finally, it should be mentioned that not all of the designs based on the arenecarboximide platform resulted in successful fluorescent probes. As observed in compound **13**, molecular modeling cannot always provide an accurate prediction of fluorescence outcomes. Such results highlight one of the biggest challenges in chemosensory devices. Although the relationship of ligand structure to the chemical and physical properties of derived host/guest complexes has become a central theme in chemosensor discovery, the proof of concept in rational design for such complexes ultimately depends on the final photophysical experiment. In many instances, especially those that combine photophysical effects such as fluorescence with host/guest interactions, this approach can frequently give unexpected yet desirable results.[44]

Acknowledgments
The author wishes to thank the many students who have contributed to this area of research. Namely and in alphabetical order, the assistance of Devi Adhikari, Haishi Cao, Amanda Carnahan, Dalia Diaz, Susan Duong, Dustin English, Lisa Eskra, Justin Heynekamp, Fang Liang, Tom McGill and Qiang Li is gratefully acknowledged. It has been their enthusiasm and dedication which provides the driving force for our ongoing research in saccharide detection. MDH thanks Professors Chris Geddes and Joseph Lakowicz for their kind invitation to contribute to this annual review. Finally, we thank NIH for several years of support via its AREA grant program.

REFERENCES

1. T.D. James, P. Linnane, S. Shinkai, Fluorescent saccharide receptors: a sweet solution to the design, assembly and evaluation of boronic acid derived PET sensors, *Chem. Commun.* 281-288 (1996).
2. T. D. James, K. R. A. S. Sandanayake, S. Shinkai, Recognition of sugars and related compounds by "reading-out" type interfaces, *Supramol. Chem.* **6**, 141-157 (1995).
3. T. D. James, K. R. A. S. Sandanayake, S. Shinkai, Saccharide sensing with molecular receptors based on boronic acid, *Angew. Chem. Int. Ed. Engl.* **35**, 1910-1922 (1996).
4. A. W. Czarnik, Chemical communication in water using fluorescent chemosensors, *Acc. Chem. Res.* **27**, 302-308 (1994).
5. A. W. Czarnik, *Fluorescent Chemosensors for Ion and Molecule Recognition.*; American Chemical Society: Washington, D.C., 1993.
6. H. Suenaga, M. Mikami, K. R. A. S. Sandanayake, S. Shinkai, Screening of Fluorescent boronic acids for sugar sensing which shows a large fluorescence change, *Tetrahedron Lett.* **36**, 4825-4828 (1995).
7. N. DiCesare, J.R. Lakowicz, Chalcone-analogue fluorescent probes for saccharides signaling using the boronic acid group, *Tetrahedron Lett.* **43**, 2615-2618 (2002).
8. N. DiCesare, J.R. Lakowicz, Fluorescent probe for monosaccharides based on functionalized boron-dipyronemethane with a boronic acid group, *Tetrahedron Lett.* **42**, 9105-9108 (2001).
9. N. DiCesare, J.R. Lakowicz, A new highly fluorescent probe for monosaccharides based on a donor-acceptor diphenyloxazole, *Chem. Commun.* 2022-2023 (2001).
10. H. Eggert, J. Frederiksen, C. Morin, J.-C. Norrild, A new glucose-selective fluorescent bis-boronic acid. first report of strong -furanose complexation in aqueous solution at physiological pH, *J. Org. Chem.* **64**, 3846-3852 (1999).
11. H. Cao, D. I. Diaz, N. DiCesare, J. R. Lakowicz, M. D. Heagy, Monoboronic acid sensor that displays anomalous fluorescence sensitivity to glucose, *Org. Lett.* **4**, 1503-1505 (2002).
12. D. P. Adhikiri, M. D. Heagy, Fluorescent chemosensor for carbohydrates which shows large change in chelation-enhanced quenching, *Tetrahedron Lett.* **40**, 7893-7896 (1999).
13. R.W. Middleton, J. Parrick, E. D. Clarke, P. Wardman, Synthesis and fluorescence of *N*-substituted-1,8-naphthalimides, J. *Heterocyclic Chem.* **23**, 849-855 (1986).
14. J. Gawronski, K. Gawronska, P. Skowronek, A. Holmén, 1,8-Naphthalimides as stereochemical probes for chiral amines: A study of electronic transitions and exciton coupling, *J. Org. Chem.* **64**, 234-241 (1999).
15. K. Nakaya, K. Funabiki, H. Muramatsu, K. Shibata, M. Matsui, *N*-aryl-1,8-naphthalimides as highly sensitive fluorescent labeling reagents for carnitine, *Dyes and Pigments* 43, 235-239 (1999).
16. S. Chang, R. E. Utecht, D. E. Lewis, Synthesis and bromination of 4-alkylamino-*N*-alkyl-1,8-naphthalimides, *Dyes and Pigments* **43**, 83-94 (1999).
17. Q. Xuhong, Z. Zhenghua, C. Kongchang, The synthesis, application and prediction of Stokes shift in fluorescent dyes derived from 1,8-naphthalic anhydride, *Dyes and Pigments* **11**, 13-20, (1989).
18. J. Gawronski, M. Kwit, K. Gawronska, Helicity Induction in a Bichromophore: A sensitive and practical chiroptical method for absolute configuration determination of aliphatic alcohols, *Org. Lett*, **4**, 4185-4188 (2002).
19. B. Ramachandram, A. Samanta, Modulation of metal-fluorophore to develop structurally simple fluorescent sensors for transition metal ions, *Chem. Commun.* 1037-1038 (1997).

20. A. Demeter, T. Bérces, L. Biczók, V. Wintgens, P. Valat, J. Kossanyi, Comprehensive model of the photophysics of *N*-phenylnaphthalimides: The role of solvent and rotational relaxation, *J. Phys. Chem.* **100**, 2001-2011 (1996).
21. R. A. Bissell, A. P. de Silva, H. Q. N. Gunaratne, P. L. M. Lynch, G. E. M. Maguire, K. R. A. S. Sandanayake, Molecular fluorescent signalling with fluor-spacer-receptor systems. Approaches to sensing and switching devices via supramolecular photophysics, *Chem. Soc. Rev.* 187-195 (1992).
22. L. M. Daffy, A. P. D. de Silva, H. Q. N. Gunaratne, C. Huber, P. L. M. Lynch, T. Werner, O. S. Wolfbeis, Arenedicarboximide Building Blocks for Fluorescent Photoinduced Electron Transfer pH Sensors Applicable with Different Media and Communication Wavelengths, *Chem. Eur. J.* **4**, 1810-1815 (1998).
23. J. A. Cowan, J. K. M. Sanders, Pyromellitimide-bridged porphyrins as model photosynthetic systems. 1 Synthesis and steady state fluorescence properties, *J. Chem. Soc., Perkin Trans. 1* 2435-2437 (1985).
24. B. M. Aveline, S. Matsugo, R. W. Redmond, Photochemical Mechanisms Responsible for the Versatile Application of Naphthalimides and Naphthaldiimides in Biological Systems, *J. Am. Chem. Soc.* **119**, 11785-11795 (1997).
25. A. Samanta, G. Saroja, Steady-state and time-resolved studies on the redox behavior of 1,8-napthalimdes in the excited state, *J. Photochem. Photobiol. A: Chem.* **84**, 19-26 (1994).
26. I. Garbtchev, Tz. Philipova, P. Meallier, S. Guittoneau, Influence of Substituents on the Spectroscopic and Photochemical Properties of Naphthalimide Derivatives, *Dyes and Pigments*, **31**, 31-34 (1996).
27. V. Wintgens, P. Valat, J. Kossanyi, A. Demeter, L. Biczok, T. Berces, Spectroscopic properties of aromatic dicarboximides. Part 4. On the modification of the fluorescence and intersystem crossing processes of molecules by electron-donating methoxy groups at different positions. The case of 1,8-naphthalimides, *New J. Chem.*, **20**, 1149-1158 (1996).
28. A calculated energy barrier of rotation for sensor and sensor-saccharide complex was determined using PC model software (version 6.0). The sensor-saccharide complex displays a higher energy barrier (15.32 kcal/mole) than free sensor (14.41 kcal/mole). Although this difference appears small, the overall strain energy was much higher 140.3 kcal/mol). On this energy difference basis, the rotational energy barrier may prevent phenyl ring rotation.
29. Y. Dromzée, J. Kossanyi, V. Wintgens, P. Valat, L. Biczók, A. Demeter, T. Bérces, Crystal and molecular structure of *N*-phenyl substituted 1,2-, 2,3- and 1,8-naphthalimides, *Z. Kristallo.* **210**, 760-765 (1995).
30. A. Weller, Exciplex and radical pairs in photochemical eletron-transfer, *Pure and Appl. Chem.* **54** 1885-1888 (1982).
31. Gibbs free energy term is divided by *–nF* where E values are in units of volts.
32. The default force field used in PCMODEL(version 8.0) is called MMX and is derived from MM2(QCPE-395, 1977) force field of N. L. Allinger, with the pi-VESCF routines taken from MMP1 (QCPE-318), also by N. L. Allinger.
33. J.H. Hageman, G.D. Kuehn, Boronic acid matrices for the affinity purification of glycoproteins and enzymes *Methods in Molecular Biology*, **11**, 45-71 (1992).
34. G.R. Kennedy, M.J. Row, The interaction of sugars with borate: an N.M.R. spectroscopic study *Carbohydrate Res.*, **28**, 13-19 (1973),
35. G. Wulff, Selective binding to polymers via covalent bonds: The construction of chiral cavities as specific receptor sites, *Pure Appl. Chem.* **54**, 2093-2102 (1982).
36. T. D. James, K. R. A. S. Sandanayake, R. Iguchi, S. Shinkai, Novel photoinduced electron-transfer sensors based on the interaction of boronic acid and amine, *J. Am. Chem. Soc.* **117**, 8982-8987 (1995).
37. K. R. A. S. Sandanayake, S.; Shinkai, Two-dimensional photoinduced electron-transfer (PET) fluorescence sensor for saccharides *Chem. Lett.* 503-504 (1995).
38. W. Yang, H. He, D. G. Drueckhammer, Computer-guided design in molecular recognition design and synthesis of a glucopyranose receptor, *Angew. Chem. Int. Ed. Engl.* **40**, 1714-1717 (2001).
39. V. V. Karnati, X. Gao, S. Gao, W. Yang, W. Ni, S. Sankar, B. Wang, A glucose-selective fluorescence sensor based on boronic acid-diol recofnition, *Bioorg. Med. Chem. Lett.* **12**, 3373-3377 (2003).
40. N. DiCesare, D. P. Adhikari, J. J. Heynekamp, M. D. Heagy, J. R. Lakowicz, Spectral properties of fluorophores combining the boronic acid group with electron donor or withdrawing groups. Implication in the development of fluorescent probes for saccharides, *J. Fluor.* **12**, 147-154 (2002).
41. M. P. Groziak, A. D. Ganguly, P. D. Robinson, Boron Heterocycles Bearing a Peripheral Resemblence to Naturally-occuring Purines: Design, Synthesis, Structures, and Properties, *J. Am. Chem. Soc.* **116**, 7597-7605 (1994).
42. H. Günther, *NMR Spectroscopy*; Georg Thieme Verlag: Stuttgart, 1987.
43. H. Noth, B. Wrackenmeyer, Nuclear Magnetic Resonance Spectroscopy of Boron Compounds, Vol 14; Springer: Berlin, 1978.
44. M. B. Francis, N. S. Finney, E. R. Jacobsen, Combinatorial approach to the discovery of novel coordination complexes, *J. Am. Chem. Soc.* **118**, 8983-8984 (1996).

PROGRESS TOWARDS FLUORESCENT MOLECULAR THERMOMETERS

Nirmala Chandrasekharan and Lisa A. Kelly*

1. INTRODUCTION

Material-based sensors have attracted an enormous amount of recent interest. Biological[1] and chemical sensors[2] are being rapidly developed to offer sensitivity and specificity for the analyte of choice. In addition, there has been growing material science interest in the creation of "smart materials,"[3, 4] capable of fast, reversible responses to environmental stimuli.

It is of broad-ranging interest to develop "smart materials" or sensors for the measurement of temperature. The measurement of temperature is ubiquitous in all fields of science, engineering and medicine. Temperature measurement accounts for 75-80% of the worldwide sensor market.[5] There is an ever-increasing variety of situations that require accurate and, in many cases, remote measurement of temperature. Conventional methods use thermocouples, thermistors or resistance thermometers. In these sensors, measurement is accomplished by means of electrical signals that are generated and converted to temperature. Temperature-dependent electrical signals will be altered in "hostile" environments where large electric or magnetic fields are present, or in chemical environments where corrosion of thermocouple junctions will occur. In many applications, it is essential to obtain temperature data by non-electrical means.

In addition, many applications require accurate measurement of the temperature distribution over a large area. In these applications, it is impractical and expensive to machine an array of thermocouple taps into the surface. In all cases, conventional approaches to temperature measurement require the sensor and electrical leads be physically connected to the object being probed. Remote temperature measurement is limited by the need to connect sensor and electrical leads to the read-out device.

* Lisa A. Kelly, Department of Chemistry and Biochemistry, University of Maryland, Baltimore County, 1000 Hilltop Circle, Baltimore, MD 21250; LKelly@umbc.edu; Phone: (410) 455-2507; Fax: (410) 455-2608.

As a means of measuring temperature, optical methods offer a viable alternative to conventional thermocouples or thermistors. The most common optical method for temperature measurement is infrared (IR) thermometry. IR thermometry makes use of the fact that all matter emits radiant energy, according to Planck's blackbody distribution. By remotely measuring the power spectrum (radiant energy density vs. wavelength), and fitting it to Planck's blackbody distribution law, the absolute temperature of an object is determined. Although IR thermometry has found wide-ranging applications, its utility is drastically reduced if the intervening medium absorbs the radiation and distorts the spectrum. As an example, water absorbs strongly in the 2000 – 8000 μm spectral region. Since this is the range in which a 200 – 300 K blackbody emits,[5, 6] IR thermometry has severe limitations if remote measurements of "ambient" temperatures are required through water vapor or aqueous media. Under these conditions, the detected emissivity of the object is unknown due to modifications from the absorbing and intervening medium.

As an alternative, temperature-sensitive coatings provide the foundation for remote and non-intrusive temperature measurement. For instance, if the material is luminescent, the emission spectral shape and/or intensity (quantum yield) may change with temperature. "Fluoroptic thermometry" offers a viable and cost-effective alternative to single point temperature measurement where temperature measurement in environments where electrical signals may be distorted. In addition, if an object is coated with a temperature sensitive material, then illuminated and imaged, remote temperature measurements and mapping are possible. Since most materials luminesce in the visible spectral region (400 – 800 nm), the light can be measured through nearly every intervening media.

Material-based temperature sensors have found broad-ranging applications in biological, chemical, and engineering applications. Fiber optic thermometry has become an ever-growing method for measuring temperature using optical fibers coated with luminescent compounds. With fiber optic arrays or bundles, it is possible to map the temperature distribution across the fiber optic area.[7] Fiber optic thermometry has found, among others, medical applications in monitoring tissue temperature during heat-induced tumor destruction, or hyperthermia. In hyperthermia, malignant tunors are heated to ca. 42.5 °C using either radiofrequency (RF) or microwave (MW) radiation[8] The treatment requires that the temperature be accurately measured and monitored during the course of the treatment. The strong electromagnetic fields from the RF or MW source will distort, in a non-predictable fashion, electrical signals from a thermocouple or thermistor.

Luminescent coatings are used in aerodynamic applications, where two-dimensional temperature fluctuations induced by frictional heating during fluid or gaseous flow are to be measured.[9, 10] Finally, temperature sensitive materials have been used in real-time polymer processing monitoring applications, where the limitations of thermocouples and infrared imaging preclude accurate temperature measurement.[11, 12, 13]

In the engineering community, a great deal of effort has been devoted to developing the instrumentation required for optical thermometry. The luminescent compounds have been predominantly from commercial sources. There have been few systematic studies devoted to developing new materials optimized for sensitivity and reversibility in temperature measurement. In the following section, we will list some of the temperature-sensitive materials that have been used to measure temperature both in solution and in solid films, and present some of our own results towards the goal of developing molecular thermometers.

2. OVERVIEW OF TEMPERATURE-SENSITIVE MATERIALS

Temperature-sensitive materials, whose color or emission is modified by temperature, offer a cost-effective method of remotely mapping temperature changes. Since the absorption and emission is usually in the visible region of the electromagnetic spectrum, they offer an alternative to IR thermometry in condensed media. There exist two main categories of temperature-sensitive materials: (i) Thermochromics are materials that exhibit a temperature-dependent and reversible color change and (ii) Thermoemissives are materials whose luminescence properties (wavelength and/or intensity) change with temperature.[5] To evaluate a specific temperature from a thermochromic material, a transmittance or reflectance measurement must be carried out. This requirement makes remote temperature monitoring difficult or impossible.

2.1 Thermochromic materials

Materials that exhibit temperature-dependent color changes have found utility ranging from the measurement of instantaneous heat transfer and fluid flow[14] to components of thermochromic inks, thermometers, and temperature indicators. Examples of these substances include: (i) inorganic materials that exhibit a color change upon a temperature dependent change in crystal structure;[15] (ii) materials that change color with temperature as a result of changes in stereochemistry[16] and (iii) materials that undergo a thermochromic semiconductor to metal transition.[17] In addition, liquid crystals are an important class of thermochromics.[14, 18] These undergo changes in reflectance as a function of temperature and are widely used in medicine and industry.

2.2 Thermoemissive materials

A variety of compounds have been used as thermoemissive materials both in solution and in solid films. In all cases, the origin of the temperature-dependent spectral changes is understood from elementary photophysics. If the luminescence is a spin-allowed fluorescence, the intensity that is observed represents the competition between the radiative and non-radiative deactivation of an excited state. Since the rate constant for activated processes (e.g. intersystem crossing) increases with temperature, fluorescence quantum yields are intrinsically temperature dependent, since they represent the competition between radiative and (activated) non-radiative decay pathways. In addition, if the luminescent excited state is subject to diffusional quenching (e.g. by molecular oxygen), both the diffusional rate constant and dissolved oxygen concentration (or polymer permeability, if the luminophore is imbedded in a matrix) may change with temperature.

Thermoemissive materials that have been used to measure surface temperature include (i) inorganic thermographic phosphors (TPs),[19, 20] and (ii) organic-based temperature-sensitive coatings. Temperature-sensitive coatings (TSCs) are composed of luminescent organic or inorganic dyes embedded in a suitable polymer matrix or binder.[10, 21] The resulting coating is applied to a surface using a brush or sprayer. When the coating is irradiated with ultraviolet or visible light, the embedded dyes provide a luminescent response that can be calibrated with temperature. TSCs have been used in aerodynamics to visualize heat transfer rates, boundary layer transition, flow separation, flow reattachment on models of transonic, supersonic and shock wind tunnels.[22]

During the past decade, the mapping of pressure and temperature distributions has been accomplished by incorporating chromophores as reporters of pressure and temperature. The compounds are embedded into an oxygen-permeable binder and used to record two-dimensional temperature and pressure maps. For example, the luminescent metal-to-ligand charge transfer (MLCT) state of ruthenium (II) complexes has been widely employed.[23, 24] Since the lifetime of these excited states is quenched by oxygen in a Stern-Volmer fashion, the luminescence intensity is a direct reporter of oxygen partial pressure on the coated surface. In a similar way, the quantum yield for luminescence is well-known to be temperature dependent. Thus, at constant pressure, the intensity has utility to map temperature distribution on a surface.[25] Likewise, porphyrin compounds possessing long-lived and luminescent excited states have been used in an equivalent way in temperature and pressure sensitive coatings.[26] In short, these examples illustrate how elementary photophysics has great utility in developing new coating and sensor technology.

Other compounds that have been shown to demonstrate temperature dependent changes in fluorescence quantum yield include acridine yellow **(1)** dissolved in a rigid saccharide glass.[27] The compound was shown to exhibit a temperature dependent reverse intersystem crossing. As predicted for an activated process, the rate constant for intersystem crossing was temperature dependent. As such, the ratio of fluorescence to phosphorescence intensities is modified with temperature. The temperature-dependent ratio of fluorescence to phosphorescence intensity was recorded and a sensitivity of 4.5%/°C was observed.

In solution, commercially available laser dyes have been used to report the temperature in alkane solvents.[28] Since the singlet-state lifetime of these laser dyes was very short, the temperature-response was independent of dissolved oxygen. The fluorescence intensity and spectral shape were correlated with temperature to map the solution temperature with an accuracy of 4°C.

As in the case of acridine yellow, other materials have been used as "ratiometric" temperature sensors. Ratiometric sensors exhibit emission in two spectral regions. The intensity of either or both may be modified with temperature. As such, the intensity ratios, rather than absolute intensity, are evaluated and correlated with temperature. For example, the fluorescence maximum and intensity of vibronic hot-band were modified by

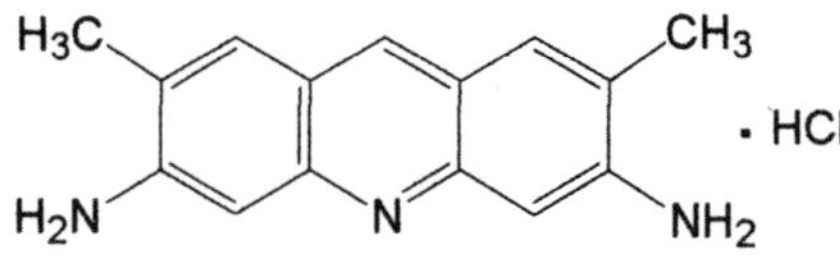

H_3C CH_3 · HCl H_2N N NH_2

(1)

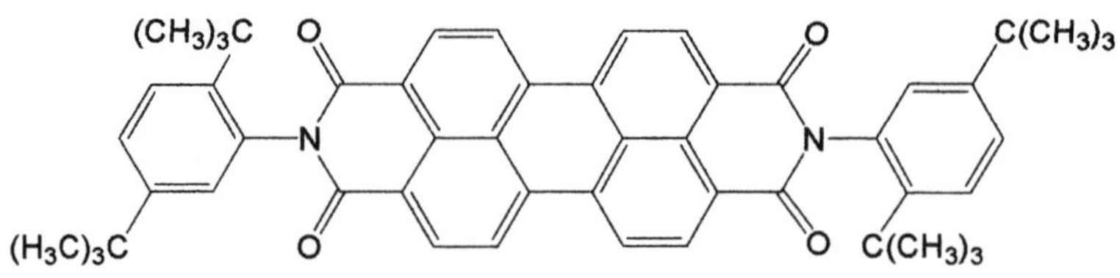

$(CH_3)_3C$ O O $C(CH_3)_3$ N N $(H_3C)_3C$ O O $C(CH_3)_3$

(2)

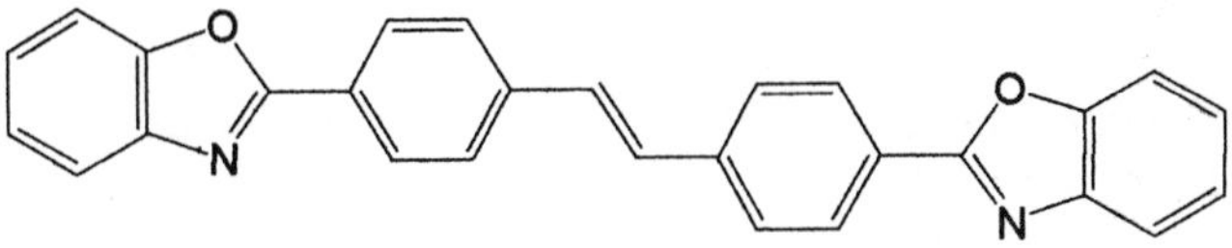

(3)

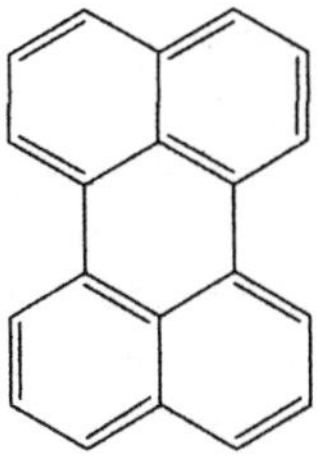

(4)

thermal population of a higher vibronic state in a perylenedicarboximide **(2)** fluorescent probe.[29]

Such thermoemissives have been used in real-time polymer processing monitoring applications, where the limitations of thermocouples and infrared imaging preclude accurate temperature measurement.[11, 12, 13] Commercially available benzoxazolyl stilbene **(3)** and perylene **(4)** have been doped into compatible polymer matrices. The compounds are resin-soluble and show outstanding thermal stability. Changes in the relative intensities of the vibronic bands were induced by temperature.[13] The temperature was measured by taking the relative intensities of two emission wavelengths (e.g. ratiometric detection). For all three dyes, the temperature is determined by measuring the intensities at two emission wavelengths. In each case, plots if the intensity ratio vs. temperature are linear, with slopes of 0.09 – 0.2% per °C.

A third approach to fluoroptical thermometry has utilized the temperature-dependent high-spin/low-spin interconversion of a nickel (II) macrocycle covalently attached to a fluorophore **(5)**. Such systems have been shown to have utility as a "molecular thermometers" in fluid solution.[30] The endothermic high-spin to low-spin equilibrium is shifted towards the low-spin form of the transition metal complex as the temperature is

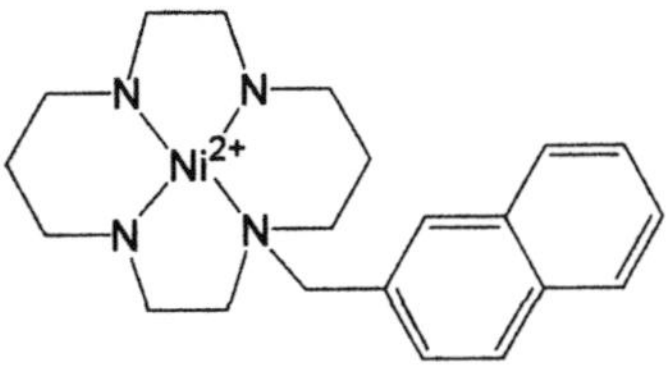

(5)

increased. When it is in the vicinity of the high-spin form, the fluorescence quantum yield of naphthalene is substantially diminished due to the presence of the paramagnetic nickel ion. As the equilibrium is shifted towards the low-spin form, the naphthalene "reports" the temperature increase as an enhancement in the luminescence quantum yield. In other words, the sensor "lights up" as it is heated.

Finally, polymers that exhibit very pronounced structural changes during heat-induced phase transitions have been labeled with a fluorophore and show very large fluorescence changes near the transition temperature. For example, poly(N-isopropylacrylamide) (PNIPAM) in water undergoes a phase transition, known as the lower critical solution temperature (LCST), at ca. 31°C.[31] Below the LCST, the polymer dissolution is enthalpically favored by hydrogen bonding with water. Above the LCST, the hydrogen bonds are broken and the negative entropic contributions dominate, making the free energy change for dissolution of the polymer positive or "uphill." At the LCST, a "cloud point" is observed and can be monitored via changes in the transmittance of the solution. When the fluorescent probe, pyrene, is attached to the polymer backbone, the phase change induces prominent changes in the dye emission at the LCST.[32, 33, 34] Fluorescently labeled PNIPAM's have been synthesized using a variety of pH-dependent, water soluble, or spin-labeled co-monomers (Figure 1).

Over the past two years, we have been investigating a new class of temperature sensitive materials that operate via another textbook example of temperature-dependent photophysics. It is well-established that the excited state of aromatic hydrocarbons form excited state dimers (excimers) or excited state complexes (exciplexes) with their ground states or appropriate electron donors in solution. Under steady-state illumination, the amount of broad-structureless excimer or exciplex emission, relative to the vibronically structured "monomer" emission changes with temperature, since both the forward and reverse (k_1, k_{-1}) reactions are activated processes (Scheme I).

As a temperature sensor, this system has an important advantage over "single-color" materials: The fluorescence is two-color, and the monomer and exciplex emission are well-separated in the visible spectrum. The temperature is measured by recording the ratio of the two emissions. Since the detection is ratiometric, the temperature calibration

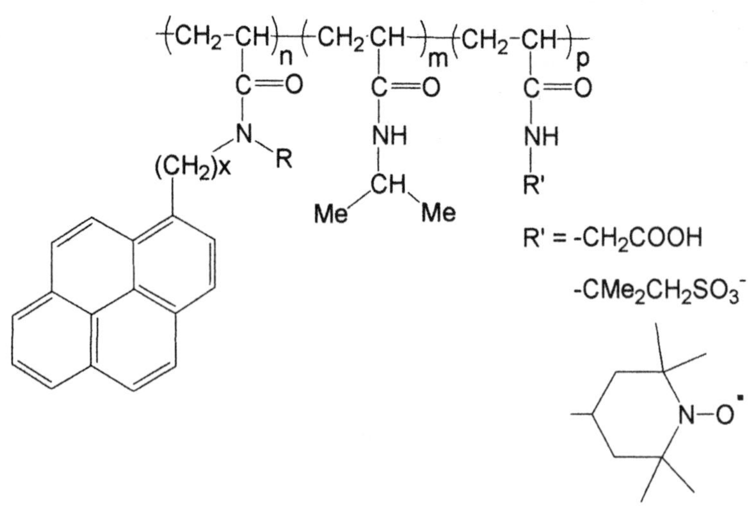

Figure 1. Structure of N-isopropylacrylamide co-polymers that are modified with a fluorescent pyrene probe and glycine,[32] alkyl sulfonate,[33] or a spin label.[34]

is independent of illumination intensity. This is advantageous, since calibrations for different light sources are unnecessary, and time-dependent intensity fluctuations for a given source are "normalized" out. The temperature dependence of the excited state dimer or complex formation, as well as the decay kinetics, is well-understood in solution.[35] Thus, the concept was deemed to have utility in preparing temperature sensitive fluorescent coatings. In this chapter, we present some of our results that demonstrate how these coatings are synthesized and characterized, and discuss how the temperature sensitivity may be optimized and quantified.

3. EXCIPLEX FORMATION IN POLYMERIC FILMS

In designing our exciplex-forming materials, we have considered several important aspects to broaden their scope of application. First, the polymer host should be hydrophobic in nature so that the coatings can be used in aqueous, biological, or humid environments. As described below, we have prepared co-polymers of styrene and an exciplex-forming donor (D in Scheme 1) as the basis of our films. To date, we have used perylene as a luminophore in our films. As an aromatic hydrocarbon, it is compatible with the host matrix so that aggregation does not occur. Finally, it is most desirable to illuminate the material in the visible spectral region, unlike alternative aromatic molecules, including pyrene or naphthalene. The scattering cross-section through intervening medium is small in the visible (compared to the ultraviolet), and cheap, blue LED's are available.

Perylene is well known to form exciplexes that show temperature dependent behavior. There are numerous solution studies of perylene and its exciplex formation with various donor species, including dimethylaniline (Scheme 2). Temperature dependent solution studies of aromatic hydrocarbons and their exciplexes with trinitrobenzene and tetrachlorophthalic anhydride have also been reported.[36] There are comparatively few reports of exciplex behavior of perylene in polymeric matrices. We found only one report of exciplex emission from perylene and N,N-diethylaniline in a styrene polymer.[37]

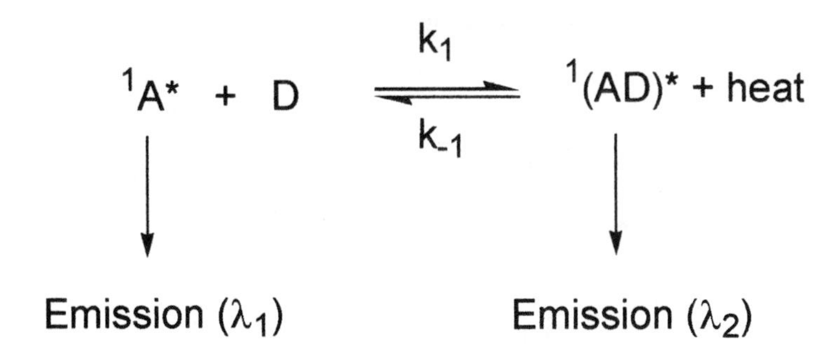

Scheme 1. Fluorescent monomer to excimer interconversion.

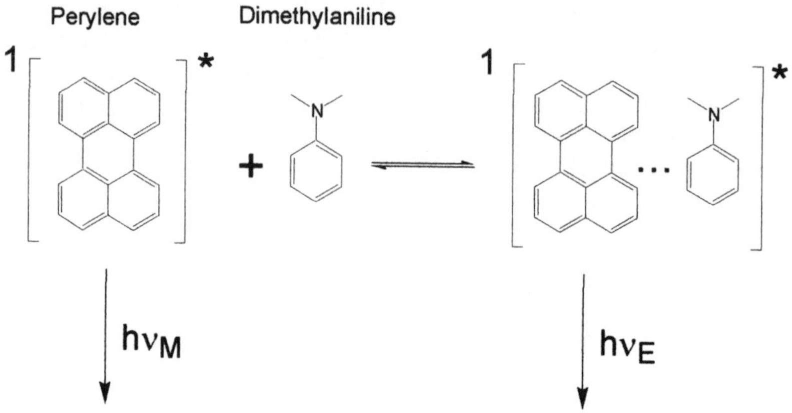

Perylene Dimethylaniline

$h\nu_M$ $h\nu_E$

Scheme 2. Exciplex formation from perylene singlet excited state and N,N-dimethylaniline.

To prepare exciplex-forming films, we have synthesized two aniline derivatives that can be co-polymerized with styrene. Our initial polymers were synthesized as co-polymers of styrene and N-allyl-N-methylaniline (**An-I**).[38] The latter compound was synthesized from N-methylaniline and allyliodide. After synthesis and purification of the aniline derivative, it was co-polymerized with styrene using free radical methods (Scheme 3).

For temperature-dependent dynamic exciplex formation to occur, we found it necessary to provide mobility of the interacting molecules within the polymer matrix. By cross-linking the polymers with divinylbenzene, cavities are introduced within the matrix. In addition, plasticizing agents are added to the films to disrupt the rigid pi-stacking of the aromatic monomer units, providing a degree of "fluidity" within the polymers. Polymers that were not cross-linked and plasticized did not show a variation in monomer to exciplex intensity with temperature. The cross-linking also prevented the polymers from melting over the temperature range being measured.

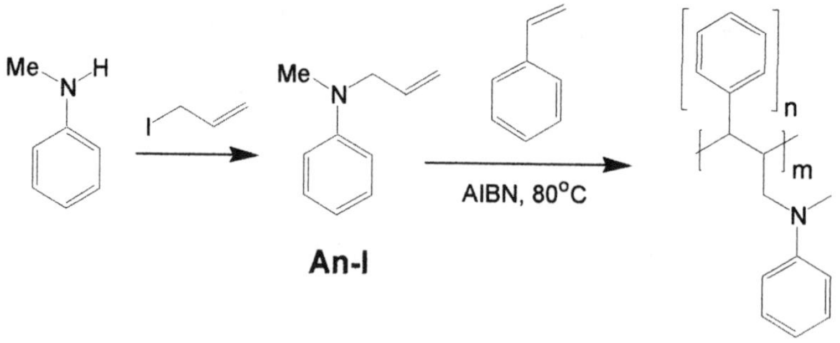

AIBN, 80°C
An-I

Scheme 3. Synthesis of styrene-N-allyl-N-methylaniline co-polymers.

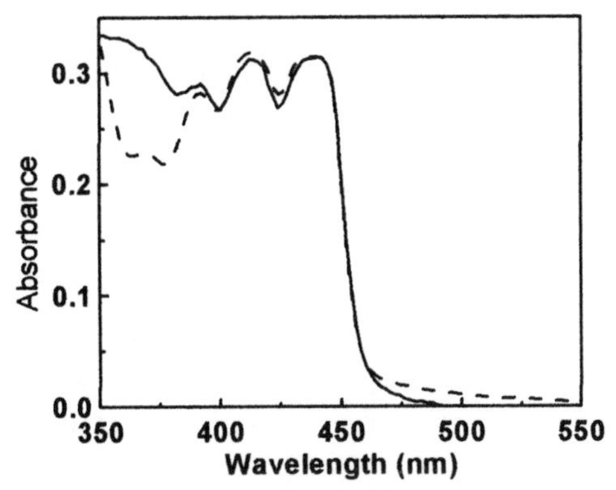

Figure 2. Absorption spectrum of 0.017 wt% of perylene and 18.96 wt% of N-allyl, N-methylaniline in a polystyrene matrix shown as a solid line. The dotted line is the absorption spectrum of the film composed of perylene alone. (Reprinted with permission from reference 38. Copyright (2001) American Chemical Society)

To prepare the material, a monomer solution containing 18.96 wt. % of N-allyl-N-methyl-aniline (**An-I**), 55.94 wt % styrene, and 0.017 wt. % perylene was co-polymerized with 5.63 wt % divinylbenzene (cross-linker), 19.27 wt % methylsalicylate (plasticizer) and 0.2 wt % AIBN as the initiator. The mixture was deaerated and heated in an oil bath at 80° C for 1 hr. In these materials, there are, on average, 3.4 styrene units per aniline.

The absorption spectrum of the film is shown in Figure 2 as a solid line. At wavelengths longer than 375 nm, it matches very well with the absorption spectrum of a film composed of the perylene luminophor alone (dashed line), with bands at 440nm, 414 nm, 392 nm and 366 nm. From the result it is concluded that the excitation light (386 nm) is absorbed exclusively by the perylene.

Upon UV excitation of the perylene, thermodynamics favors exciplex formation with the aniline derivative. The exciplex is stabilized by partial electron transfer from the lone electron pair on the aniline nitrogen to the partially vacant pi molecular orbital of the perylene excited state. The sample fluoresces green at room temperature when illuminated with a UV handlamp. When the sample is heated, the fluorescence becomes predominantly blue.[38] When we first made this observation, the green to blue reversible fluorescence change that accompanied sample heating/cooling was found to be intense enough to be visually discernable.

3.1 Performance of Temperature-Sensitive Materials Prepared from Styrene and N-methyl-N-allylaniline (An-I)

In developing temperature-sensitive materials, it is crucial to fully characterize the sensitivity, reversibility and reproducibility of each new system. Steady-state fluorescence spectroscopy is used to quantify and verify the response. The samples are mounted onto a custom-made aluminum sample block that is wrapped in a resistive heating tape. The temperature was recorded by mounting a K–type (Chromel-Alumel) thermocouple towards the center of the film. The temperature was maintained within ± 0.2 °C for every spectrum taken. Temperature is cycled over a heat/cool run and the emission spectra recorded at fixed temperature intervals for a given film. The sensitivity is determined by recording the variation of exciplex to monomer emission intensity as a function of temperature. The sensitivity is reported as the ratio of intensities of the two spectral regions as a function of temperature. To obtain the best value of the intensity from each individual species, the emission spectra are fitted as a sum of Gaussians, and the weights of the Gaussians used as a measure of the monomer and exciplex populations at each temperature.[38] The reversibility was assessed by recording the emission spectra over three heat/cool cycles. Reproducibility was assessed by performing measurements on several different films of the same composition.

The results on the first system we have studied are shown in Figure 3. On controlled heating of the film, the intensity of the bands at 463 nm and 475 nm increases. An isoemissive point at 543 nm is observed. A concomitant intensity decrease in the long-wavelength region of the spectrum is observed. The temperature response of the material, defined as the percent change in the ratio of blue/green intensity, was found to be 1% per °C. A temperature change of 2 °C could be detected with the spectrometer and scan rate used. In addition, as seen from Figure 3, the spectra show outstanding reversibility with successive heat/cool cycle.

3.2 Interpreting the Temperature Response

To understand our results, we turn to the simple kinetic scheme for exciplex formation and decay shown in Scheme 4. There are two regimes in which a change in the excimer/monomer intensity ratio will exhibit a temperature dependence. From the kinetic scheme shown in Scheme 4, the ratio of intensities from exciplex to monomer has been derived.[39] This ratio depends on the individual rate constants as expressed in Eq. (1).

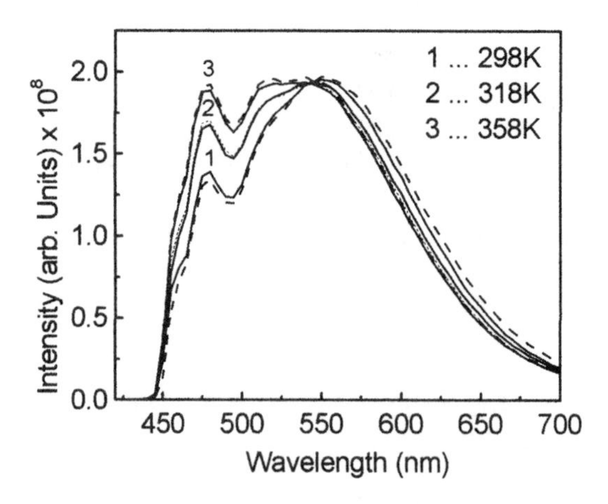

Figure 3. Plot of the emission spectra over a temperature cycle. The spectra recorded while heating are indicated by solid lines and those recorded while cooling are given by broken lines for (1) 298; (2) 318 K; (3) 358 K. (Reprinted with permission from reference 38. Copyright (2001) American Chemical Society)

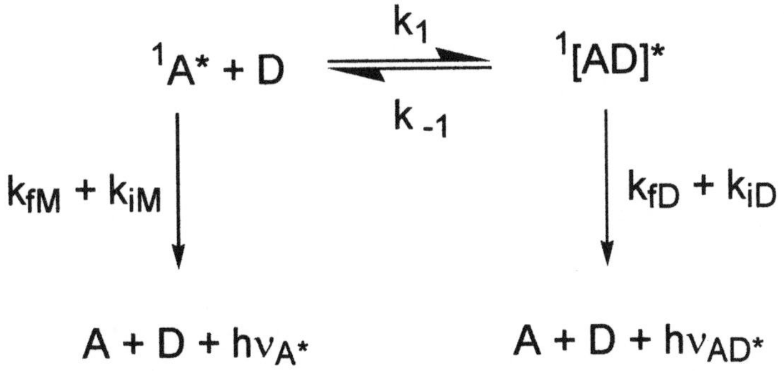

$$^{1}A^{*} + D \underset{k_{-1}}{\overset{k_1}{\rightleftharpoons}} {}^{1}[AD]^{*}$$

$$^{1}A^{*} + D \xrightarrow{k_{fM} + k_{iM}} A + D + h\nu_{A^*} \qquad {}^{1}[AD]^{*} \xrightarrow{k_{fD} + k_{iD}} A + D + h\nu_{AD^*}$$

Scheme 4. Kinetic scheme for exciplex formation and decay. In the scheme the unimolecular rate constants k_{fM} and k_{fD} are those for radiative decay of the monomer and exciplex, while k_{iM} and k_{iD} are the rate constants for non-radiative decay of the monomer and exciplex.

$$\frac{I_{AD^*}}{I_{A^*}} = \frac{k_{fD}}{k_{fM}} \frac{k_1[D]}{k_{fD} + k_{iD} + k_{-1}} \tag{1}$$

In Eq. (1), the rate constants k_1 and k_{-1} are for activated processes. As such, the steady-state populations of exciplex and monomer are expected to be temperature dependent. The temperature dependence of the processes shown in Scheme 4 has been derived and is schematically illustrated in Figure 4.[39] Figure 4 depicts two limiting and linear regions.

In the low temperature limit (high values of 1/T), exciplex formation is slow relative to the intrinsic deactivation of the excited-state of the monomer (predominantly fluorescence for aromatic hydrocarbons). In this limit, k_{-1} and k_{iD} are much smaller than k_{fD}, and Eq. (1) reduces to Eq. (2).

$$\text{Low T-behavior:} \quad \frac{I_{AD^*}}{I_{A^*}} = \frac{k_{fD} k_1 [D]}{k_{fM}} \exp\left(\frac{-E_{A_1}}{RT}\right) \tag{2}$$

where E_{A1} is the activation energy for exciplex formation. In the high temperature limit, where the equilibrium shown in Scheme 4 is rapidly established relative to decay of the exciplex (k_1, $k_{-1} >> k_{fD} + k_{iD}$), Eq. (1) reduces to:

$$\text{High T-behavior:} \quad \frac{I_{AD^*}}{I_{A^*}} = \frac{k_{fD}}{k_{fM}} \frac{k'_1 [D]}{k'_{-1}} \exp\left(\frac{-(E_{A_1} - E_{A_{-1}})}{RT}\right) \tag{3}$$

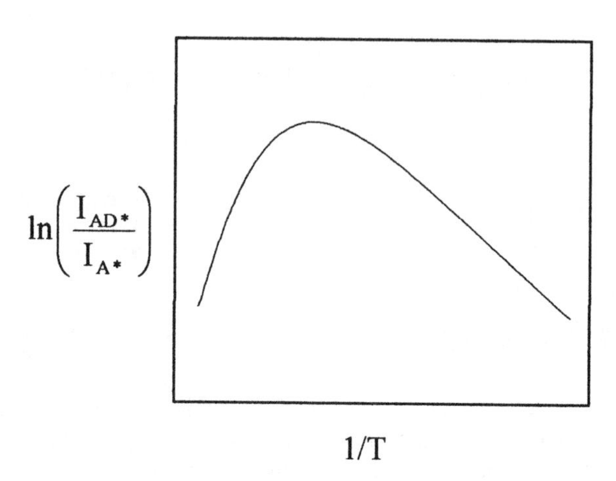

Figure 4. Predicted temperature dependence (based on Scheme 4) of exciplex to monomer emission intensities

where E_{A-1} is the activation energy for monomer production from the exciplex.

Both Eqs. (2) and (3) predict a logarithmic dependence of the ratio of exciplex to monomer intensities to reciprocal temperature. In Eq. (3), the difference in activation energies (E_{A1}-E_{A-1}) is equal to the free enthalpy (ΔH) of exciplex formation. Since the exciplex formation is an exothermic process, Eq. (3) predicts that ratio of exciplex to monomer concentrations (and thus intensities) should decrease with increasing temperature. We expect the interconversion of A* to (AD*) to be modeled simply by the thermal population of the upper state (A*). In this model, the ratio of luminescences corresponding to each state will be fitted to the van't Hoff Equation (Eq. 4).

$$\ln\left[\frac{\left(\frac{I_{(AD^*)}}{I_{A^*}}\right)_{T_2}}{\left(\frac{I_{(AD^*)}}{I_{A^*}}\right)_{T_1}}\right] = \frac{-\Delta H}{R}\left(\frac{1}{T_2} - \frac{1}{T_1}\right) \qquad (4)$$

In Eq. (4), the intensities (I), are proportional to the fluorescence quantum yields of monomer (A*) and exciplex (AD*) and are experimentally obtained from the steady-state spectra at T_1 and T_2. Since exciplex formation is enthalpically downhill ($\Delta H < 0$), the Equation predicts that the slope of Eq. (1) will be positive.

From the spectra shown in Figure 3, it is clear that the equilibrium depicted in Scheme 1 is shifted towards the monomer as the temperature is increased. The isoemis-

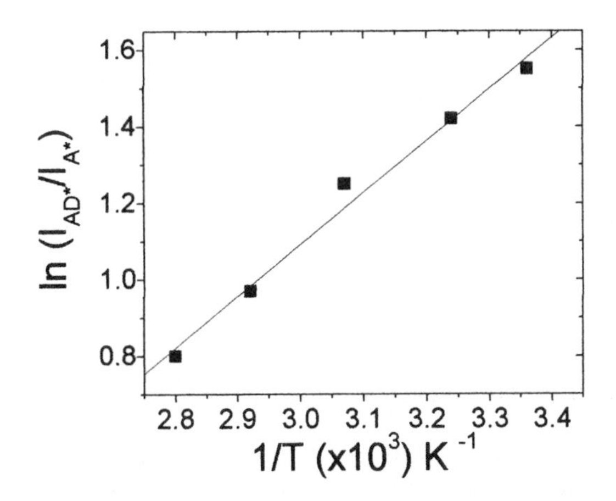

Figure 5. Intensity data from Figure 3 plotted according to Eq. (1). The intensities represent the weights of the individual guassians used to model the monomer and excimer emission bands.[38]

sive point is evidence for the simple two-state, fast equilibrium model shown in Scheme 4. The increase in monomer intensity is attributed to the thermal population of A* on the excited state potential energy surface. This represents the high temperature regime, where the monomer/exciplex equilibrium is rapidly established relative to deactivation of either species.

The ratios of these intensities were plotted according to Eq. (4). The plot is shown in Figure 5. From the slope of the line, the enthalpy of exciplex (AD*) formation was determined ($\Delta H = -2.6 \pm 0.2$ kcal/mole). The existence of an isoemissive point is consistent with the lack of temperature dependence in the fluorescence spectrum of a polystyrene film containing only perylene.

3.3 Temperature-Sensitive Materials Prepared from Styrene and Para-N,N-(dimethylamino)styrene (An-II)

More recently, we have been using a new class of co-polymers, where the anchor point of the aniline onto the polymer backbone is in the para position relative to the dimethylamino electron donating group. We have found this co-monomer to be more compatable with styrene in preparing the co-polymers. As such, we have been able to explore a broader range of compositions, to investigate how polymer structure and composition governs the observed temperature sensitivity and reversibility. In addition, the new polymers exhibit substantially improved temperature sensitivity. Finally, the "direction" of the temperature-dependent change is reversed. Unlike the polymers prepared using **An-I**, the ratio of exciplex to monomer is found to increase with temperature. Details are discussed below.

p-(N,N-dimethylamino)styrene (**An-II**) was co-polymerized with styrene and cross-linker. Preparation of the aniline derivative (**An-II**), via the Wittig reaction, was modi-

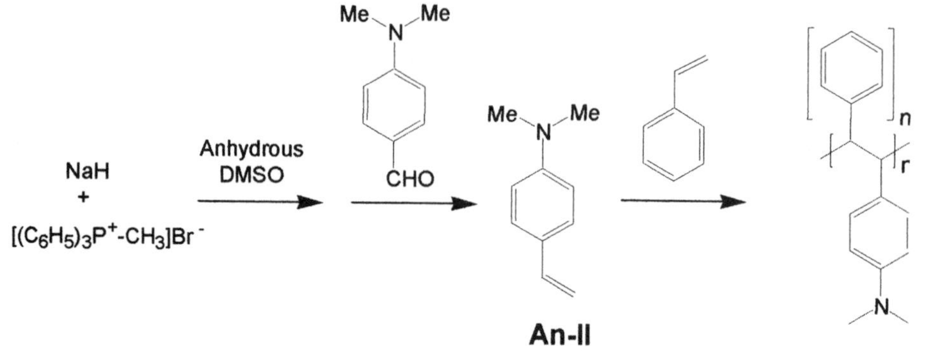

NaH + $[(C_6H_5)_3P^+\text{-}CH_3]Br^-$ → (Anhydrous DMSO) → (p-Me₂N-C₆H₄-CHO) → An-II → (styrene) → copolymer

Scheme 5. Synthesis of styrene-p-(N,N-dimethylamino)styrene (**An-II**) co-polymers.

fied from a published report.[40] Synthesis of the monomer and polymer are shown in Scheme 5. Perylene was encapsulated in the material and the temperature response was investigated.

The temperature-dependent fluorescence spectra of this material are shown in Figure 6. In Figure 6, the peaks occurring at 463, 475nm and the shoulder at 525 nm may be assigned to the perylene monomer. The emission maxima for the perylene monomer are the same as those shown in Figure 3. On controlled heating of the film, two changes are observed: (i) a small red-shift in the exciplex emission and (ii) a substantial increase in the ratio of exciplex to monomer emission. The magnitude of ratiometric change is clearly seen in the inset of Figure 6, where the spectra are normalized at 475 nm.

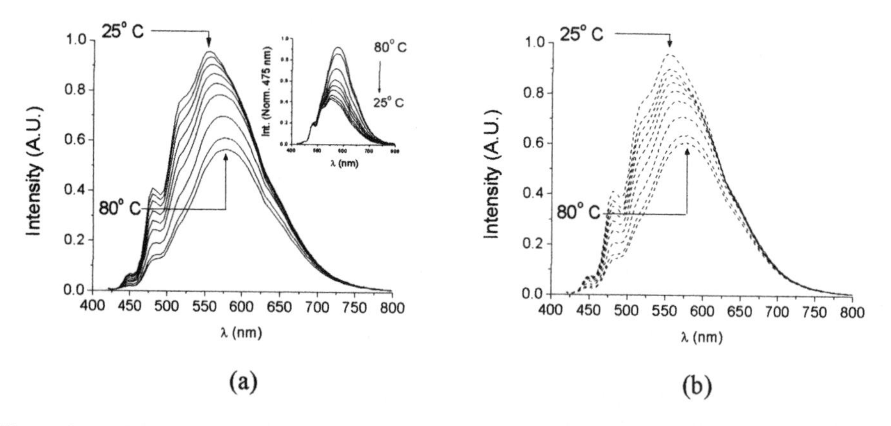

Figure 6. (a) Emission spectra observed upon 410 nm excitation as a pellet of a copolymer of styrene and p-(N,N-dimethylamino)styrene (**An-II**) was heated. The polymer was synthesized using free-radical polymerization of a co-monomer solution containing 39.4 % styrene, 20.8 % An-II and 37.6 % dibutylphthalate (percentages are weight percentages). Within the polymer there are, on average, 2 styrene monomer units per aniline. Spectra shown in (b) are those recorded as the same sample is cooled back to 25 °C.

3.4 Interpreting the Temperature Response

When exciplex formation occurs within the **An-II**-styrene copolymers, the ratio of exciplex to monomer is found to increase with temperature, unlike the polymer film prepared from **An-I**. The data are consistent with slow repopulation of the monomer excited state (k_{-1} in Scheme 4), and the interconversion of A* to (AD*) is modeled as an activated process (low temperature limit described above). In this case, the exciplex population (and thus intensity) is expected to increase with temperature. In Eq. 3, the activation energy for exciplex formation from the perylene excited state (E_{A1}) may be obtained. Shown in Figure 7 are the data extracted from the spectra shown in Figure 6. The data fit the model well, and an activation energy for exciplex formation of 3.0 kcal/mole is obtained.

3.5 Effect of Polymer Composition on Sensitivity.

Recently, we have been investigating the effect of polymer composition and preparation to elucidate the best conditions for optimum sensitivity and stability of our exciplex-forming polymer films. Composition effects on the fluorescence emission spectrum and temperature sensitivity have been demonstrated (unpublished work). A typical sample was prepared by encapsulating perylene in a poly[styrene-co-*p*-(N,N-Dimethylamino)styrene] matrix. Free radical polymerization was initiated using 2,2'-azobisisobutyronitrile (AIBN) in the presence of dissolved perylene. After the polymerization, the vial was cut and the pellet was removed. After washing the pellet with methanol, it was heated to 100°C for 1 hour to remove any unreacted monomer and methanol solvent.

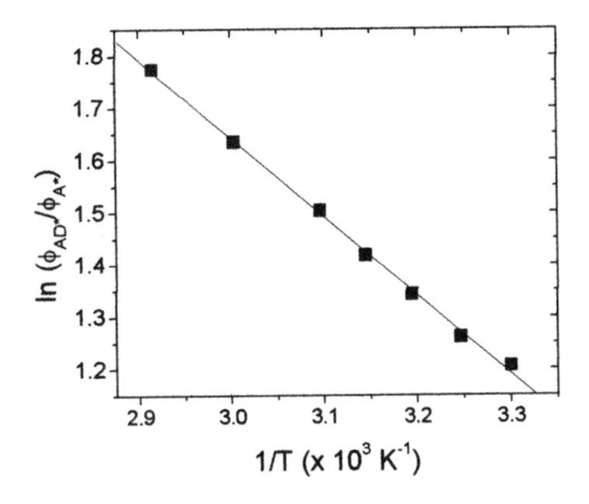

Figure 7. Intensity data from Figure 6 plotted according to the low-temperature limiting case.

To take advantage of the two-color ratiometric response of exciplex-forming polymers, we have found the polymer composition to be extremely important. Specifically, the amount of aniline that is added into the polymer must be such that an observable population of both monomer and exciplex is measured. If the polymer contains too much aniline, exciplex formation (via both static and dynamic quenching) is rapid and competes effectively with monomer deactivation. If the polymer matrix is too dilute in aniline co-monomer, only emission from the perylene monomer is observed. While absolute intensity changes may be observed in both of these cases, temperature sensing, via changes in emission intensity ratios, is not viable. For our aniline-styrene co-polymers, we have found that using molar ratios of the styrene to aniline monomers in the range of 2 – 6 gives observable emission from both the monomer and exciplex in casted films. The spectra shown in Figure 6 represent the most aniline-rich polymer we have studied. The material shows emission from both perylene monomer and exciplex, but the latter is clearly dominant.

Shown in Figure 8 are the temperature-dependent emission spectra of a polymer that is made using ca. 2.5 times less aniline (relative to styrene) in the co-monomer solution. At 25°C, there is clearly less observed exciplex emission relative to that shown in Figure 6. When the spectra are normalized (Figure 8 inset), there is a measurable increase in exciplex to monomer intensity with temperature. However, the magnitude of change, over the 55o temperature range that was used, is substantially smaller than that shown in Figure 6.

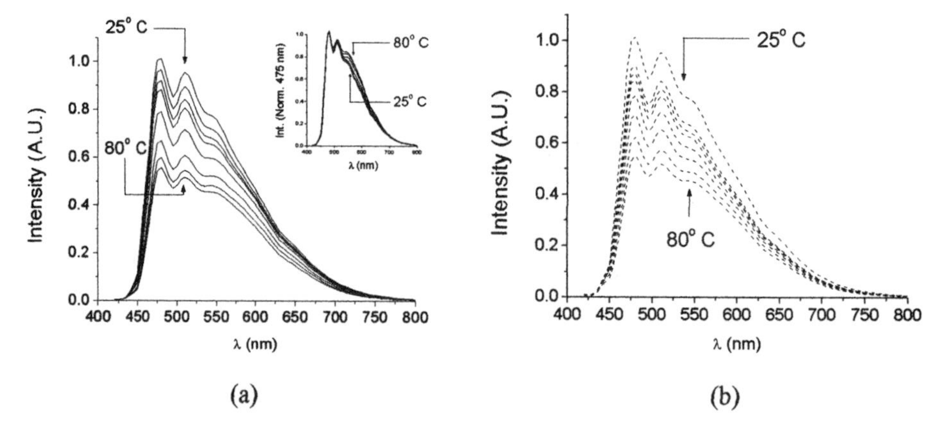

Figure 8. (a) Emission spectra observed upon 410 nm excitation of a pellet of a copolymer of styrene and p-(N,N-dimethylamino)styrene (**An-II**) was heated. The polymer was synthesized using free-radical polymerization of a co-monomer solution containing 48.4 % styrene, 11.8 % **An-II** and 37.6 % dibutylphthalate (percentages are weight percentages). Within the polymer there are, on average, ca. 5 styrene monomer units per aniline. The spectra shown in (b) are those recorded as the same sample is cooled back to 25 °C.

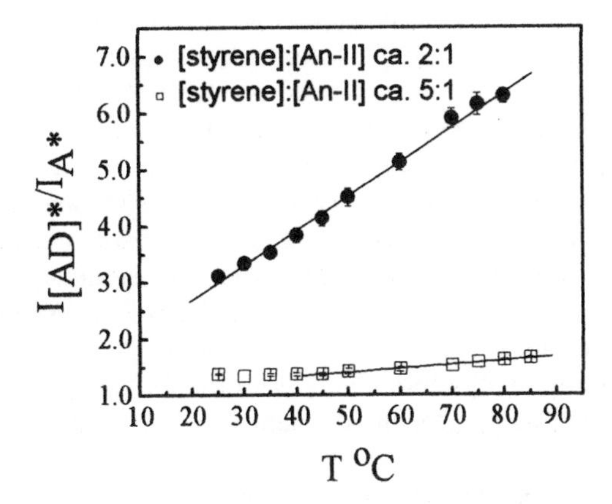

Figure 9. Temperature sensitivity of a polymer film containing perylene and two different ratios of styrene and p-(N,N-dimethylamino)styrene co-monomers. Monomer and exciplex intensities were evaluated as the integrated areas in the 420 – 525 nm and 525 – 800 nm ranges, respectively

A range of compositions has been synthesized and the temperature sensitivity has been quantified for each material. The details will be published elsewhere. However, as a general trend, we conclude that for our co-polymers of styrene and para-N,N-dimethylaminostyrene, optimum temperature sensitivity is obtained as the amount of aniline co-monomer within the polymer increases. Plots of the temperature sensitivity for the two materials whose emission spectra are shown in Figures 6 and 8 are shown in Figure 9. Detailed materials characterization studies are on-going to correlate how variation in the composition affects e.g. the glass transition temperature of the polymer. Qualitatively, we believe the origin of the effect is in the "fluidity" of the polymer. Due to the strong pi-stacking interactions of the styrene sub-units, polystyrene films are very hard. As the concentration of the aniline co-monomer increases, it is anticipated that the dimethylamino groups should disrupt the styrene-styrene interactions and allow more mobility of the interacting groups within the polymer matrix. This is also the role played by the dibutylphthalate that is added to plasticize the polymer. However, the two polymers whose temperature sensitivity is shown in Figure 9 contain identical amount of the plasticizer. In summary, the temperature sensitivity of these polymers is clearly modified by changing the composition. However, such changes will affect both the material properties and the exciplex photophysics.

4. PROGRESS TOWARDS EXCIPLEX-FORMING MOLECULAR THERMOMETERS.

As demonstrated in this chapter, exciplex-forming polymers are ideal as molecular thermometers. As functionalized polymers, films can be casted onto a fiber optic tip for fiber optic thermometry applications. In addition, the polymers can be used as clear, fluorescent coatings to map temperature distributions across an illuminated and imaged surface. We have presented some results showing that subtle structural variations within the polymer can lead to large changes in how the materials respond to temperature. The systems are quite interesting from a thermodynamic and kinetics perspective. As summarized in this chapter, changing the orientation of the electron donating group on the aniline co-monomer causes the "direction" of response to reverse. Although the results are interpreted using simple physical chemistry models developed for diffusional processes, the dynamics are certainly much more complex in the polymer matrix. Thus, there is much more to be learned about the microdynamics within these temperature sensitive materials.

To date, we have demonstrated a linear temperature response (as shown in Figure 9, circles) with our exciplex-forming polymers. From shown in Figure 9, our best materials to date exhibit a sensitivity of ca. 6% per degree. With this sensitivity, we can accurately measure temperature changes to a precision of a few tenths of a degree. Ultimately, the precision of measurement will depend upon the specific illumination and detection system used.

5. PROSPECTS ON DEVELOPING NEW MOLECULAR THERMOMETERS

In conclusion, there continue to be endless opportunities to design sensors, both chemical and physical, using elementary principles of photophysics. As presented in the Introduction to this chapter, there are numerous applications where conventional temperature measurement techniques are not viable. Thus, it is crucial for the continued development of material-based temperature sensors for use in homogeneous solution, heterogeneous media, and as coated films. While we have explored only a single exciplex-forming system Coupling these principles with material science opens new avenues for sensors of physical properties such as temperature. We must simply capitalize on what is well-understood in solution, and couple it to material science. From our initial pursuits, it is clear that "what is well-understood in solution" does not hold true when the phenomena are studied in synthetic materials (e.g. Why do the two categories of materials discussed in this article exhibit strikingly different temperature dependences?). Clearly, there is a great deal of fundamental understanding that needs to be gained in studying these materials. At the same time, practical and functional materials can be developed for sensor applications.

By coupling well-known photophysical effects with polymer chemistry, there are an infinite number of new materials that can be made. Modifications in the polymer building blocks and size can alter (i) the viscosity of polymer solutions; (ii) the dielectric in the vicinity of the polymer; and (iii) the mobility of the interacting units, just to name a few. All of these factors will inevitably play a role in optimizing the temperature sensitivity of new classes of materials. Finally, there are numerous other photophysical processes that rely on temperature-dependent intermolecular interactions. Any one of these processes can form the basis for developing new molecular thermometers.

6. REFERENCES

1. Turner, A. P. F., Karube, L., and Wilson, G. S., Eds. *Biosensors. Fundamentals and Applications.* **1987**, Oxford University Press: Oxford, UK.
2. Janata, J. and Josowicz, M., "Chemical Sensors" *Analytical Chemistry*, **1998**, *70*: p. 179R - 208R.
3. Cao, W., Cudney, H. H., and Waser, R., "Smart Materials and Structures" *Proceedings of the National Academy of Sciences USA*, **1999**, *96(15)*: p. 8330-8331.
4. Miller, J. S., *Encyclopedia of Smart Materials*, Ed. J. Harvey. **2000**, New York: John Wiley and Sons.
5. Childs, P. R. N., Greenwood, J. R., and Long, C. A., "Review of Temperature Measurement" *Reviews of Scientific Instruments*, **2000,** *71*: p. 2959-2978.
6. Wagner, J. R., van Lier, J. E., Decarroz, C., Berger, M., and Cadet, J., "Photodynamic Methods for Oxy Radical-Induced DNA Damage" *Methods in Enzymology*, *1990*, **186**: p. 502 - 520.
7. Wickersheim, K. A. and Sun, M. H., "Fluoroptic Thermometry" *Medical Electronics*, **1987**: p. 84 - 91.
8. Jia, D., Zhao, L., Cui, S., and Lin, Y., "Optical Fiber Fluorescent Thermometry for Electromagnetic Induced Heating in Medicine Treatment" *Proceedings SPIE*, **2002**, *4916*: p. 95 - 97.
9. Liu, T., Campbell, B., and Sullivan, J. P., "Heat Transfer Measurement on a Waverider at Mach 10 Using Fluorescent Paint" *Journal of Thermophysics and Heat Transfer*, **1995**, *9(4)*: p. 605-611.
10. Hubner, J., Carroll, B., Schanze, K., Ji, H., and Holden, M., "Temperature and Pressure Sensitive Paint Measurements in Short Duration Hypersonic Flow" *AIAA Journal*, **1998**, *99-0388*.
11. Bur, A. J., Vangel, M. G., and Roth, S. C., "Fluorescence Based Temperature Measurements and Applications to Real-Time Polymer Processing" *Polymer Engineering and Science*, **2001**, *41:* p. 1380 - 1389.
12. Bur, A. J. and Roth, S. C., "Temperature Monitoring of Capillary Rheometry Using a Fluorescence Technique" *Antec*, **2001**: p. 3071 - 3075.
13. Bur, A. J., Vangel, M. G., and Roth, S., "Temperature Dependence of Fluorescent Probes for Applications to Polymer Materials Processing" *Applied Spectroscopy*, **2002**, *56:* p. 174 - 181.
14. Babinsky, H. and Edward, J. A., "Automatic Liquid Crystal Thermography for Transient Heat Transfer Measurements in Hypersonic Flow" *Experiments in Fluids*, **1996**, *21(4)*: p. 227-236.
15. Belsky, V. K., Fernandez, V., Zavodnik, V. E., Diaz, I., and Martinez, J. L., "Crystal Structure and Thermochromism of Bisguanidinium Tetrachlorocuprate Dihydrate at 123 and 293K" *Crystallography Reports*, **2001**, *46(5)*: p. 779-785.
16. Jaw, H. R. C., Mooney, M. S., Novinson, T., Kaska, W. C., and Zink, J. I., "Optical Properties of the Thermochromic Compounds Ag_2HgI_4 and Cu_2HgI_4" *Inorganic Chemistry*, **1987**, *26(9)*: p. 1387-1391.
17. Maaza, M., Bouziane, K., Maritz, J., McLachlan, D. S., Swanepool, R., Frigerio, J. M., and Every, M., "Direct Production of Thermochromic VO_2 Thin Film Coatings by Pulsed Laser Ablation" *Optical materials*, **2000**, *15(1)*: p. 41-45.
18. Smith, C. R., Sabatino, D. R., and Praisner, T. J., "Temperature Sensing with Thermochromic Liquid Crystals" *Experiments in Fluids*, **2001**, *30(2)*: p. 190-201.
19. Allison, S. W. and Gillies, G. T., "Remote Thermometry with Thermographic Phosphors: Instrumentation and Applications" Reviews of Scientific Instruments, **1997**, *68*: p. 2615-2650.
20. Ozawa, L. and Jaffe, P. M., "Mechanism of Emission Color Shift with Activator Concentration in Eu^{3+} Activated Phosphors" *Journal of the Electrochemical Society*, **1971**, *118*: p. 1678.
21. Gallery, J., Gouterman, M., Callis, J., Khalil, G., Mclachlan, B., and Bell, J., "Luminescent Thermometry for Aerodynamic Measuerements" *Reviews of Scientific Instruments*, **1994**, *65(3)*: p. 712-720.
22. Morris, M. J., Benne, M. E., Crites, R. C., and Donovan, J. F. "Aerodynamic Measurements Based on Photoluminescence" In *31st Aerospace Sciences Meeting and Exhibit.* **1993**, Reno, NV.
23. Shen, Y., Bedlek-Anslow, J. M., Hubner, J. P., Carroll, B. F., Ifju, P. G., and Schanze, K. S., "Advances in Pressure-, Temperature-, and Strain-Sensitive Paints" *Polymer Preprints*, **2002**, *43*: p. 69 - 70.
24. Hai-Feng, J., Shen, Y., Hubner, J. P., Carroll, B. F., Schmehl, R. H., Simon, J. A., and Schanze, K., "Temperature-Independent Pressure-Sensitive Paint Based on a Bichromophoric Luminophore" *Applied Spectroscopy*, **2000**, *54*: p. 856 - 863.
25. *http://hawk.mae.ufl.edu/bfc/Projects/PSP_TSP/psp_tsp.html*
26. Gouterman, M., "Oxygen Quenching of Luminescence of Pressure Sensitive Paint for Wind Tunnel Research" *Journal of Chemical Education*, **1997**, *74*: p. 697-702.

27. Fister, J. C., Rank, D., and Harris, J. M., "Delayed Fluorescence Optical Thermometry" *Analytical Chemistry*, **1995**, *67*: p. 4269 - 4275.
28. Bai, F. and Melton, L. A., "High-Temperature, Oxygen-Resistant Molecular Fluorescence Thermometers" *Applied Spectroscopy*, **1997**, *51*: p. 1276 - 1280.
29. Schrum, K. F., Williams, A. M., Haerther, S. A., and Ben-Amotz, D., "Molecular Fluorescnece Thermometry" *Analytical Chemistry*, **1994**, *66*: p. 2788 - 2790.
30. Engeser, M., Fabbrizzi, L., M., L., and Sacchi, D., "A Fluorescent Molecular Thermometer Based on the Nickel (II) High-Spin/Low -Spin Interconversion" *Chemical Communications*, **1999**: p. 1191-1192.
31. Heskins, M. and Guillet, J. E., *Journal of Macromolecular Science, Chemstry A2*, **1968**: p. 1441.
32. Principi, T., Goh, C. C. E., Liu, R., C. W., and Winnik, F., M., "Solution Properties of Hydrophobically Modified Copolymers of N-Isopropylacrylamide and N-Glycine Acrylamide: A Study by Microcalorimetry and Fluorescence Spectroscopy" *Macromolecules*, **2000**, *33*: p. 2958 - 2966.
33. Mizusaki, M., Morishima, Y., and Winnik, F. M., "Hydrophobically Modified Poly(Sodium 2-Acrylamido-2-Methylpropanesulfonate)s Bearing Octadecyl Groups: A Fluorescence Study of Their Solution Properties in Water" *Macromolecules*, **1999**, *32*: p. 4317 - 4326.
34. Winnik, F. M., Ottaviani, M. F., Bossman, S. H., Pan, W., Garcia-Garibay, M., and Turro, N. J., "Phase Separation of Poly(N-Isopropylamide) in Water: A Spectroscopic Study of a Polymer Tagged with a Fluorescent Dye and a Spin Label" *Journal of Physical Chemistry*, **1993**, *97*: p. 12998 - 13005.
35. Birks, J. B., Lumb, M. D., and Munro, I. H., *'Excimer' Fluorescence V.* "Influence of Solvent Viscosity and Temperature" *Proceedings of the Royal Society of London. Series A*, **1964**, *280*: p. 289 - 297.
36. Birks, B. J., *Photophysics of Aromatic Molecules*. **1970**: John Wiley and Sons, Ltd.
37. Kurokawa, K., Honnma, T., Jinbo;, S., Y. H., and Nakamura, K., "Exciplex Emission from Perylene in Styrene Polymer" *Journal of Photopolymer Science and Technology*, **1990**, *3*: p. 37-41.
38. Chandrasekharan, N. and Kelly, L. A., "A Dual Fluorescence Temperature Sensor Based on Perylene/Exciplex Interconversion" *Journal of the American Chemical Society*, **2001**, *124*: p. 9898-9899.
39. Birks, J. B., Dyson, D. J., and Munro, I. H., "'Excimer' Fluorescence II. Lifetime Studies of Pyrene Solutions" *Proceedings of the Royal Society of London. Series A*, **1963**, *275*: p. 575 - 588.
40. Yasudo, M., Harano, K., and Kanematsu, K., "High Peri- and Regiospecificity of Phencyclone: Kinetic Evidence of the Frontier-Controlled Cycloaddition Reaction and Molecular Structure of the Cycloadduct" *Journal of Organic Chemistry*, **1980**, *45(4)*: p. 659 - 664.

FROM MOLECULAR LUMINESCENCE TO INFORMATION PROCESSING

Stéphane Content, A. Prasanna de Silva and David T. Farrell[a]

1. INTRODUCTION

Luminescence and information processing may appear to belong to different worlds, but many links can be discovered if we only care to look. Luminescence has a long history of being open to quenching, as might be expected of a phenomenon originating from an excited state which can be easily perturbed (Lakowicz, 1999; Valeur, 2001). Neutralization of the quenching agent naturally reinstates the luminescence (de Silva et al., 1997). So reversible switching of molecule-based light signals is not hard to achieve. Being highly visual, such signals are easily interfaced with humans even though they arise in a far smaller molecular domain. The ease of detection of light signals means that even one molecule will suffice, under appropriate conditions (Brasselet and Moerner, 2000). Switching of a light signal between "on" and "off" situations allows binary digital interpretations, i.e. 1 and 0 respectively.

An easy way of achieving such luminescence switching is via photoinduced electron transfer (PET) (Bissell et al., 1993). PET (when thermodynamically feasible) can outrun luminescence, especially in situations where intramolecular quenching is arranged. This is the case with "lumophore-spacer-receptor" systems. In one manifestation, the receptor serves as the electron donor and the lumophore is the electron acceptor (Figure 1). Of course, photoexcitation of the lumophore launches the entire process. The receptor module is vital to our story since this is where the chemical input is received. Naturally, this occupation of the receptor alters its electron donor characteristics. In the ideal case, the electron donor ability is greatly reduced. So the PET process becomes unfeasible in a thermodynamic sense and luminescence wins out. Ionic inputs can exert large influences primarily through their ability to hold back an electron electrostatically. Photoionic systems with ionic inputs and photonic outputs therefore represent one of the commonest types of molecular luminescent switches.

[a] Stéphane Content, Analytical Chemistry Department, Procter & Gamble Eurocor, Temselaan 100, 1853 Grimbergen, Belgium. E-mail: content.s@pg.com. A. Prasanna de Silva and David T. Farrell, School of Chemistry, Queen's University, Belfast BT9 5AG, Northern Ireland. E-mail: a.desilva@qub.ac.uk, d.farrell@qub.ac.uk

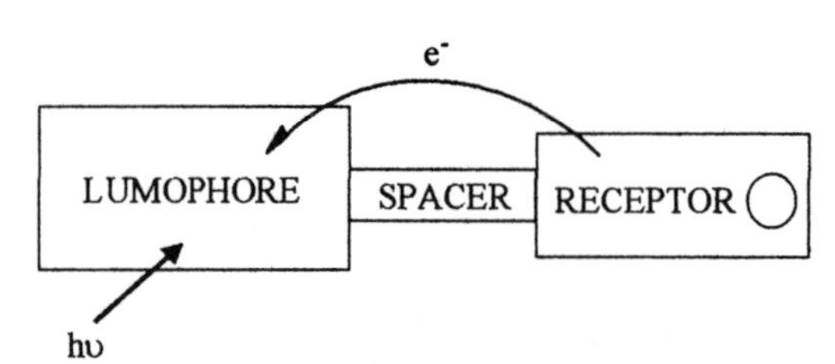

Figure 1. The "lumophore-spacer-receptor" system and its principle of operation.

The previous paragraphs outlined how a molecule- based light output can be controlled by a chemical input. Such control can be of increasing degrees of complexity. Simple switches are then replaced by various binary logic functions (Ben-Ari, 1993). Some of these control patterns are shown in the form of truth tables in Figure 2. These are labelled with their logic type or, in complex cases, with the number of the molecular structure where the logic function is seen. For instance, a two-input AND logic function produces an "on" output only if both inputs are "on" themselves. Molecules that exhibit such logic functions (Balzani et al., 2003) are clearly involved in information processing as seen in an electronics context (Malvino and Brown, 1993; Mitchell, 1995).

Two-input logic

		AND	OR	NOR	NAND	INH	XOR
IN1	IN2	OUT	OUT	OUT	OUT	OUT	OUT
0	0	0	0	1	1	0	0
0	1	0	1	0	1	1	1
1	0	0	1	0	1	0	1
1	1	1	1	0	0	0	0

Three-input logic

			INH	EOR	**24**	**25/26**
IN1	IN2	IN3	OUT	OUT	OUT	OUT
0	0	0	0	0	0	1
0	1	0	0	0	1	0
1	0	0	0	0	0	1
1	1	0	1	0	0	0
0	0	1	0	0	0	1
0	1	1	0	1	1	0
1	0	1	0	1	1	0
1	1	1	0	1	1	0

Figure 2. Some truth-tables for logic gates.

2. A COMPARATIVE STUDY

It is educational to consider an early semiconductor electronics implementation of a two-input AND logic gate in some detail (Figure 3) (Hughes, 1987). Diode-diode-logic (DDL), as its name implies, builds logic gates from diodes. A diode only passes current in one direction when a positive voltage is driving it. Application of a zero or negative voltage causes no current to flow (in the opposite direction). The diode arranges such a one-way traffic thanks to an internal electric field set up at the junction between two semiconductors carrying majority charge carriers of opposite sign (p- and n-).

Interestingly, related electric field effects can control electron movement in molecule-based systems (de Silva et al., 1995).

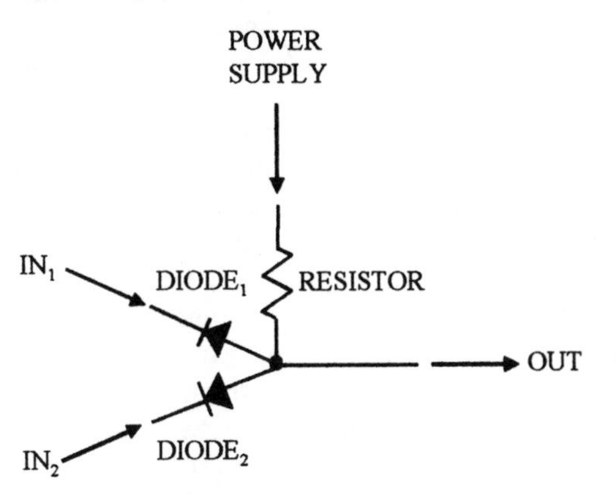

Figure 3. Circuit diagram for a diode-diode-logic AND gate.

The AND gate of Figure 3 employs 5 V as "on", 0 V as "off" for both input$_1$, input$_2$ and output. Importantly, the power supply voltage is also 5 V. If input$_1$ is "off" (0 V i.e. earth) there is 5 V dropping across diode$_1$ (right to left) since the power supply is connected to its right hand side. Such a positive voltage drop causes the diode to conduct an electric current to the earth via the input$_1$ connection. Thus the output will sense the earth, i.e. it will read 0 V ("off"). The same output situation applies if input$_2$ is "off". On the other hand if both input$_1$ and input$_2$ are "on" (5 V), both diodes have 0 V drop across them. So they both cease to conduct. Now the output will clearly sense the voltage (5 V) of the power supply. This is the "on" situation of the output. The resistor serves to limit current drawn through all components and prevent burnout.

Now let's compare this situation with that of a molecular photoionic AND logic gate **1** - in fact the first case in the primary literature (de Silva et al., 1993). In schematic terms the "lumophore-spacer$_1$-receptor$_1$-spacer$_2$-receptor$_2$" system (Figure 4) is a clear extension of the "lumophore-spacer-receptor" model described in Figure 1. As before, each receptor is chosen to engage in a PET process with the lumophore acting as the acceptor. Occupation of a given receptor with corresponding input, i.e. receptor$_1$ with ion$_1$, stops its PET process. However, luminescence will not appear until both PET processes are halted i.e. until

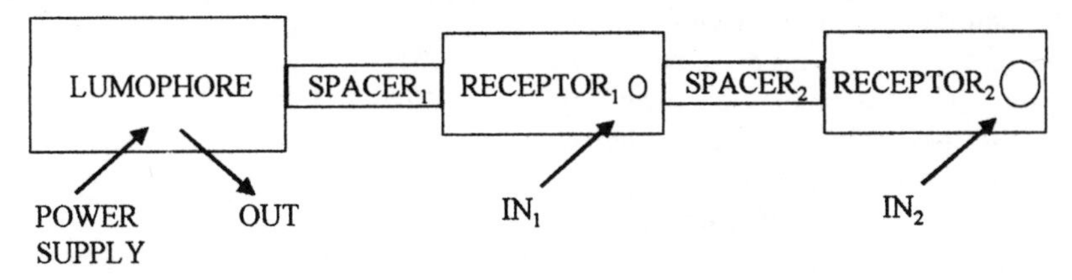

Figure 4. The "lumophore-spacer$_1$-receptor$_1$-spacer$_2$-receptor$_2$" system.

receptor$_1$ binds ion$_1$ and receptor$_2$ binds ion$_2$. Of course, this requires each of the two receptors to be essentially specific in recognizing its target ion. Thankfully this can be arranged since only one potential interferent is present in each case.

Both the semiconductor electronic and molecular photoionic cases route the power supply to the output unless the former is diverted. In both cases, each input serves to block one diversion channel. These are important functional similarities. However there are crucial differences. Figure 3 deals with electronic voltages (0 V when “off” and 5 V when “on”) for inputs and output. The power supply is 5 V as well. Hence it is no wonder that semiconductor electronic logic devices can be easily integrated by passing the output from one gate to the input of another. On the other hand, Figure 4 deals with chemical species (ions) as inputs and photons as output, with photons as the power supply. Therefore serial integration of logic functions of the photoioinic type requires extra care. Another well-publicized difference is that of size. Semiconductor gate dimensions appear to be stranded at around 100 nm after remarkable downsizing over the past three decades. In contrast, the molecular photoionic gates are functional at the 1 nm scale. The system in Figure 3 is put together by the integrating 7 wires, 1 resistor and 2 diodes (each made by fusing p- and n- semiconductors) - 12 components in all, whereas the system in Figure 4 is synthesized as a single molecular entity. In fact these photoionic logic gates are wireless. The inputs and power supply find their own way to the correct parts on the molecular device whereas the output directly interfaces with human observers. So it is clear that molecular photoionic logic gates have a significant degree of component integration in the electronic sense built-in. It is also clear that luminescent molecular information processing must develop in the future according to its strengths and weaknesses vis-à-vis the semiconductor electronic approach.

A selection of luminescent molecular systems which behave as two-input logic gates are now illustrated in some detail. The discussion of AND logic is continued in the first of these sections.

3. AND LOGIC

While we have demonstrated AND logic with ‘receptor$_1$-spacer$_1$-fluorophore-spacer$_2$-receptor$_2$’ PET systems employing simple ion inputs (de Silva, 1997), more complex chemical species have been used by Cooper and James (2000). They reported a system **2** binding a sugar diol (with a boronic acid) and an ammonium ion (with an azacrown ether) which is a selective sensor for glucosamine at near-physiological pH. **2** can also be employed as a fluorescent AND logic gate if a saccharide and ammonium ions are applied as the two inputs. We can also imagine a situation where the unprotonated glucosamine is one input for the AND gate **2** and protons the other, at a concentration such that the D-glucosamine but not the azacrown ether is protonated.

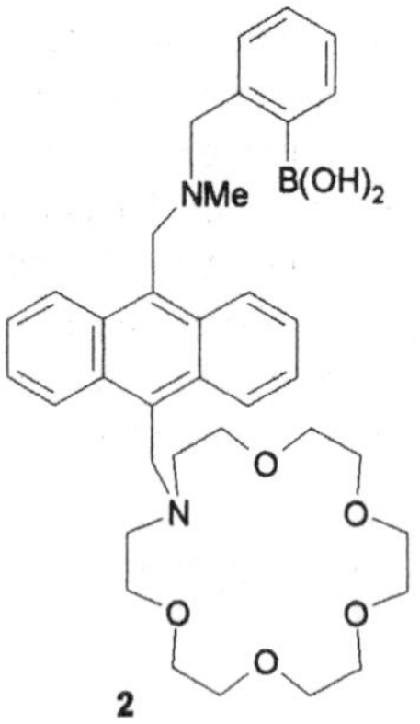

NMe B(OH)$_2$ O N O O O O **2**

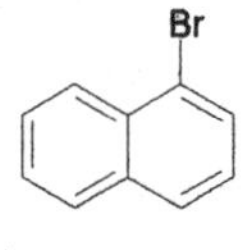

Br **3**

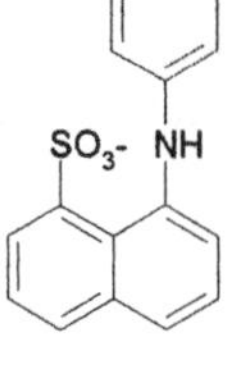

SO_3- NH **4**

The success of PET–based multi-receptor systems as AND gates allows us to discover other approaches to luminescent AND logic action even though they may not have been recognized as such at the time of publication. The luminescence behaviour of phosphor **3** is one such example. Its phosphorescence is usually quenched in aerated fluid solution at room temperature. Nocera's group has encapsulated **3** within glucosyl β-cyclodextrin and capped the complex with t-butanol (Ponce et al., 1993). These two chemical species can be viewed as the inputs. Now phosphorescence is easily observed. There are older examples of cyclodextrin-enhanced phosphorescence in the literature (Bolt et al., 1982) but these tend to be air-sensitive. Another pair of examples comes from the not-unrelated field of cyclophanes. The cyclophane macrocycle is produced from a linear precursor by complexation with a metal ion such as Zn^{2+} (Schneider and Ruf, 1990) or Ca^{2+} (Cole et al., 1992). The macrocycle can then host **4** which fluoresces once it is suitably shielded from water. The excited state of **4** is of the TICT (Twisted Intramolecular Charge Transfer) type (Kosower and Huppert, 1986; Rettig, 1994). The two chemical inputs can be viewed as the linear cyclophane precursor and the metal ion. The output is fluorescence and the AND logic gate itself is **4**.

AND logic activity can also be obtained with light and chemical inputs and fluorescence output, e.g. **5** (Pina et al., 1999). Ultraviolet irradiation (light input 1) converts the thermodynamically stable trans-form **5** to the cis-isomer **6**. Importantly, **6** cyclizes to **7** in the presence of H^+ (chemical input 1). The fluorescence emission of **7** is the output. **7** cannot be produced by photo-irradiation nor H^+ alone. Diederich has also reported work along this direction (Gobbi et al., 1999).

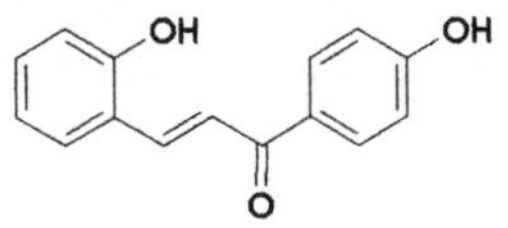

OH OH 5 OH O OH 6 OH O+ 7 CO_2^- CO_2^- N O CO_2^- CN N N MeO_2S 9 CO_2Et EtHN O N^+HEt 8

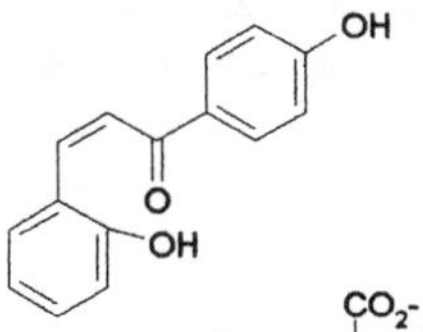

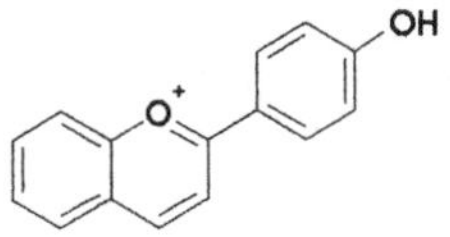

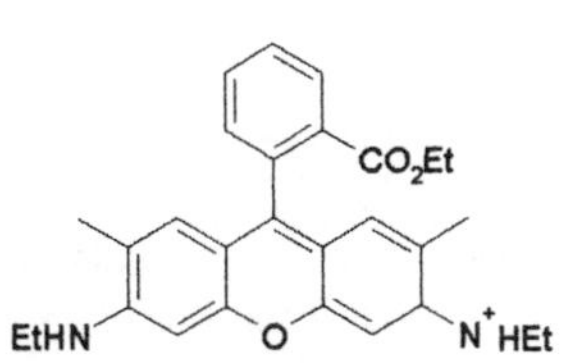

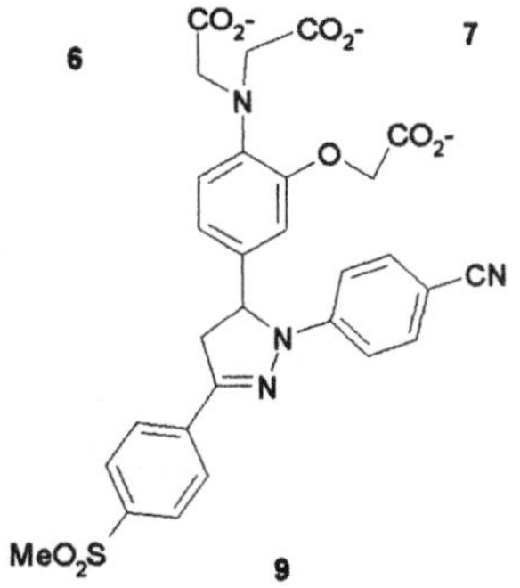

Proteins can also yield AND logic with luminescence output. To mention a recent case, Konermann's group find that cytochrome c shows AND logic activity, as well as others (Deonarine et al, 2003). The native protein has quenched fluorescence (output 0) from tryptophan since it is quite close to a quenching heme. Chemical denaturants can be used as the inputs in order to unfold the protein so that the tryptophan is able to fluoresce (output 1) by moving away from the heme. The inputs are urea and H^+ but at judiciously chosen low concentrations such that both must be present simultaneously if denaturing is to occur within the time-scale of the experiment.

Molecular AND logic can also arise from other imaginative schemes, even though the instrumentation necessary may be expensive when compared with simple luminescence experiments (which in the limit require no instruments at all). Two-photon fluorescence (TPF) (output 1) can be observed in **8** if a high-power laser is used for excitation (Remacle et al., 2001). Another way to achieve this is to employ two low-power laser beams (the two light inputs both 1). Naturally the power levels of the low-power beams must be selected judiciously. Clearly, the provision of only one or none of the laser beams will not yield TPF (output 0).

4. OR LOGIC

If a single receptor within a PET system is promiscuous towards two chemical species, there is potential for building an OR logic gate especially if the luminescence output is similarly high when "on". Similar species-binding constants are not required. Ca^{2+} or Mg^{2+} can bind to the amino acid receptor within **9** (de Silva et al., 1994). The ion-induced conformational change in the receptor is similar in the two cases. This causes loss of conjugation between the amine and the benzene ring. Loss of conjugation translates to a reduced electron transfer activity. So the PET process is virtually halted in both cases, leading to similar luminescence quantum yields.

10 is an interesting OR gate because various transition metal ions serve to switch 'on' fluorescence, though the fluorescence enhancement factors are quite diverse (Ghosh et al., 1996). Transition metal ions normally quench fluorescence very efficiently by a combination of PET, heavy atom effects, electronic energy transfer (EET) and paramagnetic effects.

5. NOR LOGIC

While NOR logic is named as a hybrid gate arising from NOT and OR components and constructed as such in semiconductor electronics (Malvino and Brown, 1993), it is not difficult to implement at the molecular scale. Of course, NOT is just one of the 16 two-input logic operations as far as George Boole is concerned (Ben-Ari, 1993). **11** (de Silva et al., 1999) has a 2,2'-bipyridyl unit which can bind either H^+ or Zn^{2+} (the

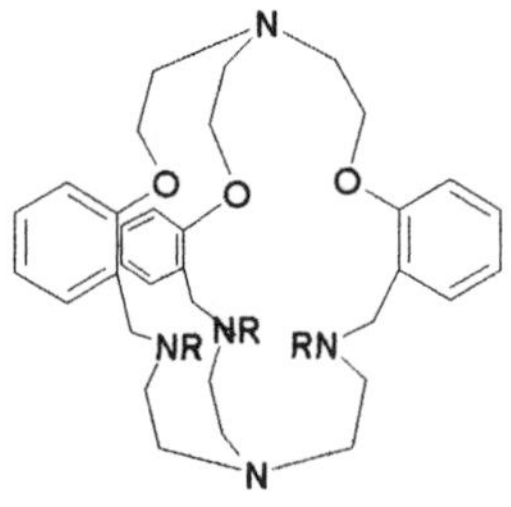

10; R = 9-Anthrylmethyl 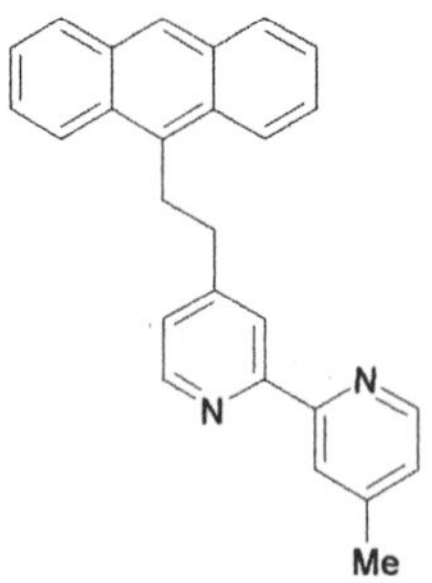 **11** 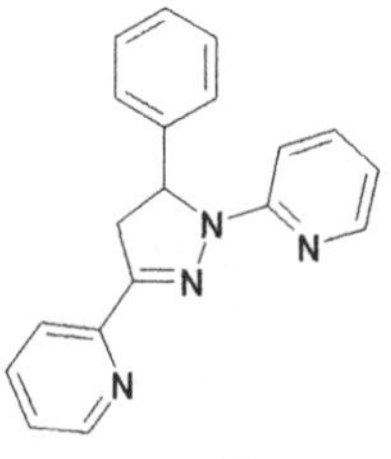 **12**

two inputs). In either case the 2,2'-bipyridyl unit becomes more electron deficient and participates in PET. Hence, the fluorescence of the anthracene unit is quenched with either or both inputs. A closely related structure is available from the work of the Fages group (Sohna Sohna et al., 1999).

Again, PET is not essential for demonstration of NOR logic activity. Protons quench the fluorescence of **12** via a vibrational loss mechanism by binding to the imine nitrogen of the pyrazoline heterocycle. Quite separately, Hg^{2+} binds to **12** which possesses a 2,2':6',2"- terpyridyl-like ligation site. Thus a non-emissive ligand to metal charge transfer (LMCT) excited state is produced owing to the redox activity of Hg^{2+}. The creation of a non-emissive LMCT state has a lot in common with PET. Thus **12** operates with H^+ and Hg^{2+} as inputs.

6. NAND LOGIC

13 was first reported as possessing AND logic activity (Iwata and Tanaka, 1994) though it is preferable now to admire it as a NAND gate. This would maintain consistency with all the other examples discussed in this review by using a positive logic convention where an "on" fluorescence signal is read as a binary output 1. The 15-crown-5 ether binds Ba^{2+}, but SCN^- ion-pairs to the Ba^{2+} via an exposed face of the crown. The SCN^- is very oxidizable so it launches a PET process to the fluorophore. Of course, the SCN^- cannot attach itself to **13** without the intermediacy of Ba^{2+}.

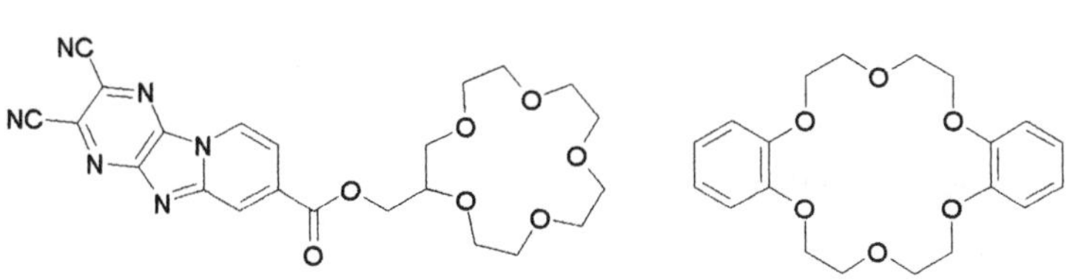

13 **14**

However, this result is not without a precedent of sorts. In an early paper, Wolfbeis and Offenbacher (1984) described the quenching of fluorescence of dibenzo-18-crown-6 ether **14** in the presence of K^+ and I^-, though the logic possibilities were not recognized at the time. The oxidizable I^- must be ion-paired to the K^+ within the crown ether as in the previous example in order to achieve this effect.

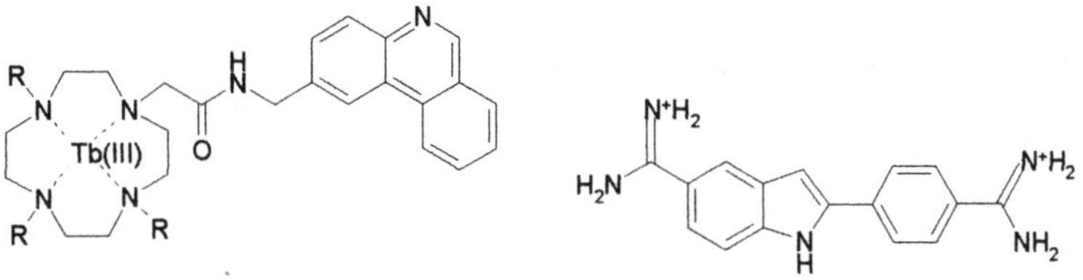

15; R = $CH_2P(Me)O_2^-$ **16**

Parker and Williams (1998) found the same logic action in the Tb^{3+} complex **15**. The lanthanide emission is switched 'off' when H^+ and O_2 are simultaneously present. Protonation of the phenanthridine sidechain lowers the energy of its triplet excited state so that it mixes with the Tb^{3+} $^5D_4$ excited state. So the metal-centred state displays an unusual O_2 sensitivity.

Akkaya and Baytekin (2000) used the common DNA binder **16** which prefers to intercalate into adenine(A)-thymine(T) base-pairs. **16** binds to a single A-T mononucleotide pair if care is taken to use an aqueous-organic solvent mixture. Small but significant changes are seen in the fluorescence spectra. A low emission intensity at a judiciously chosen wavelength (455 nm) is seen only when both A and T are present.

DNA oligomers can also show NAND logic activity when chemical inputs and fluorescence output are cleverly used (Saghatelian et al., 2003). **17** is a single-stranded 16-mer oligonucleotide with a carboxyfluorescein fluorophore at the 3'-terminus. A complementary 16-mer (**18**) is involved as the first input. The common DNA intercaland **19** forms the second input. Selective excitation of the carboxyfluorescein fluorophore gives strong emission (output 1) when one or both inputs are absent. When both inputs are applied, hybridization occurs between **17** and **18**. Also, **19** intercalates into the resulting duplex within quenching range of the carboxyfluorescein group. The quenching mechanism is EET. So the fluorescein emission is quenched (output 0). **17** can also show AND logic with different inputs. Longer DNA oligomers are used by the Stojanovic group as inputs and the logic device itself. Fluorescence outputs still provide the human communication (Stojanovic et al., 2002).

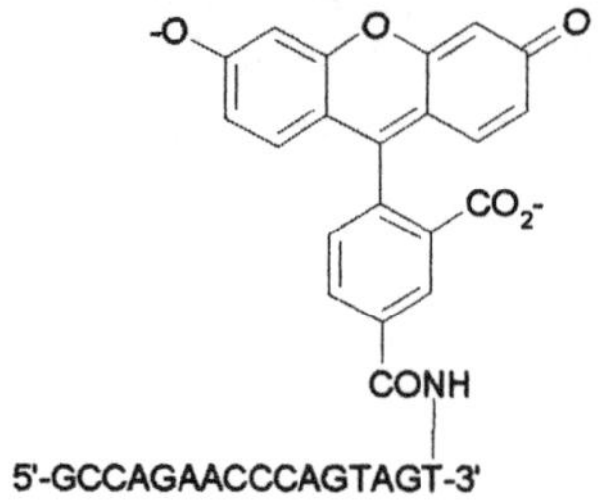

5'-GCCAGAACCCAGTAGT-3'

17

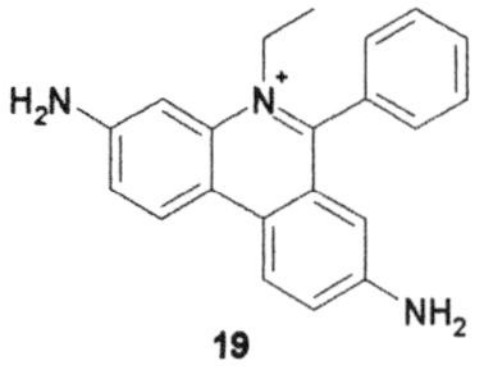

19

3'-CGGTCTTGGGTCATCA-5"

18

7. INHIBIT LOGIC

Phosphorescence output has been particularly productive in displaying INHIBIT (INH) logic. The Tb^{3+} complex **20** (Gunnlaugsson et al., 2000) provides a nice example. O_2 (the first input) quenches the delayed line-like emission, giving an output 0 regardless of the other inputs. The organic triplet state of the ligand which sensitizes the metal emission is easily waylaid by O_2. The second input (H^+) shifts the absorption band of **20** to match the excitation wavelength. Luminescence (output 1) is only observed if O_2 is absent (input 0) and acid is present (input 1). An older three-input INHIBIT system is best discussed here, since it uses organic phosphorescence. Several processes must be considered for **21** (de Silva et al., 1999) before phosphorescence can be seen. A PET process from the alkoxyaniline part of the amino acid-based receptor to the bromonaphthalene phosphor must be stopped by Ca^{2+} (the first input) binding. A bimolecular triplet-triplet annihilation involving two excited phosphors needs to be blocked by β-cyclodextrin (the second input) encapsulation. Phosphorescence of **21** is easily quenched by oxygen (the third, and disabling, input) in the present case whether the other inputs are present or not. We note that simple re-assignment of inputs for these two-input INHIBIT molecules will lead to logic gates of the REVERSE IMPLICATION type. For absorption-based versions of these, the reader is referred to some recent work (de Silva and McClenaghan, 2002).

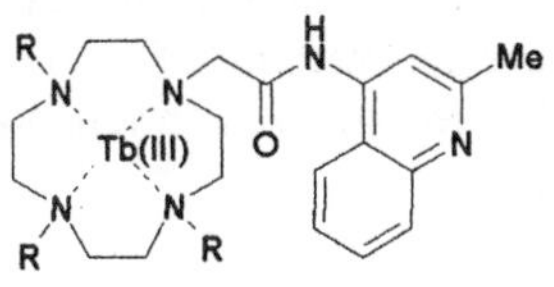

20; R = $CH_2P(Me)O_2^-$

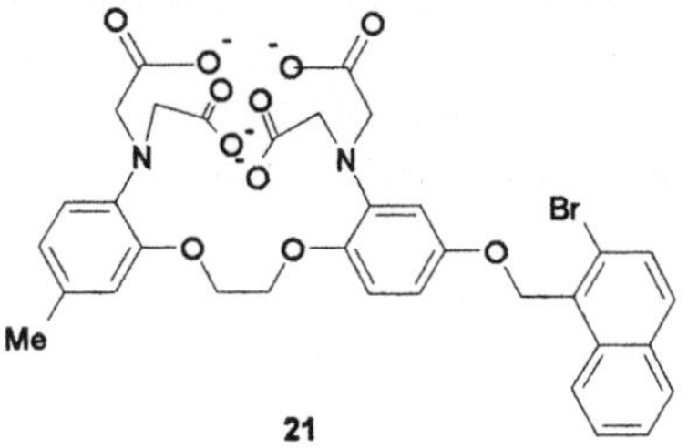

21

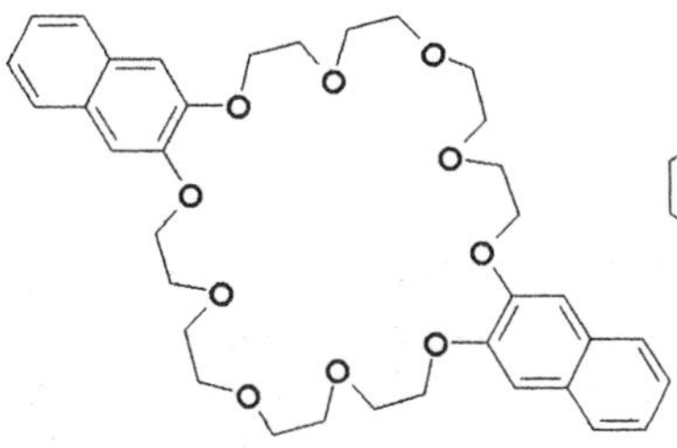

22

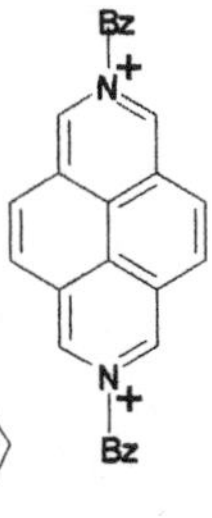

23

8. XOR LOGIC

[2]pseudorotaxanes have been persuaded to show XOR logic by the Stoddart-Balzani combine (Credi et al., 1997). Threading/dethreading processes of a macrocyclic host (**22**) and rod-like guest (**23**) complex are involved. In the presence of acid or base (the two inputs), the complex threading is not permitted and an output 1 is obtained in the form of emission at 343 nm. In neutral solution (i.e. either no acid or base, or both added simultaneously in stoichiometric amounts) the non-emissive 1:1 complex forms and an output 0 is obtained. A more general approach to molecular XOR logic is now available (de Silva and McClenaghan, 2000) but this does not use luminescence as yet.

XOR logic is also embedded within **8** (Remacle et al., 2001). While two-photon fluorescence (TPF) of **8** showed AND logic, conventional one-photon fluorescence output satisfies XOR logic under carefully defined conditions. Either of two low-power laser beams (either input 1) produces normal fluorescence emission (output 1). Of course, no emission results (output 0) if both laser beams are switched "off" (both inputs 0). When both laser beams are switched "on" (both inputs 1), normal fluorescence is diverted to TPF which is not under observation anyway. Such extraction of multiple logic activities from the same (structurally simple) molecule with different luminescence spectroscopy protocols is to be applauded.

9. MORE COMPLEX MOLECULAR LOGIC

The Bologna and Lisbon teams have shown how to extend their AND logic gate (Pina et al., 1999) to three-input enabled OR (EOR) logic (Roque et al., 1999). Their AND logic system used protons and ultraviolet photons as the two inputs on **5** to produce a fluorescence output. Now **5** is located in interfacial regions of negatively charged sodium dodecyl sulfate (SDS) detergent micelles. So **5** feels heightened proton densities due to this local electric field. So now rather low bulk proton densities can be used to deliver the low local pH value necessary to trigger the conversion of **5** into **6** and then to **7** (in the presence of ultraviolet light, of course). Protons serve as $input_1$, SDS micelles serve as $input_2$ and ultraviolet light the enabling $input_3$.

Another three-input system **24** (Ji et al., 2000) which is derived from one of our older AND gates (de Silva et al., 1997) uses an azacrown ether to target K^+ ($input_2$) or H^+ ($input_1$). A calixarene moiety targets Cs^+ ($input_3$). Cs^+, but not K^+, causes fluorescence enhancement of **24** in acid medium. K^+, but not Cs^+, causes fluorescence enhancement In basic medium. This behaviour corresponds to the truth table in Figure 2. Figure 5 shows the relative complexity of the combinational logic array which is emulated by this rather small molecule.

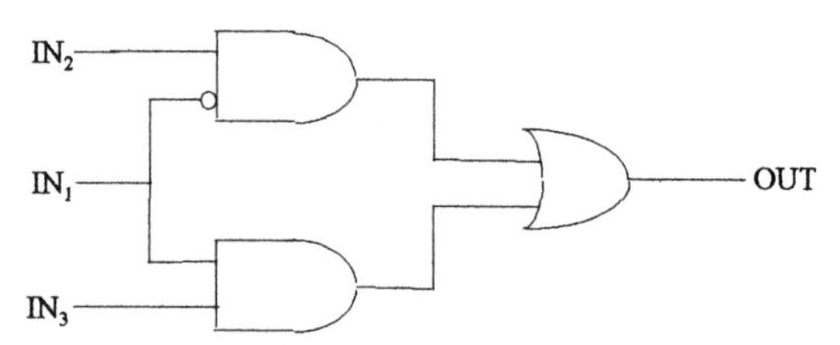

Figure 5. The gate array emulated by the logic behaviour of **24**.

Raymo and Giordani's work on three-state systems with acid-sensitive photochromic spiropyrans such as **25** also leads to very interesting three-input logic arrays (Raymo and Giordani, 2001) but several of these are based on absorbance output and hence outside the scope of this review. Nevertheless, nice applications with luminescence output are available since three molecular states with characteristic absorption spectra can be externally manipulated. For instance, the simple passage of the emission from fluorophore **26** through the three-state system leads to the logic pattern shown in Figure 2 (Raymo and Giordani, 2001). Ultraviolet light, visible light and protons form $input_1$, $input_2$ and $input_3$ respectively. The use of multiple fluorophores emitting in different wavelength regions leads to different logic patterns (Raymo and Giordani, 2002).

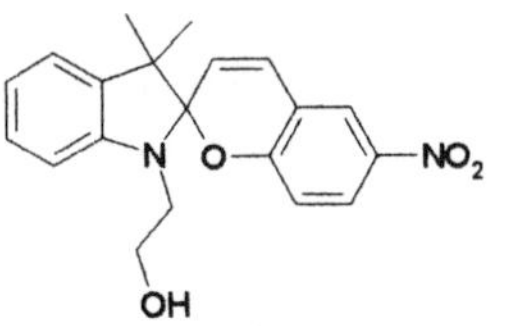

Even two-input logic systems can achieve relatively complex effects if two of these gates are run in parallel. For instance, parallel AND and XOR gates ((de Silva and McClenaghan, 2000)) will yield molecular number processing – a humanly relevant enterprise. The molecular half-adder, just like its electronic equivalent – will perform the following numerical manipulations. 0 + 0 = 0, 0 + 1 = 1, 1 + 0 = 1 and 1 + 1 = 2 (or 10 in binary). Its limitation, of course, is that numbers larger than decimal 2 are not recognized. Nevertheless, it is clearly recognizable as a molecular computational device and therefore a very important milestone to pass. Fluorescence is the output from AND gate **27**, but transmitted light is the output from XOR gate **28**. AND gate **27** works in the same way as **1**. While the **22-23** system was a nice demonstrator of XOR logic, it was difficult to use it as a component of a half-adder owing to the neutralization of its inputs. On the other hand, **28** provides an approach which is not only general but also smoothly compatible with AND gates such as **27**. Its absorption band moves one way or the other along the wavelength axis in the presence of only one or other input (Ca^{2+} or H^+) but not both. The simultaneous presence of Ca^{2+} and H^+ causes two opposite effects which cancel out. The transmittance must be monitored at the judiciously chosen wavelength of 390nm. Then it provides the second bit of the two-bit number '$output_1output_2$' – the 'sum' digit. The fluorescence output from the AND gate **27** provides the first bit of the two-bit number '$output_1output_2$' – the 'carry' digit. The **27-28** pair was the first example of small molecules performing numerical computation outside of human brains.

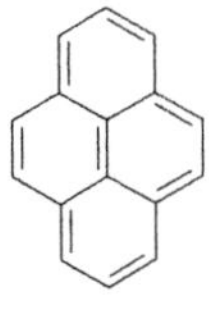

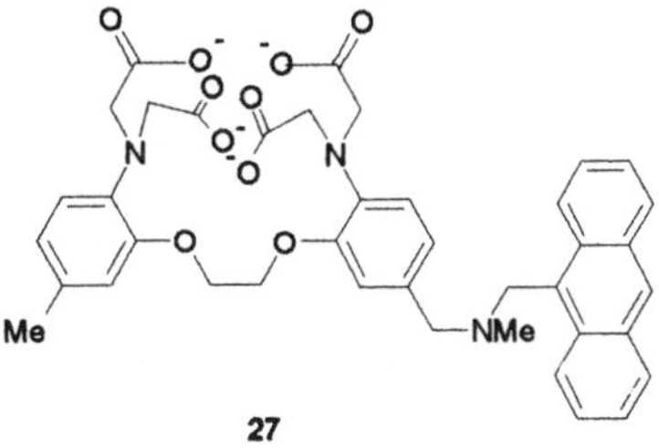

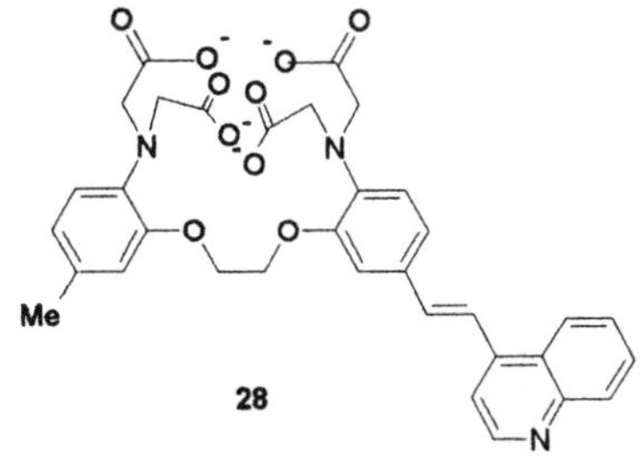

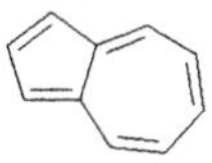

The fact that **8** (Remacle et al., 2001) featured in the sections on AND as well as XOR logic is a clear indication that entire half-adders can be found in far simpler (and cheaper) molecules than the **27** and **28** pair, provided that sophisticated luminescence techniques are accessible. The elegance increases further when we note that **29** can do what **8** does, but in a lower energy range. So the one-photon fluorescence emission of **8** (which is the first XOR output) can be fed into **29** via its absorption band by EET. Such coupling of half-adders is tantalizingly close to achieving a molecular full-adder – a molecular adding machine with no glass ceiling at the decimal number 2.

10. CONCLUSION

Photochemical principles naturally lead to ways of switching molecular luminescence according to patterns which represent various binary logic operations. While this short review mainly chose to illustrate some of the two-input logic gates, it must be noted that virtually all the one- and two-input logic gates (Ben-Ari, 1993) can now be implemented in this way. The progress in tackling three-input logic is also remarkable. Though imaginable, none of this was reasonably achievable just over a decade ago. Photochemically oriented science will continue to be a pathfinder in molecular information processing.

11. ACKNOWLEDGEMENT

We appreciate the support of the Department of Employment and Learning of Northern Ireland, Procter and Gamble Eurocor, European Union (HPRN-CT-2000-00029) and Invest NI (RTD COE 40).

12. REFERENCES

Balzani, V., Venturi, M., and Credi, A., 2003, *Molecular Devices and Machines*, Wiley-VCH, Weinheim, Ch.9.

Baytekin, H. T., and Akkaya, E. U., 2000, A molecular NAND gate based on Watson-Crick base pairing, *Org. Lett.* **2**: 1725-1727.

Ben-Ari, M., 1993, *Mathematical Logic for Computer Science*, Prentice-Hall, Hemel Hempstead.

Bissell, R. A., de Silva, A. P., Gunaratne, H. Q. N., Lynch, P. L. M., Maguire, G. E. M., McCoy, C. P., and Sandanayake, K. R. A. S., 1993, Fluorescent PET (photoinduced electron-transfer) sensors, *Top. Curr. Chem.* **168**: 223-264.

Bolt, J. D., and Turro, N. J., 1982, New probes for surfactant solutions - phosphorescent labeled detergents, *Photochem. Photobiol.* **35**: 305-310.

Brasselet, S., and Moerner, W. E., 2000, Fluorescence behaviour of single-molecule pH sensors, *Single Mol.* **1**: 17-23.

Cole, K. L., Farran, M. A., and deShayes, K., 1992, A new water-soluble cyclophane host that is organized by calcium-binding, *Tetrahedron Lett.* **33**: 599-602.

Cooper, C. R., and James, T. D., 2000, Synthesis and evaluation of D-glucosamine-selective fluorescent sensors, *J. Chem. Soc. Perkin Trans. 2* 963-969.

Credi, A., Balzani, V., Langford, S. J., and Stoddart, J. F., 1997, Logic operations at the molecular level. An XOR gate based on a molecular machine, *J. Am. Chem. Soc.*, **119**: 2679-2681.

Deonarine, A. S., Clark, S. M., and Konermann, L., 2003, Implementation of a multifunctional logic gate based on folding/unfolding transitions of a protein, *Future Generation Computer Systems* **19**: 87-97.

de Silva, A. P., Gunaratne, H. Q. N., and McCoy, C. P., 1993, A molecular photoionic and gate based on fluorescent signaling, *Nature* **364**: 42-44.

de Silva, A. P., Gunaratne, H. Q. N., and Maguire, G. E. M., 1994, Off-on fluorescent sensors for physiological levels of magnesium-ions based on photoinduced electron-transfer (PET), which also behave as photoionic OR logic gates, *J. Chem. Soc., Chem. Commun.* 1213-1214.

de Silva, A. P., Gunaratne, H. Q. N., Habib-Jiwan, J. -L., McCoy, C. P., Rice T. E., and Soumillion J. -P., 1995, New fluorescent model Compounds for the study of photoinduced electron transfer; The influence of a molecular electric field, *Angew. Chem. Int. Ed. Engl.* **34**: 1728-1731.

de Silva, A. P., Gunaratne, H. Q. N., Gunnlaugsson, T., Huxley, A. J. M., McCoy, C. P., Rademacher, J. T., and Rice, T. E., 1997, Signaling recognition events with fluorescent sensors and switches, *Chem. Rev.* **97**: 1515-1566.

de Silva, A. P., Gunaratne, H. Q. N., and McCoy, C. P., 1997, Molecular photoionic AND logic gates with bright fluorescence and "off-on" digital action, *J. Am. Chem. Soc.* **119**: 7891-7892.

de Silva, A. P., Dixon, I. M., Gunaratne, H. Q. N., Gunnlaugsson, T., Maxwell, P. R. S., and Rice, T. E., 1999, Integration of logic functions and sequential operation of gates at the molecular-scale, *J. Am. Chem. Soc.*, **121**: 1393-1394.

de Silva, A. P., and McClenaghan, N. D., 2000, Proof-of-principle of molecular-scale arithmetic, *J. Am. Chem. Soc.* **122**: 3965-3966.

de Silva, A. P., and McClenaghan, N. D., 2002, Simultaneously multiply-configurable or superposed molecular logic systems composed of ICT (internal charge transfer) chromophores and fluorophores integrated with one or two ion receptors, *Chem. Eur. J.* **8**: 4935-4945.

Ghosh, P., Bharadwaj, P. K., Mandal, S., and Ghosh, S., 1996, Ni(II), Cu(II), and Zn(II) cryptate-enhanced fluorescence of a trianthrylcryptand: A potential molecular photonic OR operator, *J. Am. Chem. Soc.* **118**: 1553-1554.

Gobbi, L., Seiler, P., and Diederich, F., 1999, A novel three-way chromophoric molecular switch: pH and light controllable switching cycles, *Angew. Chem. Int. Ed.* **38**: 674-678.

Gunnlaugsson, T., MacDonail, M., and Parker, D., 2000, Luminescent molecular logic gates: the two-input (INH) function, *Chem. Commun.*, 93-94.

Hughes, E., *Electrical Technology*, Longman, Harlow, 1987.

Iwata, S., and Tanaka, K., 1995, A novel cation and anion recognition host having pyrido[1',2'/1,2]imidazo[4,5-b]pyrazine as the fluorophore, *J. Chem. Soc., Chem. Commun.* 1491-1492.

Ji, H. F., Dabestani, R., and Brown, G. M., 2000, A supramolecular fluorescent probe, activated by protons to detect cesium and potassium ions, mimics the function of a logic gate, *J. Am. Chem. Soc.* **122**: 9306-9307.

Kosower, E. M., and Huppert, D., 1986, Excited-state electron and proton transfers, *Ann. Rev. Phys. Chem.* **37** : 127-156.

Lakowicz, J. R., 1999, *Principles of Fluorescence Spectroscopy*, 2nd ed., Plenum, New York.

Malvino, A. P., and Brown, J. A., 1993, *Digital Computer Electronics*, Glencoe, Lake Forest, 3rd edn.

Mitchell, R. J., 1995, *Microprocessor Systems: An introduction*, Macmillan, London.

Parker, D., and Williams, J. A. G., 1998, Taking advantage of the pH and pO(2) sensitivity of a luminescent macrocyclic terbium phenanthridyl complex, *Chem. Commun.* 245-246.

Pina, F., Maestri, M., and Balzani, V., 1999, Photochromic flavylium compounds as multistate/multifunction molecular-level systems, *Chem. Commun.* 107-114.

Ponce, A., Wong, P. A., Way, J. J., and Nocera, D. G., 1993, Intense phosphorescence triggered by alcohols upon formation of a cyclodextrin ternary complex, *J. Phys. Chem.* **93**: 11137-11142.

Raymo, F. M., and Giordani, S. 2001, Signal processing at the molecular level, *J. Am. Chem. Soc.* **123**: 4651-4652.

Raymo, F. M., and Giordani, S. 2001, Digital communication through intermolecular fluorescence modulation, *Org. Lett.* **3**: 1833-1836.

Raymo, F. M., and Giordani, S. 2002, Multichannel digital transmission in an optical network of communicating molecules, *J. Am. Chem. Soc.* **124**: 2004-2007.

Remacle, F., Speiser, S., and Levine, R. D., 2001, Intermolecular and intramolecular logic gates, *J. Phys. Chem. B* **105**: 5589-5591.

Rettig, W., 1994, Photoinduced charge separation via twisted intramolecular charge-transfer states, *Top. Curr. Chem.* 169 : 253-299.

Roque, A., Pina, F., Alves, S., Ballardini, R., Maestri, M., and Balzani, V., 1999, Micelle effect on the 'write-lock-read-unlock-erase' cycle of 4 '-hydroxyflavylium ion, *J. Mater. Chem.* **9**: 2265-2269.

Saghatelian, A., Völcker, N. H., Guckian, K. M., Lin, V. S. -Y., and Ghadiri, M. R., 2003, DNA-based photonic logic gates: AND, NAND, and INHIBIT, *J. Am. Chem. Soc.* **125**: 346-347.

Schneider, H. J., and Ruf, D., 1990, Host-guest chemistry .24. A synthetic allosteric system with high cooperativity between polar and hydrophobic binding-sites, *Angew. Chem. Int. Ed. Eng.* **29**: 1159-1160.

Sohna Sohna, J. -E., Jaumier, P., and Fages, F., Zinc(II)-driven fluorescence quenching of a pyrene-labelled bis-2,2 '-bipyridine ligand, 1999, *J. Chem. Res.* 134-135.

Stojanovic, M. N., Mitchell, T. E., and Stefanovic, D., 2002, Deoxyribozyme-based logic gates, *J. Am. Chem. Soc.* **124**: 3555-3561.

Valeur, B., 2001, *Molecular Fluorescence*, Wiley-VCH, Weinheim.

Wolfbeis, O.S., and Offenbacher, H., 1984, The effects of alkali cation complexation on the fluorescence properties of crown ethers, *Monatsh. Chem.* **115**: 647-654.

ZINC FLUORESCENT PROBES FOR BIOLOGICAL APPLICATIONS

Tomoya Hirano, Kazuya Kikuchi, Tetsuo Nagano

1. INTRODUCTION

Fluorescent probes, which allow visualization of cations,[1-3] anions,[4] small molecules[5,6] or enzyme activity[7-10] in living cells by fluorescence microscopy, are useful tools for studying biological systems.[11] A central achievement has been the development of probes for calcium ion (Ca^{2+}), which has many physiologically important roles in biological systems. The presently available range of small-molecular probes, such as Fura-2,[1] and Fluo-3,[2] and green fluorescent protein (GFP)-based probes, such as cameleon,[3] should be useful for clarifying the physiological functions of Ca^{2+} (Figure 1).

Zinc (Zn^{2+}) is the second most abundant transition metal cation, and several grams of Zn^{2+} is contained in the adult human body. Many proteins contain Zn^{2+} as a structural component.[12,13] Over 300 enzymes, including carbonic anhydrase and alcohol dehydrogenase, use this cation in a catalytic role, and some transcriptional factors require Zn^{2+} for binding the target genes. In addition to such protein-bound Zn^{2+}, free or loosely bound (labile, chelatable) Zn^{2+} exists at high concentration especially in brain,[14] pancreas[15] and spermatozoa,[16] and can be visualized by a staining method.[17,18] In brain, labile Zn^{2+} exists at a concentration of several mM in the vesicles of presynaptic neurons, and is released by synaptic activity or depolarization, modulating the functions of certain ion channels and receptors. In pancreas, Zn^{2+} is co-stored with insulin in secretory vesicles of pancreatic ß-cells, and is released when insulin is secreted.[19] Zn^{2+} has also been reported to regulate aspects of neuronal degeneration, e. g., in ischemia[20] and seizure.[21] Although many reports describe the significance of Zn^{2+} in biological systems, its mechanisms of action are poorly understood.

Thus, as is the case for Ca^{2+}, fluorescent probes for detecting labile Zn^{2+} are needed to clarify the function of Zn^{2+}. Previous reports suggest that fluorescent probes for Ca^{2+} or Mg^{2+} can also be used to detect Zn^{2+}.[22-24] However, these methods are useful only when

* Tomoya Hirano, Kazuya Kikuchi, Tetsuo Nagano, Graduate School of Pharmaceutical Sciences, The University of Tokyo, 7-3-1 Hongo, Bunkyo-ku, Tokyo, 113-0033 Japan.

the concentration of Ca^{2+} or Mg^{2+} shows little change, and calibration of the Ca^{2+} or Mg^{2+} concentration is required. For those reasons, fluorescent probes which can selectively detect Zn^{2+} are desirable, and many such probes have already been reported. Here, we would like to describe some of the fluorescent probes for Zn^{2+}, and discuss their usefulness for clarifying the physiological functions of Zn^{2+}.

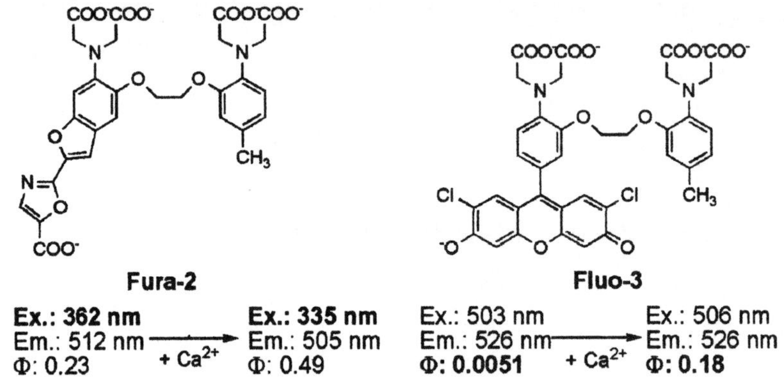

Fura-2

Ex.: 362 nm → (+ Ca^{2+}) Ex.: 335 nm
Em.: 512 nm → Em.: 505 nm
Φ: 0.23 → Φ: 0.49

Fluo-3

Ex.: 503 nm → (+ Ca^{2+}) Ex.: 506 nm
Em.: 526 nm → Em.: 526 nm
Φ: 0.0051 → Φ: 0.18

Figure 1. Structures of Ca^{2+} fluorescent probes
(Ex.: excitation wavelength, Em.: emission wavelength, Φ: quantum yield)

2. FLUORESCENT PROBES BASED ON QUINOLINE STRUCTURE

The first fluorescent probe for Zn^{2+} was *p*-toluenesulfoamide quinoline (TSQ),[25] whose fluorescence intensity was selectively increased by the addition of Zn^{2+}, but not by other biologically important cations, such as Ca^{2+} and Mg^{2+} (Figure 2).

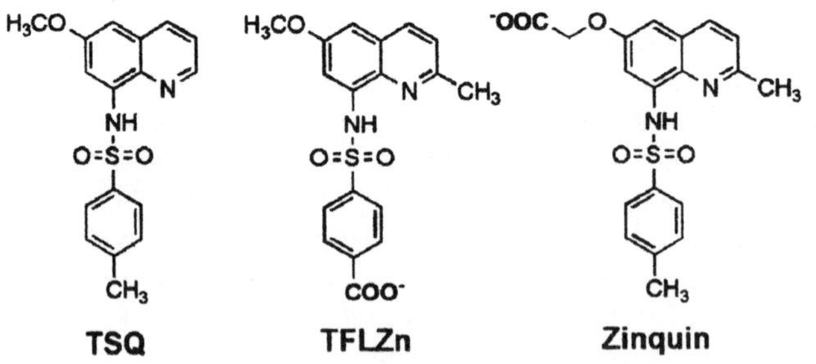

TSQ TFLZn Zinquin

Figure 2. Structures of TSQ and its derivatives

TSQ can form a water-insoluble 1:2 Zn^{2+}-complex in which the sulfonamide is deprotonated, and is mainly used for histochemical staining of Zn^{2+} in tissue sections.[21] For detection of Zn^{2+} in living cells, water-soluble derivatives have been developed by carboxylation of TSQ, thereby improving the solubility at neutral pH.[26,27] Among these derivatives, Zinquin can be used to monitor Zn^{2+} concentration in living cells.[26-29] By using Zinquin ethyl ester, the change of intracellular concentration of Zn^{2+} in hepatocytes or thymocytes can be directly detected by fluorescence microscopy, and by extracellular addition of Zinquin, secretion of Zn^{2+} from pancreatic ß-cells can similarly be detected. Recently, other derivatives of TSQ, whose fluorescence properties are improved by substitution of various functional groups onto the quinoline structure, have also been developed.[30-32]

3. FLUORESCENT PROBES BASED ON FLUORESCEIN STRUCTURE

Although quinoline-based fluorescent probes are useful because of their selectivity and applicability to living cells, they are not ideal, because the excitation wavelength is in the ultraviolet range, which may cause cell damage, and the measurement is subject to interference by autofluorescence, for example from pyridine nucleotides.[11]

Fluorescein is one of the most widely used fluorophores in biological experiments, for example, as a label in immunofluorescence studies, and is advantageous in that it has a high quantum yield of fluorescence in aqueous solution, and its excitation wavelength is in the visible range. Thus, fluorescein is a favorable fluorophore for Zn^{2+} fluorescent probes. By using macrocyclic polyamines as an acceptor for Zn^{2+}, ACFs were developed by our group.[33] At neutral pH, ACF shows almost no fluorescence, and the addition of Zn^{2+} induces fluorescence. The selectivity and sensitivity are relatively high, but the complexation with Zn^{2+} is kinetically slow, and thus it is difficult to monitor rapid changes of Zn^{2+} concentration. Another nitrogen-containing chelator, dipicolylamine, was also used as an acceptor. To our knowledge, Newport Green DCF was the first fluorescein-based probe for Zn^{2+}.[34] It coordinates Zn^{2+} via three nitrogens of dipicolylamine, and its dissociation constant is a few μM; thus, its affinity for Zn^{2+} is relatively weak, and it cannot be used to detect low concentrations of Zn^{2+}. Other probes using dipicolylamine are the Zinpyr group.[35,36] These probes form an 1:1 complex, with a dissociation constant in the sub nM range at pH 7, i.e., 2 orders magnitude lower than that of dipicolylamine (70 nM at pH 7), because of the additional coordination by phenolic oxygen. On further addition of Zn^{2+}, Zinpyr can also form 1:2 complexes. Although Zinpyrs can be used for intracellular imaging of Zn^{2+}, they also fluoresce via protonation of the benzylic nitrogen at around neutral pH, resulting in high background fluorescence. Recently, Molecular Probes, Inc. released novel type fluorescence probes, FluoZin group.[34] Its chelator structure resembles the BAPTA (*O,O'*-bis(2-aminophenyl)ethylene-glycol-*N,N,N',N'*-tetraacetic acid) structure, which is used as a chelator in fluorescent probes for Ca^{2+} (Figure 1). FluoZin-1 has about a half of the BAPTA structure; its dissociation constant with Zn^{2+} is 7.8 μM and that with Ca^{2+} is 8 mM. A much higher affinity probe of the FluoZin group, FluoZin-3, has one acetic acid removed from the BAPTA structure.[37] Its K_d value with Zn^{2+} is 15 nM, but its affinity with Ca^{2+} is not so weak: on addition of 100 μM Ca^{2+}, its fluorescence intensity is increased by up to about a half of the increase in the case Zn^{2+} addition,[37] suggesting that it would be useful only when Ca^{2+} concentration is kept in a low range.

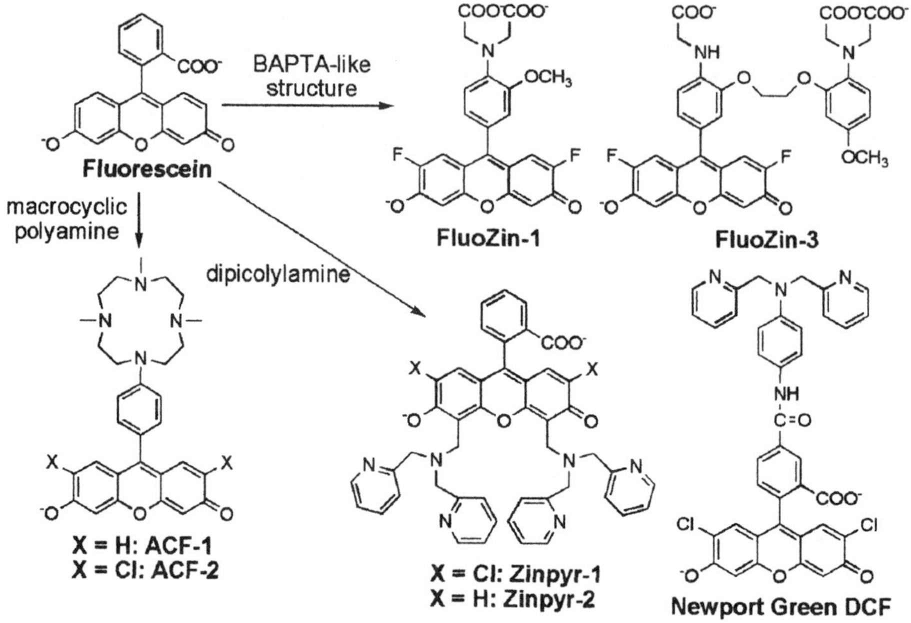

Figure 3. Structures of fluorescein-based probes for Zn^{2+}

4. FLUORESCENT PROBES BASED ON PEPTIDES OR PROTEINS

In addition to the small-molecular fluorescent probes described in sectiions **2** and **3**, peptide- and protein-based fluorescent probes have also been reported. Fluorophore-labeled peptides containing a zinc-finger motif, with the Zn^{2+} binding site Cys_2/His_2 or Cys/His_3, were developed. One of them, peptide 1 in Figure 4, and its improved derivatives are peptides labeled with a dansyl group or other environment-sensitive fluorophore, which fluoresces under more hydrophobic conditions.[38-40] The conformational change induced by the coordination with Zn^{2+} causes the fluorescence intensity to increase due to the fluorophore being placed in a more hydrophobic environment. Another example is a peptide labeled with two fluorophores, fluorescein as a donor and lissamine as an acceptor; this detects Zn^{2+} on the basis of the change in the efficiency of fluorescence resonance energy transfer (FRET) between two fluorophores (peptide 2 in Figure 4).[41] Upon binding of Zn^{2+} with the peptide, the conformational change brings the two fluorophores closer together, and enhances the efficiency of FRET, increasing the lissamine fluorescence. A fluorophore-labeled metallothionein probe, whose fluorescence property is changed by coordination with Zn^{2+}, has also been reported.[42,43] However, these probes have to be used under reducing conditions to avoid peptide oxidation, especially at cysteine.

Thompson *et al.* have reported fluorescent sensing systems consisted of apo-carbonic anhydrase (apoCA) and fluorescent aryl sulfonamide (dansyl amide or the derivatives of 2-oxa-1,3-diazole-4-sulfoamide), which were originally used as inhibitors of carbonic anhydrase. In the absence of Zn^{2+}, the fluorescent aryl sulfonamide cannot bind to apoCA. When Zn^{2+} is added, apoCA coordinates with Zn^{2+}, which allows binding of the fluorescent aryl sulfonamide to the holoenzyme. This binding results in changes of fluorescence intensity, emission wavelength, lifetime and anisotropy, which make it possible to detect the change of Zn^{2+} concentration.[44-46] This system has high sensitivity and selectivity, and can detect Zn^{2+} extracellularly released from hippocampal slices,[47] although its reversibility is a little problematic, in that the dissociation rate of Zn^{2+} from CA is so slow that this system cannot be used to take continuous measurements, and would be difficult to apply to intracellular studies. Kimua *et al.* reported a small-molecular fluorescent probe, dansylamidoethylcyclen, which was designed on the basis of the tight binding between carbonic anhydrase and arylsulfonamide discussed above.[48-50] Its fluorescence intensity is increased by coordination with Zn^{2+} via five nitrogens, four from macrocyclic polyamine, and one from deprotonated sulfonamide. Although its excitation wavelength is in the undesirable ultraviolet range, which is similar to that of TSQ derivatives (see **2**), its K_d value is relatively low, 5.5×10^{-13} (M) at pH 7.8, suggesting that this probe would have high sensitivity.

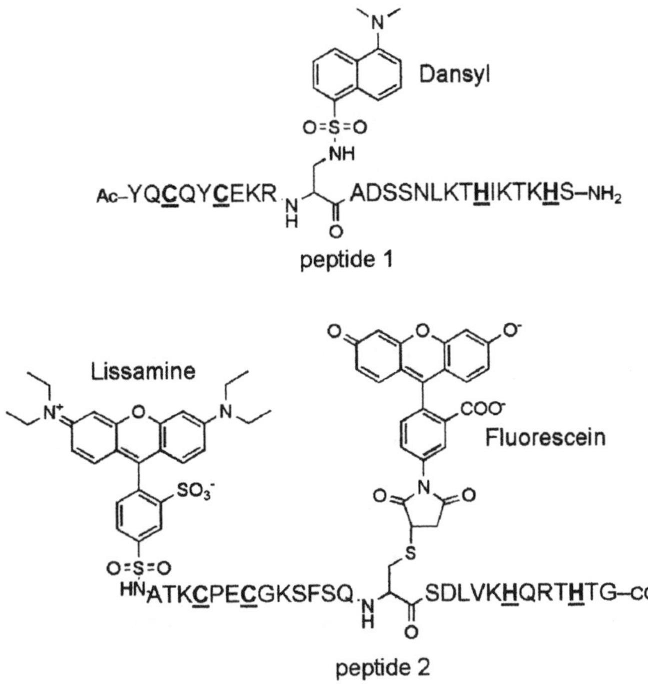

Figure 4. Some examples of peptide-based probes for Zn^{2+}.

Other reported protein-based probes, which can detect Zn^{2+}, include mutated green fluorescent proteins (GFP). These probes would be useful because they can be introduced noninvasively into the cells by transfection, and be targeted to specific tissues, organelles or cellular localizations. The usefulness of this methodology was proved by a study using cameleon,[3] which is a GFP-based fluorescent probe for Ca^{2+}. Kristian et al. introduced a mutation at the dimerization interface, which was identified from X-ray structure of GFP, of the cyan and the yellow variants of GFP (CFP and YFP), which are the donor and acceptor of a FRET system respectively.[51] This mutation was designed so that Zn^{2+} promotes dimerization between the mutated CFP and YFP resulting in an increase of FRET efficiency. Another example is a GFP mutant designed to bind Zn^{2+} via its chromophore, resulting in enhancement of its fluorescence intensity.[52] Although mutated GFP-based probes are potentially advantageous, the probes reported to date have disadvantages such as low sensitivity or a relatively slow complexation rate. Thus, further improvement is necessary before GFP-based probes can be used to monitor changes of Zn^{2+} in living cells.

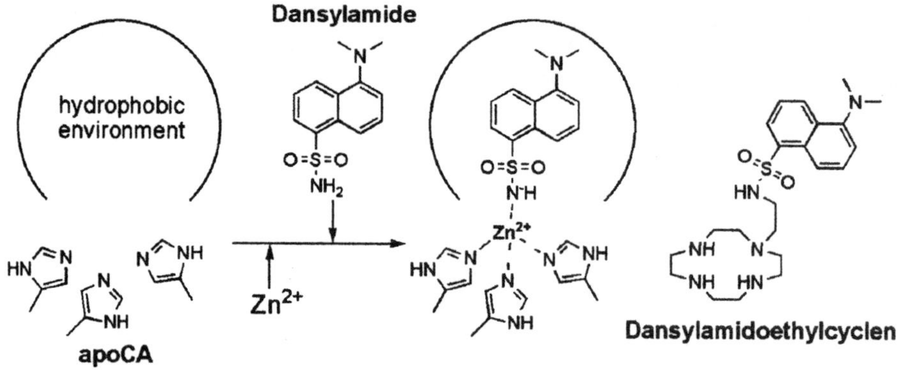

Figure 5.

5. DEVELOPMENT AND CHEMICAL PROPERTIES OF ZnAFs AS FLUORESCENT PROBES

Derivatives of fluorescein that are amino-substituted at the benzoic acid moiety exhibit weak fluorescence, but when the amino group is converted to a less electron-donating group, such as an amide group, fluorescence with a high quantum yield is obtained. This off/on switching of the fluorescence is controlled by photo-induced electron transfer (PET) between the xanthene ring, which is an electron acceptor and fluorophore, and the benzoic acid moiety, which is an electron donor.[53,54] The fluorescence off/on switching should depend on the highest occupied molecular orbital (HOMO) levels of the benzoic acid moiety. On the basis of this concept, our group has developed a range of fluorescent probes: **diaminofluoresceins** (DAFs) for nitric oxide (NO),[55,56] 9-[2-[3-(carboxy)-9,10-**diphenyl**]**anthryl**]-6-hydroxy-3*H*-**xanthen**-3-ones (DPAXs)[57] and 9-[2-[3-(carboxy)-9,10-**dimethyl**]**anthryl**]-6-hydroxy-3*H*-**xanthen**-3-ones (DMAXs)[53] for singlet oxygen ($^1O_2$). When DAFs react with NO, the diamino group is transformed to a triazole, resulting in lowering of the HOMO level of the benzoic acid moiety, and the fluorescence intensity is increased because of the hindrance of PET.

Based on this concept, we have developed novel fluorescent probes for Zn^{2+}. As an acceptor for Zn^{2+}, we chose TPEN (*N,N,N',N'*-tetrakis(2-pyridylmethyl)ethylenediamine) derivatives. TPEN is used as a heavy metal chelator, whose affinity for biologically important cations such as calcium or magnesium is relatively weak, and it has high cell permeability because of its high lipophilicity and low protonation constants (pK_a: 7.19, 4.85, 3.32, 2.85) compared with aliphatic amines. We synthesized several products consisting of fluorescein and TPEN derivatives, and some of them, ZnAFs, whose acceptor for Zn^{2+} is *N,N*-bis(2-pyridylmethyl)ethylenediamine directly attached to the benzoic acid moiety via nitrogen of aliphatic amine, exhibited good properties as Zn^{2+} probes. In the absence of Zn^{2+} both ZnAF-1 and ZnAF-2 showed little fluorescence, with quantum yields of only 0.022 and 0.023, respectively, while the addition of Zn^{2+} induced fluorescence enhancement by 17-fold for ZnAF-1 and 51-fold for ZnAF-2 (Figure 7, Table 1).[58]

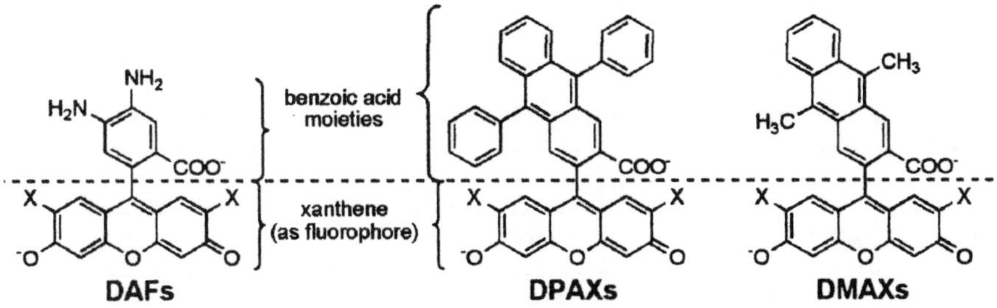

Figure 6. Structures of fluorescent probes for nitric oxide (NO), DAFs, and for singlet oxygen ($^1O_2$), DPAXs and DMAXs

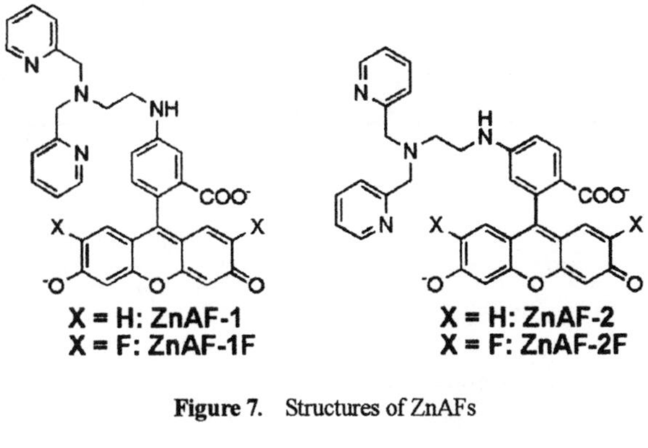

Figure 7. Structures of ZnAFs

The fluorescence intensity of the Zn^{2+} complex with ZnAF-1 or ZnAF-2 is decreased below pH 7.0 owing to the characteristics of fluorescein, the fluorophore of the probe; its fluorescence intensity decreases under acidic conditions as a result of protonation of the phenolic hydroxyl group of fluorescein, whose pK_a value is 6.43.[82] So, we also designed and synthesized ZnAF-1F and ZnAF-2F with electron-withdrawing fluorine at the ortho-position of the phenolic hydroxyl group to lower the pK_a value, and thereby obtain stable fluorescence under near-neutral conditions. Figure 8 shows that the florescence intensity of the Zn^{2+} complex of ZnAF-2F hardly changed above pH 6.0, and the pK_a value of the phenolic hydroxyl group of fluorescein was calculated to be 6.2 for both ZnAF-1 and ZnAF-2, and 4.9 for both ZnAF-1F and ZnAF-2F.[59] So, the fluorescence of the Zn^{2+} complex with ZnAF-1F or ZnAF-2F is not subject to interference by pH changes under near-neutral or slightly acidic conditions. The enhancement of the fluorescence intensity of ZnAFs might be controlled by the coordination of proton or metal cations with the nitrogen directly attached to the benzoic acid moiety, resulting in a change of HOMO levels. The pK_a value of this pivotal nitrogen for fluorescence seems to be as low as 5.0, since ZnAFs do not fluoresce upon protonation above this pH. Under strongly acidic conditions this nitrogen is protonated, but the fluorescence would be quenched because the phenolic hydroxyl group of fluorescein would also be protonated, resulting in formation of the non-fluorescent lactone form of fluorescein. This insensitivity to pH of the fluorescence is extremely useful for applications to living cells, in which pH changes can be caused by various biological stimuli.

Table 1. Chemical properties of ZnAFs with and without Zn^{2+}.[a]

	Free		+ Zn^{2+}	
Compound	ε[b]	Φ[c]	ε[b]	Φ[c]
ZnAF-1	7.4×10^4 (489)	0.022	6.3×10^4 (492)	0.21
ZnAF-2	7.8×10^4 (490)	0.023	7.6×10^4 (492)	0.32
ZnAF-1F	7.7×10^4 (489)	0.004	7.0×10^4 (492)	0.17
ZnAF-2F	7.4×10^4 (490)	0.006	7.3×10^4 (492)	0.24

[a]All data were obtained at pH 7.4 (100 mM HEPES buffer, I = 0.1 ($NaNO_3$)). [b]ε stands for extinction coefficient ($M^{-1}cm^{-1}$). Measured at the respective λ_{max}, which is shown in parentheses (nm). [c]Φ stands for quantum yield, determined using Φ of fluorescein (0.85) in 0.1 N NaOH as a standard.

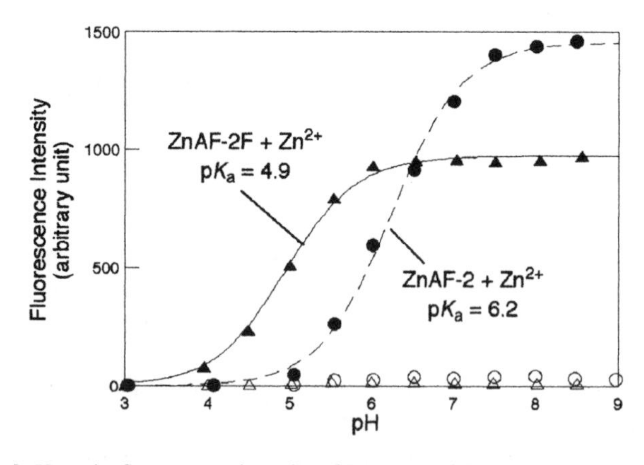

Figure 8. Effect of pH on the fluorescence intensity of ZnAF-2 and ZnAF-2F. Open circle: 1 μM ZnAF-2, closed circle: 1 μM ZnAF-2 + 1 μM Zn^{2+}, open triangle: 1 μM ZnAF-2F, closed triangle: 1 μM ZnAF-2F + 1 μM Zn^{2+}.

Upon addition of various concentrations of Zn^{2+}, the fluorescence intensity of ZnAF (1 μM) linearly increased up to a 1:1 [ZnAF] / [Zn^{2+}] ratio, and the fluorescence and absorption spectra did not change between 1 and 100 μM Zn^{2+} addition. Furthermore, a Job's plot analysis revealed that maximum fluorescence was obtained at 1:1 ratio. These data suggested that ZnAF forms a 1:1 complex with Zn^{2+}. Apparent dissociation constants were determined as 0.78 nM (ZnAF-1), 2.7 nM (ZnAF-2), 2.2 nM (ZnAF-1F) and 5.5 nM (ZnAF-2F) (Table 2), suggesting that ZnAFs can quantitatively measure the concentration of Zn^{2+} around the nM range, which is sufficient sensitivity for application in biological samples. The association (k_{on}) and dissociation (k_{off}) rate constants of ZnAFs at pH 7.4 (100 mM HEPES Buffer, I = 0.1 ($NaNO_3$)), 25°C are also shown in Table 2. The k_{on} values of ZnAFs were almost the same, 3 ~ 4 x 10^6 $M^{-1}s^{-1}$, implying that the complexation is sufficiently fast to detect an increase of Zn^{2+} concentration within a few hundred milliseconds.

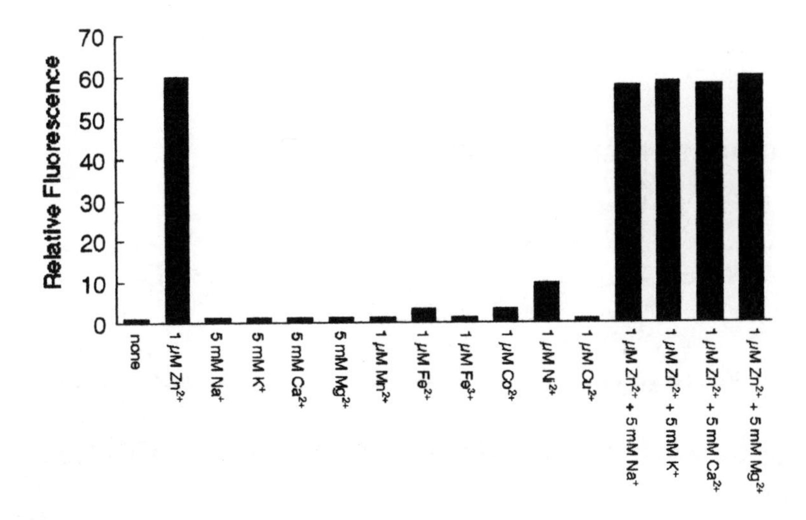

Figure 9. Relative fluorescence intensity of 1 μM ZnAF-2F in the presence of various cations. These data were measured at pH 7.4 (100 mM HEPES buffer, I = 0.1 ($NaNO_3$)).

Figure 9 shows the cation selectivity of ZnAF-2F; other ZnAFs have almost the same selectivity. Several cations which exist at high concentration in living cells, Ca^{2+}, Mg^{2+}, Na^+ and K^+, did not enhance the fluorescence intensity even at high concentration (5 mM). These results are due to the poor complexation of alkali metals or alkaline earth metals with the chelator of ZnAFs. These cations also did not interfere with the Zn^{2+}-induced fluorescence enhancement. Among first-row transition metal cations, Fe^{2+}, Co^{2+} and Ni^{2+} induced a slight enhancement of the fluorescence intensity, and Cu^{2+} quenched the fluorescence. Like TPEN, ZnAFs probably form complexes with these transition metal cations, but the fluorescence is weakened or quenched because of electron or energy transfer between the metal cation and fluorophore. Among ZnAFs, only ZnAF-1 showed almost no increase of its fluorescence intensity upon addition of Cd^{2+}. Its selectivity against Cd^{2+} was much higher than that of other reported fluorescent probes for Zn^{2+}, including other ZnAFs, and the mechanism of this selectivity is under study now.

Table 2. Apparent dissociation constants (K_d), association and dissociation rate constants (k_{on} and k_{off}) of ZnAFs in 100 mM HEPES buffer (pH 7.4, I = 0.1 ($NaNO_3$)) at 25°C.

Compound	K_d (nM)	k_{on}[a] ($M^{-1}s^{-1}$)	k_{off} (s^{-1})
ZnAF-1	0.78	4.3×10^6	3.4×10^{-3}
ZnAF-2	2.7	3.1×10^6	8.4×10^{-3}
ZnAF-1F	2.2	3.5×10^6	7.7×10^{-3}
ZnAF-2F	5.5	3.2×10^6	1.8×10^{-2}

[a]These data were measured under pseudo-first order conditions. Final concentrations: 1 μM ZnAFs; 50 μM $ZnSO_4$.

6. BIOLOGICAL APPLICATIONS OF ZnAFs

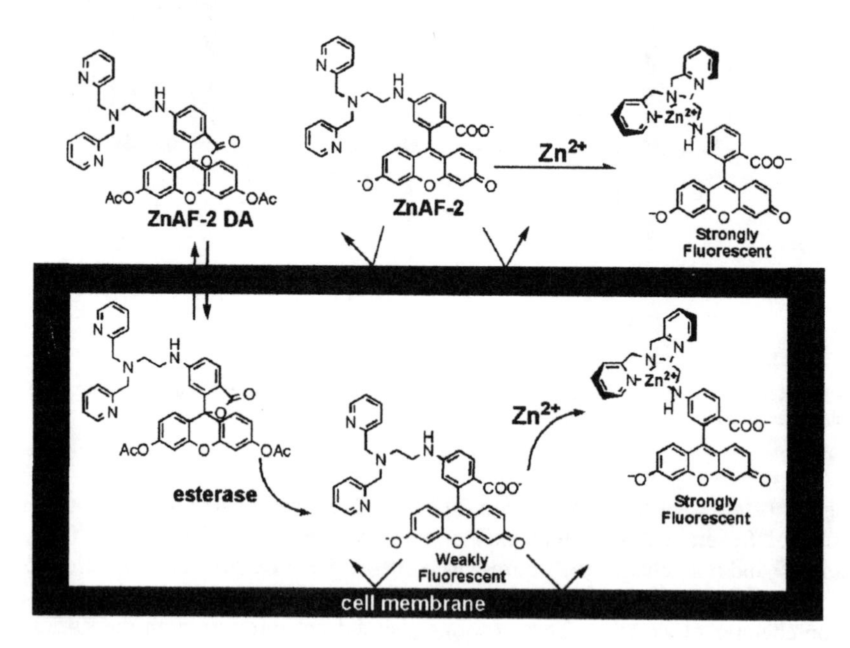

Figure 10.

ZnAFs themselves almost does not permeate cell membrane, so by extracellular addition of ZnAFs, extracellular Zn^{2+} can be monitored (Figure 10). In addition, cell membrane-permeable derivatives, ZnAF-2 DA (diacetate) and ZnAF-2F DA were developed.[58,59] ZnAF DA is more lipophilic, permeates well into cells, where it is transformed to ZnAF by esterase in the cytosol, and the dye is retained in the cells for a long time (Figure 10).

Cultured cells (RAW 264.7) were incubated with ZnAF-2 DA-containing PBS (phosphate buffered saline) to allow intracellular accumulation of ZnAF-2. The cells were washed, then Zn^{2+} and the Zn^{2+} ionophore, pyrithione (2-mercaptopyridine *N*-oxide) were added extracellularly (at 5min in Figure 12), which induced an increase of intracellular Zn^{2+}. As a result, the fluorescence in the intracellular regions (Figure 12d, 1–3) was increased, although the extracellular fluorescence (Figure 12d, 4) showed almost no change. The intracellular fluorescence was decreased by subsequent addition of the cell-permeable chelator TPEN at 20 min. These results suggested that ZnAF-2 DA can be used to monitor the changes of intracellular Zn^{2+}.

In addition to the cultured cell experiments, we applied ZnAF-2 DA to hippocampal slices. Acute rat hippocampal slices were incubated with ZnAF-2 DA for dye loading, and the fluorescence image of a dye-loaded slice is shown in Figure 11A. The fluorescence signal was detected in dentate hilus (DH), stratum lucidum (SL) and a small portion of stratum oriens (SO) but not in stratum radiatum (SR) or stratum pyramidale (SP), and this corresponds well with the localization of intracellular Zn^{2+} detected by other staining methods, such as Timm's method. The fluorescence signal was eliminated by the addition of TPEN, as shown in Figure 11D. Thus, ZnAF-2F DA can be used to detect intracellular labile Zn^{2+} in hippocampal slices.

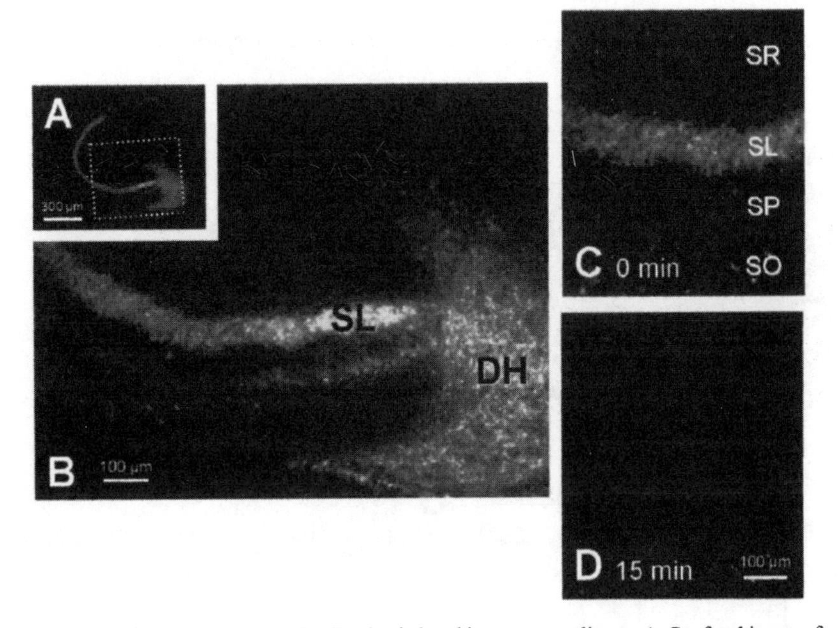

Figure 11. Fluorescence images of ZnAF-2 DA-loaded rat hippocampus slices. A, Confocal image of a hippocampal slice loaded with ZnAF-2 DA. B, Confocal image of the boxed region in A at higher magnification. ZnAF-2 images of the CA3 area in a hippocampal slice were obtained immediately before (C) and 15 min after bath application of 25 µM TPEN (D). DH: dentate hilus; SL: stratum lucidum; SO: stratum oriens; SR: stratum radiatum; SP: stratum pyramidale.

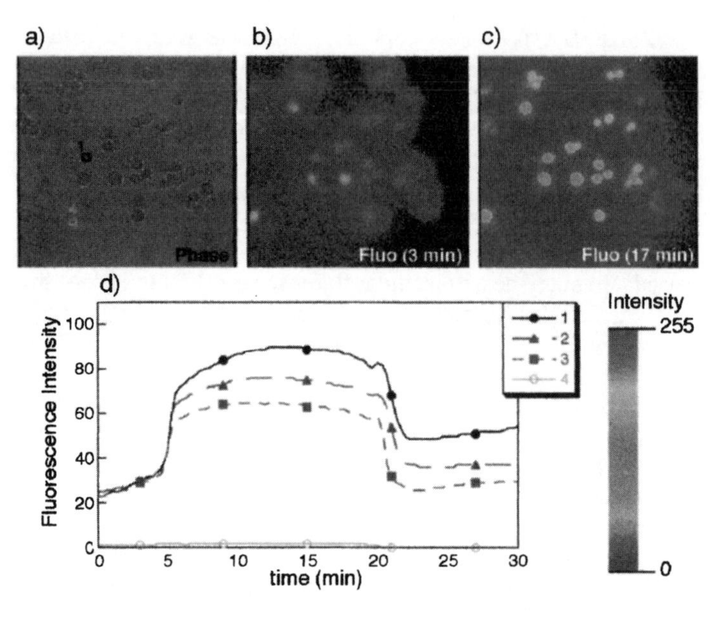

Figure 12. (See color insert section) a) Bright-field and b) c) fluorescence images of RAW 264.7 loaded with ZnAF-2 DA. The cells were incubated with 10 μM ZnAF-2 DA for 0.5 hours at 37°C. Then the cells were washed with PBS and the fluorescence excited at 470-490 nm was measured at 20 s intervals. At 5 min, 5 μM pyrithione and 50 μM $ZnSO_4$ were added to the medium, and 100 μM TPEN was added at 20 min. Fluorescence images are shown in pseudo-color and correspond to the fluorescence intensity data in d), which shows the average intensities of the corresponding areas (1 – 3: intracellular region, 4: extracellular region).

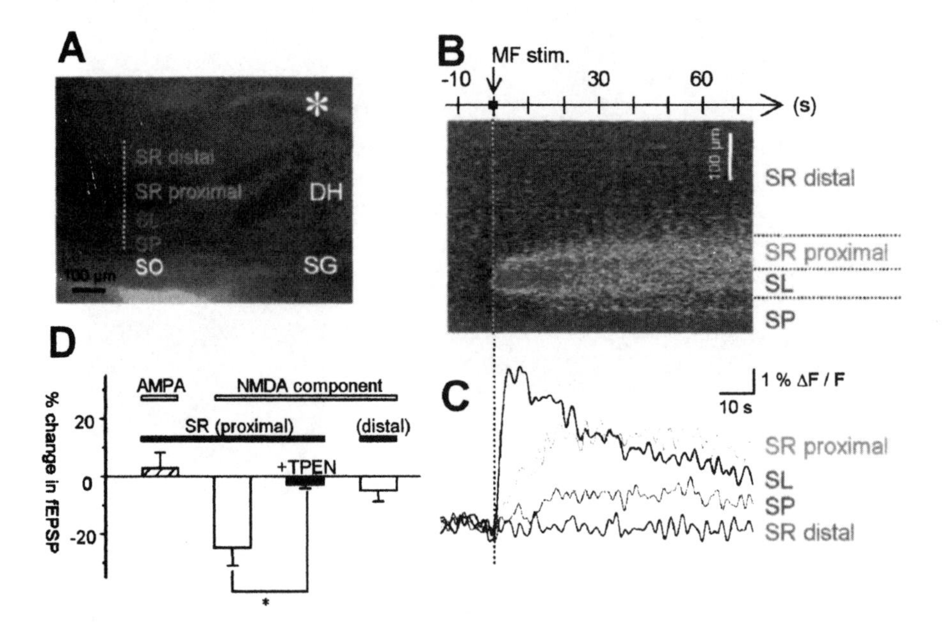

Figure 13. (See color insert section). Extracellular Zn^{2+} release, diffusion and heterosynaptic inhibition of

NMDA receptors in stratum radiatum after MF activation. A, An image of the dentate-CA3 area of a hippocampal slice perfused with ZnAF-2. The confocal ZnAF-2 signal is shown as a green-colored scale, superimposed on a transmitted beam image. Bipolar electrodes (•) were placed in the stratum granulosum (SG) to stimulate the MFs. The dotted line marks the transect of illumination during line-scan imaging. B, Line-scan image of ZnAF-2 taken at the points indicated in A. The temporal resolution was 1 s per line. "Hotter" colors correspond to increased $[Zn^{2+}]_o$ on an arbitrary pseudo-color scale. C, Data extracted from the image in B, along the time axis. Each point in time is the average %ΔF/F value across the spatial axis of the region separated by the horizontal dotted lines in B, i.e., the stratum radiatum far from stratum lucidum (SR distal, brown), the stratum radiatum near stratum lucidum (SR proximal, green), stratum lucidum (SL, red) and stratum pyramidale (SP, blue). The MFs were tetanized at 100 Hz for 2 s (MF stim.) at the time indicated by the vertical dotted line. D, Summary data for the effect of MF stimulation on AMPA and NMDA responses in proximal and distal stratum radiatum (SR). The ordinate indicates the average change in fEPSPs in associational/commissural fiber-CA3 pyramidal cell synapses 15 s after MF stimulation (100 Hz for 2 s). *$p < 0.01$; Student's *t*-test. Data are means ± SEM of . 5-7 slices.

As mentioned in section **1**, labile Zn^{2+} plays many physiologically important roles, especially in the central nerve system (CNS), where it is mainly stored in the synaptic vesicles of excitatory synapses, particularly the synaptic terminals of hippocampal mossy fibers (MFs), and is co-released with neurotransmitters in response to synaptic activity.[60,61] Zn^{2+} is known to modulate postsynaptic neurotransmitter receptor activity. For example, it inhibits *N*-methyl-D-aspartate (NMDA) receptors and γ-aminobutyric acid receptors, and potentiates α-amino-3-hydroxy-5-methyl-4-isoxazolepropionic acid (AMPA) receptors.[62] Zn^{2+} is also able to enter into the cells through ligand-gated channels, e.g., NMDA receptor channels,[63] Ca^{2+}-permeable AMPA/kainate receptor channels[64] and voltage-dependent Ca^{2+} channels,[22] and may influence various intracellular signaling pathways.[65-67] Despite numerous studies on Zn^{2+} action in the CNS, the physiological significance of synaptically released Zn^{2+} is still largely unknown. One reason is that the spatiotemporal Zn^{2+} dynamics during synaptic activity remains unclear to date. We therefore applied ZnAF-2 to monitor extracellularly released Zn^{2+}, to clarify its function. When high-frequency stimulation was delivered to the MFs, the concentration of extracellular Zn^{2+} was immediately elevated in the stratum lucidum (SL), and this was followed by a slight increase in the stratum radiatum adjacent to the stratum lucidum (SR proximal), but not in the distal area of stratum radiatum (SR distal) (Figure 13B, C).[68] Electrophysiological analyses revealed that NMDA-receptor-mediated synaptic responses ($fEPSP_{NMDA}$) in CA3 proximal stratum radiatum were inhibited in the immediate aftermath of MF activation and that this inhibition was no longer observed in the presence of a Zn^{2+}-chelating agent (Figure 13D), which indicates that Zn^{2+} serves as a heterosynaptic mediator.

We also focused on the function of Zn^{2+} in neuronal degeneration during ischemia. Transient forebrain ischemia results in the degeneration of specific populations of neurons. The hippocampus is one of the most vulnerable regions, and pyramidal cells in the CA1 region are especially vulnerable. This neuronal degeneration occurs a few days after the ischemia, and is known as delayed neuronal death.[69] We applied ZnAF-2 DA to rat hippocampus slices and monitored the change of intracellular Zn^{2+} in the early phase of ischemia. As a result, a rapid and transient increase of intracellular Zn^{2+} was detected, especially in the CA1 region. Because the CA1 region is especially vulnerable to ischemia, and excess intracellular Zn^{2+} causes cell damage,[65-67] this increase of intracellular Zn^{2+} may initiate the selective neuronal degeneration that occurs after ischemic insult.

7. OTHER TYPES OF FLUORESCENT PROBES

Recently other types of fluorescent probes have also been reported. One particularly interesting type is probes which enable ratiometric imaging. Ratiometric imaging is a technique that involves observing the changes in the ratio of fluorescence intensities at two wavelengths. Compared with the measurement of the fluorescence intensity at one wavelength, this method reduces artifacts by minimizing the influence of extraneous factors, such as the changes of the probe concentration and excitation light intensity. Based on the structure of Fura-2, which is a fluorescent probe suitable for ratiometric measurement of Ca^{2+}, novel fluorescent probes have been developed for ratiometric measurement of Zn^{2+}: FuraZin-1,[34] whose chelator structure is that of FluoZin-1, and ZnAF-R2,[70] whose chelator structure is that of ZnAF-2 (Figure 14). Their selectivity for Zn^{2+} over other cations and their dissociation constants are similar to those of their relatives, but these probes are applicable to ratiometric measurement of Zn^{2+} in living cells.

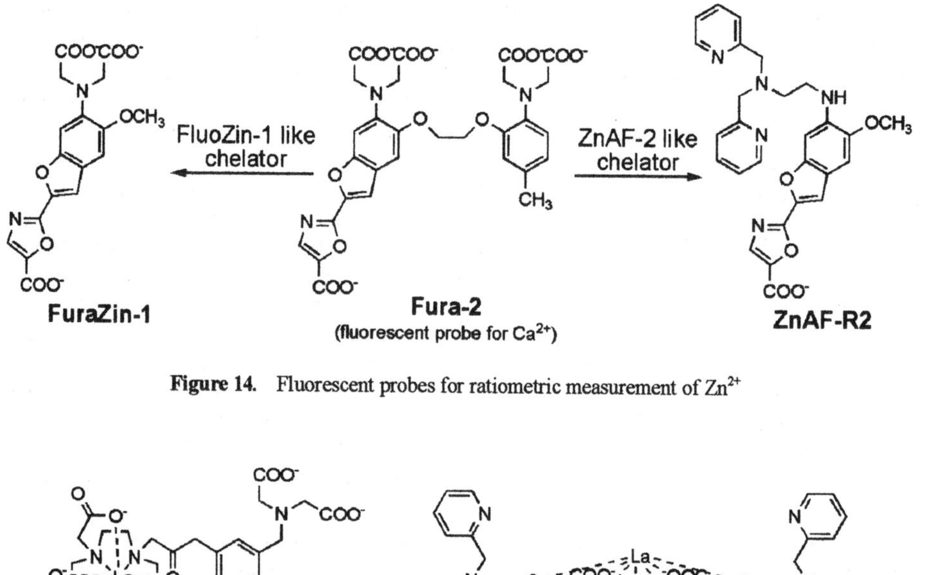

Figure 14. Fluorescent probes for ratiometric measurement of Zn^{2+}

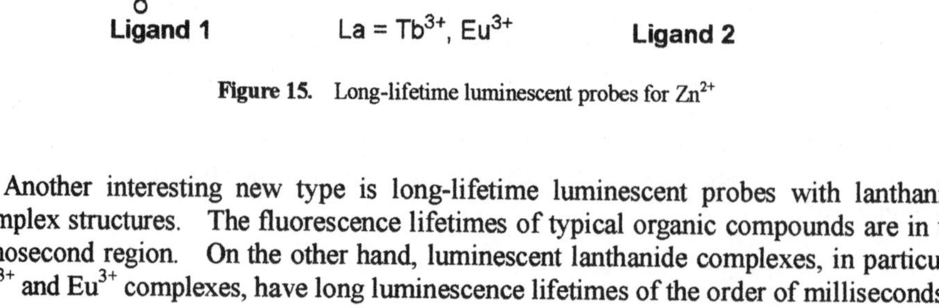

Figure 15. Long-lifetime luminescent probes for Zn^{2+}

Another interesting new type is long-lifetime luminescent probes with lanthanide complex structures. The fluorescence lifetimes of typical organic compounds are in the nanosecond region. On the other hand, luminescent lanthanide complexes, in particular Tb^{3+} and Eu^{3+} complexes, have long luminescence lifetimes of the order of milliseconds.[71]

These long-lived luminescent compounds have the advantage that short-lived background fluorescence and scattered light quickly decay to negligible levels, when a pulse of excitation light is applied and the emitted light is collected after an appropriate delay time. Figure 15 shows the structures of long-lifetime probes. Upon the addition of Zn^{2+}, the emission intensity of Ligand 1 is increased for both Tb^{3+} and Eu^{3+} complexes, and thier lifetimes are in the millisecond range. This Zn^{2+}-induced emission enhancement operates through a photo-induced electron transfer (PET) process.[72] Ligand 2 was very recently developed by our group, and its emission is also increased by the addition of Zn^{2+}. This increase is thought to be derived from a change of efficiency of intramolecular energy transfer from the pyridyl group to the lanthanide ion.[73] Although these probes have not yet been applied to the detection of Zn^{2+} in biological samples, this methodology may have the potentiality to provide a more sensitive and selective detection system for Zn^{2+}.

8. CONCLUSION ~FOR CLARIFYING THE PHYSIOLOGICAL FUNCTIONS OF Zn^{2+}~

Besides the probes discussed above, other probes for detecting Zn^{2+} have been reported, and their number is growing fast (recent examples: reference 74-78). Overall, however, our ZnAFs appear to be most useful, because of their sensitivity, selectivity and ability to readily apply to monitor both intracellular and extracellular Zn^{2+} concentrations of living cells or tissue slices. However, ZnAFs are not necessarily appropriate in all experiments. Thus, in studies for clarifying the physiological function of Zn^{2+}, which probe should be used? To answer this question, we should consider both of the chemical and biological properties of the probes. For example, to measure high concentrations of Zn^{2+} quantitatively, high-affinity probes, such as ZnAF-2, would not be suitable, because ZnAF-2 can quantitatively measure the concentration of Zn^{2+} around the nM range (0.1 nM – 10 nM order), but above this range, it maximally fluoresces, and can detect change of Zn^{2+} only qualitatively. In such a case, low-affinity probes, whose dissociation constant is around the expected concentration of Zn^{2+}, should be used. The kinetic parameters, such as association and dissociation rates constant, are also important. In order to measure rapid changes of Zn^{2+} concentration, such as those associated with neuronal activity, probes with sufficiently fast kinetic parameters should be used. In addition to these chemical parameters, biological parameters, for example, cell-membrane permeability, intracellular localization, and the toxicity of the excitation light and the probes themselves to the cells, should be considered. Lipophilic probes can generally permeate through the cell membrane, but tend to accumulate in membrane structures, while cationic compounds tend to accumulate in mitochondria. The purpose of the measurement is also important. For example, to study the distribution of Zn^{2+} in biological samples with high resolution and no requirement for temporal resolution, a staining method, such as Timm's staining, measured by electron microscopy would be most suitable.

Recently, we[59,68] and others[79-81] have used fluorescent probes and fluorescence microscopy to examine the changes of the concentration and distribution of endogenous Zn^{2+}, and their correlation with the physiological functions, such as memory and neuronal degeneration, especially in central nerve system, where levels of labile Zn^{2+} are high. The further development of fluorescent probes, optical instruments and methodology should yield much information about the mechanisms of Zn^{2+} functions.

9. REFERENCES

1. G. Grynkiewicz, M. Poenie, and R. Y. Tsien, A new generation of Ca^{2+} indicators with greatly improved fluorescence properties, *J. Biol. Chem.* **260,** 3440-3450 (1985).
2. A. Minta, J. P. Y. Kao, and R. Y. Tsien, Fluorescent indicators for cytosolic calcium based on rhodamine and fluorescein chromophores, *J. Biol. Chem.* **264,** 8171-8178 (1989).
3. A. Miyawaki, J. Llopis, R. Heim, J. M. McCaffery, J. A. Adams, M. Ikura, R. Y. Tsien, Fluorescent indicators for Ca^{2+} based on green fluorescent proteins and calmodulin, *Nature,* **388,** 882-887 (1997).
4. S. Mizukami, T. Nagano, Y. Urano, A. Odani, and K. Kikuchi, A fluorescent anion sensor that works in neutral aqueous solution for bioanalytical application, *J. Am. Chem. Soc.* **124,** 3920-3925 (2002).
5. K. Setsukinai, Y. Urano, K. Kikuchi, T. Higuchi, and T. Nagano, Fluorescence switching by O-dearylation of 7-aryloxycoumarins. Development of novel fluorescence probes to detect reactive oxygen species with high selectivity. *J. Chem. Soc. Perkin Trans. 2.* 2453-2457 (2000).
6. K. Setsukinai, Y. Urano, K. Kakinuma, H. J. Majima, and T. Nagano, Development of novel fluorescence probes that can reliably detect reactive oxygen species and distinguish specific species. *J. Biol. Chem.* **278,** 3170-3175 (2003).
7. S. Mizukami, K. Kikuchi, T. Higuchi, Y. Urano, T. Mashima, T. Tsuruo, and T. Nagano, Imaging of caspase-3 activation in HeLa cells stimulated with etoposide using a novel fluorescent probe, *FEBS Lett.* **453,** 356-360 (1999).
8. Y. Kawanishi, K. Kikuchi, H. Takakusa, S. Mizukami, Y. Urano, T. Higuchi, and T. Nagano, Design and synthesis of intramolecular resonance-energy transfer probes for use in ratiometric measurements in aqueous solution, *Angew. Chem. Int. Ed.* **39,** 3438-3440 (2000).
9. H. Takakusa, K. Kikuchi, Y. Urano, S. Sakamoto, K. Yamaguchi, and T. Nagano, Design and synthesis of an enzyme-cleavable sensor molecule for phosphodiesterase activity based on fluorescence resonance energy transfer, *J. Am. Chem. Soc.* **124,** 1653-1657 (2002).
10. H. Takakusa, K. Kikuchi, Y. Urano, H. Kojima, and T. Nagano, A novel design method of ratiometric fluorescent probes based on fluorescence resonance energy transfer switching by spectral overlap integral, *Chem. Euro. J.* **9,** 1479-1485 (2003).
11. W. T. Mason, *Fluorescent and luminescent probes for biological activity, 2nd ed.* (Academic Press, London, UK, 1999)
12. B. L. Vallee, and K. H. Falchuk, The biochemical basis of zinc physiology, *Physiol. Rev.* **73,** 79-118 (1993).
13. J. J. R. F. da Silva, and R. J. P. Williams, in: *The biological chemistry of elements: The inorganic chemistry of life, 2nd ed.* (Oxford UP, New York, 2001), pp. 315-339.
14. C. J. Frederickson, and A. I. Bush, Synaptically released zinc: Physiological functions and pathological effects, *Biometals* **14,** 353-366 (2001).
15. P. D. Zalewski, S. H. Millard, I. J. Forbes, O. Kapaniris, A. Slavotinek, W. H. Betts, A. D. Ward, S. F. Lincoln, and I. Mahadevan, Video image-analysis of labile zinc in viable pancreatic-islet cells using a specific fluorescent-probe for zinc, *J. Histochem. Cytochem.* **42,** 877-884 (1994).
16. P. D. Zalewski, X. Jian, L. L. L. Soon, W. G. Breed, R. F. Seamark, S. F. Lincoln, A. D. Ward, and F. Z. Sun, Changes in distribution of labile zinc in mouse spermatozoa during maturation in the epididymis assessed by the fluorophore Zinquin, *Reprod. Fertil. Dev.* **8,** 1097-1105 (1996).
17. F. Timm, Zur histochemie der schwermetalle das sulfid-silberverfahren, *Dtsch. Z. Gesamte. Gerichtl. Med.* **46,** 706–711 (1958).
18. S. de Biasi, and C. Bendotti, A simplified procedure for the physical development of the sulphide silver method to reveal synaptic zinc in combination with immunocytochemistry at light and electron microscopy, *J. Neurosci. Methods* **79,** 87-96 (1998).
19. W. -J. Qian, C. A. Aspinwall, M. A. Battiste, R. T. Kennedy, Detection of secretion from single pancreatic ß-cells using extracellular fluorogenic reactions and confocal fluorescence microscopy, *Anal. Chem.* **72,** 711-717 (2000).
20. J. -Y. Koh, S. W. Suh, B. J. Gwag, Y. Y. He, C. Y. Hsu, and D. W. Choi, The role of zinc in selective neuronal death after transient global cerebral ischemia, *Science* **272,** 1013–1016 (1996).
21. C. J. Frederickson, M. D. Hernandez, and J. F. McGinty, Translocation of zinc may contribute to seizure-induced death of neurons, *Brain Res.* **480,** 317–321 (1989).
22. D. Atar, P. H. Backx, M. M. Appel, W. D. Gao, and E. Marban, Excitation-Transcription coupling mediated by zinc influx through voltage-dependent calcium channels, *J. Biol. Chem.* **270,** 2473-2477 (1995).
23. T.J.B. Simons, Measurement of free Zn^{2+} ion concentration with the fluorescent-probe Mag-Fura-2 (Furaptra), *J. Biochem. Biophys. Methods* **27,** 25-37 (1993).
24. S. L. Sensi, L. M. T. Canzoniero, S. P. Yu, H. S. Ying, J.-Y. Koh, G. A. Kerchner, and D. W. Choi,

Measurement of intracellular free zinc in living cortical neurons: Route of entry, *J. Neurosci.* **17**, 9554-9564, (1997).
25. C. J. Frederickson, E. J. Kasarskis, D. Ringo, and R. E. Frederickson, A quinoline fluorescence method for visualizing and assaying the histochemically reactive zinc (bouton zinc) in the brain. *J. Neurosci. Methods* **20**, 91-103 (1987).
26. P. D. Zalewski, I. J. Forbes, and W. H. Betts, Correlation of apoptosis with change in intracellular labile Zn(II) using Zinquin (2-methyl-8-*p*-toluenesulphonamido-6- quinolyloxy)acetic acid , a new specific fluorescent-probe for Zn(II), *Biochem. J.* **296**, 403-408 (1993).
27. T. Budde, A. Minta, J. A. White, and A. R. Kay, Imaging free zinc in synaptic terminals in live hippocampal slices, *Neuroscience* **79**, 347-358 (1997).
28. P. Coyle, P. D. Zalewski, J. C. Philcox, I. J. Forbes, A. D. Ward, S. J. Lincoln, I. Mahadevan, and A. M. Rofe, Measurement of zinc in hepatocytes by using a fluorescent-probe, Zinquin - relationship to metallothionein and intracellular zinc, *Biochem. J.* **303**, 781-786 (1994).
29. P. D. Zalewski, I. J. Forbes, R. F. Seamark, R. Borlinghaus, W. H. Betts, S. F. Lincoln, and A. D. Ward, Flux of intracellular labile zinc during apoptosis (gene-directed cell death) revealed by specifiv chemical probe, Zinquin, *Chem. Biol.* **1**, 153-161 (1994).
30. M. C. Kimber, I. B. Mahadevan, S. F. Lincoln, A. D. Ward, and E. R. T. Tiekink, The synthesis and fluorescent properties of analogues of the zinc(II) specific fluorophore Zinquin ester, *J. Org. Chem.* **65**, 8204-8209 (2000).
31. M. C. Kimber, I. B. Mahadevan, S. F. Lincoln, A. D. Ward, and W. H. Betts, A preparative and spectroscopic study of fluorophores for zinc(II) detection, *Aust. J. Chem.* **54**, 43-49 (2001).
32. D. A. Pearce, N. Jotterand, I. S. Carrico, B. Imperiali, Derivatives of 8-hydroxy-2-methylquinoline are powerful prototypes for zinc sensors in biological systems, *J. Am. Chem. Soc.* **123**, 5160-5161 (2001).
33. T. Hirano, K. Kikuchi, Y. Urano, T. Higuchi, and T. Nagano, Novel zinc fluorescent probes excitable with visible light for biological applications, *Angew. Chem. Int. Ed.* **39**, 1052-1054 (2000).
34. R. P. Haugland, *Handbook of fluorescent probes and research products, 9th ed.* (Molecular Probes, Inc. Eugene, OR, 2002).
35. G. K. Walkup, S. C. Burdette, S. J. Lippard, and R. Y. Tsien, A new cell-permeable fluorescent probe for Zn^{2+}. *J. Am. Chem. Soc.* **122**, 5644-5645 (2000).
36. S. C. Burdette, G. K. Walkup, B. Springler, R. Y. Tsien, and S. J. Lippard, Fluorescent sensors for Zn^{2+} based on fluorescein platform: synthesis, properties and intracellular distribution, *J. Am. Chem. Soc.* **123**, 7831-7841 (2001).
37. K. R. Gee, Z. -L. Zhou, W. -Q. Qian, R. Kennedy, Detection and imaging of zinc secretion from pancreatic ß-cells using a new fluorescent zinc indicator, *J. Am. Chem. Soc.* **124**, 776-778 (2002).
38. G. K. Walkup; and B. Imperiali, Design and evaluation of a peptidyl fluorescent chemosensor for divalent zinc, *J. Am. Chem. Soc.* **118**, 3053–3054 (1996).
39. G. K. Walkup; and B. Imperiali, Fluorescent chemosensors for divalent zinc based on zinc finger domains. Enhanced oxidative stability, metal binding affinity, and structural and functional characterization, *J. Am. Chem. Soc.* **119**, 3443–3450 (1997).
40. G. K. Walkup; and B. Imperiali, Stereoselective synthesis of fluorescent α-amino acids containing oxine (8-hydroxyquinoline) and their peptide incorporation in chemosensors for divalent zinc, *Org. Chem.* **120**, 3053–3054 (1998).
41. H. A. Godwin, and J. M. Berg, A fluorescent zinc probe based on metal-induced peptide folding, *J. Am. Chem. Soc.* **118**, 6514-6515 (1996).
42. L. L. Pearce, R. E. Gandley, W. P. Han, K. Wasserloos, M. Stitt, A. J. Kanai, M. K. McLaughlin, B. R. Pitt, and E. S. Levitan, Role of metallothionein in nitric oxide signaling as revealed by a green fluorescent fusion protein, *Proc. Nat. Acad. Sci.* **97**, 477-482 (2000).
43. S. H. Hong, and W. Maret, A fluorescence resonance energy transfer sensor for the beta-domain of metallothionein, *Proc. Nat. Acad. Sci.* **100**, 477-482 (2003).
44. R. B. Thompson, and E. R. Jones, Enzyme-based fiber optic zinc biosensor, *Anal. Chem.* **65**, 730-734 (1993).
45. R. B. Thompson, and M. W. Patchan, Lifetime-based fluorescence energy-transfer biosensing of zinc, *Anal. Biochem.* **227**, 123-128 (1995).
46. R. B. Thompson, B. P. Maliwal, and C. A. Fierke, Expanded dynamic range of free zinc ion determination by fluorescence anisotropy, *Anal. Chem.* **70**, 1749-1754 (1998).
47. R. B. Thompson, W. O. Whetsell, B. P. Maliwal, C. A. Fierke, and C. J. Frederickson, Fluorescence microscopy of stimulated Zn(II) release from organotypic cultures of mammalian hippocampus using a carbonic anhydrase-based biosensor system, *J. Neurosci. Methods* **96**, 35-45 (2000).
48. T. Koike, T. Watanabe, S. Aoki, E. Kimura, and M. Shiro, A novel biomimetic zinc(II)-fluorophore, dansylamidoethyl- pendant macrocyclic tetraamine 1,4,7,10-tetraazacyclododecane (cyclen), *J. Am. Chem.*

Soc. **118**, 12696-12703 (1996).

49. T. Koike, T. Abe, M. Takahashi, K. Ohtani, E. Kimura, and M. Shiro, Synthesis and characterization of the zinc(II)-fluorophore, 5-dimethylaminonaphthalene-1-sulfonic acid 2-(1,5,9-triazacyclododec-1-yl)ethyl amide and its zinc(II) complex. *J. Chem. Soc. Dalton Trans.* 1764-1768 (2002).
50. E. Kimura, S. Aoki, E. Kikuta, and T. Koike, A macrocyclic zinc(II) fluorophore as a detector of apoptosis. *Proc. Nat. Acad. Sci.* **100**, 3731-3736 (2003).
51. K. K. Jensen, L. Martini, and T. W. Schwartz, Enhanced fluorescence resonance energy transfer between spectral variants of green fluorescent protein through zinc-site engineering, *Biochemistry* **40**, 938-945 (2001).
52. D. P. Barondeau, C. J. Kassmann, J. A. Tainer, E. D. Getzoff, Structural chemistry of a green fluorescent protein Zn biosensor, *J. Am. Chem. Soc.* **124**, 3522-3524 (2002).
53. K. Tanaka, T. Miura, N. Umezawa, Y. Urano, K. Kikuchi, T. Higuchi, and T. Nagano, Rational design of fluorescein-based fluorescence probes, mechanism-based design of a maximum fluorescence probe for singlet oxygen. *J. Am. Chem. Soc.* **123**, 2530-2536 (2001).
54. T. Miura, Y. Urano, K. Tanaka, T. Nagano, K. Ohkubo, and S. Fukuzumi, Rational design principle for modulating fluorescence properties of fluorescein-based probes by photoinduced electron transfer, *J. Am. Chem. Soc.* in press.
55. H. Kojima, N. Nakatsubo, K. Kikuchi, S. Kawahara, Y. Kirino, H. Nagoshi, Y. Hirata, and T. Nagano, Detection and imaging of nitric oxide with novel fluorescent indicators: Diaminofluoresceins, *Anal. Chem.* **70**, 2446-2453 (1998).
56. H. Kojima, Y. Urano, K. Kikuchi, T. Higuchi, Y. Hirata, and T. Nagano, Fluorescent indicators for imaging nitric oxide production. *Angew. Chem. Int. Ed.* **38**, 3209-3212 (1999).
57. N. Umezawa, K. Tanaka, Y. Urano, K. Kikuchi, T. Higuchi, and T. Nagano, Novel fluorescent probes for singlet oxygen, *Angew. Chem. Int. Ed.* **38**, 2899-2901 (1999).
58. T. Hirano, K. Kikuchi, Y. Urano, T. Higuchi, and T. Nagano, Highly zinc-selective fluorescent sensor molecules suitable for biological applications. *J. Am. Chem. Soc.* **122**, 12399-12400 (2000).
59. T. Hirano, K. Kikuchi, Y. Urano, and T. Nagano, Improvement and biological applications of fluorescent probes for zinc, ZnAFs, *J. Am. Chem. Soc.* **124**, 6555-6562 (2002).
60. S. Y. Assaf, and S. H. Chung, Release of endogenous Zn^{2+} from brain-tissue during activity, *Nature* **308**, 734-736, (1984).
61. G. A. Howell, M. G. Welch, and C. J. Frederickson, Stimulation-induced uptake and release of zinc in hippocampal slices. *Nature* **308**, 736-738 (1984).
62. R. Dingledine, K. Borges, D. Bowie, and S. F. Traynelis, The glutamate receptor ion channels, *Pharmacol. Rev.* **51**, 7-61 (1999).
63. J. Y. Koh, and D. W. Choi, Zinc toxicity on cultured cortical-neurons - Involvement of N-methyl-D-aspartate receptors, *Neuroscience* **60**, 1049-1057 (1994).
64. H. Z. Yin, and J. H. Weiss, Zn^{2+} permeates Ca^{2+} permeable AMPA kainate channels and triggers selective neural injury, *Neuroreport* **6**, 2553-2556, (1995).
65. S. L. Sensi, H. Z. Yin, S. G. Carriedo, S. S. Rao, and J. H. Weiss, Preferential Zn^{2+} influx through Ca^{2+}-permeable AMPA/kainate channels triggers prolonged mitochondrial superoxide production, *Proc. Nat. Acad. Sci.* **96**, 2414-2419 (1999).
66. J. A. Park, and J. Y. Koh, Induction of an immediate early gene egr-1 by zinc through extracellular signal-regulated kinase activation in cortical culture: Its role in zinc-induced neuronal death, *J. Neurochem.* **73**, 450-456 (1999).
67. K. M. Noh, Y. H. Kim, and J. Y. Koh, Mediation by membrane protein kinase C of zinc-induced oxidative neuronal injury in mouse cortical cultures, *J. Neurochem.* **72**, 1609-1616 (1999).
68. S. Ueno, M. Tsukamoto, T. Hirano, K. Kikuchi, K. M. Yamada, N. Nishiyama, T. Nagano, N. Matsuki, Y. Ikegaya, Mossy-fiber Zn^{2+} spillover modulates heterosynaptic N-methyl-D-aspartate receptor activity in hippocampal CA3 circuits. *J. Cell Biol.* **158**, 215-220 (2002).
69. T. Kirino, Delayed neuronal death in the gerbil hippocampus following ischemia, *Brain Res.* **239**, 57-69 (1982).
70. S. Maruyama, K. Kikuchi, T. Hirano, Y. Urano, and T. Nagano, A novel, cell-permeable, fluorescent probe for ratiometric imaging of zinc ion, *J. Am. Chem. Soc.* **124**, 10650-10651 (2002).
71. M. Li and P. R. Selvin, Luminescent polyaminocarboxylate chelates of terbium and europium: the effect of chelate structure, *J. Am. Chem. Soc.* **117**, 8132-8138 (1995).
72. O. Reany, T. Gunnlaugsson, and D. Parker, A model system using modulation of lanthanide luminescence to signal Zn^{2+} in competitive aqueous media, *J. Chem. Soc. Perkin Ttrans. 2* 1819-1831 (2000).
73. K. Hanaoka, K. Kikuchi, H. Kojima, Y. Urano and T. Nagano, Selective detecting of zinc ions with novel luminescent lanthanide probes, *Angew. Chem. Int. Ed.* in press.
74. L. Prodi, F. Bolletta, M. Montalti, and N. Zaccheroni, Searching for new luminescent sensors: Synthesis

and photophysical properties of a tripodal ligand incorporating the dansyl chromophore and of its metal complexes, *Eur. J. Inorg. Chem.* 455-460 (1999).

75. P. J. Jiang, L. Z. Chen, J. Lin, Q. Liu, J. Ding, X. Gao, Z. J. Guo, Novel zinc fluorescent probe bearing dansyl and aminoquinoline groups, *Chem. Comm.* 1424-1425 (2002).
76. T. W. Kim, J. H. Park, J. I. Hong, Zn^{2+} fluorescent chemosensors and the influence of their spacer length on tuning Zn^{2+} selectivity, *J. Chem. Soc. Perkin Trans. 2* 923-927 (2002).
77. S. C. Burdette, S. J. Lippard, The rhodafluor family. An initial study of potential ratiometric fluorescent sensors for Zn^{2+}, *Inorg. Chem.* **41**, 6816-6823 (2002).
78. S. C. Burdette, C. J. Frederickson, W. M. Bu, S. J. Lippard, ZP4, an improved neuronal Zn^{2+} sensor of the Zinpyr family, *J. Am. Chem. Soc.* **125**, 1778-1787 (2003).
79. Y. Li, C. J. Hough, S. W. Suh, J. M. Sarvey, and C. J. Frederickson, Rapid translocation of Zn^{2+} from presynaptic terminals into postsynaptic hippocampal neurons after physiological stimulation, *J. Neurophysiol.* **86**, 2597-2604 (2001).
80. Y. Li, C. J. Hough, C. J. Frederickson, and J. M. Sarvey, Induction of mossy fiber -> CA3 long-term potentiation requires translocation of synaptically released Zn^{2+}, *J. Neurosci.* **21**, 8015-8025 (2001).
81. S. L. Sensi, D. Ton-That, P. G. Sullivan, E. A. Jonas, K. R. Gee, L. K. Kaczmarek, and J. H. Weiss, Modulation of mitochondrial function by endogenous Zn^{2+} pools, *Proc. Nat. Acad. Soc.* **100**, 6157-6162 (2003).
82. W.-C. Sun, K. R. Gee, D. H. Klaubert, and R. P. Haugland, Synthesis of fluorinated fluoresceins, *J. Org. Chem.* **62**, 6469–6475 (1997).

USE OF FLUORESCENCE SPECTROSCOPY TO MONITOR PROTEIN-MEMBRANE ASSOCIATIONS

Suzanne F. Scarlata[1]

1. INTRODUCTION

The binding of proteins to membranes mediates many physiological processes such as endocytosis, endosomal recycling and signal transduction (see [1]). Membrane association serves to concentrate proteins on a quasi-two dimensional surface promoting interactions between protein partners. Membrane binding can be reversible, such as the binding of phospholipases and phosphodiesterases, or irreverisible, such as the binding and assembly of HIV-Gag proteins on the plasma membrane of host cells. Proteins may associate only on the surface of the membrane, penetrate into the head group region or incorporate into the hydrocarbon interior. Membrane binding can be non-specific, such as for MARCKS [2] or depend of the presence of particular lipids, such is the case for PLCδ [3].

There are few methods that allow one to study membranes and membrane proteins in real time. Fluorescence spectroscopy is now routinely used to monitor the interactions of proteins with membranes, the fusion of membranes with each other, and the association between membrane-bound proteins under model and cellular conditions. Membrane fusion and the association between membrane proteins have been the subject of recent reviews [4-6], and so here we will focus on the fluorescence-based methods to monitor the binding of proteins to membranes. Almost all of the methods described here can be used for the association of purified proteins to model membranes or cellular membrane preparations. In complex solutions, then membrane binding maybe followed by attaching a fluorescent-tagged antibody to the protein of interest, or by monitoring green fluorescent proteins (GFP)- or GFP analogs chimeras of the protein of interest.

Fluorescence spectroscopy allows sensitive steady state and kinetic measurements

[1] Suzanne F. Scarlata, Dept. of Physiology & Biophysics, SUNY Stony Brook, Stony Brook, NY 11794-8661, U.S.A

of protein-membrane associations. The steady state and on- and off-rates of these associations can be monitored by nearly every fluorescence method including changes in intensity, lifetime, emission energy, anisotropy and fluorescence correlation. Protein association can be followed by probes located in the lipid, the protein or both. While these methods are only briefly mentioned here, a comprehensive description of these methods can be found in the textbook of Lakowicz [7]. The ionic nature of membrane surface, and the hydrocarbon nature of its interior results in a dramatic environmental change for many probes and allows for detection by these many fluorescence methods. Here, we will first review the membrane properties that give rise changes in protein fluorescence or changes in the membrane that occur upon protein association. We will then discuss the methods used to follow the association of proteins to membranes from the perspective of the lipids, and then of the protein.

2. OVERVIEW OF MEMBRANE PROPERTIES

Lipid bilayers have hydrocarbon interiors of low polarity sandwiched between head group regions that contain polar and ionic groups. Natural membranes have ionic surfaces and contain 10-20% anionic lipids [8]. There are several fluorescence probes that are sensitive to the surface potential and have been used to measure this potential under different environmental conditions (see [9]). The highly charged membrane surface contributes greatly to the changes in the emission properties of fluorophores upon protein association. Further, the extent of fluorescence changes that a particular probe will undergo will depend on the charge of the membrane surface. Because of the anionic nature of membrane surfaces, the driving force for the association of many proteins to membrane surfaces is electrostatic, and many membrane binding proteins have a cluster of basic residues that target their host protein to negatively charged membranes [2 8]. The binding energy from electrostatic interactions is highly dependent on the local ionic strength, and will weaken considerably as the salt concentration is raised allowing dissociation at sufficient ionic strengths. Thus, using fluorescence to follow the reversible electrostatic binding of a protein to a membrane requires a probe that is not sensitive to changes in ionic strength, or careful control studies of these background changes.

Contributing to the changes in probe emission seen upon membrane-protein associations is the curvature of the membrane surface. Small, unilamellar vesicles prepared by extrusion or more commonly by sonication, are typically ~250 Å and have twice the surface area and 70% of the lipids in the outer leaflet making their surfaces very curves as compared to large and giant unilamellar vesicles, which are are relatively flat (see [10]). The curvature of small unilamellar vesicles causes solvent exposure of less ionic/polar groups underlying the surface thus resulting in stress of the membrane surface. Surface curvature allows for protein interactions with less charged groups in the membrane to relieve this stress. If protein-membrane association is being monitored by charged-induced changes in fluorescence, then curvature would result in reduced changes in probe emission upon the membrane association. Curved membranes tend to be fusinogenic and frequently the addition of proteins with basic residues will reduce electrostatic repulsion between anionic bilayers and promote vesicle fusion. It is notable that curvature also promotes penetration of the protein into the bilayer interior which would alter emission changes as discussed below.

While the use of small, unilamellar vesicles does present experimental difficulties as described above, their study is directly relevant to many biochemical processes, such as directing the binding of proteins during the fusion events that occur during endocytosis and vesicle trafficking (see [4]). The geometric factors that contribute to curvature are given are out of the scope of this review but we note that phosphatidyethanolamine (PE) lipids whose width of their head groups is less than their hydrocarbon chains under most physiological conditions, promote bilayer curvature and tend to promote non-specific binding (see [10]).

Another factor that may contributes to the changes in probe fluorescence that occurs upon the association of proteins to membrane is the presence of lipid, and/or protein domains. The surfaces of biological membranes are far from uniform. Irregularities of the membrane surface can be caused by integral membrane proteins that protrude from the surface. Membrane head groups have varying size and their lipid chains have varying length which can either cause lipids to protrude from or sink into the membrane surface. Over the past several years it has become increasingly apparent that lipid domains on membrane surfaces called lipid rafts will form independent of protein. Lipid rafts were first described as the domains that remain after detergent solubilization of cell membranes (for review see [11,12]). Model membranes containing high concentrations of cholesterol will separate into `rafts' of liquid-ordered phase (L_o) lipids, having reduced fluidity and extended chains, that float in a sea of liquid disordered phase (L_d) phase lipids. Fluorescence methods have been used to determine the phase diagram of raft-forming mixtures in model systems [13]. In biological membranes, lipid rafts have been proposed to localize proteins into signal transduction complexes, or sequester and internalize protein receptors. Although the size and lifetime of lipid rafts is not yet clear, the combined evidence suggests that they are very small and short-lived (see [12]). Therefore, it is not clear whether rafts would significantly alter the surface structure of natural membranes enough to affect the association of proteins to the membrane surface, although it is probable to affect the localization of membrane probes (see below).

A subset of detergent-insoluble domains are caveolae, which are flask-shaped invaginations on the plasma membrane visualized by electron microscopy composed primarily of caveolin proteins (for review see [14,15]). These domains are large in size and could easily carry out functions such as protein organization and receptor sequestration. In regards to membrane association, for proteins that do not bind to caveolae, their presence has the effect of reducing the membrane area in which the protein can bind thus causing the apparent binding constant to be lower. In the case where proteins specifically bind to caveolae, then changes in fluorescence result from protein interactions with caveolae proteins as opposed to protein-lipid interactions. The binding curves will thus reflect the association constant between the proteins. If a protein has both lipid and caveolae protein affinity, then the changes in fluorescence would result from lipid and caveolae protein associations, and

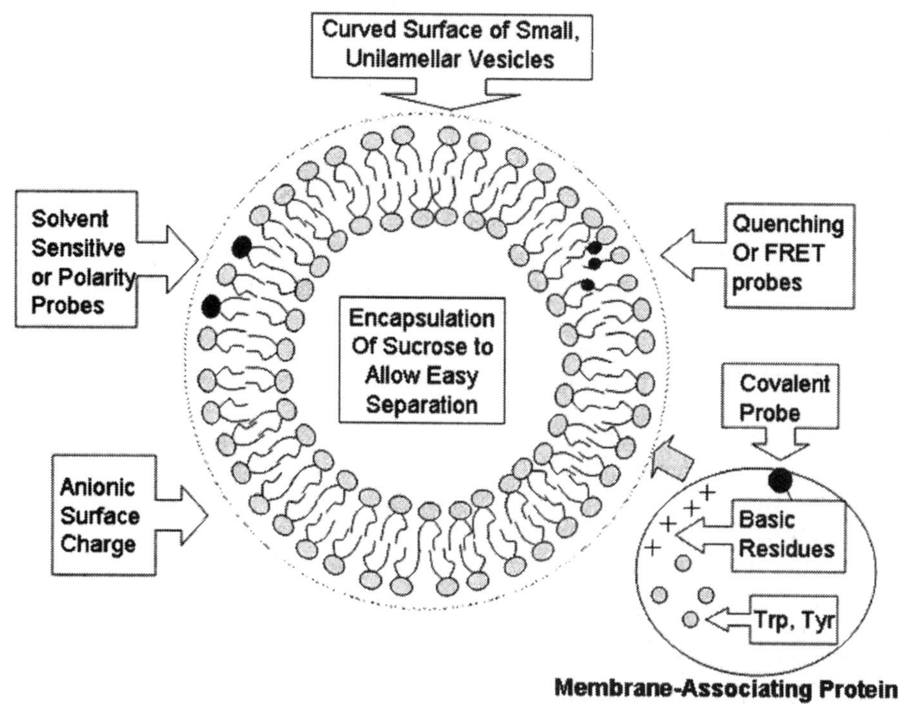

Figure 1 Overview of key membrane surface properties and different fluroescence methods that can be used to monitor protein association

the apparent association constant would consist of the membrane partition coefficient as well as the apparent dissociation constant between the proteins, and this latter value would depend on whether the protein associates directly to the caveolae from the solvent or whether the protein binds first to the membrane and then laterally associates to caveolae.

The repercussions of the non-uniform surfaces of lipid and protein domains in regards to protein-membrane associations is that these structures may cause membrane probes to partition either inside of outside of the domains making their membrane distribution non-uniform. Thus, detection of protein association to lipid domains that contain a reduced amount of membrane probe would be reduced, whereas protein association to the domains where membrane probes reside would be enhanced. Also, following the discussion above, the preferential partitioning of a membrane associating protein would yield stronger apparent membrane affinities due to the effective increase in membrane areas due to non-binding domains.

3. MONITORING PROTEIN-LIPID ASSOCIATION FROM THE MEMBRANE'S PERSPECTIVE

There are many fluorescence-based methods that follow the binding of proteins to membrane surface, and some of these are depicted in Figure 1. Experimentally, these studies are conduced by starting with a solution of lipid membranes doped with a small amount (i.e. less than one mol percent) of probe, so as to not perturb the integrity of the membrane, and add increasing amounts of protein, although often this method require a prohibitive amounts of protein. It is thus more practical to start with a small amount of protein and titrate in labeled lipid, and compare to controls that substitute buffer for protein. While very low lipid concentrations may be missed using this method, the membrane partition coefficient is readily obtained (see below).

There is a large assortment of membrane probes commercially available and the spectral properties and environmental sensitivity are listed in the Molecular Probes, Inc. catalog (http://www.molecularprobes.com), and their properties described in Lakowicz [7]. These fall into two major groups: those that reside on the membrane surface and those that reside in the membrane interior. Surface-localized probes may position themselves at slightly different depths in different membrane environment and so control studies should be done to eliminate fluorescence changes due to probe position from those due to protein binding. Many membrane surface probes are very sensitive to the change in local dielectric when a protein displaces water on the membrane surface as it binds, and these probes usually undergo a shift to higher energies and an increase in quantum yield. The most common polarity sensitive probe is Laurdan, which has a substituted naphthalene head group that is anchored to the membrane surface by a hydrocarbon (lauryl) chain. Some surface probes are quenched by water and so protein binding results in de-quenching and an increase in quantum yield, which is the case for dansyl probes. It is important to keep in mind that in fluid phase bilayers, the probes undergo rapid lateral diffusion and so more insensitive probes will only transmit changes in the membrane surface when a substantial amount of protein is bound.

Incorporation of the probes varies with probe type. Detergent-like probes having a single hydrocarbon chain will usually incorporate into preformed bilayers by simple addition, or by a brief low energy sonication step. Since detergent probes could greatly perturb the structure of the bilayer, their mole percent concentration should be as low as possible. In contrast, lipid-like probes with two hydrocarbon chains must be incorporated into membranes by mixing the probe with lipid in organic solvent before preparing model membranes or by using phospholipid transfer proteins to incorporate the probes into preformed membranes.

Probes that reside in the hydrocarbon interior can be used to monitor the surface association of proteins by fluorescence resonance energy transfer to or from a label on the protein. Hydrocarbon-localized probes are more often used to monitor the properties of the lipid bilayers itself (i.e. fluidity, fusion, phase separations and transitions) and can also be used to monitor the penetration of membrane proteins, either by energy transfer, or a collision quenching using brominated or spin-labeled groups or by fluorescence anisotropy since protein penetration tends to reduce the fluidity of the membrane .

When using probes attached in the middle to lower half of a long hydrocarbon chain, it is possible for the chain to bend back such that the probe localizes on the membrane surface. This unanticipated result sometimes happens when incorporating detergent or fatty acid probes into preformed bilayers, or when the probe is much more polar than the bilayer interior. Although this probe configuration would still be expected to be sensitive to protein association, it may be disruptive to the packing of the lipid chains in the membrane [16].

4. MONITORING LIPID-PROTEIN ASSOCIATION FROM THE PROTEIN'S PERSPECTIVE

Viewing the association of protein to membranes can be accomplished by monitoring changes in the intrinsic fluorescence of the protein, or by the changes in a covalently attached probe (see Figure 1). Typically, these studies are done by starting with a dilute solution of protein and titrating in unlabeled lipid. Membrane association can be easily followed by changes in fluorescence as lipid is added.

The emission intensity of intrinsic fluorophores is usually very sensitive to membrane binding since Trp and Tyr residues tend to be interfacial (see [17]). The interfacial position of Trp and Try residues of membrane binding proteins suggests that these residues are solvent exposed before membrane association and will undergo a shift from a water environment to an ionic environment upon membranes binding. The Trp and Tyr side chains tend to be quenched by the charged groups on the membrane surface and thus binding of proteins with interfacial Trp/Tyr can be followed by the quenching of intrinsic fluorescence. Of course, protein binding using this method will depend on the location of the intrinsic fluorophores. On example of this is the protein phospholipase Cβ which is a 134 kDa protein that binds to membranes primarily through the N-terminal pleckstrin homology domain. This protein undergoes an ~30% decrease upon binding to membranes that contain 33% anionic lipids and 67% zwitterionic lipids of POPC, and an ~40% to vesicles composed of 100% anionic lipids [18]. Since this quenching is static, no significant changes in fluorescent lifetime are seen. Membrane binding also results in an observed shift to lower energies presumably due to the quenching of the surface-exposed interfacial residues. Membrane binding of this protein is driven by its N-terminal pleckstrin homology domain which is ~120 residues in size [19]. This domain contains a single buried Trp that is completely insensitive to membrane association in contrast to the large quenching of the intrinsic fluorescence of whole enzyme.

In some cases, membrane binding of a protein results in an increase in emission intensity and a shift to higher energies. These changes are indicative of penetration of the protein into the lipid matrix. There are several methods to verify this idea such as surface pressure measurements using monolayers, or collisional quenching of the protein to brominated or spin-labeled probes residing at different membrane depths, or similar fluorescence resonance energy transfer methods to probes such as the series of commercially available anthroyloxy fatty acids where the probes are attached from the 2 to 16 positions, although the spacial resolution of this latter method is not very good.

Proteins can be labeled with a variety of probes that are sensitive to membrane association and sensitivity will depend on the type of probe as well as the location of its site of attachment (see Molecular Probes Handbook (www.molecularprobes.com)). Labeling sites are typically on external groups making the chances of reacting a group on or close to the membrane binding site good. To insure that the presence of the label does not affect the binding affinity, different types of covalent attachment should be tested (i.e. labeling on Cys side chains or primary amines), or the membrane binding affinity using this method should match the value obtained using membrane probes.

Long-lived fluorescent probes can also be used to measure membrane binding through changes in steady state or time-resolved fluorescence anisotropy since the protein undergoes a large increase in rotation volume as it binds to the much larger lipid vesicle. An experimental drawback of this method is that the polarizers needed for these measurements reject a large portion light and is thus difficult to use when the emitter is weak as compared to the background scattering signal from the lipid vesicles.

5. LIMITATION OF METHODS AND ANALYSIS

The simplest method to determine the strength of the binding of a protein to membrane surfaces is to present the results in terms of a membrane binding partition coefficient (K_p). The partition coefficient refers to the lipid concentration at which the fraction of bound protein is equal to the fraction of unbound. Since the fraction of bound and unbound is usually taken from the changes in some fluorescence property as lipid is added and it is possible that there is a silent, non-binding population which may present itself as smaller than expected changes in the particular fluorescence parameter. To determine whether all or only a portion of a protein binds to a membrane, a second technique, usually a separation-based method, should be used. Some methods may include encapsulating sucrose or another heavy compound into the interior of the bilayer, adding protein and sedimenting the lipids and measuring the loss of protein in the supernatent or the gain of protein in the lipid pellet (e.g. [18]). Other separation methods, such as the use of spin tubes containing a separating filter, dialysis, or gel filtration chromatography also work well.

While this review described methods for equilibrium binding,, most methods described can be use in a rapid mixing chamber inserted into a fluorometer to measure the on-rate. For many membrane -associating proteins, this rate is rapid and may require changes in the protocol to obtain resolvable data, such as increases in solvent viscosity and lowering the temperature, etc. It should be noted that these changes may affect the membrane properties; for example, adding an osmolyte may result in vesicle rupture due to relieve the osmotic pressure, and so any changes to the solvent must be carefully carried out. On the other hand, membrane dissociation is usually much slower and can be done by dilution of the protein-lipid solution by buffer so that the species are below the K_p, or displacing the protein from the membrane through the addition of another species.

One complication that may occur is protein aggregation due to surface crowding when the protein concentration is much higher than the lipid concentration as may be the case in the beginning of a lipid titration. For this reason, it is recommended that protein concentrations are kept much lower than the initial additions of lipid (e.g. 1 μM lipid added to a 10 nM solution of protein), or lipid protein solutions be used, which may require the addition of a covalent label.

The membrane binding curve is usually fit to hyperbolic function using package software which directly gives the lipid concentration where the ratio of bound/free protein is one. The hyperbolic fit reflects the simplest model for membrane binding which defines the membrane partition coefficient (K_p) as $K_p = [P_b]/[P_f] * 1/L$) where L equals the total amount of lipid (see [10]). It should be kept in mind that only the outer leaflet of the bilayer is available for binding, and so the amount of total lipid needed to bind protein must be more than twice the reported value of K_p. This twofold difference is true for large and giant unilamellar vesicles, while the amount of lipid in the outer leaflet of small, unilamellar vesicles is ~10% lower due to curvature. Of course, the available lipid surface for multilamellar vesicles is much lower. This simple binding model is applicable in many cases but note that it does not account for electrostatic interactions between charged proteins or other processes such as aggregation, etc. However, when other processes are involved, such as membrane binding followed by protein-protein association on the membrane surface, then other methods, such as fluorescence resonance energy transfer between the proteins, must to used. In a few cases, a protein will bind specifically to a particular lipid head group. In these cases membrane binding must be treated as a enzyme-substrate association, and the binding constant would be a biomolecular association rater than a membrane partition coefficient. An example of this behavior is the association of the pleckstrin homology domain of phospholipase Cδ to phosphatidylinositol 4,5 bisphosphate [20]. This association involves ionic interactions as well as several hydrogen bonding giving the interaction strong specificity. In fact, this protein will only bind strongly to membranes if phosphatidylinositol 4,5 bisphosphate is present.

6. CONCLUDING STATEMENTS

This review is written to convey the versatility of fluorescence methods to monitor the association of proteins to membranes. Almost all types of fluorescence techniques can be used for general binding studies, and simple corroborative methods should be used to determine whether a significant non-binding population exists. When other mechanisms besides simple associations are operative, such as protein insertion into the lipid matrix, or protein-protein associations on the membrane, then methods such as quenching and energy transfer may be employed.

7. ACKNOWLEDGMENTS

The author would like to acknowledge the support of N.I.H. (GM53132). The author would also like to acknowledge the contribution of a large number of primary fluorescence references which were not specifically cited and hope that these authors are not offended.

8. REFERENCES

1. Alberts, B.; Bray, D.; Lewis, J.; Raff, M.; Roberts, K.; Watson, J. *Molecular Biology of the Cell*; (Garland: New York, 1994).
2. Kim, J.; Mosior, M.; Chung, L. A.; Wu, H.; McLaughlin, S. Binding of peptides with basic residues to membranes containing acidic phospholipids,*Biophys J.***60**, 135-48(1991).
3. Rebecchi, M.; Boguslavsky, V.; Boguslavsky, L.; McLaughlin, S. Phosphoinositide-specific phospholipase C-delta 1: effect of monolayer surface pressure and electrostatic surface potentials on activity,*Biochemistry*, **31**, 12748-53 (1992).
4. Jahn, R.; Grubmuller, H. Membrane Fusion,*Curr.Opin.Chem.Biol.*, **14**, 488-95 (2002).
5. Scarlata, S. Determination of the strength and specificity of membrane-bound protein association using fluorescence spectroscopy: Application to the G protein - phospholipase C system. In *Methods in Enzymology*; Hildrebrandt, I. a., Ed.; Academic Press: New York, 2002; Vol. 345C; pp 306-27.
6. Scarlata, S.; Dowal, L. The use of green fluorescent proteins to view associations between phospholipase Cbeta and G protein subunits in cells. In *Methods in Molecular Biology*; Smrcka, A., Ed.; Humana Press Inc: Totowa, NJ, 2003; Vol. 237: G protein signaling: methods and protocols; pp in press.
7. Lakowicz, J. *Principles of Fluorescence Spectroscopy, Second Edition*; (Plenum: New York, 1999).

8 McLaughlin, S. The electrostatic properties of membranes.,*Annual Review of Biophysics & Biophysical Chemistry*, **18**, 113-36 (1989).

9. Singh, Y.; Gulyani, A.; Bhattacharya, S. A new ratiometric fluorescence probe as strong sensor of surface charge of lipid vesicles and micelles, *FEBS Letters*, **541**, 132-6 (2003).
10. Gennis, R. B. *Biomembranes: Molecular Structure and Function*; (Springer-Verlag: New York, 1989).
11. Brown, D. A.; London, E. Functions of lipid rafts in biological membranes, *Annu.Rev.Cell Dev.Biol.*, **14**, 111-36 (1998).
12. Edinin, M. The state of lipid rafts: from model membranes to cells, *Annu.Rev.Biophy.Biomolec.Struc.*, **32**, in press (2003).
13. Xu, X.; London, E. The effect of sterol structure on membrane lipid domains reveals how cholesterol can induce lipid domain formation, *Biochem.*, **39**, 843-49 (2000).
14. Okamoto, T.; Schlegel, A.; Scherer, P. E.; Lisanti, M. P. Caveolins, a family of scaffolding proteins for organizing "preassembled signaling complexes" at the plasma membrane, *J.Biol.Chem.* **273**, 5419-22 (1998).
15. Anderson, R. G. The caveolae membrane system,*Annu.Rev.Biochem.*, **67**, 199-225

(1998).
16. Kaiser, R.; London, E. Determination of the depth of BODIPY probes in model membranes by parallax analysis of fluorescence quenching, *Biochim.Biophys.Acta*, **1375**, 13-22 (1998).
17. Wimley, W. C.; White, S. H. Experimentally determined hydrophobicity scale for proteins at membrane interfaces, *Nature Structural Biology*, **3**, 842-8 (1996).
18. Runnels, L. W.; Jenco, J.; Morris, A.; Scarlata, S. Membrane binding of phospholipases C-beta 1 and C-beta 2 is independent of phosphatidylinositol 4,5-bisphosphate and the alpha and beta gamma subunits of G proteins, *Biochemistry*, **35**, 16824-32 (1996).
19. Wang, T.; Pentyala, S.; Rebecchi, M.; Scarlata, S. Differential association of the pleckstrin homology domains of phospholipases C-β1, C-β2 and C-δ1 with lipid bilayers and the βγ subunits of heterotrimeric G proteins,*Biochemistry*, **38**, 1517-27 (1999).
20. Garcia, P.; Gupta, R.; Shah, S.; Morris, A. J.; Rudge, S. A.; Scarlata, S.; Petrova, V.; McLaughlin, S.; Rebecchi, M. J. The pleckstrin homology domain of phospholipase C-delta 1 binds with high affinity to phosphatidylinositol 4,5-bisphosphate in bilayer membranes, *Biochemistry*, **34**, 16228-34 (1995).

APPLICATION OF GREEN FLUORESCENT PROTEIN-BASED CHLORIDE INDICATORS FOR DRUG DISCOVERY BY HIGH-THROUGHPUT SCREENING

A. S. Verkman*, Peter M. Haggie and Luis J.V. Galietta

SUMMARY

High-throughput screening of small molecule collections has become a widely used approach in modern drug discovery. Compounds with activity in a primary screen using automated assays are evaluated and optimized for development of drugs for clinical use. Our lab has been interested in discovering inhibitors and activators of the cystic fibrosis transmembrane conductance regulator protein (CFTR), a cAMP-activated chloride channel which when defective causes the genetic disease cystic fibrosis. Development of a robust and sensitive primary screen is of central importance in drug discovery by high-throughput screening. We review the development and optimization of a green fluorescent protein-based cell assay for measurement of CFTR halide transport in epithelial cells. Screening of compound collections has yielded CFTR inhibitors with potential application as antidiarrheals, and CFTR activators with potential application as correctors of defective mutant CFTRs in cystic fibrosis.

A.S. Vekman and P.M. Haggie, Departments of Medicine and Physiology, Cardiovascular Research Institute, University of California, San Francisco, CA, 94143-0521, USA; L.J.V. Galietta, Laboratorio di Genetica Molecolare, Istituto Giannina Gaslini, 16148 Genova, Italy

1. INTRODUCTION

Drug discovery by high-throughput screening has become the paradigm worldwide for identification of compounds with specified target activity. Large collections of synthetic drug-like small molecules or natural compounds are screened individually or in small groups using an assay designed to identify active compounds with high efficiency and reliability. Compounds found to be active in an initial primary screen, or 'hits', are further evaluated for their potency, specificity, toxicity, structure-activity relationship, and *in vivo* efficacy and pharmacology. One or a small number of compounds with favorable profiles, or 'leads', are then optimized by synthetic organic chemistry and intensively screened as potential new drugs. Fluorescence-based primary screening is particularly attractive because of its rapidity, suitability for automation, and potential for high specificity and selectivity.

Our lab has been interested in developing inhibitors and activators for the Cystic Fibrosis Transmembrane Conductance Regulator (CFTR) protein. The CFTR gene was identified in 1989 as the genetic basis of the hereditary lethal disease cystic fibrosis (1). CFTR functions as a cAMP-activated Cl^- channel in the apical plasma membrane of epithelial cells in the airways, sweat duct, testis, pancreas, intestine and other fluid-transporting tissues (2). CFTR is a large transmembrane glycoprotein containing two 6-helix membrane-spanning domains, each followed by a nucleotide binding domain (NBD), with a regulatory (R) domain linking the first NBD and the second membrane-spanning domain (3). CFTR activation primarily involves ATP binding and hydrolysis at NBDs, and phosphorylation of multiple R-domain sites. In cystic fibrosis, defective epithelial cell fluid transport produces chronic lung infection and slow deterioration in lung function, as well as pancreatic insufficiency, meconium ileus, and male infertility (2). There are now >1000 CFTR mutations associated with cystic fibrosis, though ~90 % of cystic fibrosis subjects have as at least one allele the ΔF508 (deletion of phenylalanine at 508 position) mutation (4). The ΔF508 mutation causes two distinct defects in CFTR that produce Cl^- impermeable cells: ***a.*** retention at the endoplasmic reticulum due to misfolding and/or defective interactions with molecular chaperones (5, 6), and ***b.*** impaired intrinsic Cl^- conductance (reduced open channel probability, refs. 7-10). Drug treatment of subjects with the ΔF508 mutation may require the combined use of two drugs to correct the two defects.

Our lab initially developed small-molecule fluorescent indicators of chloride and iodide that have been used to study anion transport in many types of cultured cell models and *in vivo* (reviewed in refs. 11, 12). Although in principle suitable for high-throughput screening, small-molecule indicators require external addition and washing, and are complicated by indicator leakage between the time of washing and assay. For this reason we focused on the development of targetable, intrinsically fluorescent indicators based on green fluorescent protein (GFP) that do not require cell loading or washing. We review here the development and optimization of a cell-based assay for CFTR activity, and provide examples demonstrating the application of the assay to the discovery of CFTR inhibitors and activators by high-throughput screening.

2. GREEN FLUORESCENT PROTEIN-BASED HALIDE INDICATORS

GFP is a genetically-encoded, intrinsically fluorescent protein of ~30 kDa isolated from the jellyfish *Aequoria victoria* (for review, see ref. 13). GFP can be genetically targeted to specific cellular sites in cell culture models and *in vivo* and has been expressed in a wide variety of organisms. Indeed, GFP has had a major impact upon cell biology with diverse applications in studies of protein location, dynamics, interactions and regulation (14, 15). One exciting application of GFP has been the engineering of real-time cellular sensors of pH, ion concentrations, second messengers, enzyme activities and other parameters (16, 17). Compared to classical chemical probes, genetically-encoded fluorescent proteins permit stable, non-invasive staining at specific subcellular sites with little or no cellular toxicity.

The GFP chromophore is autocatalytically generated by the post-translational cyclization and oxidation of three residues encoded within the primary sequence (13, 18). Amino acids throughout wild type GFP, including the chromophore, have been mutated to generate an array of fluorescent proteins with altered spectral and biophysical properties. Yellow fluorescent protein (YFP), the GFP variant from which our halide sensors are derived, was generated using a rational mutagenic strategy based upon crystallographic data (19). Aromatic amino acids were introduced at Thr203 to extend the π-system of the chromophore, lower its excited state energy and consequently increase emission wavelength. YFP contains tyrosine at position 203 as well as mutations to increase folding efficiency (S65G, V68L, S72A). Maximum fluorescence emission of YFP is at 528 nm, ~20 nm longer than that of GFP-S65T.

High-resolution crystal structures of several GFP variants have been solved and found to be similar; GFPs are cylindrical (~40 × 20 Å) and composed of an 11-standed ß-barrel that encloses the chromophore (19-21). Although not directly accessible to solvent, the phenolic group within the chromophore of most GFP variants is pH sensitive. Acidic environments quench fluorescence by protonation of the chromophore (22). GFPs, including YFP, have thus been exploited to non-invasively measure pH *in vivo* (23-25). The pK_a value of a fluorescent protein, which determines the pH range over which the protein is sensitive, is influenced by amino acids in the vicinity of the chromophore. The phenolic portion of the chromophore is located near the His148 imidazole ring structure, where there are irregularities in the ß-strand structure. Backbone atoms from residues 144-150 do not form hydrogen bonds with backbones of adjacent residues (165-170); instead the ring structure of His148 participates in a hydrogen bond network. The chromophore of YFP is also ~1 Å closer to the protein surface when compared to other GFPs, a consequence of the introduced π-π stacking (20). To investigate the effect of solvent accessibility on the YFP chromophore the crystal structure of mutant YFP-H148G was determined (20). YFP-H148G remained fluorescent despite the presence of a large cavity in the vicinity of the chromophore that opened the ß-strand structure permitting solvent access to the chromophore (20).

The structure of YFP-H148G prompted investigation into the potential fluorescence modulation of YFPs by small solvent molecules. It was determined that halides and nitrate could quench the fluorescence of YFPs but not GFP-S65T (26). This initial study indicated that the pK_a values of YFP chromophores changed by up to 1.8 pH units for 0–400 mM chloride and that fluorescence was differentially quenched by various halides. The H148Q mutant of YFP demonstrated characteristics most suitable for cell based assays of chloride flux (27).

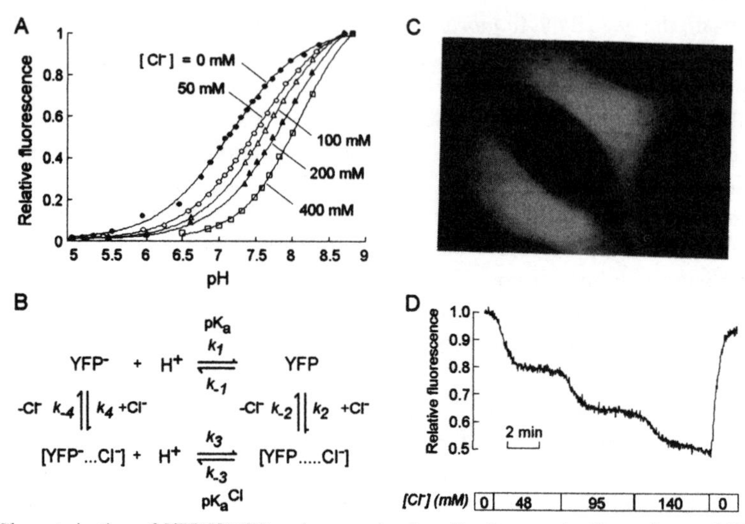

Figure 1. Characterization of YFP-H148Q and expression in cell cultures. **A.** Dependence of YFP-H148Q fluorescence on pH at indicated Cl^- concentrations. **B.** Kinetic scheme for YFP-H148Q interaction with Cl^- deduced from kinetic data. Rate constants: $k_1 = 1.4 \times 10^7\ s^{-1}$; $k_{-1} = 2.7 \times 10^7\ s^{-1}$; $k_2 = 5.6 \times 10^8\ M^{-1}s^{-1}$; $k_{-2} = 1.7 \times 10^7\ s^{-1}$; $k_3 = 3.5 \times 10^6\ s^{-1}$; $k_{-3} = 2.3 \times 10^6\ s^{-1}$; $k_4 = 3.6 \times 10^7\ M^{-1}s^{-1}$; $k_{-4} = 9.3 \times 10^6\ s^{-1}$; relative fluorescence: $[YFP^-] = 1$; $[YFP^-...Cl^-] = 0.3$; YFP and $[YFP...Cl^-] = 0$. **C.** Confocal fluorescence micrograph of YFP-H148Q transfected fibroblasts expressing CFTR showing cytoplasmic and nuclear staining. **D.** Intracellular fluorescence Cl^- titration at pH 7.4 using high K^+ buffer containing ionophores and indicated concentrations of Cl^-. Adapted from ref. 27.

Fluorescence titrations of purified recombinant YFP-H148Q indicated a pK_a of ~7 in the absence of Cl^- that increased to ~8 at 150 mM Cl^- (Fig. 1A) (27). At pH 7.5, YFP-H148Q fluorescence decreased with increasing chloride and iodide concentration such that the protein was 50 % quenched by 100 mM Cl^- and 21 mM I^-. The anion selectivity sequence for YFP-H148Q quenching ($F^- > I^- > NO_3^- > Cl^- > Br^-$) suggested the strong binding of weakly hydrated chaotropic ions. YFP-H148Q was insensitive to large ions (including gluconate, sulphate, phosphate and isothionate) (27) and the K_d for a particular halide decreased with pH (28) implying that quenching was influenced by steric issues and the ionic state of the chromophore. A static quenching mechanism involving Cl^- binding to YFP-H148Q was established by biophysical methods. The YFP-H148Q fluorescence lifetime was insensitive to chloride, whereas YFP-H148Q molar absorbance decreased with increasing chloride concentration. Crystallographic studies of YFP-H148Q in the presence of iodide identified a discrete binding site for halides (28). Bound iodide was shown to interact with the chromophore and the phenol group of T203Y although additional amino acids contributed to stabilize the YFP-H148Q/halide interaction. As such, halide binding to YFP-H148Q effectively stabilizes the deprotonated form of the chromophore mimicking a lower pH at the chromophore.

Stopped-flow fluorescence analysis was used to establish the kinetics and mechanism of chloride quenching of YFP-H148Q. In the absence of chloride, YFP-H148Q fluorescence changed rapidly ($t_{1/2}$ < 10 ms) in response to pH changes, whereas the fluorescence response was biexponential in the presence of 100 mM chloride with fast ($t_{1/2}$ < 10 ms) and slower ($t_{1/2}$ ~ 100 ms) components. Chloride dissociation and association with the deprotonated chromophore (pH 8) was relatively slow with $t_{1/2}$ ~ 200 ms, but faster ($t_{1/2}$ ~70 ms) for the protonated chromophore (pH 6.4). A kinetic model incorporating these quantitative results was deduced from the kinetic analysis, containing four equilibria involving YFP-H148Q protonation and chloride binding (Fig. 1B). The rapid kinetics of these changes in fluorescence permitted application of the YFP-H148Q to the generally much slower cell-based assays of chloride flux.

YFP-H148Q was expressed in tissue culture cell lines to test its utility in cell-based assays of chloride concentration (Fig. 1C). As reported for other GFP variants, YFP-H148Q fluorescence was observed throughout the cell cytoplasm and nucleus. *In vivo* calibration experiments using ionophore-treated cell cultures indicated that the pH and chloride (Fig. 1D) sensitivities of YFP-H148Q were similar in cells and aqueous solutions. The utility of YFP-H148Q as a cellular Cl^- / halide indicator to follow cAMP-stimulated, CFTR-mediated halide fluxes was tested and protocols were established based upon the ability of CFTR to conducts certain anions (Cl^-, NO_3^-, I^-; ref. 29) that differentially quenched YFP-H148Q fluorescence (27). In cells expressing CFTR and bathed in forskolin-containing solutions to activate CFTR, replacement of solution Cl^- by gluconate or nitrate (to drive Cl^- efflux) produced small changes (~15%) in cellular fluorescence consistent with the expected unquenching of YFP-H148Q determined *in vitro*. To increase the magnitude of fluorescence response for assays of anion transporters like CFTR, we exploited the observations that YFP-H148Q fluorescence is substantially more sensitive to I^- than to Cl^- and that anion channels conduct I^- efficiently. Large changes in fluorescence intensity of ~50 % were observed in Cl^-/I^- exchange protocols (Fig. 1D) (27). This substantial signal change is comparable to that observed for LZQ, the best chemical halide indicator developed to date (30). Cl^-/I^- exchange has been used in cell-based high-throughput screens for CFTR inhibitors and activators as described further below (31-33).

Although a good probe for some cell-based assays of Cl^-/halide flux mediated by ion channels we sought to discover variants of YFP-H148Q with increased sensitivities (34). In particular, the anion conductance of the ΔF508-CFTR channel is greatly reduced when compared to wild type CFTR (7-10), requiring a more sensitive YFP-H148Q variant for primary screening. Based upon structural data (28) random mutations were introduced in six hydrophobic residues lining the YFP-H148Q halide binding site in an attempt to modify the polarity and/or size of the cavity and thus halide binding affinities (Fig. 2) (34). Degenerate primers were used to generate YFP-H148Q libraries containing mutations in the residue pairs val150/ile152, val163/phe165, and leu201/tyr203. The mutagenesis procedure generated a diverse library as indicated by the varying intensities of colonies generated (Fig. 2, *left, middle*). An efficient screening protocol was developed to measure halide sensitivities of the expressed YFP mutants. Fluorescent bacterial colonies were distinguished by illumination with ultraviolet light and transferred to 96-well microplates. Colonies were grown, replicated to agar plates, and efficiently lysed *in situ* by digestion with lysozyme followed by repeated freeze-thaw cycles.

Approximate K_d values for Cl^- and I^- binding were then determined using a fluorescence plate reader.

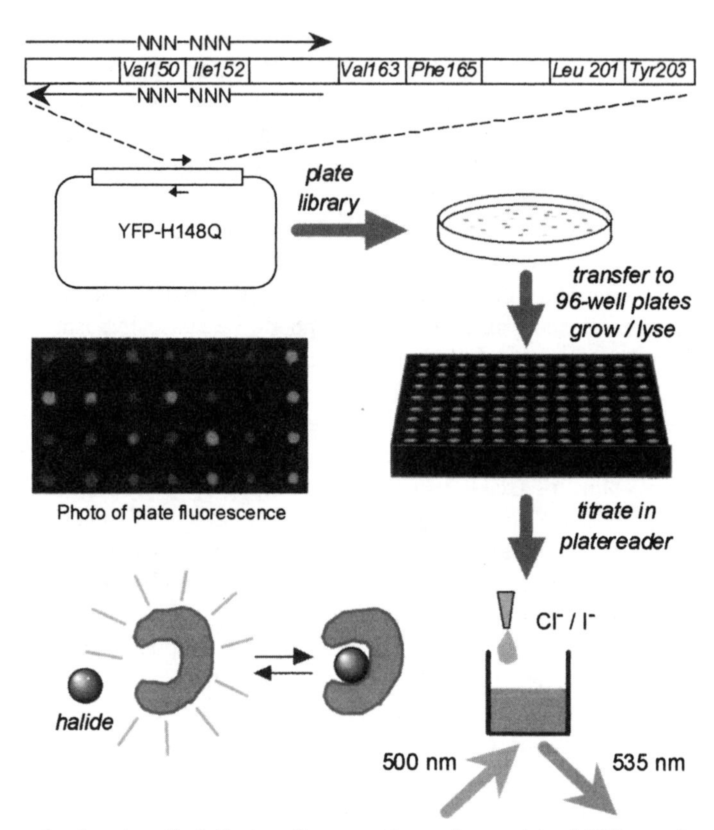

Figure 2. (See color insert section) Strategy for generating and screening of YFP mutational libraries. Indicated pairs of amino acids were randomly mutated using appropriate primers. Transformed bacterial colonies were transferred to 96–well microplates for growth, replication to agar plates, lysis and screening. Left, middle: Photograph of a section of a replicate agar plate showing bacteria with differing amounts of fluorescence. Adapted from ref. 34.

Screening of >1000 colonies from the libraries yielded YFP-H148Q mutants with significantly different Cl^- and I^- affinities compared to YFP-H148Q (34). The mutants V150T, I152L/Y, V163T/L, V150A/I152L, V163A/F165Y and V163T/F165Y had K_d values for I^- of <15 mM in the initial screen. We characterized the I152L mutant further because of its low K_d for I^- of 3 mM (Fig. 3). Dissociation constants for Cl^-, I^- and NO_3^- at cytoplasmic pH, determined by fluorescence titrations using purified recombinant YFP-H148Q/I152L, were 85, 10 and 2 mM, respectively (Fig. 3A, *left*). As was determined for YFP-H148Q (27), indicator pK_a decreased with increasing $[Cl^-]$ (Fig. 3A, *right*); for I152L, the pK_a was 6.95 in the absence of Cl^-, increasing to 7.70 and 7.89 in the presence of 75 and 150 mM Cl^-, respectively. Stopped-flow fluorescence analysis indicated a rapid response of YFP-H148Q/I152L fluorescence to changes in Cl^-

concentration (Fig. 3A, *inset*). Fluorescence lifetime analysis and absorption spectroscopy confirmed a static quenching mechanism of YFP-H148Q/I152L by halides as was determined for YFP-H148Q.

YFP-H148Q/I152L was introduced into a mammalian expression vector to test its applicability in cellular anion exchange measurements. Replacement of 20 mM Cl^- by I^- produced a slow decline in fluorescence due to basal CFTR activity, which increased rapidly with addition of the cAMP agonist forskolin (Fig. 3B). The maximum fluorescence decrease of ~50 % was much greater than that of <10 % in identical experiment performed using YFP-H148Q. Further, given the significant sensitivity of I152L to NO_3^-, CFTR activity could also be studied using a Cl^-/NO_3^- exchange protocol. Replacement of 100 mM Cl^- by NO_3^- in exchange experiments with YFP-H148Q/I152L transfected cells generated a fluorescence decrease of ~50 %, similar to that obtained for Cl^-/I^- protocols with YFP-H148Q.

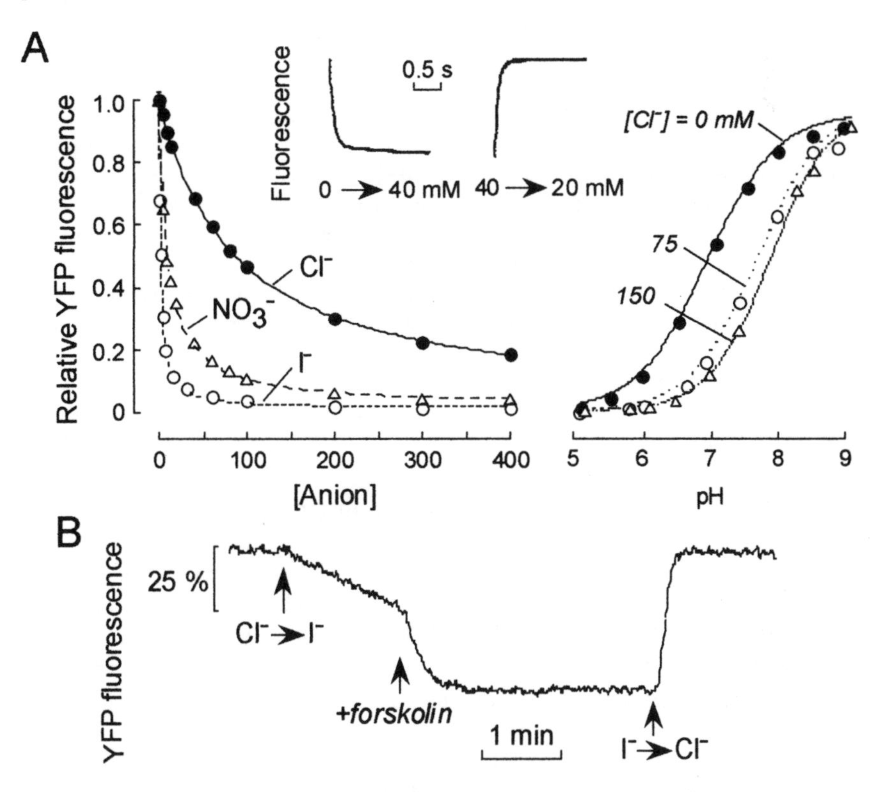

Figure 3. Characterization of purified YFP-H148Q/I152L and application to cell-based assay of CFTR halide transport. **A.** Titration of YFP-H148Q/I152L with Cl^-, I^- and NO_3^- done at cytoplasmic pH (*left*). Fluorescence pH titrations of YFP-H148Q/I152L done at indicated Cl^- concentrations (*right*). Cl^- association and dissociation kinetics of indicated YFP-H148Q/I152L measured by stopped-flow fluorimetry (*inset*). **B.** Time course of fluorescence in response to exchange of 20 mM Cl^- for I^- and addition of forskolin in CFTR-expressing cells transiently transfected with YFP-H148Q/I152L. Adapted from ref. 34.

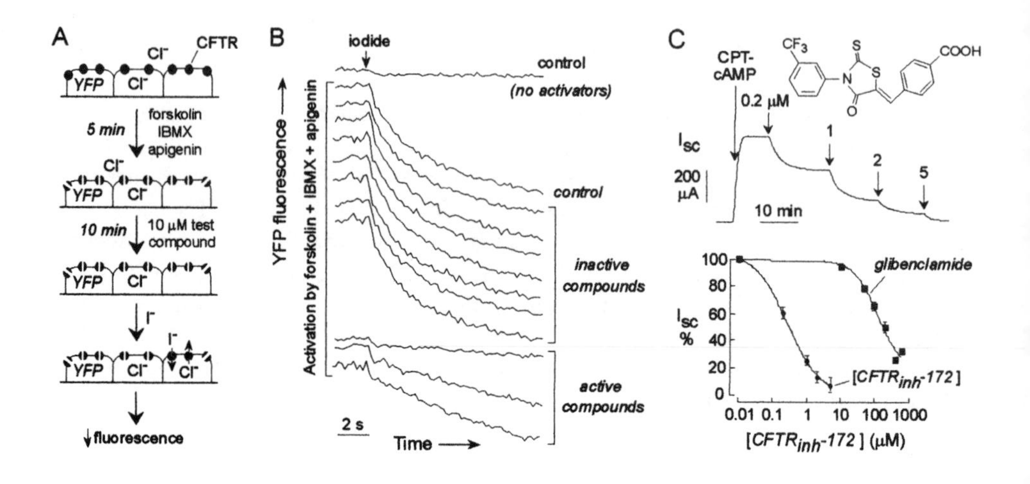

Figure 4. Identification of CFTR inhibitors by high-throughput screening. **A.** Screening protocol. Cells co-expressing CFTR and a halide-sensitive yellow fluorescent protein were plated in 96-well microplates. After 24-48 hours, culture medium was removed and cells were stimulated by an agonist mixture (forskolin, apigenin, IBMX) to fully activate CFTR. After addition of test compound, I^- influx was induced by adding an I^- containing solution. **B.** Representative original fluorescence data from individual wells of a 96–well plate showing controls (no activators, no test compound) and test wells. **C.** (*top*) Structure of the thiazolidinone $CFTR_{inh}$-172, and inhibition of short-circuit current in permeabilized FRT cells expressing human CFTR after stimulation by 100 μM CPT-cAMP. (*bottom*) Dose-inhibition data for $CFTR_{inh}$-172 and glibenclamide . Adapted from ref. 32.

3. CFTR INHIBITORS

CFTR inhibition has been proposed as a therapy for some forms of secretory diarrhea (35). The intestine of CFTR null mice does not secrete fluid in response to cholera toxin (36), and a diarylsulfonylurea CFTR inhibitor prevented Cl^- secretion in porcine mucosa (37). Also, selective CFTR inhibitors should be useful to probe CFTR function *in vitro* and *in vivo*. The CFTR blockers used previously, such as glibenclamide, suffer from lack of specificity and potency (38), requiring high micromolar concentrations at which they block other types of anion and cation channels (39-41).

To identify new CFTR inhibitors, we recently screened a collection of 50,000 drug-like molecules (32). Fisher Rat Thyroid (FRT) cells co-expressing human CFTR and the YFP-H148Q halide sensor were stimulated by a CFTR-activating cocktail and then subjected to an iodide gradient (Fig. 4A). Iodide addition produced a prompt decrease in cell fluorescence after CFTR activation (Fig. 4B). Inhibitors ('active compounds') were identified from a reduction in the (negative) fluorescence slope. The best inhibitor identified by screening and subsequent optimization was the 2-thioxo-4-thiazolinone compound $CFTR_{inh}$-172 (Fig. 4C, top). $CFTR_{inh}$-172 inhibited CFTR function at submicromolar concentrations, ~500 times better than glibenclamide studied under similar conditions (Fig. 4C, middle and bottom). Further analysis showed voltage-independent inhibition of CFTR Cl^- conductance with prolonged mean channel closed time and without change in unitary conductance. $CFTR_{inh}$-172 did not affect several other Cl^- channels (calcium and volume-activated) or ABC transporters (MDR-1, SUR). Rodent pharmacology studies indicated low toxicity, a large volume of distribution with slow elimination by renal glomerular filtration, no metabolism, and enterohepatic circulation with accumulation in bile and intestine (42).

The antidiarrheal efficacy of the thiazolidinone $CFTR_{inh}$-172 was tested in a mouse closed ileal loop model (42). Injection of cholera toxin into loops produced fluid secretion over 6 hours after a slow onset. A single intraperitoneal injection of $CFTR_{inh}$-172 (just after cholera toxin infusion) reduced fluid accumulation by ~90%, with 50% inhibition at ~5 µg $CFTR_{inh}$-172. Orally-administered $CFTR_{inh}$-172 was also effective in blocking intestinal fluid secretion after oral cholera toxin in an open-loop model. Finally, $CFTR_{inh}$-172 inhibited fluid secretion after *E. coli* STa toxin, as well as cAMP and cGMP-stimulated Cl^- currents in human intestine. $CFTR_{inh}$-172 is thus a potentially useful lead compound for drug treatment of enterotoxin-mediated and possibly other secretory diarrheas.

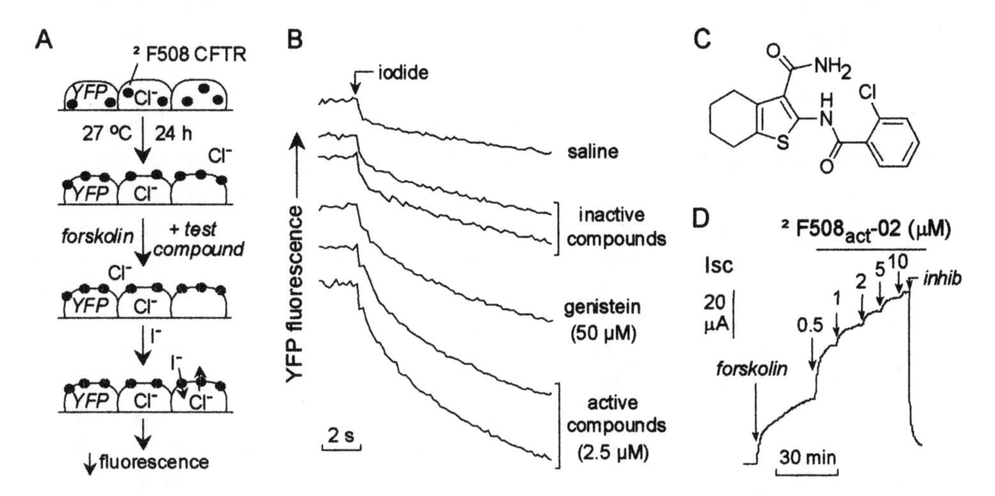

Figure 5. ΔF508-CFTR activators identified by high-throughput screening. **A.** High-throughput screening procedure. Cells co-expressing ΔF508-CFTR and the halide-sensitive fluorescent protein YFP-H148Q/I152L were grown for 24 h at 27 °C (to give plasma membrane ΔF508-CFTR expression). After washing, test compounds (2.5 µM) and forskolin (20 µM) were added, and I^- influx was assayed from the time course of YFP-H148Q/I152L fluorescence after adding I^- to the external solution. **B.** Representative time courses of YFP-H148Q/I152L fluorescence in control wells (saline, negative control; 50 µM genistein, positive control) with examples of inactive and active test compounds. **C.** Chemical structure of the most potent tetrahydrobenzothiophene. **D.** Transepithelial short-circuit current (I_{sc}) in FRT cells expressing ΔF508-CFTR showing responses to 20 µM forskolin and potentiation by the tetrahydrobenzothiophene $\Delta F508_{act}$-02. Where indicated, the CFTR inhibitor $CFTR_{inh}$-172 (5 µM) was added. Adapted from ref. 46.

4. CFTR ACTIVATORS

Different types of organic molecules have been found to increase the activity of wild type and mutant CFTRs, possibly by a direct interaction mechanism (8, 9, 38, 43-45). However, many of these compounds, such as genistein, apigenin, and IBMX, are not selective and require high concentrations to be effective. In an initial project to discover new classes of CFTR modulators with improved potency and selectivity, we screened 60,000 diverse drug-like compounds (33). Compounds were tested on FRT cells co-expressing human wild type CFTR and a halide-sensitive YFP sensor. Primary screening consisted of short-term stimulation of cells with 10 μM test compound and 0.5 μM forskolin followed by I^- challenge. The screen yielded 57 strong activators (greater activity than reference compound apigenin), most of which were unrelated in chemical structure to known CFTR activators. Secondary analysis of the strong activators included analysis of CFTR specificity, forskolin requirement, transepithelial short-circuit current, activation kinetics, dose-response, toxicity and mechanism. Three compounds, the most potent being a dihydroisoquinoline, activated CFTR indirectly by increasing cellular cAMP. Fourteen compounds activated CFTR without cAMP elevation or phosphatase inhibition, suggesting direct CFTR interaction. This proof-of-principle study showed the utility of high-throughput screening in identifying novel classes of potent CFTR activators.

As mentioned in the introduction, ΔF508 is the most common CFTR mutation causing cystic fibrosis. We have been working on identifying correctors of defective ΔF508-CFTR gating and intracellular processing. To identify correctors of defective ΔF508-CFTR gating (also called 'potentiators'), we designed a high-throughput screen using stably transfected FRT cells that co-express ΔF508-CFTR and the halide sensitive indicator YFP-H148Q/I152L (46). A cell clone was selected that showed consistent expression of ΔF508-CFTR protein at the plasma membrane when incubated at low temperature. A collection of 100,000 diverse drug-like small molecules was screened for activation of halide transport in these cells by incubation at 27 °C for 24 h to permit ΔF508-CFTR plasma membrane targeting, followed by 15 min incubation with forskolin and test compounds (Fig. 5A). Rapid I^- influx was found for some of the 100,000 test compounds (Fig. 5B). We identified >30 compounds that potentiated ΔF508-CFTR Cl^- channel activity with submicromolar affinity, with most compounds belonging to 6 distinct chemical classes that are unrelated structurally to known CFTR activators or inhibitors.

A secondary library of >1000 compounds with structural similarity to each class of ΔF508-CFTR potentiators was screened to establish structure-activity relationships and to identify the best compounds for further analysis. Seventeen tetrahydrobenzothiophenes were identified giving good ΔF508-CFTR activation with K_d down to 60 nM. Further analysis showed rapid and reversible ΔF508-CFTR activation, with strong Cl^- current activation in short-circuit experiments in the transfected FRT cells (Fig. 5C) and the natively ΔF508-CFTR expressing human bronchial epithelial cells. Experiments to identify correctors of defective ΔF508-CFTR intracellular processing are in progress.

5. SUMMARY

Fluorescence assays have become the preferred approach for drug discovery by high-throughput screening because of their sensitivity, robustness, technical simplicity, and relatively low cost. In the example described here, halide-sensitive green fluorescent protein mutants were optimized for cell-based kinetic assays, and applied to the discovery of new inhibitors of wild type CFTR and activators of a defective CFTR mutant (ΔF508) causing the genetic disease cystic fibrosis. Potent inhibitors and activators were discovered using the fluorescence-based screen and are being optimized for potential clinical applications.

6. REFERENCES

1. J.M. Rommens, M.C. Iannuzzi, B. Kerem, M.L. Drumm, G. Melmer, M. Dean, R. Rozmahel, J.L. Cole, D. Kennedy, N. Hidaka, M. Zsiga, M. Buchwald, J.R. Riordan, L.C. Tsui, and F.S. Collins, Identification of the cystic fibrosis gene: chromosome walking and jumping, *Science 245*, 1059-1065 (1989).
2. J.K. Pilewski and R.A. Frizzell, Role of CFTR in airway disease, *Physiol. Rev. 79*, S215-S255 (1999).
3. D.N. Sheppard and M.J. Welsh, Structure and function of CFTR chloride channel, *Physiol. Rev. 79*, S23-S45 (1999).
4. J.L. Bobadilla, M.J. Macek, J.P. Fine, and P.M. Farrell, Cystic fibrosis: a worldwide analysis of CFTR mutations - correlation with incidence data and application to screening, *Hum. Mutat. 19*, 575-606 (2002).
5. G.M. Denning, M.P. Anderson, J.F. Amara, J. Marshall, A.E. Smith, and M.J. Welsh, Processing of mutant cystic fibrosis transmembrane conductance regulator is temperature-sensitive, *Nature 358*, 761-764 (1992).
6. R.R. Kopito, Biosynthesis and degradation of CFTR, *Physiol. Rev. 79*, S167-S173 (1999).
7. W. Dalemans, P. Barbry, G. Champigny, S. Jalle, K. Dott, D. Dreyer, R.G. Crystal, A. Pavirani, J.P. Lecocq, and M. Lazdunski, Altered chloride ion channel kinetics asssociated with the DF508 cystic fibrosis mutation, *Nature 354*, 526-528 (1991).
8. C.M. Haws, I.B. Nepomuceno, M.E. Krouse, H. Wakelee, T. Law, Y. Xia, H. Nguyen, and J.J. Wine, DF508-CFTR channels: kinetics, activation by forskolin and potentiation by xanthines, *Am. J. Physiol. 270*, C1544-1555 (1996).
9. T.C. Hwang, F. Wang, I.C. Yang, and W.W. Reenstra, Genistein potentiates wild-type and DF508-CFTR channel activty, *Am. J. Physiol. 273*, C988-C998 (1997).
10. F. Wang, S. Zeltwanger, S. Hu, and T.C. Hwang, Deletion of phenylalanine 508 causes attentuated phosphorylation-dependent activation of CFTR chloride channels, *J. Physiol. 524*, 637-648 (2000).
11. M.K. Mansoura, J. Biwersi, M.A. Ashlock, and A.S. Verkman, Fluorescent chloride indicators to assess the efficacy of CFTR delivery, *Hum. Gene Ther. 10*, 861-875 (1999).
12. A.S. Verkman and S. Jayaraman, Fluorescent indicator methods to assay functional CFTR expression in cells, *Meth. Mol. Med. 70*, 187-196 (2002).

13. R.Y. Tsien, The green fluorescent protein, *Annu. Rev. Biochem. 67*, 509-544 (1998).
14. J. Lippincott-Schwartz, E. Snapp, and A. Kenworthy, Studying protein dynamics in living cells, *Nat. Rev. Mol. Cell Biol. 2*, 444-456 (2001).
15. A.S. Verkman, Solute and macromolecule diffusion in cellular aqueous compartments, *Trends Biochem. Sci. 27*, 27-33 (2002).
16. P. Haggie and A.S. Verkman, GFP sensors, *Topics in Fluorescence*, vol. *9*, In Press (2003).
17. J. Zhang, R.E. Campbell, A.Y. Ting, and R.Y. Tsien, Creating new fluorescent probes for cell biology, *Nat. Rev. Mol. Cell Biol. 3*, 906-918 (2002).
18. B.G. Reid and G.C. Flynn, Chromophore formation in green fluorescent protein, *Biochemistry 36*, 6786-6791 (1997).
19. M. Örmo, A.B. Cubitt, K. Kallio, L.A. Gross, R.Y. Tsien, and S.J. Remington, Crystal structure of the *Aequorea victoria* green fluorescent protein, *Science 273*, 1392-1395 (1996).
20. R.M. Wachter, M.-A. Elsliger, K. Kallio, G.T. Hanson, and S.J. Remington, Structural basis of spectral shifts in the yellow-emission variants of green fluorescent protein, *Structure 6*, 1267-1277 (1998).
21. R.M. Wachter, B.A. King, R. Heim, K. Kallio, R.Y. Tsien, S.G. Boxer, and S.J. Remington, Crystal structure and photodynamic behaviour of the blue emission variant Y66H/Y145F of green fluorescent protein, *Biochemistry 36*, 9759-9765 (1997).
22. M.-A. Elsliger, R.M. Wachter, G.T. Hanson, K. Kallio, and S.J. Remington, Structural and spectral repsonse of green fluorescent protein variants to changes in pH, *Biochemistry 38*, 5296-5301 (1999).
23. M. Kneen, J. Farinas, Y. Li, and A.S. Verkman, Green fluorescent protein as a noninvasive intracellular pH indicator, *Biophys. J. 74*, 1591-1599 (1997).
24. J. Llopis, J.M. McCaffery, A. Miyawaki, M.G. Farquhar, and R.Y. Tsien, Measurement of cytosolic, mitochondrial, and Golgi pH in single living cells with green fluorescent proteins, *Proc. Natl. Acad. Sci. USA 95*, 6803-6808 (1998).
25. G. Miesenböck, D.A.D. Angelis, and J.E. Rothman, Visualizing secretion and synaptic transmission with pH-sensitive green fluorescent proteins, *Nature 394*, 192-195 (1998).
26. R.M. Wachter and S.J. Remington, Sensitivity of the yellow variant of green fluorescent protein to halides and nitrate, *Curr. Biol. 9*, R628-R629 (1999).
27. S. Jayaraman, P. Haggie, R.M. Wachter, S.J. Remington, and A.S. Verkman, Mechanism and cellular application of a green fluorescent protein-based halide sensor, *J. Biol. Chem. 275*, 6047-6050 (2000).
28. R.M. Wachter, D. Yarbough, K. Kallio, and S.J. Remington, Crystallographic and energetic analysis of binding of selected anions to the yellow variants of green fluorescent protein, *J. Mol. Biol. 301*, 157-171 (2000).
29. D.C. Dawson, S.S. Smith, and M.K. Mansoura, CFTR: mechanism of anion conductance, *Physiol. Rev. 79*, S47-S75 (1999).
30. S. Jayaraman, L. Teitler, B. Skalski, and A.S. Verkman, Long-wavelength iodide-sensitive fluorescent indicators for measurement of functional CFTR expression in cells, *Am. J. Physiol. 277*, C1008-1018 (1999).

31. L.J. Galietta, M.F. Springsteel, M. Eda, E.J. Niedzinski, K. By, M.J. Haddadin, M.J. Kurth, M.H. Nantz, and A.S. Verkman, Novel CFTR chloride channel activators identified by screening of combinatorial libraries based on flavone and benzoquinolizinium lead compounds, *J. Biol. Chem. 276*, 19723-19728 (2001).
32. T. Ma, J.R. Thiagarajah, H. Yang, N.D. Sonawane, C. Folli, L.J. Galietta, and A.S. Verkman, Thiazolidinone CFTR inhibitor identified by high-throughput screening blocks cholera toxin-induced intestinal fluid secretion, *J. Clin. Invest. 110*, 1651-1658 (2002).
33. T. Ma, L. Vetrivel, H. Yang, N. Pedemonte, O. Zegarra-Moran, L.J. Galietta, and A.S. Verkman, High-affinity activator of cystic fibrosis transmembrane conductance regulator (CFTR) chloride conductance identified by high-throughput screening, *J. Biol. Chem. 277*, 37235-37241 (2002).
34. L.J. Galietta, P.M. Haggie, and A.S. Verkman, Green fluorescent protein-based halide indicators with improved chloride and iodide affinities, *FEBS Lett. 499*, 220-224 (2001).
35. K. Kunzelmann and M. Mall, Electrolyte transport in mammalian colon: mechanism and implications for disease, *Physiol. Rev. 82*, 245-289 (2002).
36. B.R. Grubb and S.E. Gabriel, Intestinal physiology and pathology in gene-targeted mouse models of cystic fibrosis, *Am. J. Physiol. 273*, G258-G266 (1997).
37. E.K. O'Donnell, R.L. Sedlacek, A.K. Singh, and B.D. Schultz, Inhibition of enterotoxin-induced porcine colonic secretion by diarylsulfonylureas in vitro, *Am. J. Physiol. 279*, G1104-G1112 (2000).
38. B.D. Schultz, A.K. Singh, D.C. Devor, and R.J. Bridges, Pharmacology of CFTR chloride channel activity, *Physiol. Rev. 79*, S109-S144 (1999).
39. G. Edwards and A.H. Weston, Induction of glibenclamide-sensitive K-current by modification of a delayed rectifier channel in rat portal vein in insulinoma cells, *Br. J. Pharmacol. 110*, 1280-1281 (1993).
40. A. Rabe, J. Disser, and E. Fromter, Cl^- channel inhibition by glibenclamide is not specific for the CFTR-type Cl^- channel, *Pflugers Arch. 429*, 659-662 (1995).
41. J. Yamazaki and J.R. Hume, Inhibitory effects of glibenclamide on cystic fibrosis transmembrane regulator swelling-activated, and Ca^{2+}-activated Cl^- channels in mammalian cardiac myocytes, *Circ. Res. 81*, 101-109 (1997).
42. J.R. Thiagarajah, T. Broadbent, E. Hsieh, and A.S. Verkman, Prevention of toxin-induced intestinal ion and fluid secretion by a small-molecule CFTR inhibitor, *Gastroenterology 126*, 511-519 (2004).
43. V. Chappe, Y. Mettey, J.M. Vierfond, J.W. Hanrahan, M. Gola, B. Verrier, and F. Becq, Structural basis for specificity and potency of xanthine derivatives as activators of the CFTR chloride channel, *Br. J. Pharmacol. 123*, 683-693 (1998).
44. B. Ilek, H. Fischer, G.F. Santos, J.H. Widdecombe, T.E. Machen, and W.W. Reenstra, cAMP-independent activation of CFTR Cl^- channel by the tyrosine kinase inhibitor genistein, *Am. J. Physiol. 268*, C886-C893 (1995).
45. S. Singh, C.A. Syme, A.K. Singh, D.C. Devor, and R.J. Bridges, Benzimidazolone activators of chloride secretion: potential therapeutics for cystic fibrosis and chronic obstructive pulmonary disease, *J. Pharmacol. Exp. Ther. 296*, 600-611 (2001).
46. H. Yang, A.A. Shelat, R.K. Guy, V.S. Gopinath, T. Ma, K. Du, G.L. Lukacs, A. Taddei, C. Folli, N. Pedemonte, L.J.V. Galietta, and A.S. Verkman, Nanomolar-

affinity small-molecular activators of DF508-CFTR chloride channel gating, *J. Biol. Chem. 278*, 35079-35085 (2003).

EXPLORING MEMBRANE MICRODOMAINS AND FUNCTIONAL PROTEIN CLUSTERING IN LIVE CELLS WITH FLOW AND IMAGE CYTOMETRIC METHODS

György Vereb, János Szöllősi, Sándor Damjanovich, and János Matkó*

1. INTRODUCTION

Fluorescence technologies have shown a spectacular spreading in 21st century biological research. This is mainly due to the enormous complexity of cells where one usually has to detect, localize or quantitatively measure a single type of molecule in a large assembly (millions) of other molecules. Fluorescence labeling may assure a high level of selectivity (dark background) and sufficient sensitivity for this purpose. A systematic development of fluorescent molecular probes in the past two decades (lipid, peptide, protein constituents of cells conjugated with various fluorophores; highly specific, fluorochrome-conjugated monoclonal or polyclonal antibodies (Fab); organelle specific probes; ion-selective and potential-sensitive fluorescent indicators; photoactivatable caged-compounds; natural reporter proteins: green fluorescent protein, GFP, and its analogues with biochemically tunable spectral properties, etc.) allow us now to label a wide variety of functional sites or molecules in living cells and even to analyze their functional responses. Among others, fluorescence technology revolutionized studies on cellular morphology and signal transduction, as well as the biophysical approaches to monitor localization, interaction and mobility of single molecules.

In addition to successful probe design, fluorescence based flow cytometry and the rapidly developing microscopic imaging techniques provided a solid technical background for studying the biology of cells at new, subcellular and even single molecule

* György Vereb, János Szöllősi, and Sándor Damjanovich, Department of Biophysics & Cell Biology, and Cell Biophysics Research Group of the Hungarian Academy of Sciences, University of Debrecen, Debrecen, Hungary, H-4012. János Matkó, Department of Immunology, Eötvös Loránd University, Budapest, Hungary, H-1117.

levels. Specific techniques are available now for detecting molecular proximity and interactions in cells, such as flow or image cytometric variations of fluorescence resonance energy transfer, FRET, or the videomicroscopy based single particle tracking (SPT) and the fluorescence correlation spectroscopy/microscopy (FCS/FCM) for monitoring mobility of different fluorescently labeled cellular components, molecules.

Flow cytometric techniques offer the advantage of rapid analysis on a large number of cells (~10^5 cells in some minutes) with a high statistical value and a possibility for analyzing heterogeneity at the population level. The new generation bench-top and research flow cytometers also offer the advantage of multiparameter analysis (increasing number of available laser-excited optical channels), increased sensitivity owed to optical improvements in photon collection, and the possibility of slow kinetic measurements at a population level. Flow cytometry, however, does not provide any information about the spatial localization of fluorescent probes, but instead measures the fluorescence intensity averaged over each cell. In contrast, microscopic techniques provide a high spatial resolution: conventional fluorescence microscopy has a ~250 nm resolution limited by diffraction of the optics, while in scanning near-field optical microscopy (SNOM) both horizontal and vertical (z-axis) resolution is improved to ~50-100 nm. Although microscopy has several further advantages in detecting molecular dynamics or kinetics of changes in distribution or intensity of fluorescent probes, it suffers from a low statistical value, especially in the case of quantitative measurements. Thus, a combined application of flow and image cytometry in resolving particular biological questions can be a very powerful approach.

Since the plasma membrane (PM) of cells represents a "platform of first contact" with different stimuli received from the extracellular milieu, with numerous invading pathogens or with drugs applied in treatments of pathological situations, extensive studies on the molecular architecture and dynamics of PM seem to be indispensable to understand the molecular background and mechanistic details of these biological processes. In the present overview we will focus on recent works from our and other laboratories using fluorescence-based flow and imaging cytometric techniques in exploring a particular biological question, namely how the molecular architecture, heterogeneity (microdomain structure) and dynamics of the plasma membrane looks like in live / intact cells. Several specific examples of our particular areas of interest (e.g. how membrane organization of multichain immunoreceptors and association pattern of ErbB receptor tyrosine kinases are affected by lipid rafts) will also be shown.

2. DYNAMIC MICROHETEROGENEITY OF THE CELL MEMBRANE: QUESTIONS ON LIVE CELLS

That most mammalian cell membranes display a considerable microheterogeneity (existence of microdomains) in their structure is known for a long time, but the mechanisms, transport and biosynthetic processes underlying their formation are still poorly understood[1-3]. The hypothesis of membrane "rafts" was introduced in 1997 by Simons and Ikonen[4] proposing ordered sphingolipid- and cholesterol-rich microdomains as autonomic, mobile molecular assemblies to coexist with fluid, glycerophospholipid-rich regions of the plasma membrane of cells. Lipid rafts can be considered as transient molecular associations between lipid and protein constituents of the PM providing a dynamic patchiness/local order in the fluid mosaic membrane[1]. The microdomain concept is widely accepted by now, and existence of rafts was confirmed by many lines of experimental evidence (e.g. biochemical data on detergent resistance, resolving membrane patchiness by high resolution fluorescence and electron microscopies, tracking by videomicroscopy the lipid and protein motions in the membrane, etc.), but some basic questions about the microdomains still remained unclear. Membrane microdomains are defined in different ways depending on the experimental approach used to detect them. Chemically they are defined on the basis of their resistance to solubilization with cold nonionic detergents (Triton-X100, Brij, Chaps, etc.)[5, 6]. Microdomains are also often defined as (co)clusters of proteins or lipids detected by optical or electron microscopic techniques, such as confocal laser scanning microscopy, fluorescence resonance energy transfer microscopy, atomic force microscopy or transmission electron microscopy in intact cells[7-12]. Microdomains may also arise from specific constraints to the lateral diffusion of mobile membrane proteins as revealed by tracking molecular motions with videomicroscopy[13]. This confinement may arise from lipid-lipid or lipid-protein interactions,[2] as well as from a "capturing" effect of the membrane skeleton[14].

Thus, an important question originating from the diversity of definitions is whether microdomains, detected and defined in the above different ways, correspond to each other, and if so, to what extent. Additionally, the experimental approaches suffer to a certain degree from some specific drawbacks, such as perturbation of the *in situ* membrane organization by the applied labels themselves, e.g. antibodies, antibody-coated beads that are too large in size or can initiate protein or lipid crosslinking. Therefore, the *in situ* size and stability (lifetime) of cell surface rafts still remains unresolved. As Edidin pointed out in a recent review[1] "the domains are now thought to be smaller and less stable" than in 1992.

Our recent studies addressing some of these questions were mainly related to exploring some basic static and dynamic features of membrane rafts in live cells with special attention to their relationship to the protein clusters (non-random protein distributions in the PM) observed by different biophysical techniques earlier (such as FRET, long range energy transfer, CLSM)[9, 15-19]. Functional consequences of the observed protein clustering were also subjects of our study.

3. FÖRSTER-TYPE RESONANCE ENERGY TRANSFER (FRET): A SPECTROSCOPIC RULER ADAPTABLE FOR MAPPING MOLECULAR LEVEL INTERACTIONS OF CELL SURFACE PROTEINS

Modern proteomics has revealed a great number of membrane proteins that are *in vitro* capable of interacting with each other. However, the extent to which this reflects their actual behavior in living cells is not clear. Henceforth it becomes more and more important to be able to detect such interactions *in situ* on the surface of cells.

A major asset in studying molecular level interactions was the application of FRET to cellular systems. In the 1970s, cell surface lectins were the first to be investigated by FRET[20, 21]. Subsequently, many molecular interactions have been subjected to FRET analysis, including – just to mention a few examples – the homo- and heteroassociations of MHC class I and class II,[22] the interleukin-2 receptor α-subunit and ICAM-1,[23, 24] the TCR/CD3 complex,[25] tetraspan molecules (CD53, CD81, CD82) and CD20 with MHC class I and class II,[17] the three subunits of the multi-subunit IL-2 receptor[26] the TNF receptor,[27] and Fas (CD95)[28]. Excellent reviews are available on the applicability of FRET to biological systems as well as descriptions and comparisons of various approaches. Only a few are quoted here[12, 29-34].

In FRET, an excited fluorescent dye, called donor donates energy to an acceptor dye, a phenomenon first described *correctly* by Förster already in 1946[35]. For the process to occur, a set of conditions has to be fulfilled:

- The emission spectrum of the donor has to overlap with the excitation spectrum of the acceptor. The larger the overlap, the higher the rate of FRET is.
- The emission dipole vector of the donor and the absorption dipole vector of the acceptor need to be close to parallel. The rate of FRET decreases as the angle between the two vectors increases. In biological situations where molecules are free to move (rotate) we generally assume that dynamic averaging takes place, i.e. during its excited state lifetime the donor assumes many possible streric positions, among them those that can yield an effective transfer of energy.
- The distance between the donor and acceptor should be between 1-10 nm.

This latter phenomenon is the basis of the popularity of FRET in biology: The distance over which FRET occurs is small enough to characterize proximity of possibly interacting molecules, under special circumstances even provide quantitative data on exact distances, and additionally information of spatial orientation of molecules or their domains. Hence the very apropos term from Stryer, who equaled FRET to a "spectroscopic ruler"[36].

The usual term for characterizing the efficiency of FRET is

$$E = \frac{R_0^6}{R_0^6 + R^6} \tag{1}$$

The rate of energy transfer is dependent on the negative 6th power of the distance R between the donor and the acceptor, resulting in a fast dropping curve when plotting E against R, centered around R_0, the distance where E=0.5, that is, where half of the photons absorbed by the donor get transferred to the acceptor. As separation between the donor and acceptor increase, E decreases, and at R=2R_0 E is already getting negligible. Conversely, as the distance reaches values below 1 nm, strong ground-state interactions or transfer by exchange interactions become dominant at the expense of FRET[37].

The occurrence of FRET has profound consequences on the fluorescence properties of both the donor and the acceptor. An additional de-excitation process is introduced in the donor, which decreases the fluorescent lifetime and the quantum efficiency of the donor, rendering it less fluorescent. The decrease in donor fluorescence (often termed donor quenching) can be one of the most facilely measured spectroscopic characteristic that indicates the occurrence of FRET. Additionally, since the acceptor is excited as a result of FRET, acceptors that are fluorescent will emit photons (proportional to their quantum efficiency) also when FRET occurs. This is called sensitized emission and can also be a sensitive measure of FRET.

The various approaches that can be used to quantitate FRET can be categorized based on the spectrofluorimetric parameter detected, and whether the donor or the acceptor is investigated:

1. Measurements based on intensity changes upon FRET
   a. donor quenching based on measuring donor with and without the acceptor
      i. based on samples labeled with and without acceptor[38]
      ii. based on photobleaching the acceptor[10, 39]

   b. measuring sensitized acceptor emission[40]
   c. measuring both donor quenching and sensitized emission
      i. measuring integrated emission in spectral bands through filters[41, 42]
      ii. measuring and fitting emission spectra[43]
2. Measurements based on changes of donor lifetime
   a. direct measurement of donor lifetime with and without the acceptor[39, 44]
   b. estimation of donor lifetime based on donor photobleaching rate [10, 45, 46]
   c. estimation of donor lifetime based on fluorescence anisotropy[47]
3. Measurements based on changes of acceptor photobleaching rate[48]
4. Measurements based on changes in fluorescence anisotropy of the donor. In this case homo-transfer between the same type of molecules (donor only) can be estimated while the fluorescence lifetime does not change[49, 50]

Most of these approaches can be applied either in the flow cytometer or in the microscope. Flow cytometric FRET (FCET) carries the advantage of examining large cell populations in a short time, and provides a FRET efficiency value averaged over the population, or on a cell by cell basis, but averaged over each cell. The microscopic approaches have the ability to provide subcellular detail and the possibility to correlate FRET values with other biological information gained from fluorescent labeling, on a pixel by pixel basis. However, the sample size is restricted, as data acquisition and processing is rather time consuming. Also, biological variation may influence the composition of the cell population selected for imaging.

4. FLUORESCENCE MICROSCOPY AS A VERSATILE TOOL TO DETECT DIFFERENT HIERARCHICAL LEVELS OF MEMBRANE PROTEIN ORGANIZATION

Membrane organization, that is, the ever changing mobility and proximity relationships of the lipid and protein components in the plasma membrane have a significant impact on essential biological processes, such as ligand-receptor recognition, receptor activation triggering cell proliferation or death, intercellular interactions, etc. A new feature of present day cell biological investigations at the cellular level is the demand for quantitative assessment of individual receptor-receptor and receptor-ligand interactions, which is not provided by classical immunological methods, such as coprecipitation. However, a selection of microscope based fluorescence methods has evolved during the past decades to quantitatively measure membrane organization and dynamics.

On the dynamic side, fluorescence recovery after photobleaching (FRAP),[51-53] single-particle tracking techniques (SPT)[54-57] and optical trapping (by laser-optical tweezers)[58-60] have created foundations for the present day concept of microdomains as areas to which mobility of membrane molecules is restricted.

Restrictions in the lateral mobility of both lipid and protein components were extensively studied by Edidin using the FRAP technique. FRAP measures the diffusion of fluorescently labeled membrane components from non-bleached areas from the cell membrane into a small bleached spot. Different spot sizes in FRAP measurements were applied in order to assess lateral diffusion parameters of different membrane proteins, mostly of MHC class I and class II molecules[59, 61]. It has been found that lateral diffusion parameters were highly dependent on the bleaching spot size. Since cell membrane diffusion rate is expected to be independent of time and the size of the observed spot, one possible and plausible explanation for these findings is the mosaic-like domain structure of lipids. These domains restrict the barrier free path (BFP) of proteins and are partly responsible for the clustered arrangements of membrane proteins.

SPT follows the "random walk" of a gold particle fixed to a cell surface protein usually visualized with a sensitive CCD camera in a microscope. Kusumi *et al.*[55] have determined a barrier free path of ~ 400 nm for the transferrin receptor and E-cadherin using SPT, while Edidin *et al.*,[59] using optical tweezers, measured ~600 nm for wild-type and ~1700 nm for mutant class I MHC glycoproteins truncated in their cytoplasmic domains. The wild-type membrane spanning isoforms showed a close value to that of the transferrin receptor indicating that for membrane spanning proteins these BFP values are more or less in the same range regardless of the protein and the method of examination.

Fluorescence correlation spectroscopy (FCS), although it has had long tradition in studying reaction kinetics and molecular interaction in solutions[62, 63] has only recently been applied to studying cellular systems in the microscope[64-66]. The method, if used with critical caution, allows the determination of absolute molecular concentration, mobility, and co-mobility in small confocal volume elements of living cells (for a comprehensive review see[67]).

As for static measurements, confocal laser scanning microscopy (CLSM)[11, 68] proved to be successful in determining the uneven cell surface distribution of various lymphocyte antigens. Scanning nearfield optical microscopy (SNOM), a method ideal for assessing localization of membrane proteins at the resolution of several 10 nm has also been gaining space in the investigation of the plasma membrane[7, 10, 69-74]. Additionally, the combination of FRET and the high submicron resolution afforded by CLSM and SNOM allows the precise subcellular spatial localization of various molecular interactions[8, 75]. Measuring FRET in the microscope is mostly based on donor photobleaching kinetics[8, 10, 45] and assesment of donor quenching by photodistruction of the acceptor[10, 39]. However, intensity-based FRET measurements were also successfully applied to detect the uneven spatial distribution of receptor associations at the subcellular level[42].

Thus, data obtained with CLSM and SNOM, evaluated using appropriate digital image processing algorithms provide valuable information about the spatial localization, colocalization and compartmentation of membrane constituents that are highly complementary to information from dynamic measurements. In general, these approaches have, to date, provided ample evidence for the domain-like distribution of lipids and proteins in biological membranes[11, 53, 56, 58, 70, 75-77].

4.1. Clusters of class I and class II MHC molecules in the plasma membrane of antigen-presenting cells: fluorescence for detecting clusters and revealing function

Major histocompatibility complex (MHC) proteins are gene products in mammalian cells functioning as "identity cards" for the immune system, as well as involved in presentation of antigenic peptides derived from internalized and processed pathogens towards T lymphocytes, the key players of the cellular immune response[78]. Therefore, the structure of the MHC-peptide complex, its organization in the PM and its interaction with the antigen-specific receptors of T cells (TCR) are all crucially important questions to understand details of the cellular immune response. Early studies on the membrane organization of MHC molecules, using biophysical techniques (FRET, and lateral

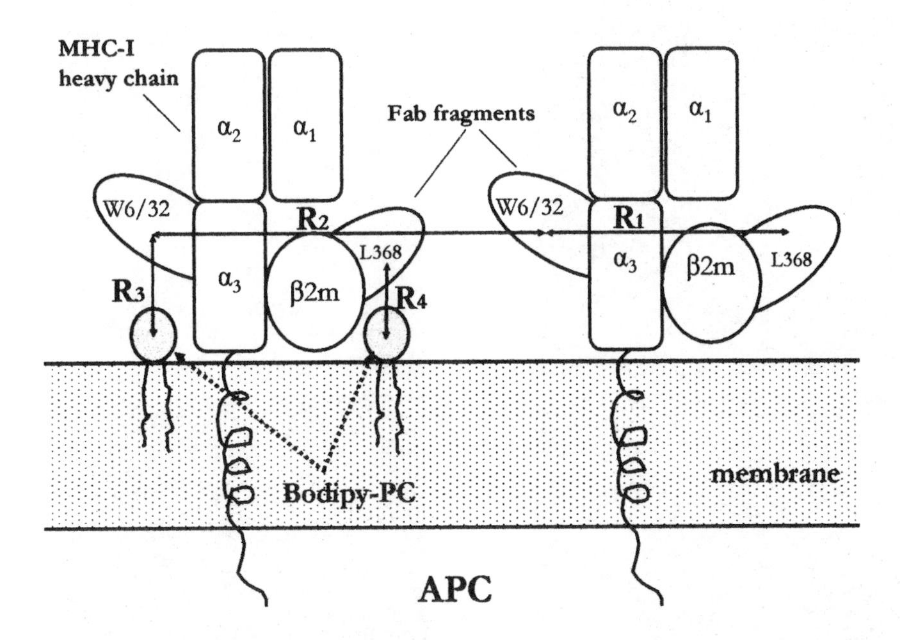

Figure 1. Mapping of antibody binding epitopes and the relative position of MHC-I membrane protein to the exofacial surface of the plasma membrane in live cells by FRET. The strategy is based on knowledge of antibody (Fab fragment) docking sites on MHC-I and that the single fluorescence labels on Fabs are located near their mass center. Labels in the lipid bilayer were expected to be distributed in the plane of polar headgroups in a unifrom fashion. FRET was analyzed by a modified Stern-Volmer method in terms of closest approach distances[84].

diffusion measurements with FRAP) indicated that formation of homo- or heteroassociates for both MHC-I and MHC-II molecules is their inherent property in the PM of different cells including B and T lymphocytes as well as fibroblasts[15, 16, 79-81]. X-ray crystallographic data also indicated this tendency of homoassociation for MHC-II gene products[82].

In order to analyze supramolecular clusters of membrane proteins involved in the contact of antigen presenting cells and T lymphocytes, we have recently used a genuine approach. *In situ* FRET data estimating intra- and intermolecular distances within and among MHC complexes and the spatial relationship of some structurally well characterized antibody-binding sites (epitopes) on MHC molecules relative to the lipid bilayer of the PM were combined with the available X-ray crystallographic models of MHC-TCR complexes[83, 84]. The scheme of FRET measurements used in this approach is shown in Fig. 1. Comparison of FRET efficiencies allowed for positioning MHC molecules relative to the plane of the membrane. In addition, a computer program generating 3D structural models of MHC, CD8 and T cell receptor molecules allowed selecting the sterically possible orientations and interactions between these three molecules matching the in situ FRET data. This modeling supported a coreceptor (CD8)-mediated formation of "molecular recognition lattice" presumably driving cytotoxic T lymphocyte activation[84]. Such strategy of exploiting *in situ* FRET measurements and crystallographic data could be useful for many other membrane receptors.

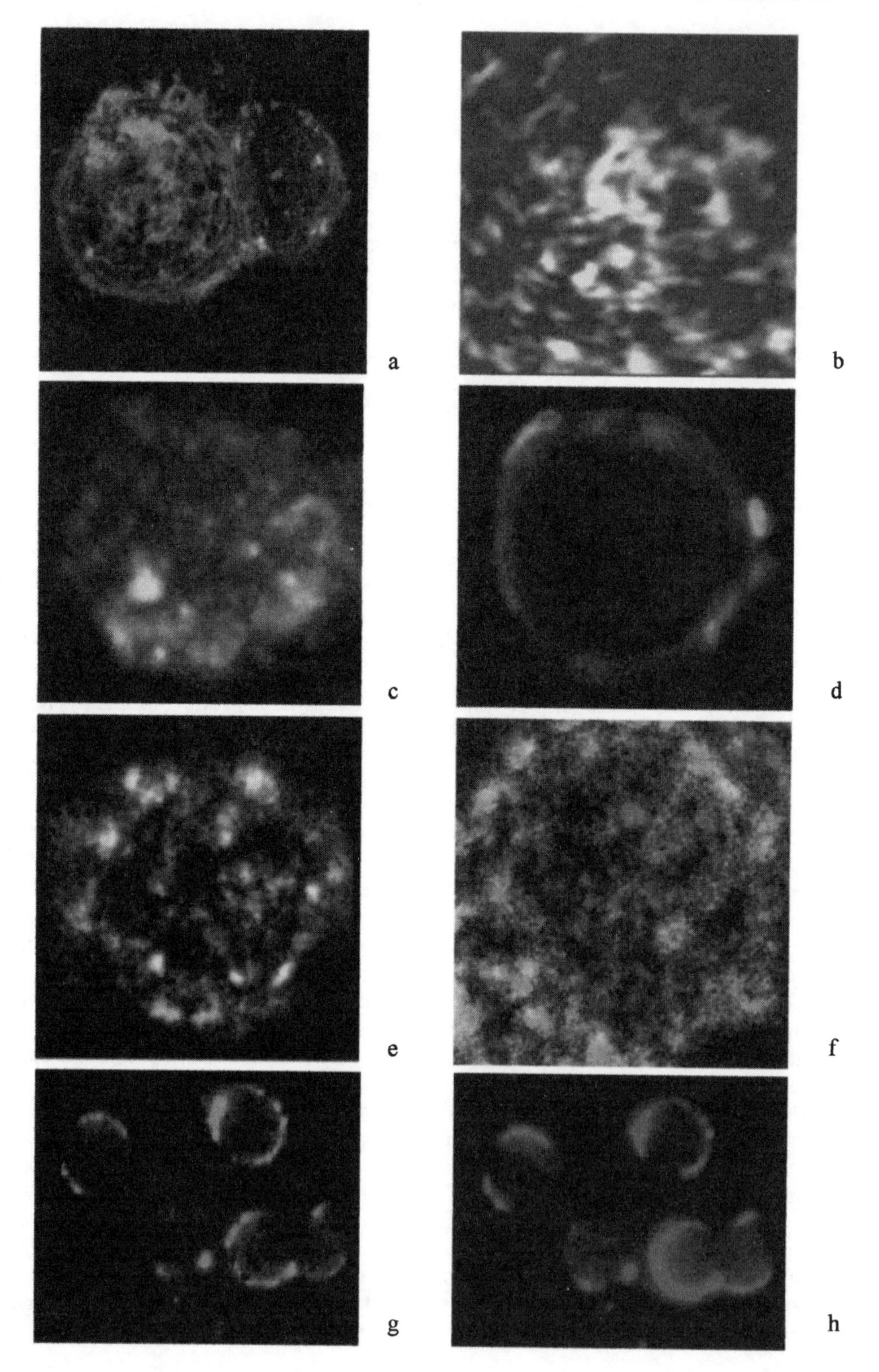
a
b
c
d
e
f
g
h

Figure 2. **(See color insert section)**. Fluorescence confocal microscopy allows analysis of submicrometer molecular colocalization of membrane constituents in live cells. (**a.**) Immunological synapse between human B cells (antigen presenting cell; green) and cytotoxic T lymphocytes (red). (**b.**) SNOM image of a human B cell surface: intact MHC-I molecules (labeled with SF-X-W6/32 mAb, green) and β2m-free MHC heavy chains (labeled with Cy5-HC10 mAb; red) are shown in a 15x15μm representative area. There is a substantial overlap (yellow color) between the two labels (cross-correlation coefficient: 0.771) suggesting high degree of molecular co-clustering of the two forms of MHC-I. (**c.**) Colocalization of GM1 gangiosides (labeled with Alexa488CTX; green) and MHC-II (I-E^d) molecules (labeled with TM-rhodamine-Ab; red) in murine B cells (3D reconstruction, overlap in yellow color). (**d.**) Central optical slice of a cell labeled as in (**c.**), but after cholesterol depletion of the PM. Note the remarkable segregation of the labeled species upon raft-disruption. (**e.**) IL-2R alpha subunit and MHC class I double labeled with red and green fluorophores for confocal microscopy. The overlaping clusters of these proteins result in yellow spots. (**f.**) IL-2R alpha (green) and the transfrerrin receptor (red) exhibit clusters of distinctly different size that hardly overlap. (**g.**) GPI-linked CD48 proteins on a T leukemia cell are tagged with specific monoclonal Alexa488-conjugated Fab and crosslinked with GAMIG to induce capping. (**h.**) IL-2R alpha subunit labeled with Cy3-conjugated anti-Tac Fab on the same cell show identical localization with CD48 indicating co-capping of the two raft-resident proteins.

The factors regulating the formation and possible rearrangements of these supramolecular clusters have been the target of many recent investigations. Earlier, FRET analysis on MHC-I molecules in human B-cells revealed that their homoassociation is sensitive to membrane cholesterol level[85]. Recently, clustering of MHC-I molecules was also found to be controlled by extracellular factors, such as the concentration of available β2-microglobulin[86]. As an *ex vivo* model for analyzing the biological function of this molecular association, T lymphocytes were selected from human peripheral blood with specificity against human JY B-lymphoblastoid cells as antigen-presenting cells. The two cell types were found to form stable immunological synapses, as detected by confocal microscopy using fluorescent immunocytochemical staining (Fig. 2a).

It was shown that exogenous β2-microglobulin, a serum component, is able to recombine with β2-microglobulin-free MHC heavy chains (expressed in the PM of activated or transformed human cells) and, as a consequence, disrupt/destabilize MHC clusters in the PM. This was convincingly reinforced by FRET microscopic and SNOM studies[86] (Fig. 2b, Fig 3). In addition to recognizing that exogenous β2-microglobulin can control MHC-clustering on antigen-presenting cells, we also wanted to understand the functional consequences of this molecular-level regulation. Application of fluorescent immunocytochemistry and flow cytometric detection allowed us to demonstrate that a reduced clustering of MHC-I on antigen-presenting cells resulted in a weaker T cell activation in the cell-conjugates, as shown by a decreased T cell receptor internalization (monitored by the decrease of Fab labeled TCR fluorescence) and by a weaker cytolytic function of cytotoxic T lymphocytes (monitored by measuring the release of luminescent europium indicator from the target cells)[86]. This latter assay implemented as a time-resolved (boxcar-gated) luminometric measurement of long lifetime Eu luminescence can also be performed with plates allowing numerous parallel samples, and has the great advantage of excluding the short lived autofluorescence background, thus providing sensitivity comparable to, or – by the use of a photochemical amplifier – even exceeding that of the classical radioactive methods used for this type of assay earlier[87].

MHC-II molecules present antigens to helper T lymphocytes and thereby control the whole process of cellular immune response. Our earlier FRET studies showed that MHC-II and MHC-I molecules are consistently co-clustered at the surface of various immunocompetent cells[17, 22]. Recent studies directed on the dynamics of raft microdomains in PMs and using the method of cell fusion have shown that while several

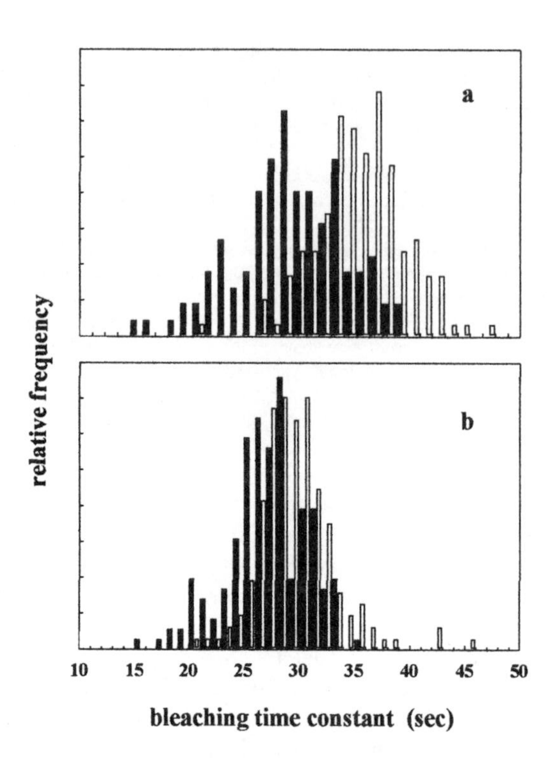

Figure 3. Kinetic measurement of donor photobleaching in human B lymphoblasts: MHC-I molecules labeled with FITC-W6/32 mAb (donor) were bleached with a shorter time constant in the absence (black histogram in a) than in the presence of acceptor (TRITC-W6/32-mAb, white histogram), indicating substantial FRET between MHC-I molecules on the nanometer scale. Treatment of the cells with β2-microglobulin greatly decreased this homoassociation as indicated by the remarkably shortened donor bleaching time constant in the presence of acceptor (white histogram, b).

protein constituents of the rafts in the PM are easily and rapidly exchanged among raft domains upon cell fusion, some protein associations, such as MHC-I/MHC-II interactions remain conserved[73]. In this study, simultaneous detection of FRET between the molecular species labeled with Fabs (indicating molecular level proximity) and the patchy distribution of certain proteins detected by SNOM was used to analyze the dynamics of lipid rafts, in terms of exchanging proteins. Fluorescence labeling strategy was the following: the same kinds of proteins were immunocytochemically labeled with green or red fluorescent dyes in two aliquots of the same cell population. After fusing the green and red cells with polyethylene-glycol (PEG), the kinetics of FRET and the spatial distribution of tagged species were followed in time. This approach was repeated for a number of membrane proteins including both raft- and non-raft ones. We could learn from this analysis that small (nanometer) scale association detected by FRET and large (μm) scale clustering detected by SNOM of the same proteins represent two distinct hierarchical levels of protein organization in the PM, consistent with earlier atomic force and electron microscopy data[8]. Dynamic exchange of components between small scale clusters was shown to follow the redistribution of small scale clusters themselves with a delay. The reasons behind this phenomenon, as well as behind the intriguing stability of heterooligomers formed by MHC-I and MHC-II molecules are still unresolved[73].

Clusters (patchy distribution) of MHC-II antigens were also detected at the surface of murine antigen-presenting cells (APCs). These clusters were associated to a large degree with lipid raft markers, such as GM1 gangliosides labeled with Alexa488-cholera toxin B subunit (Fig. 2c). Disruption of membrane microdomains (rafts) by cholesterol depletion resulted in the separation of MHC-II and GM-1-positive clusters at the submicron level (Fig. 2d) and also in a simultaneous decrease of FRET efficiency (from 5.5±1.3% to 0.8±1.0%) characteristic for MHC-II homoassociation[88]. Such modifications of the APC membranes resulted also in a decreased efficiency of antigen-presentation and consequential T cell activation, suggesting that homo-clusters of MHC-II molecules on APCs may regulate the threshold of activation of T cells encountering them. Part of the microscopically observable MHC-II clusters are associated with raft microdomains in the PM, while another fraction of MHC-II molecules was found associated with so called tetraspan domains insensitive to cholesterol depletion on APCs[17, 89, 90].

From these examples it is clear that fluorescent technologies have provided many important molecular details for better understanding the mechanism of antigen presentation and the function of immunological synapses,[91] and at the same time highlighted some details of raft-dynamics in the PM. A future challenge for modern fluorescent microscopic techniques is to spatio-temporally resolve complex intracellular and intercellular biological structures and processes, such as synapses, junctions, vesicle targeting, or protein recycling.

4.2. Compartmentation of cytokine receptors by lipid rafts: advanced microscopies and FRET assess molecular arrangements and membrane compartmentation

Cytokines regulating immune responses have their specific, private receptor, but may also share public receptors with other cytokines. Interleukin-2 (IL-2) secreted by T lymphocytes when stimulated with antigen or mitogens is essential for T cell growth[92, 93]. The private receptor for IL-2 is the IL-2Rα subunit, exhibiting relatively low affinity for IL-2 compared to the IL-2Rαβγ heterotrimer, which is considered a fully functional receptor[94]. It was shown in a fluorescence resonance energy transfer study that the IL-2R α, β and γ subunits are preassembled on the surface of unstimulated Kit 225 K6 T lymphoma cells, and cannot, therefore, be considered as a transient signaling assembly[26]. In addition, there is evidence that the IL-15 receptor α subunit, which shares the β and γ subunits with IL-2Rα, can also form preassembled supramolecular structures with IL-2Rβ and the "common" γ chain[95-97]. Interestingly, the colocalization of IL-2R subunits was significantly modulated by binding of relevant interleukins: IL-15 loosened the molecular linkage between the α and γ subunits, while IL-2 treatment caused the α-β-γ "triangle" to tighten up[26]. FRET measurements[68] have also shown that all three IL-2R subunits were in the molecular proximity of both MHC-I and MHC-II glycoproteins as well as the GPI-linked CD48 antigen[98, 99]. Nonetheless, it was still unclear how IL-2Rα is organized in the PM and how it is recruited to the less abundant β and γ chains to form the functionally active receptor.

Fluorescence resonance energy transfer measurements can detect molecular association only in the 1 – 10 nm range. This limitation could be successfully overcome by combining scanning force microscopy (SFM) and electron microscopy (EM) as complementary methods to fluorescence spectroscopy[8, 100, 101]. Application of EM and SFM made possible the discovery of a new, higher hierarchical level of receptor clustering in lymphoid cells, that was confirmed under hydrated conditions, using tapping mode SFM[8]. The observation that nm-scale islets of MHC class I molecules in the cell membrane are organized into micrometer-sized "island groups" was soon extended for MHC class II[101] and for the IL-2 receptor alpha subunit and the transferrin receptor[11]. With these methods, direct visualization of the receptor patterns was achieved, making statistical evaluation of randomness (or non-randomness) relatively easy. Such calculations were carried out for MHC class I and II, the IL-2 receptor alpha subunit and the transferrin receptor (TrfR). Besides concluding that their plasma membrane distribution was different from the Poissonian expected for randomly scattered molecules, it was also possible to estimate the average size of higher order molecular clusters (or "island groups")[11]. These data were in the range of 400-800 nm for the various receptors, which is in good correlation with the mean barrier free path obtained for transmembrane proteins using single particle tracking or optical tweezers[55, 59]. Confocal laser scanning microscopy of fluorescently labeled live and fixed cells followed by surface reconstruction also confirmed the existence of such supramolecular receptor clusters of the IL-2Rα and the TrfR on lymphoid cells (Fig. 2e,f.). For the confocal images, the two-dimensional autocorrelation function $G_{(\rho,\phi)}$ was calculated to reveal cluster sizes[11]:

$$G_{(\rho,\varphi)} = \left\langle F_{(r,\Theta)} \cdot F_{(r+\rho,\Theta+\varphi)} \right\rangle, \tag{2}$$

where $F_{(r,\Theta)}$ is the normalized pixel fluorescence at radius r and angle Θ relative to the center of the image, and the angle brackets indicate summation over the whole domain of the r radius and Θ angle for each ρ, and φ change of radius and angle. The autocorrelation image was calculated by taking the inverse Fourier-transform of the two-dimensional power spectrum matrix of the original images, and since no anisotropy was seen or expected in the images, the rotation invariant autocorrelation $G_{(\rho)}$ was calculated by averaging $G_{(\rho,\phi)}$ over the range $0 \leq \varphi < 2\pi$. $G_{(\rho)}$ was fitted to the equation

$$G_{(\rho)} = \sum_i A_i e^{-\left(\frac{\rho}{R_i}\right)^2}, \tag{3}$$

where the R_i characteristic radii serve as an adequate measure of the mean size (half-width at the 1/e height of a Gaussian distribution) of each class of clusters distinguishable on the basis of its size. Cluster radii of IL-2Rα were essentially equal (R_i~400 nm) to those seen in EM analysis of the immunogold label distribution[11]. In these experiments, very similar cluster sizes were found for other molecules that are preferentially localized to lipid rafts, such as MHC-I, MHC-II, and CD48, the latter being a typical example of the GPI-anchored proteins. Patches of these molecules also showed a great degree of colocalization with the IL-2Rα (Fig. 2e.). Furthermore, crosslinking either CD48 or IL-2Rα with specific antibodies from mouse and then polyclonal anti-mouse IgG not only caused capping of the receptor targeted, but also resulted in the spatio-temporally coordinated aggregation, co-capping of the other molecule (Fig. 2g,h.). Interestingly, clusters of the constitutively expressed transferrin receptor, which based on detergent resistance analysis is excluded from rafts,[102] were only about half the size (R_i~200 nm) and showed hardly any overlap with IL-2R clusters as judged form the cross correlation coefficient (C) of the two images being close to 0 (Fig. 2f.). C was calculated as

$$C = \sum_i \sum_j \left(x_{i,j} - \langle x \rangle\right)\left(y_{i,j} - \langle y \rangle\right) \Big/ \sqrt{\sum_i \sum_j \left(x_{i,j} - \langle x \rangle\right)^2 \sum_i \sum_j \left(y_{i,j} - \langle y \rangle\right)^2} \tag{4}$$

where $x_{i,j}$ and $y_{i,j}$ are fluorescence pixel values at coordinates i,j in images x (e.g. the green channel) and y (e.g. the red channel), and the theoretical maximum is $C = 1$ for identical images.

Treatment of cells with methyl-β-cyclodextrin[103] that depleted membrane cholesterol, as well as *in situ* complexation of cholesterol with the fungal antibiotic Filipin III[104] increased cluster sizes determined form the autocorrelation function for all raft-associated proteins to about double, and also caused a blurring of spot boundaries. However, the size and appearance of TrfR domains has not changed significantly[11].

A clear advantage of the above-mentioned spectroscopic and microscopic methods in the investigation of membrane protein association is that they do not require disruption of the integrity of the plasma membrane and thus provide evidence that indeed microscopic equivalents of rafts do exist. Based on the notion that rafts are essentially membrane units formed from transport vesicles fusing to the membrane, one would expect that their size is very small[49, 105]. In photonic force microscopic experiments, it was determined that raft size is less then 50 nm in diameter[106] representing about 3500 sphyngolipid molecules. This indicates that membrane patches observed in fluorescence microscopy, bearing raft marker proteins and/or lipids, are probably aggregates of these basic building blocks. In some instances, these larger aggregates are not observed in resting cells,[107, 108] and can only be seen upon cross-linking the "unit rafts"[107, 109]. In other cases, as detailed above, cells in their native state present surface patches of sub-micron size, identifiable as rafts based on their composition[11, 110].

Following up these morphological data in further experiments, it consequently turned out that the disassembly by cholesterol depletion or *in situ* complexation not only destroys the morphology of rafts,[11] but also impairs their organizing and signaling capabilities in lymphocytes[68]. On the nm scale, the effect of cholesterol depletion was a significant 50% decrease of FRET efficiency between IL-2Rα and MHC-II, while association of IL-2Rα with CD48 was characterized by a gross drop in FRET efficiency from 12.6±1.9 to 2.3±1.5. As expected from the morphological studies, association between TrfR and IL-2Rα was not affected significantly by depletion of PM cholesterol. Strikingly, the FRET values for this receptor pair were in the 10% range as well, in spite of the virtually insignificant colocalization of their visualized clusters, which is best explained by the interaction of these receptors at the borders of clusters, as well as the notion that probably there exists a fraction of IL-2Rα that is not raft-associated. This latter is supported by the results of immunoblotting the detergent resistant and detergent soluble fractions of these cells[68].

In spite of the probable existence of this fraction, there is evidence for the raft-bound fraction of IL-2Rα being the active signaling entity. Disruption of rafts by depleting PM cholesterol causes a marked decrease of IL-2 evoked tyrosine phosphorylation as assessed from Western blots and enhanced chemiluminescence reaction. Furthermore, using flow cytometry of *in situ* immunofluorescently labeled intracytoplasmic phosphoproteins Stat3 and Stat5 that are specifically phosphorylated during IL-2 signalization, it was shown that IL-2 induced increase in Stat3 and Stat5 phosphorylation levels is greatly inhibited by cholesterol depletion [68]. Thus it is reasonable to hypothesize that lipid rafts function not only to compartmentalize the IL-2R α chain, but also to organize in a dynamic manner its access to the other IL-2R subunits, β and γ, to make it functional.

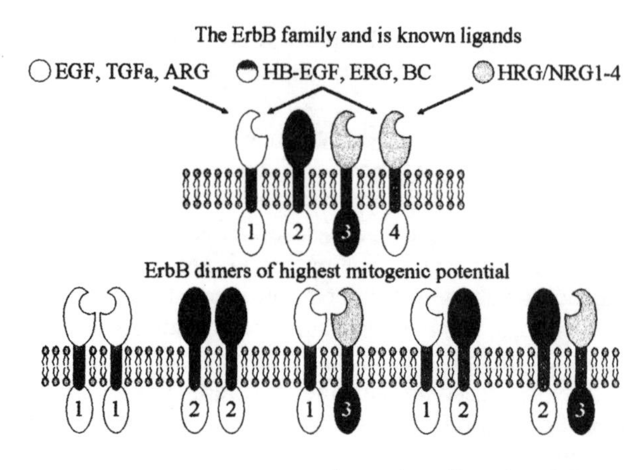

Figure 4. Members of the ErbB family are designated ErbB1 (commonly known as the EGF receptor), ErbB2 (or neu), ErbB3 and ErbB4. Their ligands belong to three groups: EGF-like ligands (EGF, TGF-α and Amphiregulin) bind to ErbB1. Heregulin-like (or neuregulin-like) ligands (Heregulin/Neuregulin 1 through 4) bind to ErbB3 and ErbB4. The third group, EGF- and heregulin-like ligands (Heparin-binding EGF, Epiregulin and Betacellulin) are capable of ligating all three ErbB receptors. Currently no physiological ligand for ErbB2 is known. The ErbB3 receptor has a defective tyrosine kinase domain. Either upon ligand binding or constitutively, ErbB kinases compose various homo- and heterodimers in the cell membrane, as revealed by FRET microscopy and FCET techniques. Many of these have outstanding mitogenic potential.

4.3. Fluorescence cytochemistry and FRET for mapping of cell surface Erb oncogene products: Erb/raft domains and cancer

The type I family of transmembrane receptor tyrosine kinases comprises four members: epidermal growths factor receptor (EGFR or ErbB1), ErbB2 (HER2 or Neu), ErbB3 and ErbB4[111-113] (Fig. 4). Within a given tissue, these receptors are rarely expressed alone, but are found in various combinations. Members of the family form homo- and heteroassociations at the cell surface. Their ligands belong to three groups: EGF-like ligands bind only to ErbB1, heregulin-like (or neuregulin-like) ligands bind only to ErbB3 and ErbB4, while EGF- and neuregulin-like ligands bind ErbB1 and ErbB4. ErbB3 shares growth factor binding specificity with ErbB4 but lacks intrinsic kinase activity. ErbB2 is an orphan receptor: no physiological ligand has been found for it. Despite this fact, ErbB2 participates actively in ErbB receptor combinations, and receptor complexes including ErbB2 appear to be more potent than other receptor combinations[111, 112, 114]. Thus, both ligand and receptor can vary tissue by tissue, and will, in part, determine the specificity and the potency of cellular signals.

Molecular scale physical associations among ErbB family members have been studied by classical biochemical,[115, 116] molecular biological, and biophysical methods[42, 46, 70]. When isolated from cells, members of the ErbB family self-associate (homoassociate) and associate with other family members (heteroassociate)[115]. However, experiments on isolated proteins are inherently unable to detect interactions in cellular environments *in vivo* and *in situ*, and cannot detect heterogeneity within or among cells. FRET between membrane proteins labeled with fluorescent antibodies or their Fab fragments measures the monomer-dimer distribution of ErbB1 receptors in fixed,[46, 117] and living cells,[117] and was also applied to detect a heterogeneous pattern of ErbB2 association in breast tumor cells[42, 118]. Fluorescence microscopy could visualize FRET within single cells with spatial resolution limited only by diffraction in the optical microscope,[10, 39, 42] allowing detailed analysis of the spatial heterogeneity of molecular interactions.

The association state of ErbB2 and how it is affected by EGF treatment in breast tumor cell lines was assessed by measuring FRET between fluorescent monoclonal antibodies and Fab fragments. There was considerable homoassociation of ErbB2 and heteroassociation of ErbB2 with EGFR in quiescent breast tumor cells. ErbB2 homosassociation was enhanced by EGF treatment in SK-BR-3 cells and in the BT474 subline BT474M1 with high tumorogenic potential, whereas the original BT474 line was resistant to this effect. These differences correlated well with EGFR expression. Experiments also revealed extensive pixel-by-pixel heterogeneity in ErbB2 homoassociation[42].

Membrane domains with erbB2 homoassociation had mean diameters of less than one micrometer[70, 118]. This suggests that single molecule interactions occur in the context of larger domains. We made these measurements using scanning near-field optical microscopy (SNOM)[10, 119, 120]. ErbB2 was concentrated in irregular membrane patches with a mean diameter of approximately 500 nm, containing up to 1000 ErbB2 molecules in non-activated SK-BR-3 and MDA453 human breast tumor cells. The mean cluster diameter increased to 600-900 nm when SK-BR-3 cells were treated with EGF, heregulin or a partially agonistic anti-erbB2 antibody. The increase in cluster size was inhibited by an EGFR-specific tyrosine kinase inhibitor. SNOM results were confirmed with CLSM on hydrated samples[70].

Lipid rafts on breast tumor cells were also investigated using a CLSM equipped with three lasers for the colocalization of the glycosphingolipid GM1 ganglioside labeled with fluoresceinated subunit B of cholera toxin (CTX-B)[107] and ErbB2 clusters labeled with Cy5-conjugated Fab fragments of the 4D5 antibody. The results suggest that ErbB2 is localized mostly in lipid rafts, similarly to ErbB1. Since stimulating ErbB2 increases the size of ErbB2 clusters[70] and lipid rafts,[75] the amount of ErbB2 concentrated in rafts is very likely related to the function of the protein. The localization of ErbB2 in lipid rafts is dynamic, since it can be dislodged from rafts by cholera toxin-induced raft crosslinking. The association properties and biological activity of ErbB2 expelled from rafts differs from that inside rafts. For example, 4D5-mediated internalization of ErbB2 is blocked in cholera toxin-pretreated cells, while its antiproliferative effect is not. These results emphasize that alterations in the local environment of ErbB2 strongly influence its association properties, which are reflected in its biological activity and in its behavior as a target for therapy[75].

5. A RAPID CYTOFLUORIMETRIC ASSAY FOR ASSESSMENT OF RAFT-ASSOCIATED MEMBRANE PROTEINS

Constituents of membrane microdomains are often identified on the basis of their resistance to cold non-ionic detergents[5]. As an alternative, they are associated to raft microdomains on the basis of their colocalization with the major lipid components of rafts, e.g. GM-1 gangliosides labeled with fluorescent cholera toxin B subunits[107, 121].

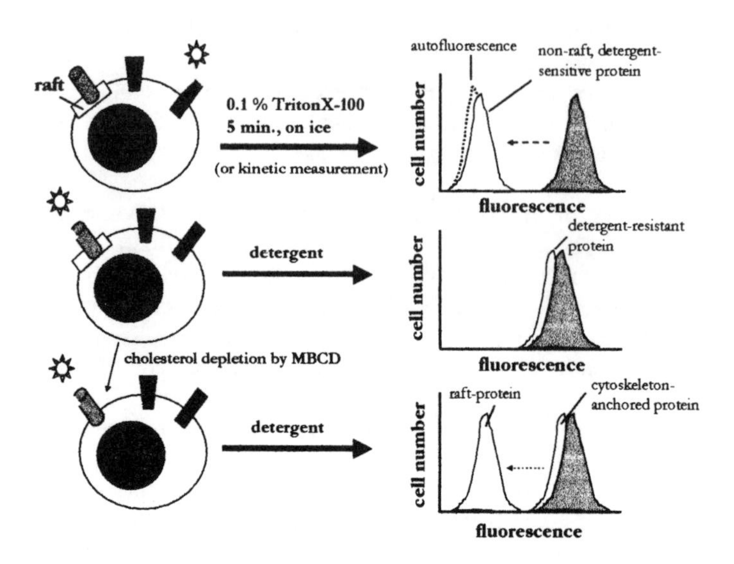

Figure 5. Principles of the flow cytometric differential detergent resistance assay (FCDRA). The assay is based on a comparative analysis of detergent-treatment induced changes in cellular fluorescence derived from specific immunocytochemical labeling in control cells and cells with depleted membrane cholesterol (for details see text).

Dynamic association to lipid rafts of most proteins is of special interest in the cell-, neuro- and immuno-biological research, where rafts are assumed to compartment cell surface receptors, signal and regulatory proteins via physical coupling or isolation both laterally or vertically in the PM,[122] this way controlling the functional responses of cells to extracellular signals. As a most widely used approach, association (constitutive or induced) of membrane proteins to raft microdomains is usually investigated by isolating detergent-resistant, „floating" (low buoyant density) membrane fractions via sucrose-gradient ultracentrifugation of cell lysates obtained with detergent (e.g. TritonX-100) solubilization[6]. This approach requires a large amount of cells (~10^8 cells/ experiment) and, besides the possible artifacts generated by the 1% detergent used for solubilization, it suffers from additional uncertainties related to the detection procedure consisting of SDS-gel electrophoresis, immunoblotting, and chemiluminescence.

As an alternative, we have recently developed a rapid and reliable assay based on the flow cytometric detection of antibody-labeled epitopes of membrane proteins that follows detergent dissolution kinetics[123]. This assay is based on the measurement of fluorescence intensity levels characteristic of an *in situ* immunofluorescently labeled cell membrane protein before, and 5 minutes after treatment with 0.1% cold Triton-X100 (or Brij, Chaps) detergent. Proteins that are sensitive to detergent extraction are dissolved easily and rapidly, and therefore show a decreased level of cellular fluorescence, while membrane proteins that are insensitive to detergent extraction due either to their raft localization or cytoskeletal anchoring remain mostly undissolved (Fig. 5). Our approach to discriminate raft-associated and non-raft proteins is based on their differential sensitivity to cholesterol-depletion (by methyl-β-cyclodextrin, MBCD). Proteins associated with glycosphingolipid- and cholesterol-rich microdomains (rafts) are dissolved by detergents easily after depletion of membrane cholesterol (raft-disruption), while proteins resistant to detergents and not dissolved upon MBCD pretreatment are presumably anchored by the cytoskeletal matrix[124]. The details of this method are under publication elsewhere. Investigating the detergent resistance of several T cell surface proteins, our data convincingly coincided with those obtained for the same proteins using the classical biochemical detergent resistance assay and microscopic analysis of co-localization[122]. This assay offers the advantages of rapid analysis, requirement for low amounts of cells (~10^4 cells per sample) and a high degree of reproducibility. In addition, in some cases the fraction of raft-associated proteins can be quantitatively estimated and induced raft-association (e.g. by ligation / crosslinking of receptors) can also be monitored in time.

Therefore, as a new member of the wide-scale methodological repertoire to study raft-localization of proteins, we propose this flow cytometric assay based on the detection of differential detergent-resistance (FCDRA) as an easy and rapid assay performable on any conventional bench-top flow cytometer.

6. ACKNOWLEDGEMENTS

This work was supported in part by grants from the Hungarian Academy of Sciences OTKA T037831, T034393, T042618, T043061, TS 040773 and from the Hungarian Ministry of Health ETT 524/2003 and 532/2003. GV is a Békésy Fellow of the Hungarian Ministry of Education.

7. REFERENCES

1. M. Edidin, Shrinking patches and slippery rafts: scales of domains in the plasma membrane, *Trends Cell Biol* **11,** 492-496. (2001).
2. R. G. Anderson and K. Jacobson, A role for lipid shells in targeting proteins to caveolae, rafts, and other lipid domains, *Science* **296,** 1821-1825. (2002).
3. W. K. Subczynski and A. Kusumi, Dynamics of raft molecules in the cell and artificial membranes: approaches by pulse EPR spin labeling and single molecule optical microscopy, *Biochim Biophys Acta* **1610,** 231-243. (2003).
4. K. Simons and E. Ikonen, Functional rafts in cell membranes, *Nature* **387,** 569-572. (1997).
5. D. A. Brown and J. K. Rose, Sorting of GPI-anchored proteins to glycolipid-enriched membrane subdomains during transport to the apical cell surface, *Cell* **68,** 533-544. (1992).
6. S. Ilangumaran, S. Arni, G. van Echten-Deckert, B. Borisch and D. C. Hoessli, Microdomain-dependent regulation of Lck and Fyn protein-tyrosine kinases in T lymphocyte plasma membranes, *Mol Biol Cell* **10,** 891-905. (1999).
7. J. Hwang, L. A. Gheber, L. Margolis and M. Edidin, Domains in cell plasma membranes investigated by near-field scanning optical microscopy., *Biophys. J.* **74,** 2184-2190. (1998).
8. S. Damjanovich, G. Vereb, A. Schaper, A. Jenei, J. Matkó, J. P. Starink, G. Q. Fox, D. J. Arndt-Jovin and T. M. Jovin, Structural hierarchy in the clustering of HLA class I molecules in the plasma membrane of human lymphoblastoid cells, *Proc Natl Acad Sci U S A* **92,** 1122-1126. (1995).
9. S. Damjanovich, J. Matkó, L. Mátyus, G. J. Szabó, J. Szöllõsi, C. Pieri, T. Farkas and R. J. Gáspár, Supramolecular receptor structures in the plasma membrane of lymphocytes revealed by flow cytometric energy transfer, Scanning Force- and Transmission electron-microscopic analyses, *Cytometry* **33,** 225-234. (1998).
10. G. Vereb, C. K. Meyer and T. M. Jovin, in: *Interacting protein domains, their role in signal and energy transduction. NATO ASI series*, edited by L. M. G. Heilmeyer Jr (Springer-Verlag, New York, 1997), pp. 49-52.

11. G. Vereb, J. Matkó, G. Vamosi, S. M. Ibrahim, E. Magyar, S. Varga, J. Szöllősi, A. Jenei, R. Gáspár, Jr., T. A. Waldmann and S. Damjanovich, Cholesterol-dependent clustering of IL-2Ralpha and its colocalization with HLA and CD48 on T lymphoma cells suggest their functional association with lipid rafts, *Proc Natl Acad Sci U S A* **97**, 6013-6018. (2000).
12. G. Vereb, J. Szöllősi, J. Matkó, P. Nagy, T. Farkas, L. Vígh, L. Mátyus, T. A. Waldmann and S. Damjanovich, Dynamic, yet structured: The cell membrane three decades after the Singer-Nicolson model, *Proc Natl Acad Sci U S A* **100**, 8053-8058. (2003).
13. M. J. Saxton and K. Jacobson, Single-particle tracking: applications to membrane dynamics, *Annu Rev Biophys Biomol Struct* **26**, 373-399. (1997).
14. A. Kusumi and Y. Sako, Cell surface organization by the membrane skeleton, *Curr Opin Cell Biol* **8**, 566-574. (1996).
15. J. Matkó, Y. Bushkin, T. Wei and M. Edidin, Clustering of class I HLA molecules on the surfaces of activated and transformed human cells, *J Immunol* **152**, 3353-3360. (1994).
16. L. Bene, M. Balázs, J. Matkó, J. Most, M. P. Dierich, J. Szöllősi and S. Damjanovich, Lateral organization of the ICAM-1 molecule at the surface of human lymphoblasts: a possible model for its co-distribution with the IL-2 receptor, class I and class II HLA molecules, *Eur J Immunol* **24**, 2115-2123. (1994).
17. J. Szöllősi, V. Horejsi, L. Bene, P. Angelisova and S. Damjanovich, Supramolecular complexes of MHC class I, MHC class II, CD20, and Tetraspan Molecules (CD53, CD81, and CD82) at the surface of a B cell line JY, *J Immunol* **157**, 2939-2946. (1996).
18. S. Damjanovich, R. Gáspár, Jr. and C. Pieri, Dynamic receptor superstructures at the plasma membrane, *Q Rev Biophys* **30**, 67-106. (1997).
19. S. Damjanovich, L. Bene, J. Matkó, L. Mátyus, Z. Krasznai, G. Szabo, C. Pieri, R. Gáspár and J. Szöllősi, Two-dimensional receptor patterns in the plasma membrane of cells. A critical evaluation of their identification, origin and information content., *Biophys Chem* **82**, 99-108. (1999).
20. S. M. Fernandez and R. D. Berlin, Cell surface distribution of lectin receptors determined by resonance energy transfer, *Nature* **264**, 411-415. (1976).
21. S. S. Chan, D. J. Arndt-Jovin and T. M. Jovin, Proximity of lectin receptors on the cell surface measured by fluorescence energy transfer in a flow system, *J Histochem Cytochem* **27**, 56-64. (1979).
22. J. Szöllősi, S. Damjanovich, M. Balázs, P. Nagy, L. Trón, M. J. Fulwyler and F. M. Brodsky, Physical association between MHC class I and class II molecules detected on the cell surface by flow cytometric energy transfer, *J Immunol* **143**, 208-213. (1989).
23. J. Szöllősi, S. Damjanovich, C. K. Goldman, M. J. Fulwyler, A. A. Aszalós, G. Goldstein, P. Rao, M. A. Talle and T. A. Waldmann, Flow cytometric resonance energy transfer measurements support the association of a 95-kDa peptide termed T27 with the 55-kDa Tac peptide, *Proc Natl Acad Sci U S A* **84**, 7246-7250. (1987).
24. J. Burton, C. K. Goldman, P. Rao, M. Moos and T. A. Waldmann, Association of intercellular adhesion molecule 1 with the multichain high-affinity interleukin 2 receptor, *Proc Natl Acad Sci U S A* **87**, 7329-7333. (1990).
25. A. De la Hera, U. Muller, C. Olsson, S. Isaaz and A. J. Tunnacliffe, Structure of the T cell antigen receptor (TCR): Two CD3 e subunits in a functional TCR/CD3 complex., *J Exp Med* **173**, 7-17. (1991).
26. S. Damjanovich, L. Bene, J. Matkó, A. Alileche, C. K. Goldman, S. Sharrow and T. A. Waldmann, Preassembly of interleukin 2 (IL-2) receptor subunits on resting Kit 225 K6 T cells and their modulation by IL-2, IL-7, and IL-15: a fluorescence resonance energy transfer study, *Proc Natl Acad Sci U S A* **94**, 13134-13139. (1997).
27. F. K. Chan, H. J. Chun, L. Zheng, R. M. Siegel, K. L. Bui and M. J. Lenardo, A domain in TNF receptors that mediates ligand-independent receptor assembly and signaling, *Science* **288**, 2351-2354. (2000).
28. R. M. Siegel, J. K. Frederiksen, D. A. Zacharias, F. K. Chan, M. Johnson, D. Lynch, R. Y. Tsien and M. J. Lenardo, Fas preassociation required for apoptosis signaling and dominant inhibition by pathogenic mutations, *Science* **288**, 2354-2357. (2000).
29. R. M. Clegg, Fluorescence resonance energy transfer, *Curr Opin Biotechnol* **6**, 103-110. (1995).
30. G. W. Gordon, G. Berry, X. H. Liang, B. Levine and B. Herman, Quantitative fluorescence resonance energy transfer measurements using fluorescence microscopy, *Biophys J* **74**, 2702-2713. (1998).
31. J. Szöllősi, S. Damjanovich and L. Mátyus, Application of fluorescence resonance energy transfer in the clinical laboratory: routine and research, *Cytometry* **34**, 159-179. (1998).
32. P. I. Bastiaens and A. Squire, Fluorescence lifetime imaging microscopy: spatial resolution of biochemical processes in the cell, *Trends Cell Biol* **9**, 48-52. (1999).
33. R. M. Clegg, FRET tells us about proximities, distances, orientations and dynamic properties, *J Biotechnol* **82**, 177-179. (2002).
34. C. Berney and G. Danuser, FRET or No FRET: A Quantitative Comparison, *Biophys J* **84**, 3992-4010. (2003).

35. T. Förster, Energiewanderung und Fluoreszenz., *Naturwissenschaften* **6**, 166-175. (1946).
36. L. Stryer and R. P. Haugland, Energy transfer: a spectroscopic ruler, *Proc Natl Acad Sci U S A* **58**, 719-726. (1967).
37. D. L. Dexter, A theory of sensitized luminescence in solids., *J. Chem. Phys.* **21**, 836-850. (1953).
38. G. Turcatti, K. Nemeth, M. D. Edgerton, U. Meseth, F. Talabot, M. Peitsch, J. Knowles, H. Vogel and A. Chollet, Probing the structure and function of the tachykinin neurokinin-2 receptor through biosynthetic incorporation of fluorescent amino acids at specific sites, *J Biol Chem* **271**, 19991-19998. (1996).
39. P. I. Bastiaens and T. M. Jovin, Microspectroscopic imaging tracks the intracellular processing of a signal transduction protein: fluorescent-labeled protein kinase C beta I, *Proc Natl Acad Sci U S A* **93**, 8407-8412. (1996).
40. Y. Suzuki, Detection of the swings of the lever arm of a myosin motor by fluorescence resonance energy transfer of green and blue fluorescent proteins, *Methods* **22**, 355-363. (2000).
41. L. Trón, J. Szöllősi, S. Damjanovich, S. H. Helliwell, D. J. Arndt-Jovin and T. M. Jovin, Flow cytometric measurement of fluorescence resonance energy transfer on cell surfaces.
Quantitative Evaluation of the transfer efficiency on a cell-by-cell basis., *Biophys J* **45**, 939-946. (1984).
42. P. Nagy, G. Vamosi, A. Bodnár, S. J. Lockett and J. Szöllősi, Intensity-based energy transfer measurements in digital imaging microscopy, *Eur Biophys J* **27**, 377-389. (1998).
43. R. M. Clegg, A. I. Murchie, A. Zechel, C. Carlberg, S. Diekmann and D. M. Lilley, Fluorescence resonance energy transfer analysis of the structure of the four-way DNA junction, *Biochemistry* **31**, 4846-4856. (1992).
44. R. M. Clegg, O. Holub and C. Gohlke, Fluorescence lifetime-resolved imaging: measuring lifetimes in an image, *Methods Enzymol* **360**, 509-542. (2003).
45. T. M. Jovin and D. J. Arndt-Jovin, in: *Cell Structure and Function by Microspectrofluorimetry.*, edited by E. Kohen and J. G. Hirschberg (Academic Press, San Diego, CA, 1989), pp. 99-117.
46. T. W. Gadella, Jr. and T. M. Jovin, Oligomerization of epidermal growth factor receptors on A431 cells studied by time-resolved fluorescence imaging microscopy. A stereochemical model for tyrosine kinase receptor activation, *Journal of Cell Biology* **129**, 1543-1558. (1995).
47. J. Matkó, A. Jenei, L. Mátyus, M. Ameloot and S. Damjanovich, Mapping of cell surface protein-patterns by combined fluorescence anisotropy and energy transfer measurements., *J Photochem Photobiol B: Biol* **19**, 69-73. (1993).
48. V. M. Mekler, A photochemical technique to enhance sensitivity of detection of fluorescence resonance energy transfer., *Photochem Photobiol* **59**, 615-620. (1994).
49. R. Varma and S. Mayor, GPI-anchored proteins are organized in submicron domains at the cell surface, *Nature* **394**, 798-801. (1998).
50. A. H. Clayton, Q. S. Hanley, D. J. Arndt-Jovin, V. Subramaniam and T. M. Jovin, Dynamic fluorescence anisotropy imaging microscopy in the frequency domain (rFLIM), *Biophys J* **83**, 1631-1649. (2002).
51. M. Edidin, Y. Zagyansky and T. J. Lardner, Measurement of membrane protein lateral diffusion in single cells, *Science* **191**, 466-468. (1976).
52. K. Jacobson, E. Elson, D. Koppel and W. Webb, Fluorescence photobleaching in cell biology, *Nature* **295**, 283-284. (1982).
53. T. J. Feder, I. Brust-Mascher, J. P. Slattery, B. Baird and W. W. Webb, Constrained diffusion or immobile fraction on cell surfaces: a new interpretation, *Biophys J* **70**, 2767-2773. (1996).
54. C. M. Anderson, G. N. Georgiou, I. E. Morrison, G. V. Stevenson and R. J. Cherry, Tracking of cell surface receptors by fluorescence digital imaging microscopy using a charge-coupled device camera. Low-density lipoprotein and influenza virus receptor mobility at 4 degrees C, *J Cell Sci* **101 (Pt 2)**, 415-425. (1992).
55. A. Kusumi, Y. Sako and M. Yamamoto, Confined lateral diffusion of membrane receptors as studied by single particle tracking (nanovid microscopy). Effects of calcium-induced differentiation in cultured epithelial cells, *Biophys J* **65**, 2021-2040. (1993).
56. Y. Sako and A. Kusumi, Compartmentalized structure of the plasma membrane for receptor movements as revealed by a nanometer-level motion analysis, *J Cell Biol* **125**, 1251-1264. (1994).
57. R. Simson, E. D. Sheets and K. Jacobson, Detection of temporary lateral confinement of membrane proteins using single-particle tracking analysis, *Biophys J* **69**, 989-993. (1995).
58. M. Edidin, S. C. Kuo and M. P. Sheetz, Lateral movements of membrane glycoproteins restricted by dynamic cytoplasmic barriers, *Science* **254**, 1379-1382. (1991).
59. M. Edidin, M. C. Zuniga and M. P. Sheetz, Truncation mutants define and locate cytoplasmic barriers to lateral mobility of membrane glycoproteins, *Proc Natl Acad Sci U S A* **91**, 3378-3382. (1994).
60. Y. Sako and A. Kusumi, Barriers for lateral diffusion of transferrin receptor in the plasma membrane as characterized by receptor dragging by laser tweezers: fence versus tether, *J Cell Biol* **129**, 1559-1574. (1995).

61. M. Edidin, Patches and fences: probing for plasma membrane domains, *J Cell Sci Suppl* **17**, 165-169. (1993).
62. D. Magde, E. L. Elson and W. W. Webb, Fluorescence correlation spectroscopy. II. An experimental realization, *Biopolymers* **13**, 29-61. (1974).
63. M. Ehrenberg and R. Rigler, Fluorescence correlation spectroscopy applied to rotational diffusion of macromolecules, *Q Rev Biophys* **9**, 69-81. (1976).
64. R. Brock, G. Vámosi, G. Vereb, and T. M. Jovin, Rapid characterization of green fluorescent protein fusion proteins on the molecular and cellular level by fluorescence correlation microscopy, *Proc Natl Acad Sci U S A* **96**, 10123-10128. (1999).
65. P. Schwille, U. Haupts, S. Maiti and W. W. Webb, Molecular dynamics in living cells observed by fluorescence correlation spectroscopy with one- and two-photon excitation, *Biophys J* **77**, 2251-2265. (1999).
66. Z. H. Zhong, A. Pramanik, K. Ekberg, O. T. Jansson, H. Jornvall, J. Wahren and R. Rigler, Insulin binding monitored by fluorescence correlation spectroscopy, *Diabetologia* **44**, 1184-1188. (2001).
67. E. L. Elson, Fluorescence correlation spectroscopy measures molecular transport in cells, *Traffic* **2**, 789-796. (2001).
68. J. Matkó, A. Bodnár, G. Vereb, L. Bene, G. Vamosi, G. Szentesi, J. Szöllősi, R. Gáspár, V. Horejsi, T. A. Waldmann and S. Damjanovich, GPI-microdomains (membrane rafts) and signaling of the multi-chain interleukin-2 receptor in human lymphoma/leukemia T cell lines, *Eur J Biochem* **269**, 1199-1208. (2002).
69. A. K. Kirsch, V. Subramaniam, G. Striker, C. Schnetter, D. J. Arndt-Jovin and T. M. Jovin, Continuous wave two-photon scanning near-field optical microscopy, *Biophys J* **75**, 1513-1521. (1998).
70. P. Nagy, A. Jenei, A. K. Kirsch, J. Szöllősi, S. Damjanovich and T. M. Jovin, Activation dependent clustering of the erbB2 receptor tyrosine kinase detected by scanning near_field optical microscopy, *J Cell Sci* **112**, 1733-1741. (1999).
71. A. Jenei, A. K. Kirsch, V. Subramaniam, D. J. Arndt-Jovin and T. M. Jovin, Picosecond multiphoton scanning near-field optical microscopy, *Biophys J* **76**, 1092-1100. (1999).
72. A. Lewis, A. Radko, N. Ben Ami, D. Palanker and K. Lieberman, Near-field scanning optical microscopy in cell biology, *Trends Cell Biol* **9**, 70-73. (1999).
73. P. Nagy, L. Mátyus, A. Jenei, G. Panyi, S. Varga, J. Matkó, J. Szöllősi, R. Gáspár, T. M. Jovin and S. Damjanovich, Cell fusion experiments reveal distinctly different association characteristics of cell-surface receptors, *J Cell Sci* **114**, 4063-4071. (2001).
74. M. Edidin, Near-field scanning optical microscopy, a siren call to biology, *Traffic* **2**, 797-803. (2001).
75. P. Nagy, G. Vereb, Z. Sebestyén, G. Horváth, S. J. Lockett, S. Damjanovich, J. W. Park, T. M. Jovin and J. Szöllősi, Lipid rafts and the local density of ErbB proteins influence the biological role of homo- and heteroassociations of ErbB2, *J Cell Sci* **115**, 4251-4262. (2002).
76. M. J. Edidin, Function by association? MHC antigens and membrane receptor complexes, *Immunol Today* **9**, 218-219. (1988).
77. R. J. Cherry, P. R. Smith, I. E. Morrison and N. Fernandez, Mobility of cell surface receptors: a re-evaluation, *FEBS Lett* **430**, 88-91. (1998).
78. R. N. Germain, MHC-dependent antigen processing and peptide presentation: providing ligands for T lymphocyte activation, *Cell* **76**, 287-299. (1994).
79. R. J. Cherry, K. M. Wilson, K. Triantafilou, P. O'Toole, I. E. Morrison, P. R. Smith and N. Fernandez, Detection of dimers of dimers of human leukocyte antigen (HLA)-DR on the surface of living cells by single-particle fluorescence imaging, *J Cell Biol* **140**, 71-79. (1998).
80. P. R. Smith, I. E. Morrison, K. M. Wilson, N. Fernandez and R. J. Cherry, Anomalous diffusion of major histocompatibility complex class I molecules on HeLa cells determined by single particle tracking, *Biophys J* **76**, 3331-3344. (1999).
81. K. Triantafilou, M. Triantafilou, K. M. Wilson and N. Fernandez, Human major histocompatibility molecules have the intrinsic ability to form homotypic associations, *Hum Immunol* **61**, 585-598. (2000).
82. J. H. Brown, T. S. Jardetzky, J. C. Gorga, L. J. Stern, R. G. Urban, J. L. Strominger and D. C. Wiley, Three-dimensional structure of the human class II histocompatibility antigen HLA-DR1, *Nature* **364**, 33-39. (1993).
83. Z. Bacsó, L. Bene, A. Bodnár, J. Matkó and S. Damjanovich, A photobleaching energy transfer analysis of CD8/MHC-I and LFA-1/ICAM-1 interactions in CTL-target cell conjugates, *Immunol Lett* **54**, 151-156. (1996).
84. R. Gáspár, Jr., P. Bagossi, L. Bene, J. Matkó, J. Szöllősi, J. Tozser, L. Fesus, T. A. Waldmann and S. Damjanovich, Clustering of class I HLA oligomers with CD8 and TCR: three-dimensional models based on fluorescence resonance energy transfer and crystallographic data, *J Immunol* **166**, 5078-5086. (2001).

85. A. Bodnár, A. Jenei, L. Bene, S. Damjanovich and J. Matkó, Modification of membrane cholesterol level affects expression and clustering of class I HLA molecules at the surface of JY human lymphoblasts, *Immunol. Lett.* **54**, 221-226. (1996).
86. A. Bodnár, Z. Bacsó, A. Jenei, T. M. Jovin, M. Edidin, S. Damjanovich and J. Matkó, Class I HLA oligomerization at the surface of B cells is controlled by exogenous beta(2)-microglobulin: implications in activation of cytotoxic T lymphocytes, *Int Immunol* **15**, 331-339. (2003).
87. C. Granberg, K. Blomberg, I. Hemmila and T. Lovgren, Determination of cytotoxic T lymphocyte activity by time-resolved fluorometry using europium-labelled concanavalin A-stimulated cells as targets, *J Immunol Methods* **114**, 191-195. (1988).
88. C. Detre, I. Gombos, B. Rethi, G. Vamosi, E. Rajnavolgyi and J. Matkó, Lipid rafts in APC and ceramide in T cell membrane modulate activation threshold and death signaling in APC (B cell)-T cell cojugates/synapses, *Immunol Lett* **87**, 103. (2003).
89. H. Kropshofer, S. Spindeldreher, T. A. Rohn, N. Platania, C. Grygar, N. Daniel, A. Wolpl, H. Langen, V. Horejsi and A. B. Vogt, Tetraspan microdomains distinct from lipid rafts enrich select peptide- MHC class II complexes, *Nat Immunol* **3**, 61-68. (2002).
90. A. B. Vogt, S. Spindeldreher and H. Kropshofer, Clustering of MHC-peptide complexes prior to their engagement in the immunological synapse: lipid raft and tetraspan microdomains, *Immunol Rev* **189**, 136-151. (2002).
91. S. K. Bromley, W. R. Burack, K. G. Johnson, K. Somersalo, T. N. Sims, C. Sumen, M. M. Davis, A. S. Shaw, P. M. Allen and M. L. Dustin, The immunological synapse, *Annu Rev Immunol* **19**, 375-396. (2001).
92. T. A. Waldmann, The structure, function, and expression of interleukin-2 receptors on normal and malignant lymphocytes, *Science* **232**, 727-732. (1986).
93. T. A. Waldmann, The interleukin-2 receptor, *J Biol Chem* **266**, 2681-2684. (1991).
94. Y. Nakamura, S. M. Russell, S. A. Mess, M. Friedmann, M. Erdos, C. Francois, Y. Jacques, S. Adelstein and W. J. Leonard, Heterodimerization of the IL-2 receptor beta- and gamma- chain cytoplasmic domains is required for signaling, *Nature* **369**, 330-333. (1994).
95. Y. Tagaya, J. D. Burton, Y. Miyamoto and T. A. Waldmann, Identification of a novel receptor/signal transduction pathway for IL-15/T in mast cells, *Embo J* **15**, 4928-4939. (1996).
96. T. A. Waldmann and Y. Tagaya, The multifaceted regulation of interleukin-15 expression and the role of this cytokine in NK cell differentiation and host response to intracellular pathogens, *Annu Rev Immunol* **17**, 19-49. (1999).
97. T. A. Waldmann, The meandering 45-year odyssey of a clinical immunologist, *Annu Rev Immunol* **21**, 1-27. (2003).
98. V. Horejsi, M. Cebecauer, J. Cerny, T. Brdicka, P. Angelisova and K. Drbal, Signal transduction in leucocytes via GPI-anchored proteins: an experimental artefact or an aspect of immunoreceptor function?, *Immunol Lett* **63**, 63-73. (1998).
99. V. Horejsi, K. Drbal, M. Cebecauer, J. Cerny, T. Brdicka, P. Angelisova and H. Stockinger, GPI-microdomains: a role in signalling via immunoreceptors, *Immunol Today* **20**, 356-361. (1999).
100. E. D. Sheets, R. Simson and K. Jacobson, New insights into membrane dynamics from the analysis of cell surface interactions by physical methods, *Curr Opin Cell Biol* **7**, 707-714. (1995).
101. A. Jenei, S. Varga, L. Bene, L. Mátyus, A. Bodnár, Z. Bacsó, C. Pieri, R. Gáspár, Jr., T. Farkas and S. Damjanovich, HLA class I and II antigens are partially co-clustered in the plasma membrane of human lymphoblastoid cells, *Proc Natl Acad Sci U S A* **94**, 7269-7274. (1997).
102. R. Xavier, T. Brennan, Q. Li, C. McCormack and B. Seed, Membrane compartmentation is required for efficient T cell activation, *Immunity* **8**, 723-732. (1998).
103. A. E. Christian, M. C. Haynes, M. C. Phillips and G. H. Rothblat, Use of cyclodextrins for manipulating cellular cholesterol content, *J. Lipid Res.* **38**, 2264-2272. (1997).
104. B. de Kruijff and R. A. Demel, Polyeneantibiotic-sterol interactions in in membranes of Acholeplasma laidlawii cells and lecithin liposomes. III. Molecular structure of the polyene antibiotic-cholesterol complexes, *Biochim. Biophys. Acta* **339**, 57-63. (1974).
105. N. M. Hooper, Detergent-insoluble glycosphingolipid/cholesterol-rich membrane domains, lipid rafts and caveolae (review), *Mol Membr Biol* **16**, 145-156. (1999).
106. A. Pralle, P. Keller, E. L. Florin, K. Simons and J. K. Horber, Sphingolipid-cholesterol rafts diffuse as small entities in the plasma membrane of mammalian cells, *J Cell Biol* **148**, 997-1008. (2000).
107. T. Harder, P. Scheiffele, P. Verkade and K. Simons, Lipid domain structure of the plasma membrane revealed by patching of membrane components, *J Cell Biol* **141**, 929-942. (1998).
108. A. K. Kenworthy, N. Petranova and M. Edidin, High-resolution FRET microscopy of cholera toxin B-subunit and GPI- anchored proteins in cell plasma membranes, *Mol Biol Cell* **11**, 1645-1655. (2000).

109. T. Friedrichson and T. V. Kurzchalia, Microdomains of GPI-anchored proteins in living cells revealed by crosslinking, *Nature* **394**, 802-805. (1998).
110. P. S. Pyenta, D. Holowka and B. Baird, Cross-correlation analysis of inner-leaflet-anchored green fluorescent protein co-redistributed with IgE receptors and outer leaflet lipid raft components, *Biophys J* **80**, 2120-2132. (2001).
111. Y. Yarden and M. X. Sliwkowski, Untangling the ErbB signalling network, *Nat Rev Mol Cell Biol* **2**, 127-137. (2001).
112. P. Nagy, A. Jenei, S. Damjanovich, T. M. Jovin and J. Szöllősi, Complexity of signal transduction mediated by ErbB2: clues to the potential of receptor-targeted cancer therapy, *Pathol Oncol Res* **5**, 255-271. (1999).
113. G. Vereb, P. Nagy, J. W. Park and J. Szöllősi, Signaling revealed by mapping molecular interactions: implications for ErbB-targeted cancer immunotherapies, *Clinical and Applied Immunology Reviews* **2**, 169-186. (2002).
114. M. X. Sliwkowski, J. A. Lofgren, G. D. Lewis, T. E. Hotaling, B. M. Fendly and J. A. Fox, Nonclinical studies addressing the mechanism of action of trastuzumab (Herceptin), *Semin Oncol* **26**, 60-70. (1999).
115. E. Tzahar, H. Waterman, X. Chen, G. Levkowitz, D. Karunagaran, S. Lavi, B. J. Ratzkin and Y. Yarden, A hierarchical network of interreceptor interactions determines signal transduction by Neu differentiation factor/neuregulin and epidermal growth factor, *Mol Cell Biol* **16**, 5276-5287. (1996).
116. M. X. Sliwkowski, G. Schaefer, R. W. Akita, J. A. Lofgren, V. D. Fitzpatrick, A. Nuijens, B. M. Fendly, R. A. Cerione, R. L. Vandlen and K. L. Carraway, 3rd, Coexpression of erbB2 and erbB3 proteins reconstitutes a high affinity receptor for heregulin, *J Biol Chem* **269**, 14661-14665. (1994).
117. T. W. Gadella, Jr., R. M. Clegg and T. M. Jovin, Fluorescence lifetime imaging microscopy: Pixel-by-pixel analysis of phase-modulation data, *Bioimaging* **2**, 139-159. (1994).
118. P. Nagy, L. Bene, M. Balázs, W. C. Hyun, S. J. Lockett, N. Y. Chiang, F. Waldman, B. G. Feuerstein, S. Damjanovich and J. Szöllősi, EGF-induced redistribution of erbB2 on breast tumor cells: flow and image cytometric energy transfer measurements, *Cytometry* **32**, 120-131. (1998).
119. A. Kirsch, C. Meyer and T. M. Jovin, in: *Proceedings of NATO Advanced Research Workshop: Analytical Use of Fluorescent Probes in Oncology, Miami, Fl. Oct. 14-18 1995*, edited by E. Kohen and J. G. Kirschberg (Plenum Press, New York, 1996), pp. 317-323.
120. E. Monson, G. Merritt, S. Smith, J. P. Langmore and R. Kopelman, Implementation of an NSOM system for fluorescence microscopy, *Ultramicroscopy* **57**, 257-262. (1995).
121. J. L. Thomas, D. Holowka, B. Baird and W. W. Webb, Large-scale co-aggregation of fluorescent lipid probes with cell surface proteins, *J Cell Biol* **125**, 795-802. (1994).
122. J. Matkó and J. Szöllősi, Landing of immune receptors and signal proteins on lipid rafts: a safe way to be spatio-temporally coordinated?, *Immunol Lett* **82**, 3-15. (2002).
123. I. Gombos, Z. Bacsó, C. Detre, G. Szabo and J. Matkó, A rapid flow cytometric assay for detecting raft association in immunecompetent cells, *Immunol Lett* **87**, 320. (2003).
124. C. P. Chia, I. Khrebtukova, J. McCluskey and W. F. Wade, MHC class II molecules that lack cytoplasmic domains are associated with the cytoskeleton, *J Immunol* **153**, 3398-3407. (1994).

RECENT ADVANCES IN SINGLE MOLECULE FLUORESCENCE SPECTROSCOPY

Jörg Enderlein*

1. INTRODUCTION

The field of single-molecule detection (SMD) or single-molecule spectroscopy (SMS) by laser induced fluorescence has seen an explosive development over the past decade. After the pioneering work of Richard Keller's group in Los Alamos (Dovichi et al., 1983; Shera et al., 1990), the number of publications on single molecule detection and spectroscopy in solution and on surfaces evolved in an exponential manner. Recently, there have been several, partially exhaustive reviews concerning the quickly evolving field of research namely (Meixner, 1998; Xie and Trautman, 1998; Weiss, 1999; Ambrose et al., 1999; Ishii and Yanagida, 2000; Keller et al., 2002; Michalet and Weiss, 2002; Moerner, 2002), and two books have been published (Rigler et al., 2001; Zander et al., 2002). The present paper intends to give an overview of the recent five years of single molecule fluorescence spectroscopy with a strong emphasis on biological applications. It is thus not intended to present a complete bibliography of SMS from the first beginnings, but is thought as an updated continuation of the reviews and books cited above. The scope of the review will be restricted to fluorescence SMS in liquids and on surfaces at room temperature, thus neglecting the broad research done on low-temperature spectroscopy (see e.g. Orrit et al., 1996; Tamarat et al., 2002; Moerner, 2002). Moreover, near-field microscopy of single molecules is also excluded because it would need the attention of a special review by itself.

The review is divided into a Techniques section, where the recent methodological advances in SMS are discussed, and two Application sections, where the diverse applications of SMS in physics, chemistry, and biology are presented. The emphasis of these latter Application sections will be on biological applications, where SMS seems to generate the biggest impact in terms of new and fundamental scientific insight. Sometimes it is difficult to clearly divide between publications dealing purely with methodological aspects of SMS, and papers describing significant new scientific results that were achieved by the application of SMS. Thus,

* Institute for Biological Information Processing 1, Forschungszentrum Jülich, D-52425 Jülich, Germany.

papers with a strong emphasis on the application of SMS but containing also new methodological developments will be discussed in the Applications sections, whereas papers with a more methodological content are mainly discussed in the Technqiues section.

2. TECHNIQUES

2.1. Probes and Markers

2.1.1. Fluorescent Dyes and Labels

A topic of considerable interest was the development of new near-infrared and infrared dyes with properties suitable for SMS, i.e. having high fluorescence quantum yield and high photostability (Arden-Jacob et al., 2001). There are two principal reasons for being interested in such dyes: firstly, scattering intensity of the excitation light, a major source of unwanted background in SMS, drops approximately off with the fourth power of the wavelength. Secondly, using NIR and IR dyes prevents the unwanted excitation of autofluorescence in biological samples. This is significant for many SMS applications in single-cell studies or for SM sensitive screening of biological fluids.

Willets et al. (2003) reported on a new class of nonlinear optical chromophores based on dicyanodihydrofuran acceptors paired with amine donors. These chromophores exhibited sufficiently large fluorescence quantum yields and stability to enable single-molecule detection in polymeric hosts. The emission spectra of these systems spanned a range from 505 to 646 nm. In contrast to conventional single-molecule fluorophores, the new molecules featured sensitivity to local rigidity, large ground-state dipole moments, and large polarizability anisotropies.

Zang et al. (2002) reported on a new molecular probe in the green/yellow spectral range which uses an intramolecular electron transfer mechanism to detect binding, local structure, and interfacial processes. The core idea is that molecular systems in which photoinduced intramolecular electron transfer (IET) from a high-energy nonbonding electron pair efficiently quenches the excited state of the chromophore form an important class of chemosensory materials. Reactions of this electron pair, with protons, metal atoms, organic electrophiles, or surfaces, lower the energy of the electron pair below the highest occupied molecular orbital of the chromophore, thus turning off IET and turning on fluorescence.

2.1.2. Quantum Dots

Although not exactly 'single-molecular', quantum dots emerged as an interesting alternative to single molecules as ultra-small fluorescent labels. Quantum dots are inorganic semiconductor nanocrystals with fluorescence properties partially similar to single molecules. They also show strong fluorescence intensity fluctuations and on/off blinking, but they have usually a much higher photostability and broader absorption spectra than fluorescence dyes, making them potentially interesting labels in many SMS applications.

Dahan et al. (2001) used the long fluorescence lifetime of CdSe quantum dot for time-gated imaging of biological samples, thus enhancing the signal-to-noise contrast by filtering out scattering and fast fluorescence background in the time domain.

Lacoste et al. (2000) reported about hybrid labels consisting of energy-transfer fluorescent beads and semiconductor nanocrystal quantum dots, that can be excited by a single laser wavelength but emit at different wavelengths. They use this multicolor emission for high-precision co-localization measurements of the labels based on their different emission colors. Chan and Nie (1998) proved the applicability of quantum dots as fluorescent labels in biological studies. Quantum dots that were labeled with the protein transferrin underwent receptor-mediated endocytosis in cultured HeLa cells, and those dots that were labeled with immunomolecules recognized specific antibodies or antigens. The same group reported on new developments in preparing highly luminescent and biocompatible CdSe quantum dots, and in synthesizing quantum-dot-encoded micro- and nano-beads (Han et al., 2001; Gao et al., 2002).

2.1.3. Color Centers

A potentially interesting alternative to dyes and quantum dots are color centers in crystalline solids, such as N-V centers in diamond (Jelezko et al., 2001). The N-V-center in diamond is a defect consisting of a substitutional nitrogen atom adjacent to a carbon-atom vacancy. The optical transition between the ^3A ground state and the ^3E excited state has a very high quantum efficiency allowing single defect spectroscopy. Small particles containing single N-V-centers could provide highly photostable and efficient fluorescent labels.

2.1.4. Metal Nanoparticles

In 2001, Lakowicz coined the term 'radiative decay engineering' describing the modification of emission properties of fluorophores or chromophores in the presence of metals, leading to enhanced photostability and optical excitation efficiency (Lakowicz, 2001). Enderlein (2000a, 2000b) made detailed calculations of these effects for fluorescently loaded nanometric metal cavities. He theoretically showed that for certain cavity parameters (radius, metal thickness) remarkable enhancement values can be achieved, although the metal is completely enclosing the fluorescing entity. The calculations clearly show that the synthesis of such metal/fluorescent composites may lead to new fluorescing labels with extraordinary optical properties.

Lee et al. (2002) report on the optical behavior of silver nanoclusters (2~8 atoms) that was completely different from that of bulk metal. These metal molecules exhibited extreme electroluminescence enhancements ($>10^4$ vs. bulk and dc excitation on a per molecule basis) when excited with specific ac frequencies, due to field extraction of electrons with subsequent reinjection and radiative recombination. Zheng and Dickson (2002) prepared and studied silver nanoclusters that were embedded within poly(amidoamine) dendrimer hosts. The dendrimer cage stabilized and solubilized the nanoclusters to yield highly stable, photoactivated single nanodots ranging in size from 2 to 8 silver atoms. These multicolored, highly fluorescent species are extremely photostable and were readily observed on the single molecule scale with weak mercury lamp excitation.

2.2. Confocal Detection of Single Molecules

2.2.1. Confocal Detection in Solution

In order to detect and to study individual fluorescent molecules, the collection of the single molecule's fluorescence emission has to be maximized while efficiently rejecting any background signal. The most efficient way of achieving that is to use high-quality optical filters, and to minimize the detection volume as much as possible.

A very convenient way of minimizing the detection volume is confocal imaging. In confocal imaging, the sample is illuminated by focusing a laser beam into a diffraction limited spot with an appropriate objective, and fluorescence light collection is done through the same objective (epifluorescence setup). Before detecting the collected light with a photoelectric detector, it is send through a confocal circular aperture. The diffraction limited focusing of the excitation laser assures minimum lateral extension of the detection volume (perpendicular to the optical axis), whereas the confocal aperture minimizes its axial extension (along the optical axis) by rejecting light emerging from above or below the in-focus plane. This is the principle of confocal detection SMS and results in a detection volume of roughly 0.5 μm diameter and 2 μm elongation. A thorough and detailed technical description of a SM sensitive multi-channel confocal detection system with pulsed excitation and time-correlated single-photon counting can be found in (Böhmer et al., 2001) and in the chapter by Böhmer and Enderlein in (Zander, 2002).

A shortcoming of the usual confocal detection setup is that one has no control of the way a molecule diffuses though the detection region. This makes quantitation of fluorescence brightness of an individual molecule impossible, and is a serious problem for application where it is necessary to detect all molecules of a given sample. One can overcome this shortcoming by directionally guiding single molecules through the small detection volume. Several solutions have been proposed for achieving that. The first approach is based on a modified flow cytometer that guides the molecules with a fluid flow through the detection region and uses the principle of sheath-flow focusing to counteract the radial diffusion of the molecules away from the flow axis, see e.g. (Demas et al., 1998). This was the method used for the first successful detection of single molecules in solution and is still used today for e.g. DNA fragment-sizing, see below. Van Orden et al. (1998) used the molecular fluorescence brightness information obtainable with the cytometer setup together with fluorescence lifetime measurements for distinguishing between different molecular species in a mixed sample.

The second approach restricts the mobility of the molecules in two directions by a capillary or a microstructure, allowing them to flow along the third direction only. Fister et al. (1998) used a microchip for guiding single molecules together with electrophoresis for transporting them through the detection region. Achieved concentration detection limits estimated at > 99% detection confidence were a few picomolar. In (Fister et al., 1999), this work was extended by cross-correlating the detection at two regions on the chip, which significantly improved the signal-to-noise ratio and thus confidence level of the detection. Dörre et al. (2001) describe a microfluidic device for SMS with multiple confocal detection regions, thus achieving a detection efficiency of nearly 60 % (3 of 5 molecules in the sample are positively detected).

A special setup for SMS near a surface on the background of high solution concentration of fluorescing molecules was presented by Levene et al. (2003). They used arrays of zero-mode waveguides consisting of sub-wavelength holes in a metal film for studying single-molecule dynamics at micromolar concentrations with microsecond temporal resolution.

Finally, Enderlein (2000a, 2000b) has proposed a novel setup for actively tracking diffusing molecules in solution via constantly probing the molecule's position with a rotationally scanning confocal detection region, and tracking the molecule by readjustment of the global position of that detection region.

2.2.2. Confocal Scanning on Surfaces

For studying molecules that are immobilized on a surface, the focused laser has to be scanned over the surface, either to record the fluorescence signal along single lines or to record a complete fluorescence image by scanning in two directions. Confocal imaging and scanning constitute the basic ingredients of a confocal laser-scanning microscope (CLSM). Scanning can also be extended into the third spatial dimension by scanning planes at different locations along the optical axis (so called *z*-scanning), resulting in a thee-dimensional fluorescence image of a sample (e.g. a dye tagged cell, or a transparent substrate with embedded fluorescent molecules). Tinnefeld et al. (2000) and Böhmer et al. (2001) described a combination of confocal scanning of surfaces with time-correlated single photon counting for SM lifetime imaging as well as monitoring fast photophysical processes. An interesting extension of this work is presented in (Tinnefeld et al., 2001b) and (Weston et al., 2002), where the authors used pulsed excitation and cross-correlation of the detected photons for directly observing so-called photon anti-bunching (vanishing cross-correlation at zero delay time), which is caused by the fact that a single molecule requires a certain time after emission to repopulate the excited state and thus being able to emit the next photon. By measuring the amplitude of the cross-correlation at zero delay time, the authors could directly count the number of discrete emitters (molecules) at a given location. Heilemann et al. (2002) used confocal lifetime-imaging for high-precision distance measurements between single fluorescent dye molecules based solely on their fluorescence lifetime, achieving a spatial resolution of less than 30 nm.

2.3. Wide-Field Microscopy

In wide-field imaging of single molecules, two principally distinct excitation methods are widely used: excitation by total-internal reflection (so called total-internal reflection fluorescence microscopy – TIRF), and direct wide-field illumination. Paige et al. (2001) compared both methods for SMS. Vacha and Kotani (2003) used alternate switching between both excitation modes for obtaining information about *three-dimensional* SM orientation by using the fact that TIRF illumination produces a strong electric-field component perpendicular to the sample and, for a given in-plane angle of the molecular dipole, provides unambiguous information on the 3D molecular dipole orientation. That it is possible to image single dye molecules even with a conventional video camera was demonstrated by (Adachi et al., 1999), where individual Cy3 molecules were used to image actin against heavy meromyosin sliding. Schütz et al. (2001) demonstrated that is possible to locate single fluorescent molecules with high precision (30-40 nm) in three dimensions employing a *z*-scan imaging technique.

An important improvement in SM imaging was the ability to simultaneously record images at different wavelengths and polarizations, for dual-color and/or polarization-resolved studies. Mörtelmayer et al. (2002) used two-color imaging for discriminating label fluorescence against autofluorescence in live cells. Cognet et al. (2002) presented a dual-color dual-polarization wide-field imaging setup, projecting simultaneously four frames at two emission colors and two polarizations onto the imaging CCD camera. Ma et al. (2000) imaged complete spectra of individual molecules by placing a grating in front of the CCD. They were able to distinguish different molecules (YOYO-I and POPO-III stained DNA, R-phycoerythrin) by these recorded spectra. Sonnleitner et al. (1999b) demonstrated successful SM imaging with two-photon excitation.

Wide-field imaging can be used to directly read out the three-dimensional orientation of the *emission* dipole of single molecules. This can be achieved by introducing optical aberrations when imaging, or by simply defocused imaging. Then, every molecule is imaged onto the CCD detector not as a sharp diffraction limited image, but as a complicated interference pattern, containing information about the angular distribution of radiation of the imaged molecule. Abberational imaging for determining molecular orientation was first demonstrated by Dickson et al. (1998) and Bartko and Dickson (1999a, 1999b). Later, Böhmer and Enderlein (2003) used defocused imaging for that purpose and presented an exact electromagnetic theory of this technique.

That it is even possible to observe single molecules near metal surfaces, where higher scattering background and significant fluorescence quenching by the metal take place, was demonstrated by Yokota et al. (1998). Their work is important for biological SMS applications when metal surface are used for biomolecule immobilization. Enderlein (2000c) gave an extended theoretical analysis of these experiments, deriving optimal experimental parameters for best possible SM detection.

2.4. Combination with Other Techniques

2.4.1. Force Microscopy

An exciting extension of fluorescence SMS is its combination with force microscopy. In force microscopy, single molecules are mechanically held and manipulated by either optical tweezers or an atomic force microscope. This allows the measurement of forces exerted by individual molecules as well as their reaction to applied external forces. Bennink et al. (1999) report the manipulation of single fluorescently stained DNA molecules with optical tweezers to study, among others, the interaction of the intercalating dye YOYO with DNA.

2.4.2. Electrophysiology

Electrophysiological patch-clamp measurements on single membrane channels have become one of the most important techniques in cell biology (Sakmann and Neher, 1984). Thus, a fascinating topic is its combination with SMS to observe, e.g., conformational changes within the membrane protein during ion or small molecule transits through the channel. Ide et al. (2002) developed an experimental apparatus for the simultaneous measurement of optical and electrical properties of single ion channels. Single molecules in bilayers were observed under an epifluorescence or an objective-type TIRF microscope. Minute currents across the bilayers were measured by a patch clamp amplifier under voltage clamped conditions. The apparatus was sensitive enough to detect the optical signals from single-fluorophores in the bilayer simultaneously with the single-channel current recording. Borisenko et al. (2003) performed electrical and optical measurements on fluorescently labeled gramicidin channels. The authors used Fluorescence Resonant Energy Transfer (FRET) to observe the formation of a gramicidine transmembrane channel, and could simultaneously detect the emerging ion flux through the channel.

2.4.2. Other Techniques

Qiu et al. (2003) used the probe of an AFM to inject electrons into individual porphyrin molecules adsorbed on an ultrathin aluminum film. Vibrational features were observed in the light-emission spectra that depended sensitively on the different molecular conformations and corresponding electronic states obtained by scanning tunneling spectroscopy. The high spatial resolution of the STM enabled the demonstration of variations in light-emission spectra from different parts of the molecule.

2.5. Data Evaluation

There are two principally different approaches for evaluating SMS data. The first set of methods can be called *fluctuation analysis* and processes the data with no explicit recourse to the *single molecule* nature of the experiment. These methods exploit the fluctuations of the measured light intensity that are caused by the fluctuating number of molecules within the detection volume. The most prominent representative of these methods is the so called *fluorescence correlation spectroscopy* (FCS), but recently new and promising techniques such as *fluorescence intensity distribution analysis* (FIDA) or *photon-counting histogramming* (PCH) have been developed. These methods will be discussed in the next two sections.

The second set of methods explicitly uses the fact that the measured data are mostly generated by single-molecule transits through the detection volume. Here, every single-molecule transit is identified and isolated within the continuous data stream, and every transit is then individually analyzed. These methods are called *burst-by-burst analysis* and are the subject of section 2.5.3.

2.5.1. Fluorescence Correlation Analysis

One of the most widely used data evaluation techniques when discussing ultrasensitive fluorescence spectroscopy is FCS. The technical basis of an FCS measurement is a confocal fluorescence detection scheme as discussed above, which has the main purpose to minimize the detection volume as much as possible. However, confocal detection can also be used for SMS without doing FCS. The actual essence of FCS is a special kind of data processing, where the measured fluorescence intensity signal $I(t)$ is correlated with a time delayed copy of itself, $I(t+\tau)$,

$$g(\tau) = \langle I(t) I(t+\tau) \rangle$$

Here, the angular brackets denote averaging over time t. The resulting autocorrelation function $g(\tau)$ contains information about translational and rotational diffusion of the fluorescent molecules, the average number of fluorescing molecules within the detection volume (i.e. chemical concentration), triplet state dynamics, possible photo-isomerization kinetics, and reaction rates between different fluorescing molecular species. It is important to understand that, although FCS lives on the small number fluctuations of molecules within the detection volume, any information about *individual* molecules is lost during the data analysis. The final result in FCS is the autocorrelation function $g(\tau)$, which is averaged over many single-molecule transits through the detection volume.

In contrast to single-channel FCS, where fluorescence is monitored at a single wavelength region only, multi-channel FCS employs two or more detection channels at different wavelength regions and subsequently applies a cross-correlation analysis. This allows for monitoring the interaction, e.g. binding dynamics, of different molecular species with different emission spectra (Bacia et al., 2002), or for looking at the dissociation kinetics of bound molecules (Kettling et al., 1998; Koltermann et al., 1998). The two challenges of multicolor detection FCS are, firstly, to achieve an exact overlap of the detection volumes at different detection wavelengths, which may be difficult due to chromatic aberrations, and secondly the necessity to use either several lasers for efficiently exciting molecules with different absorption wavelengths, or using a femtosecond pulsed laser for two-photon excitation of fluorescence, exploiting the broad two-photon absorption spectra of most molecules (Heinze et al., 2000; Rarbach et al., 2001; Heinze et al., 2002). An interesting alternative to two-color FCS is time-resolved FCS, where the different fluorescence lifetimes of different molecular species are used for implementing a multi-channel FCS (Böhmer et al., 2002).

FCS can also be used for immobile molecules when scanning the sample instead of waiting for the diffusion of molecules through the detection volume. Amediek et al. (2002) used scanning FCS together with two-channel detection and cross-correlation for co-localization studies. In principle, this is very similar to applying a single-molecule coincidence analysis (Winkler et al., 1999).

A critical aspect of FCS is that it heavily relies on the correctness of the assumed models used for data evaluation, particularly the made assumptions about molecular diffusion and the geometry of the detection region. The final result of an FCS experiment is the autocorrelation function that is condensed out of a large file of recorded fluorescence intensities. When applying a model to that autocorrelation function, it is often difficult to distinguish whether something like anomalous diffusion of the observed molecules has caused a deviation of the autocorrelation from the ideal model, or whether some technical artifact was responsible for that deviation. For example, Balakrishnan (2000) showed the influence of spatial curvature effects on FCS for molecules diffusing within membranes. Chirico et al. (2002) modeled the small bias on molecule diffusion introduced by weak optical trapping, an effect that was earlier observed by Osborne et al. (1998). Egner et al. (1998) made a detailed study of the aberration effects introduced by even small refractive index mismatches, which may be a problem for correct FCS data evaluation when working in biological samples, although Ganic et al. (2000) could show that such aberrational effects may be reduced when using two-photon excitation. Fradin et al. (2003) showed that FCS measurements inside cells can lead to erroneous values of the diffusion coefficient if the influence of membrane dynamics is not recognized. Gennerich and Schild (2000) demonstrated that FCS in small cytosolic compartments can lead to gross errors in diffusion coefficients, if confinement effects are not correctly taken into account.

Although FCS works only if single or only few molecules are present within the detection volume at the same time, it is essentially a statistical method that relies on averaging over many single molecule transits through the detection volume. It is thus an intermediate between a bulk measurement and 'true' SM measurements, where indeed only individual molecules are observed and studied. The number of papers dealing with FCS is at least as large as the rest of all fluorescence SMS publications, and there are several comprehensive reviews (Edman, 2000; Schwille, 2001; Webb, 2001; Hess et al., 2002) and a book publication (Rigler and Elson, 2001) concerned with that method. We will thus mostly exclude FCS from further discussion in this review.

2.5.2. Fluorescence Fluctuation Analysis

FCS is mostly used to obtain information about the average number of fluorescent molecules within the detection volume and the diffusion coefficient of these molecules. A recent and related method for evaluating SMS data is the so called fluorescence intensity distribution (FIDA) (Kask et al., 1999; Kask et al., 2000) or photon counting histogram (PCH) (Chen et al., 1999a; Chen et al., 1999b; Müller et al., 2000) analysis. In that method, one histograms the frequency of occurrence of the observed fluorescence photons per given time interval. It is obvious that this histogram depends on the average number of molecules within the detection volume, and on the apparent fluorescence brightness of the molecules (as defined by their photophysical properties, the excitation light intensity distribution, and the fluorescence-light collection efficiency distribution). The striking advantage of this method, when compared with FCS, is that it addresses molecular brightness instead of molecular diffusion, a parameter that changes much more from molecular species to molecular species than the diffusion coefficient. Thus, it is the method of choice when analyzing mixtures of several different molecular species, as well as for studying e.g. dimerization processes. Moreover, molecular brightness is a more robust molecular parameter than the diffusion coefficient, as was shown by Chen et al. (2002) when studying *in vivo* fluorescently loaded HeLa cells. A potentially strong application of the FIDA or PCH method is the elucidation of the stoichiometry of macromolecular complexes or the number of binding sites on a macromolecule (Chen et al., 2000). Kask et al. (1999) developed a powerful data evaluation method for FIDA and applied it to the simultaneous determination of concentrations and specific brightness values of several fluorescent species in solution. They demonstrated the capability of the method by studying the hybridization of 5'-(6-

carboxytetramethylrhodamine)-labeled (TAMRA-labeled) and unlabeled complementary oligonucleotides and the subsequent cleavage of the DNA hybrids by restriction enzymes. An extension of FIDA is two-dimensional FIDA (2D-FIDA) (Kask et al., 2000), using additional polarization resolved or two-color detection for applying FIDA on two detection channels. As an example of polarization studies by 2D-FIDA, binding of TAMRA-labeled theophylline to an anti-theophylline antibody was studied, and as an example of two-color 2D-FIDA, binding of TAMRA-labeled somatostatin-14 to the human type-2 high-affinity somatostatin receptors in vesicles was studied. Palo et al. (2000) extended FIDA for including the impact of diffusion on the fluorescence intensity distributions at varying sampling times (fluorescence intensity multiple distribution analysis – FIMDA). Palo et al. (2002) integrated fluorescence lifetime analysis into FIDA for achieving higher accuracy in the analysis of multi-analyte samples. Peleg et al. (2001) used a variant of FIDA for probing conformational changes of the β_2 adrenergic receptor which causes quenching of the bound fluorophore (rhodmaine 6G). They found at least two distinct substates for the native adrenergic membrane receptor.

2.5.3. Burst-by-Burst Analysis

In burst-by-burst analysis, the recorded fluorescence data are first processed to determine the transits of individual molecules through the detection volume, and subsequently the detected photons associated with each transit are individually anaylzed. Moreover, burst finding allows for exclusion of extraneous events from subsequent analysis, thus improving the effective signal-to-noise ratio. Special attention in SM burst-by-burst analysis has been paid to fluorescence lifetime evaluations. Lifetime measurements have the advantage to be technically, at least today, simple to implement. The method usually used for SM lifetime measurements is the so-called time-correlated single-photon counting technique, which times the delay between fluorescence photon detection events and the pulses of a pulsed excitation laser. The prerequisites are a pulsed laser with sufficiently high-repetition rate, and a time-resolved photon counting electronics together with a fast detector (usually a single-photon avalanche diode). A detailed description of such a detection system can be found e.g. in (Böhmer et al., 2001). In contrast to multi-color excitation and detection schemes, no additional optical excitation and/or detection channels are necessary for measuring the fluorescence lifetime. Thus, fluorescence lifetime is a convenient parameter for distinguishing and identifying single molecules. Several papers describe the successful implementation of this method, both from hardware (Becker et al., 1999; Böhmer et al., 2001) as well as data evaluation point of view. Fries et al. (1998) used fluorescence lifetime for selectively counting single molecules and thus determining chemical concentrations in solution. Herten et al. (2000) used fluorescence lifetime and spectral information for distinguishing single fluorescently labeled nucleotides. Knemeyer et al. (2003) used the same technique for discriminating signals from fluorescently labeled nucleotides against autofluorescence in live cells. Maus at al. (2001) made an experimental comparison between maximum-likelihood and non-linear least square fitting of SM fluorescence decay times, demonstrating that maximum-likelihood is superior to non-linear least square at low levels of photon detection. Enderlein and Sauer (2001) showed that a maximum-likelihood estimator yields the best possible algorithm for distinguishing molecules by their lifetime and experimentally exemplified this concept on a mixture of four different dyes. Dahan et al. (1999) studied the possibility to identify and to classify single molecules on the basis of their fluorescence polarization (fluorescence polarization anisotropy) as well as emission color (in a single pair FRET system).

The burst-by-burst analysis of SMS data was driven to its limits by the research of Seidel and co-workers by using up to four detection channels. Eggeling et al. (1998a, 2001) and Kühnemuth and Seidel (2001) describe the general strategy to identify and quantify sample molecules in dilute solution employing a specific burst analysis which they denote ***multi-parameter fluorescence detection*** (MFD). They used pulsed excitation and time-correlated single-photon counting to simultaneously monitor the evolution of four fluorescence parameters: intensity, lifetime, anisotropy, and spectral signature. Schaffer et al. (1999) applied these ideas for identifying single molecules based on their fluorescence lifetime *and* anisotropy.

3. APPLICATIONS

3.1. Photophysics

3.1.1. Basic Molecular Photophysics

A topic of continued and fundamental interest is the so called blinking of single molecules and fluorescent Nanoparticles: In many experiments on such entities it has been observed that the stream of emitted photons is interrupted by dark intervals. This phenomenon is a true single-particle effect because in an ensemble the various members emit independently and uncorrelated which generally leads to some constant average intensity of the fluorescence. Single emitters may pass through cycles of full, intermediate or no emission. While the phenomenon appears to be quite universal, the physical mechanisms causing the intensity fluctuations are very diverse and widespread. They include quantum jumps between states of different multiplicity, spectral shifts due to fluctuating environments, or trapping of charge carriers (Basché, 1998). Köhn et al. (2002b) studied the on/off kinetics in fluorescence intensity traces of single organic dye molecules and found that the occupation of the triplet state principally causes the off periods in some cyanine dyes structurally related to the frequently used dyes DiI or Cy3. Comparing recorded traces for various cyanine dyes bearing different atoms at the 1- and 1'-positions, they demonstrated the consequences of the internal heavy atom effect on the on/off dynamics of single organic dye molecules. Tinnefeld et al. (2001a) carried out an extensive study of SM photophysics on surfaces with a lifetime-imaging confocal scanning microscope. They investigated and compared the photophysical parameters of single oxazine (JA242), rhodamine (JF9), and carbocyanine (Cy5) derivatives adsorbed on glass surfaces under air-equilibrated conditions, monitoring fluorescence intensity, lifetime, and color. They observed discrete jumps in fluorescence intensity from single molecules which lacked spectral diffusion, and changes in radiative lifetime with correlation times (triplet lifetimes) spanning several orders of magnitude (from 2 s for the rhodamine derivative up to several seconds for the oxazine dye) and amplitude. They could not find a direct correlation between the radiative decay time and spectral fluctuations. Vargas et al. (2002) observed transitions of fluorescent Rhodamine 6G dye molecules into metastable dark states with lifetimes of several seconds. The samples were protected with different organic thin films and were characterized using atomic force microscopy. They found indications that oxygen migration or polarity changes are responsible for the transitions.

Weston and Buratto (1998) and Weston et al. (1998) studied the photophysics of single DiIC(12) molecules adsorbed on a silica surface. They observed a new type of rapid intensity fluctuations between an „on" level and a „dim" level distinct from on/off blinking. They attributed these fluctuations to small motions of the nuclear coordinates of the adsorbed molecule. In addition, molecules that showed these fast fluctuations also had emission spectra blue-shifted from the emission spectrum of the ensemble, indicating that this behavior is sensitive to the binding site of the molecule. Weston et al. (1999) extended these measurements by controlling air pressure. They found that triplet lifetimes were significantly longer and far more easily observed at moderate vacuum pressures than in air. Additionally, photobleaching rates decreased by over an order of magnitude in vacuum.

Blum et al. (2001) studied the photophysics of two different perylenediimide dyes with a spectrally resolving confocal detection setup. They observed several discrete, spectroscopically different states of the molecules, which were associated with different molecular conformations. Remarkably, they were able to identify vibronic structures within the SM fluorescence spectra, which is usually impossible within inhomogeneously broadened bulk spectra. The same technique was used to study the fluorescence properties of individual DsRed molecules, an important new fluorescent protein in the red spectral region (Blum et al. 2002). Hofkens et al. (2001b) observed conformational changes (twisting) in single dye molecules by taking fluorescence emission spectra of individual molecules. Within the low energy component of the distribution, two different vibronic shapes of the emission spectrum were seen, corresponding to two different twists of the dye molecules around their long axis.

Ying and Xie (1998) presented an extended SMS study of the light-harvesting allophycocyanin trimer (APC) from cyanobacteria. They performed measurements of fluorescence spectra, lifetimes, intensity trajectories, and polarization modulation. The intensity trajectories and polarization modulation experiments indicate reversible exciton trap formation within the three quasi-independent pairs of strong interacting α84 and β84 chromophores in APC, as well as photobleaching of individual chromophores. Comparison experiments under continuous-wave and pulsed excitation revealed a two-photon mechanism for generating exciton traps and/or photobleaching, which involves exciton-exciton annihilation. Zehetmayer et al. (2002) studied phycoerythrocyanin (PEC) as part of the light harvesting system of cyanobacteria. They used spectrally resolved detection, double resonance excitation, and polarization sensitive SMS. The studies identified the *trans* isomer of the phycoviolobilin chromophore as a short-lived dark state of monomeric PEC. The experiments revealed also that more than one-half of the PEC molecules exhibit an energy transfer behaviour significantly different from the bulk. These heterogeneities persisted on a time scale of several seconds and were attributed to minor shifts in the spectra of the chromophores.

Van Sark et al. (2000) performed a comparative study of the photophysics of Cy5 molecules and (CdSe)ZnS quantum dots by using both wide-field imaging as well as spectrally resolved confocal scanning microscopy. The difference in observed blinking behavior of single Cy5 molecules and (CdSe)ZnS quantum dots could be understood by their physical origin, i.e. the presence of a single dark state in case of Cy5 vs. a distribution of dark states in case of (CdSe)ZnS. This intrinsic difference resulted in higher probabilities to find a quantum dot in a short-lived (<0.02 s) or long-lived (>0.2 s) state than a Cy5 molecule.

Eggeling et al. (1997) studied the photophysics of the coumarine dye coumarin-120 and presented evidence for the importance of two-step photobleaching via excitation of the first excited singlet and the triplet state. These photobleaching studies were subsequently extended to several coumarine and rhodamine dyes (Eggeling et al. 1998b). In 1999, these authors presented an exhaustive review of the photophysical parameters of many dyes used for SM studies (Eggeling et al., 1999). Weber et al. (1999) used confocal microscopy in conjunction with polarization analysis, fluorescence spectroscopy, time-resolved detection, and excitation saturation for studying the photophysics of Rhodamine dyes at the glass-air interface without and with additional cover of a thin protective polymer film. They found that, under the polymer layer, the SM fluorescence was more stable than at the glass-air interface. Wennmalm and Rigler (1999b) investigated photobleaching of TMR molecules that were linked by an oligonucleotide to a glass surface. The found an exponential distribution of bleaching times. Deschenes and Vanden Bout (2002) studied the photobleaching behavior of Rhodamine 6G in polymethylmethacrylate (PMMA) under vacuum. A four-level system is used to model the photobleaching rate, and a higher triplet excited state is found to be the predominant reactive state for photobleaching. In contrast, Dittrich and Schwille (2001) found that the population of triplet states does not appear to be responsible for the limited emission rate with *two-photon* excitation. Rather, photobleaching pathways via the formation of radicals seem to be most significant. Viteri et al. (2003) investigated the connection between molecule mobility and photostability for sol-gel encapsulated DiI molecules. Both tumbling and fixed molecules exhibited nonuniform photostability, indicative of the very heterogeneous guest-host interactions within each subgroup. It was found that the fixed molecules exhibited a higher photostability than the tumbling molecules. Köhn et al. (2001) monitored contact ion pair formation of rhodamine dyes on glass surfaces by observing a blue shift in the SM emission spectra. Hübner et al. (2001) studied the influence of oxygen on the triplet state dynamics of individual DiI molecules using a special photon-counting CCD camera setup, enabling multichannel time-correlated single photon counting. They found that, in the presence of oxygen, the triplet state lifetime decreases from several tens of milliseconds down to fractions of a millisecond, whereas no changes of the intersystem crossing quantum yield or the fluorescence lifetime were observed.

3.1.2. Photophysical Characterization of Fluorescent Dyes and Proteins

An important topic in SMS is the exact photophysical characterization of fluorescent dyes and proteins that can be used as labels in biological or chemo-analytical SMS applications. Schmidt et al. (2002) presented a collection of photophysical data, in particular photostability, for many dyes used in SMS. Buschmann et al. (2003) presented a thorough photophysical study of 13 commercially available dyes in the red and near-infrared spectral region. They studied the influence of solvent polarity and viscosity as well as the addition of detergents on the fluorescence properties. Also, fluorescence quenching upon binding to biotin/streptavidin was investigated, which is important when using the dyes as labels in some biological applications.

The green fluorescent protein (GFP) of the bioluminescent jellyfish *Aequorea* and its mutants have gained widespread usage as an indicator of structure and function within cells. Moerner et al. (1999) and Peterman et al. (1999) studied phenolate ion mutants of GFP and observed complex blinking and fluctuating behavior, a phenomenon that is hidden in measurements on large ensembles. In these studies, Moerner et al. (1999) compared the SMS performance of internal reflection microscopy with scanning confocal microscopy, whereas Peterman et al. (1999) investigated the influence of several parameters such as pH on blinking. Zumbusch and Jung (2000) presented a comprehensive review of SM experiments with GFP as a fluorophore. The main emphasis of their paper was the reviewing of investigations of GFP's photophysics with the different available SMS techniques. Harms et al. (2001a) determined spectral and photophysical characteristics of several autofluorescent proteins and compared them to flavinoids to test their applicability for single-molecule microscopy in live cells. They found that the yellow-fluorescent protein mutant eYFP is superior compared to all other fluorescent proteins for single-molecule studies *in vivo*. Cognet et al. (2002) carried out a comparative study of autofluorescent proteins for SMS applications in biology, including several GFP mutants, eYFP, citrine, and DsRed. Cotlet et al. (2001a) used SMS for elucidating photo-conversion processes of DsRed. Lounis et al. (2001) studied the photostability of DsRed and found that DsRed may be slightly superior to GFP as a fluorescent label as long as the DsRed tetramerization is not an issue. Cotlet et al. (2001b) studied oligomerization and collective photophysics on the same protein and eGFP. A recent review of the results of photophysical SMS characterization of fluorescent proteins is given by Moerner (2002).

Oiwa et al. (2000) presented a technical study on the fluorescence properties of cyanine-conjugated nucleotides upon interaction with myosin, comparing SM results with ensemble studies. Wazawa et al. (2000) investigated the fluorescence spectra of a single fluorophore attached to a single protein molecule (myosin subfragment-1) in aqueous solution using TIRF microscopy. They attributed the observed spectral fluctuation to slow spontaneous conformational changes in the protein.

3.1.3. Fluorescence Resonant Energy Transfer

Fluorescence resonant energy transfer (FRET) is one of the most powerful techniques in SMS applications in biology (Weiss, 2000). This technique, capable of measuring distances on the 2- to 8-nm scale, relies on the distance-dependent energy transfer between a donor fluorophore and acceptor fluorophore. In biological applications, it is widely used for studying the colocalization of different molecules, or for investigating dynamical changes within the structure of one and the same molecule. The first measurement of energy transfer between a single donor and a single acceptor (single pair FRET or spFRET) was reported by Ha et al. (1996).

Deniz et al. (1999) used DNA as a rigid spacer for synthesizing a series of constructs with varying intramolecular donor-acceptor spacing. The constructs were used to measure the mean and distribution width of spFRET efficiencies as a function of distance. The group continued its work on spFRET on studying the fluorescence fluctuations observed with a FRET-pair attached to protein-staphylococcal nuclease (Ha et al., 1999a), being able to monitor conformational fluctuations on the millisecond timescale (Ha et al., 1999c). The first successful application of spFRET to monitoring the conformational change of a single molecule was presented in (Ha et al., 1999b), where conformational changes of individual three-helix junction RNA molecules induced by the binding of ribosomal protein S15 or Mg^{2+} ions were studied. An anomalously broad distribution of RNA conformations at intermediate ion concentrations was observed that was attributed to foldability differences among RNA molecules. A summary of fundamental aspects as well as several biological applications of spFRET is given by Deniz et al. (2001). In an extended study, Dietrich et al. (2002) used several different acceptor-donor-pairs for spFRET studies with DNA as the space molecule. They found that none of the dyes can be observed as a free rotor. Additionally, they observed strong fluorescence quenching of the donor that could not be attributed solely to FRET, but also to electron transfer via the DNA double-helix. Similarly, Ying et al. (2000) studied FRET on doubly labeled DNA molecules. Their fluorescence lifetimes and anisotropy measurements also suggested non-negligible fluorophore-DNA interaction. Lee et al. (2001) studied FRET between single TMR and azulene molecules. They found some deviation of the SM lifetime distributions from that predicted by classical Förster theory. In 1999, Ishii et al. (1999) showed that spFRET can principally be used to detect the structural changes in biomolecules. They observed strong changes in FRET efficiency from Cy3 to Cy5 attached to α–tropomyosin upon denaturatuation of that protein, demonstrating that it is possible to detect the assembly-disassembly of individual protein molecules as well as conformational changes occurring within a single protein molecule.

3.1.4. Dendrimers and Multichromophors

An extended study of the photophysics of dendrimeric multi-chromophoric systems was given by Yip et al. (1998). The analysis of the recorded fluorescence intensity fluctuations yielded information about „blinking" due to triplet bottlenecks, spectral diffusion due to environmental fluctuations, and interchromophoric energy transfer. Classification of the relevant photophysical processes was aided by SM wavelength-resolved emission spectroscopy and two-color excitation spectroscopy, as well as stochastic simulations. Hofkens et al. (1998, 2000) and Gensch et al. (1999) investigated the on/off-blinking dynamics of the fluorescence intensity of single dendrimeric molecules with multiple chromophores, using a scanning confocal microscope. Hofkens et al. (2001a) explained the observed collective on/off jumps in the fluorescence intensity traces of the dendrimers by assuming the simultaneous presence of both a radiative trap (energetically lowest chromophoric site) and a non-radiative trap (triplet state of one chromophore) within one individual dendrimer. It was shown that an analogues scheme could explain the collective on/off jumps in the fluorescence intensity traces of the photosynthetic pigment β-phycoerythrin. In (Gensch et al., 2001), polarization resolved detection together with polarization modulated excitation was used for elucidating energy hopping of the excited state energy within the multichromophoric system. Köhn et al. (2001) studied the photophysics of a host-guest system that consists of a second-generation polyphenylene dendrimer and the cyanine dye pinacyanol, and Köhn et al. (2002a) reported on temporarily fluctuating FRET efficiencies in a multichromophoric system, where all donor and acceptor molecules used were cyanine dyes. Vosch et al. (2001) focused their studies on the influence of structural and rotational isomerism on the triplet blinking of individual dendrimer molecules. They identified two distinct subsets of isomers of the studied bi-chromophoric system, which showed profoundly different photophysical properties and resulted in different probabilities of collective on/off jumps. They argued that an increased yield of intersystem crossing is the result of a lower singlet/triplet energy gap in the isomer containing two chromophores that interact in a dimer-like fashion. Tinnefeld et al. (2002) studied the photophysics of a dendrimer containing four chromophores by monitoring antibunching together with lifetime and polarization information. Antibunching (vanishing cross-correlation of the detected photons at zero delay time) occurs when only a single molecule is observed: the single emitter needs some time to become re-excited after having emitted a photon. Thus, the authors were able to directly prove that the multichromophoric system behaves like a single emitter. Tinnefeld et al. (2003)

studied collective blinking in a bichromophoric system (short peptide labeled with TMR and Cy5). They observed photoinduced reverse intersystem crossing ($T_1 \rightarrow T_n \rightarrow S_1 \rightarrow S_0$) in single Cy5 molecules. Moreover, even when the Cy5 fluorophore was in the triplet state, it continued to act as an energy transfer acceptor. This demonstrated that singlet-triplet energy transfer occurs, and can lead to efficient quenching of the fluorescence of the whole bichromophoric system. Lounis and Moerner (2000) used the antibunching property of SM fluorescence for devising a highly defined single photon source, a topic of considerable interest within the context of quantum cryptography. They realized a controllable source of single photons using optical pumping of a single molecule in a solid. Triggered single photons were produced at a high rate, whereas the probability of simultaneous emission of two photons was nearly zero.

Yu et al. (2000) investigated the photochemistry of a conjugated polymer and found a strong dependence of its photophysics on the presence of oxygen, which increased singlet exciton quenching dramatically. Spectroscopy on isolated single molecules in polycarbonate films that excluded oxygen revealed two distinct polymer conformations. Time-resolved single-molecule data demonstrated that one of these conformation exhibits a „landscape" for intramolecular electronic energy relaxation with a „funnel" that contains a singlet exciton trap at the bottom. The exciton traps can be converted to exciton quenchers by reaction with oxygen. They concluded that conformationally-induced directed energy transfer is a critical dynamical process responsible for many of the distinctive photophysical properties of conjugated polymers.

3.1.5. Light-Harvesting Complex

Light Harvesting II Complexes (LH2) are found peripheral to the Light Harvesting I Complexes (LH1) and Reaction Centers (RCs) in purple photosynthetic bacteria and funnel solar energy towards the LH1/RC core complex where the first chemical processes of photosynthesis take place. Bopp at al. (1999) studied LH2 on mica at room temperature with polarized light excitation. In their study, they observed slowly fluctuating elliptic deformations of the circular LH2 molecule. Tietz et al. (2001) performed spectroscopic and polarization measurements on single LH2 complexes of higher plants. They found that monomeric complexes emitted roughly linearly polarized fluorescence light thus indicating the existence of only one emitting state. This observation was explained by efficient triplet quenching restricted to one chlorophyll *a* (Chl *a*) molecule or by rather irreversible energy transfer within the pool of Chl *a* molecules. Gerken et al. (2002) studied photobleaching of single LH2b complexes that were immobilized at a surface with a histidine-tag, thus allowing the examination of the complexes in an aqueous environment. Most complexes showed photobleaching in one step, indicating coupling between the monomeric subunits of the LH2b leading to an energy transfer between adjacent subunits. Saga et al. (2002) used fluorescence emission spectroscopy for investigating supramolecular LH complexes (chlorosomes) from the green filamentous photosynthetic bacterium *Chloroflexus aurantiacus* on a quartz plate. All the fluorescence bands from bacteriochlorophyll-*c* aggregates of a single chlorosome showed a similar peak position and also a similar bandwidth.

3.2. Single molecules as fluorescent probes

An important issue is the site-specific and reliable labeling of biomolecules with synthetic fluorescent dyes or fluorescent proteins. Kapanidis and Weiss (2002) reviewed and discussed strategies for site-specific *in vitro* and *in vivo* labeling of biomolecules.

3.2.1. Single Molecules as Environmental Probes

Wang et al. (1998) probed the heterogeneity of silicate and polymer films with SM Rhodamine B fluorescence. They found that silicate materials are highly inhomogeneous in comparison to polymer films. Similarly, Ye et al. (1998) studied individual crystal violet molecules within poly(methyl methacrylate) (PMMA) and found a bi-modal distribution of fluorescence decay times which they attributed to a bi-modal site distribution within the polymer film. Talley and Dunn (1999) used SM fluorescence for probing the local environment of lipid membranes. They found that the observed large intensity fluctuations of SM emission were strongly coupled to the lipid environment surrounding the probe molecule. The monolayer and bilayer results were most consistent with a mechanism for intensity fluctuations driven by small twisting motions in the probe molecule that modified its emission properties. Trabesinger et al. (2000) used SM imaging for monitoring the transition of a phase-separated polymer blend into a molecular dispersion induced by solid-state tensile deformation. Statistical analysis of conjugated polymer cluster sizes as a function of the degree of matrix deformation showed that phase-separated domains of conjugated polymer transform into smaller clusters and single molecules as the degree of matrix deformation increases. Concomitantly, the conjugated guest molecules tended to adopt the preferential orientation of the surrounding matrix. Weston and Goldner (2001) used confocal detection with modulated excitation polarization for monitoring SM reorientations in thin polymer films. They found that molecules reorient with higher frequency and a broader distribution of jump rates in progressively thinner polymer films. Bartko et al. (2002) studied slow reorientations of individual molecules in glassy polymers by measuring SM orientations with fluorescence imaging. Brasselet and Moerner studied the pH-sensitivity of fluorescent pH-sensing molecules on a SM level. They observed, near the pK_a-value of the used molecules, a much broader heterogeneity of molecule fluorescence than could be expected by simple noise considerations. They attributed this heterogeneity as a sign of the heterogeneity of the host matrix (agarose gel). Hou et al. (2000) used single nile-red molecules for locally probing (polarity and rigidity) thin poly(vinyl alcohol) (PVA) and PMMA films, by recording fluorescence spectra of individual molecules. The spectral data were analyzed using the dependence of the Nile Red charge-transfer transition on the properties of the surrounding medium. Hou and Higgins (2002) used the same method for systematically looking at the influence of water content within the polymer films. Mei et al. (2003) studied single metal-free porphyrin cytochrome-*c* and Zn porphyrin cytochrome-*c* molecules within a trehalose matrix using a polarization resolving confocal microscope. They observed large angular motions of the molecules on time scales up to many seconds.

3.2.2. Diffusion in Solutions, Membranes, and Cells

A straightforward application of SM fluorescence imaging is following the diffusion of fluorescently tagged molecules within artificial or cellular lipid membranes. Harms et al. (1999) used anisotropy imaging for studying lateral and rotational diffusion of rhodamine-labeled lipids on supported phospholipid membranes. Schütz et al. (1998) exploited SM imaging to study colocalization of two different ligand molecules on an individual receptor in the cell membrane. The use of dual-wavelength imaging allowed discrimination between isolated and colocalized ligands with an accuracy of 40 nm. When the two ligands came very close together, spFRET was observed, proving co-localization on the molecular level. Schütz et al. (2000a, 2000b) employed binding of fluorescence labeled hongotoxin to the voltage gated potassium channel K_V 1.3 for visualizing the 3-dimensional positions of the channels within the plasma membrane of human T lymphocytes *in vivo*. Successive imaging of the same cell allowed the determination of 3D trajectories of individual ion channels, showing a very low mobility close to immobilization. Schütz et al. (2000c) used SM imaging of Cy5-labeled lipids for studying microdomains in the cell membrane of living human coronary artery smooth muscle cells. Their results indicated the existence of the hypothesized lipid rafts in the cell membranes. Harms et al. (2001b) used fusion proteins of L-type Ca^{2+}-channels with YFP for studying channel aggregation and motion in the cell membrane. Sonnleitner et al. (1999a) investigated the mobility of single phospholipids in free-standing and supported membranes. They identified different types of diffusional motion. Burden and Kasianowicz (2000) studied diffusion of fluorescently labeled lipids in unsupported lipid membranes. They observed biased diffusion of the lipids in dependence on excitation intensity and position with respect to the exciting laser focus due to weak optical trapping of the fluorophor's dipole by the focused excitation laser, an effect that can be important for the exact evaluation of SM tracking studies in membranes. Ke and Naumann (2001) presented a technique for imaging single molecules at the air-water interface. They used this technique for studying the diffusion of fluorescently labeled molecules within phospholipid monolayers. Vrljic et al. (2002) observed the translational motion of glycosyl phosphatidylinositol-linked and native I-E^k class II MHC membrane proteins in the plasma membrane of ovary cells, finding no strong evidence for significant confinement of either linked or native proteins in the plasma membrane of the cells.

Recently, a growing number of publications deal with SM observations within cells. Goulian and Simon (2000) imaged the diffusion of *R*-phycoerythrin within nucleoplasm and cytoplasm of mammalian cells. The observed diffusion dynamics of the fluorophores showed anomaleous diffusion behavior that cannot be described by a single diffusion constant. Byassee et al. (2000) observed fluorescence of externally tagged probes within cells. They obtained data on three types of fluorescent probes at spatially resolved locations (e.g., cytoplasm and nucleus) inside human HeLa cells. First, the iron transport protein transferrin labeled with TMR underwent rapid receptor-mediated endocytosis, and single transferrin molecules were detected inside living cells. Second, the cationic dye Rhodamine 6G entered cultured cells by a potential-driven process, and single molecules were observed as intense photon bursts when they moved in and out of the intracellular laser beam. Third, the authors reported results on synthetic oligonucleotides that were tagged with a fluorescent dye and were taken up by living cells via a passive, nonendocytotic pathway. Kubitschek et al. (2000) imaged and tracked the diffusion of single GFP molecules in solution in three dimensions, being able to follow the molecules up to 10 μm along the optical axis. This work was important for subsequent SM tracking studies within eukaryotic cells (Kues et al. 2001a; Kubitschek, 2002). The same group used the fluorescently labeled splicing factor uridine-rich small nuclear ribonucleoprotein (U snRNPs) to track its diffusion within the cell nucleus of 3T3 cells (Kues et al., 2001b; Kues and Kubitschek, 2002). Seisenberger et al. (2001) were able to directly visualize the infection pathway of single viruses in living cells, each labeled with only one fluorescent dye molecule. Diffusion trajectories with high spatial and time resolution showed the existence of various modes of motion of adeno-associated viruses (AAV) during their infection pathway into living HeLa cells. The real-time visualization of the infection pathway of single AAVs showed a much faster infection than was generally observed so far. Iino et al. (2001) imaged single GFPs linked to the cytoplasmic carboxyl terminus of E-cadherin in mouse fibroblast cells, using an objective-type TIRF microscope. They found indication of E-cadherin oligomerization prior to its assembly at cell-cell adhesion sites. The presence of many oligomers greater than dimers on the free surface suggested that these greater oligomers are the basic building blocks for the two-dimensional cell adhesion structures (adherens junctions). Watanabe and Mitchison (2002) analyzed spatial regulation of actin polymerization and depolymerization *in vivo* within lamellipodia (thin, veil-like extensions at the edge of cells). They tracked single molecules of actin fused to GFP. They observed basal polymerization and depolymerization throughout lamellipodia with largely constant kinetics, and that polymerization was promoted within one micron of the lamellipodium tip. Most of the actin filaments in the lamellipodium were generated by polymerization away from the tip. A similar method was applied by Zicha et al. (2003) to track one GFP molecule after localized photobleaching by using another GFP-variant as a colocalized reference. By visualizing the ratio of bleached to total molecules, the authors found that actin was delivered to protruding zones of the leading edge of the cell at speeds that exceeded 5 μm per second.

The ability to detect rare events on a large background made SMS an interesting tool for adsorption/desorption studies at surfaces. Grama et al. (2001) used SM imaging for studying the diffusion-limited adsorption/desorption kinetics of titin molecule at a glass surface. Similarly, Wirth and Swinton (1998, 2001), Wirth et al. (2003), and Ludes and Wirth (2002) studied the adsorption/desorption kinetics of DiI at interfaces as used in chromatography. The capability of SMS to detect rare events allowed them to detect rare adsorption modes where DiI molecules adsorbed to specific sites on the surface with unusual long residence times. Schuster et al. (2000) imaged single molecules within ultrathin films of ethylene glycol. They showed that the diffusion of dye molecules becomes slower near the interface as compared to the bulk value, which was attributed to anomalous diffusion due to attachment and detachment of molecules at the surface. In (Schuster et al., 2002), they demonstrated how to employ spot size of imaged molecules for determining SM diffusivity near a surface.

Seebacher et al. (2002) used confocal microscopy for studying the translational motion of single terrylenediimide molecules in a mesostructured molecular sieve. Only isotropic motion was found indicating that the channel-like motion lies below the optical resolution limit of the experiment. Hashimoto et al. (2003) applied TIRF microscopy together with single-photon counting to study lateral diffusion of single DiI molecules adsorbed at a dodecane-water interface in the absence and presence of surfactants. Hu and Lu (2003) employed SM nanosecond anisotropy to study the tethered protein motion of a T4 lysozyme molecule on a biologically compatible surface under water. They observed dynamic inhomogeneities of the rotational diffusion dynamics, i.e., diffusion rate fluctuation, because of interactions between the proteins and the surface, although the long time average of the rotational diffusion occurred to be the same among different molecules. Moreover, they found that tethered proteins stay predominately in solution, suggesting that the use of tethered proteins on modified glass surfaces under water is a reasonable way to study protein dynamics in solution.

An interesting application of confocal SM detection for mapping microfluidic flow profiles from slow laminar flow to fast near-turbulent flow was described by Shelby and Chiu (2003). They used a photo-activated fluorophore, nanosecond-duration photolysis pulses from a nitrogen laser, and high-sensitivity single-molecule detection with Ar^+ laser excitation. SM detection was necessary both because of the short time delay (submicrosecond) between laser photolysis and fluorescence detection and the fast transit times (as low as 10 ns) of the fluorescent molecules across the diffraction-limited beam waist of the Ar^+ laser focus. The technique is capable of high-resolution three-dimensional mapping and analysis of a wide range of velocity profiles in confined spaces that measure a few micrometers in dimension. Zheng and Yeung (2002) studied the motion of DNA molecules in capillary electrophoresis (CE) with applied Poiseuille flow. They found unexpected radial migration of the molecules: When Poiseuille flow was applied from the cathode to the anode, DNA molecules moved toward the center of the capillary, forming a narrow, highly concentrated zone. Conversely, when the flow was applied from the anode to the cathode, DNA molecules moved toward the walls, leaving a DNA-depleted zone around the axis. They discuss potential applications and consequences of the effect for flow-cytometer based SMD and capillary electrophoresis.

3.2.3. *Linear and Rotational Molecular Motors*

ATP synthase of mitochondria, chloroplasts, and bacteria catalyzes ATP synthesis coupled with a transmembrane proton flow. The enzyme consists of a membrane-embedded, proton-conducting portion (F_0) and a protruding portion (F_1) in which catalytic sites for ATP synthesis/hydrolysis exist. The isolated F_1 portion has ATPase activity; hence, it is often called F_1-ATPase. It is composed of five different subunits with a stoichiometry of $\alpha_3\beta_3\gamma\delta\varepsilon$. The $\alpha_3\beta_3\gamma$ subcomplex is the minimum ATPase-active complex, which has catalytic features similar to F_1-ATPase. In the crystal structure, the central γ subunit is surrounded by an $\alpha_3\beta_3$ cylinder where three α and three β subunits are arranged alternately, and the six nucleotide binding sites are located at the α/β subunit interfaces. The first direct observation of the rotation of a bacterially derived F_1-ATPase molecule was achieved by attaching a fluorescently labeled actin filament to the moving rotor (γ subunit) of the complex, and imaging the rotation of the attached actin filament by video microscopy (Noji et al., 1997; Yasuda et al., 1998). With this technique, even substeps of 90 ° were observed within the 120 ° rotations that characterize the catalytic steps of the enzyme (Yasuda et al., 2001; Hirono-Hara et al., 2001). The same technique was used by Hisabori et al. (1999) to study the rotation of the γ subunit of chloroplast F_1-ATPase, coming to the conclusion that that the rotation of the γ subunit in the F_1-motor is a ubiquitous phenomenon in all F_1-ATPases, in prokaryotes as well as in eukaryotes. Kaim et al. (2002) reported on the first observation of intersubunit rotation in fully coupled single F_0F_1 molecules (Na^+-translocating ATP synthase of *Propionigenium modestum*) during ATP synthesis or hydrolysis. They used polarization resolved confocal microscopy to monitor the rotation of the fluorescently labeled γ subunit after reconstituting the enzyme into proteoliposomes and subsequently applying a diffusion potential. During ATP hydrolysis, stepwise rotation of the labeled γ subunit was found in the presence of 2 mM NaCl, but was absent without the addition of Na^+ ions. Moreover, upon the incubation with the F_0-specific inhibitor dicyclohexylcarbodiimide the rotation was severely inhibited. Masaike et al. (2000) gave a review on SMS of F_1-ATPase and the found results. A striking result found in these studies was the determined efficiency of energy conversion, reaching nearly 100%. In a clever variation of the actin filament method, Pänke et al. (2001) and Cherepanov and Junge (2001) used the measured curvature of the actin filament for deducing torque values during turnings of the ATPase rotor. Börsch et al. (1998) took retarded diffusion of single ATPase molecules upon binding of nucleotides as an indication of the changed ATPase conformation after binding.

Muscle contraction is powered by the interaction of the molecular motor myosin with actin. Funatsu et al. (1995) developed an SM assay by refining epifluorescence and TIRF microscopy. By using fluorescently labeled (Cy3) ATP molecules, individual ATP turnover reactions of single myosin molecules were monitored. In (Funatsu et al., 1997), this method was combined with an optical trap to measure elemental steps of single kinesin molecules. Similarly, Kitamura et al. (1999) combined mechanical manipulation of single myosin subfragment-1 molecules using a scanning probe with visualization of these subfragments by using fluorescent labels for monitoring the motion of myosin along actin filaments. They found that myosin moves along an actin filament with single mechanical steps of approximately 5.3 nanometers; groups of two to five rapid steps in succession often produce displacements of 11 to 30 nanometers during just one biochemical cycle of ATP hydrolysis. Warshaw et al. (1998) measured orientational states in the smooth muscle myosin light chain domain during the process of motion generation. Fluorescently labeled turkey gizzard smooth muscle myosin was prepared by removal of endogenous regulatory light chain and re-addition of the light chain labeled with a dye. The measurements provided direct evidence that the myosin light chain domain adopts at least two orientational states during the cyclic interaction of myosin with actin, a randomly disordered state, most likely associated with myosin whereas weakly bound to actin, and an ordered state in which the light chain domain adopts a finite angular orientation whereas strongly bound after the powerstroke. Miyamoto et al. (2000), using both *in vitro* motility and SM motility assays, studied the influence of local anesthetics on kinesin. In the SM assay it was found that the local anesthetic tetracaine inhibited the motility of individual kinesin molecules in a dose-dependent manner. The concentrations of the anesthetics that inhibited the motility of kinesin correlated well with those blocking fast anterograde axoplasmic transport (FAAT), which is responsible for transport of membranous organelles, vesicles, or protein complexes along microtubules. They concluded that the charged form of local anesthetics directly and reversibly inhibits kinesin motility in a dose-dependent manner, and it is the major cause of the inhibition of FAAT by local anesthetics. Conibear et al. (1998) determined the degree of mechanochemical coupling in actomyosin by simultaneously measuring, with imaging TIRF microscopy, the sliding velocity and the associated ATP turnover at the single filament level by using the two labels Cy3 and Cy5. Okada and Hirokawa (1999) were able to show that even a single-headed myosin molecule is able to move along a microtubule for more than 1 µm without detaching, in contradiction to the so-called „walking model" of myosin motion. Rock et al. (2000) describe a combination of gliding filament and optical trap assays to observe the action of myosin motors. In the gliding filament assay, myosin molecules are randomly adsorbed to a microscope coverslip. Once the coverslip is blocked with bovine serum albumin (BSA) to prevent further, nonspecific protein adsorption, actin filaments, typically labeled with TMRphalloidin, are bound to the myosin molecules, and motility buffer containing ATP is introduced, leading to an observable motion of the actin filaments. Rock et al. (2001) use this technique for studying the working of myosin VI. They found a remarkably large step size and highly irregular stepping for this motor.

Forkey et al. (2001) gave a detailed description of SM fluorescence polarization and its applications to the study of protein structural dynamics. In particular, they used this method to study the three-dimensional structural dynamics of myosin (Corrie et al., 1999; Forkey et al., 2003). Corrie et al. (1999) used a bifunctional rhodamine dye for labeling myosin light-chain domains in such a way that the rhodamine has a fixed orientation with respect to the protein. This allowed them to monitor, via fluorescence polarization measurements, the relative orientation of the light-chain domain with respect to actin filaments in muscle cells. Forkey et al. (2003) observed back and forth tilting of fluorescent calmodulin light chains as the myosin molecule processively translocated along actin filaments. The results provided evidence for lever arm rotation of the calmodulin-binding domain in myosin V, and supported a 'hand-over-hand' mechanism for the translocation of double-headed myosin V molecules along actin filaments. To a similar conclusion came Yildiz et al. (2003), who observed also the motion of fluorescently labeled myosin V, finding a step size of myosin motion that alternated between $37 + 2x$ nm and $37 - 2x$, where x was the distance along the direction of motion between the dye and the midpoint between the two heads. Yanagida et al. (2000a, 2000b) presented two comprehensive reviews of SMS studies on the actomyosin motor system.

Kinesin is a dimeric motor protein that can move along a microtubule for several microns without releasing (termed processive movement). The two motor domains of the dimer are thought to move in a coordinated, hand-over-hand manner. A region adjacent to kinesin's motor catalytic domain (the neck) contains a coiled coil that is sufficient for motor dimerization and has been proposed to play an essential role in processive movement. Recent models have suggested that the neck enables head-to-head communication by creating a stiff connection between the two motor domains, but also may unwind during the mechanochemical cycle to allow movement to new tubulin binding sites. Romberg et al. (1998) investigated the role of that neck region by mutating the neck coiled coil and imaging the motion of the GFP fused mutant kinesin. Their observations contradict models in which extensive unwinding of the coiled coil is essential for movement. Moreover, deletion of the neck coiled coil decreased processivity 10-fold, but did not abolish it. Pierce and Vale (1998) studied the processivity of the non-conventional monomeric *Caenorhabditis elegans* kinesin unc104 and the sea urchin heteromeric kinesin KRP85/95. Both motors were fused to green fluorescent protein, and the fusion proteins were tested for processive ability using a SM fluorescence imaging. Neither unc104-GFP nor KRP85/95-GFP exhibited processive movement, although both motors were functional in multiple motor microtubule gliding assays. The results suggested that these motors have low duty cycles and that high processivity may not be required for efficient vesicle transport. Friedman and Vale (1999) used a SM assay (employing rhodamine labeled microtubules) that measures the motility of kinesin unattached to a surface. They showed that full-length kinesin binds microtubules and moves about ten times less frequently and exhibits discontinuous motion compared with a truncated kinesin lacking a tail. Mutation of either the stalk hinge or neck coiled-coil domain activated motility of full-length kinesin, indicating that these regions are important for tail-mediated repression. The results suggest that the motility of soluble kinesin in the cell is inhibited and that the motor becomes activated by cargo binding. Peterman et al. (2001) used monofunctional and bifunctional labels conjugated to the kinesin motor domain for determining the orientation of kinesin bound to microtubules in the presence of the nonhydrolyzable AMP analog AMP-PNP. They found close agreement with previous models derived from cryo-electron microscopy. By comparing the polarization anisotropy of monomeric and dimeric kinesin constructs bound to microtubules in the presence of AMP-PNP they deduced a mechanochemistry that requires a state in which both motor domains of a kinesin dimer bind simultaneously with similar orientation with respect to the microtubule. Shimizu et al. (2000) studied the motor activity of several mutant kinesin molecules for elucidating the role of different protein sections on correct microtubule binding, ATPase activity, and processive motion. The SM assay revealed defects in distinct processes in kinesin's mechano-chemical cycle. Dennis et al. (1999) used rhodamine conjugated tubulins for elucidating the working of kinesin in a synthetic environment (highly oriented poly(tetrafluoroethylene) films). It was found that the polymer films functionalized with the kinesin are able to direct the motion of microtubules in straight lines along the films' orientation axes. Inoue et al. (2001) investigated the motility of truncated one-headed kinesin molecules, using TIRF microscopy. They found that a single kinesin head has basal motility, but coordination between the two heads is necessary for stabilizing that basal motility to the normal level of kinesin processivity.

3.2.4. Enzymology

A fruitful field of SMS application is the study of the reactivity if single enzyme molecules. This is especially convenient when the substrate or product of the enzyme is itself fluorescent, or if the fluorescence of the enzyme itself changes during action. Lu et al. (1998) observed the turnovers of single cholesterol oxidase molecules by monitoring the emission from the enzyme's fluorescent active site, flavin adenine dinucleotide (FAD). Statistical analyses of single-molecule trajectories revealed a significant and slow fluctuation in the rate of cholesterol oxidation by FAD. The static disorder and dynamic disorder of reaction rates, which are essentially indistinguishable in ensemble-averaged experiments, were determined separately. A molecular memory phenomenon, in which an enzymatic turnover was not independent of its previous turnovers because of a slow fluctuation of protein conformation, was evidenced by spontaneous spectral fluctuation of FAD. Sytnik et al. (1998) studied the functional activity of single ribosomal complexes by observing a growing peptide labeled at its *N*-terminus with TMR. SM detection revealed dynamics on the scale of seconds at the ribosomal peptidyl transferase center. Lyon et al. (1998) developed an integrated epifocal and evanescent-wave optical microscope for real-time observations and bond cleavage studies of single DNA molecules. Large genomic DNA was stretched in a laminar flow stream and immobilized on a polylysine-coated glass surface by strong electrostatic interactions. Taking advantage of the elastic nature of double-stranded DNA, the authors measured DNA relaxation events that were triggered by two phosphodiester bond breaks, thus watching the enzymatic cleavage of DNA. Edman et al. (1999) studied the enzymatic activity of horseradish peroxidase, using the non-fluorescent substrate dihydrorhodamine 6G which after oxidation yields the highly fluorescing Rhodamine 6G. The experiments showed that thermodynamic fluctuation phenomena on a wide range of time scales affect enzyme activity. Bagshow and Conibear (2000) used Cy3-labeled nucleotides for studying the myosin enzymatic activity. They present a critical evaluation of possible artifacts when using such an assay for SM enzymatic studies. Zhuang et al. (2000a) studied the catalysis by and folding of individual *Tetrahymena thermophila* ribozyme molecules. The dye-labeled and surface-immobilized ribozymes used were shown to be functionally indistinguishable from the unmodified free ribozyme in solution. The enzymatic reaction was monitored by watching the disappearance of SM fluorescence after cleavage of the dye-labeled oligonucleotide substrate. Using spFRET allowed the detection of a reversible local folding step in which a duplex docks and undocks from the ribozyme core. From SM fluorescence intensity traces, rate constants were determined and the transition state characterized. A rarely populated docked state, not measurable by ensemble methods, was observed. In the overall folding process, intermediate folding states and multiple folding pathways were observed. In addition to observing previously established folding pathways, a pathway with a folding rate constant of 1 per second was discovered. Rothwell et al. (2003) used multi-parameter SMS to study the heterogeneity of HIV-1 reverse transcriptase. They found three structurally distinct forms of reverse transcriptase/nucleic acid complexes in solution, two enzymatically productive and one non-productive.

A completely different approach towards SM enzyme activity was taken by Craig and Dovichi (1998). In their work, they used a combination of capillary electrophoresis and ultrasensitive fluorescence detection for detecting the fluorescing reaction products of a single enzyme, thus indirectly measuring its activity. They found significant heterogeneity in enzyme activity (β-galactosidase), which they attributed to several distinct conformational states of the protein. With a similar method, Polakowski et al. (2000) could show that highly purified molecules of bacterial alkaline phosphatase generate identical activity; structurally identical molecules behave identically. In contrast, the glycosylated mammalian enzyme demonstrated a complex isoelectric focusing pattern and had a dramatic molecule-to-molecule variation in activity and activation energy. Glycosylation affected both the kinetics and energetics of this enzymatically catalyzed reaction. Shoemaker et al. (2003) used the same technique for studying the activity of β-galactosidase after crystallization. They found a significantly different activity distribution for molecules obtained from crystals than that of the enzyme used to grow the crystals.

Recent reviews on SM enzymology are given by Xie (2001, 2002) and by Xie and Lu (1999).

3.2.5. Molecular Binding and Interaction

Molecular binding and interaction studies are a broad field for applications of SMS. Harada et al. (1999) describe SM binding studies between Cy3-labeled RNA polymerase and a DNA double strand that was suspended in solution between two optically trapped beads. They found different association/dissociation rates of the polymerase for promoter associated sequences and other sequences, an AT/GC content dependency of the binding frequency, and also an influence of DNA strain. Brasselet et al. (2000) used autocorrelation, cross-correlation and polarization information for studying the binding kinetics of Ca^{2+} ions with the Ca^{2+}-binding protein calmodulin. Calmodulin was fused with two different fluorescent proteins, which were forming a FRET pair and being thus sensitive reporters of the calmodulin's conformation. Kaseda et al. (2000) used total internal reflection fluorescence microscopy for directly observing the interaction between dextran and TMR-labeled glucosyltransferase I (GTF) of *Streptococcus sobrinus*. They found a first order kinetics for the binding and indication that sucrose accelerated the dissociation of GTF from dextran. Sako et al. (2000a, 2000b) studied dimerization of the epidermal growth factor receptor (EGFR) in cell membranes using fluorescently labeled EGF. SM tracking revealed that the predominant mechanism of dimerization involves the formation of a cell-surface complex of one EGF molecule and an EGFR dimer, followed by the direct arrest of a second EGF molecule, indicating that the EGFR dimers were probably preformed before the binding of the second EGF molecule. EGF-EGFR dimerization was also checked by SM FRET measurements. Use of a monoclonal antibody specific to the phosphorylated (activated) EGFR revealed that the EGFR becomes phosphorylated after dimerization. Taguchi et al. (2001) analyzed the dynamics of the chaperonin (GroEL)-cochaperonin (GroES) interaction. In the presence of ATP and non-native protein, binding of GroES to the immobilized GroEL occurred at a rate that is consistent with bulk kinetics measurements. However, the release of GroES from GroEL occurred after a lag period (3 s) that was not recognized in earlier bulk-phase studies. This observation suggests a new kinetic intermediate in the GroEL-GroES reaction pathway. Ueda et al. (2001) described their application of SM imaging to follow binding of individual cyclic AMP (cAMP) molecules to heterotrimeric guanine nucleotide-binding protein coupled receptors on the surface of living *Dictyostelium discoideum* cells. They found uniform distribution and rapid diffusion of binding sites on the cell. The probabilities of individual association and dissociation events were greater for receptors at the anterior end of the cell. Agonist-induced receptor phosphorylation had little effect on any of the monitored properties, whereas G protein coupling influenced the binding kinetics. Daniel et al. (2002) used TMR conjugated to a synthetic peptide containing the sequence-specific DNA binding domain of Tc3 transposase for studying conformational changes of that peptide as well as its binding kinetics to DNA. They could use the fact that the TMR fluorescence brightness increased several fold upon peptide binding on DNA.

3.2.6. Conformational Dynamics of Polymers and Biomolecules

One of the most exciting and possibly most promising applications of SMS is the study of conformational dynamics of individual biomolecules. In principle, the motion of rotational and linear molecular motors or the action of enzymes that were discussed in the previous sections are also connected with conformational changes of these molecules. Here, mostly conformational studies on elongated polymer molecules will be considered. The core technique for these investigations is to label the polymer with a large number of either intercalating or covalently bound dye molecules, and to watch the conformational dynamics via conventional videomicroscopy. This, of course, presupposes that the polymer molecule under study is, at least in its elongated form, larger than the resolution of the microscope. Smith and Chu (1998) studied individual fluorescently stained DNA molecules at thermal equilibrium when exposed to an elongational flow producing a high strain rate. The flow was turned on suddenly so that the entire evolution of molecular conformation could be observed without initial perturbations. It occurred that the rate of stretching of individual molecules was highly variable and depended on the molecular conformation that developed during stretching. This variability was due to a dependence of the dynamics on the initial, random equilibrium conformation of the polymer coil. Smith et al. (1999) continued this work on the elongation of DNA in fluid flow by determining the probability distribution for the molecular extension as a function of shear rate for two different polymer relaxation times. In contrast to the behavior in pure elongational flow, the average polymer extension in shear flow did not display a sharp coil-stretch transition. Large, aperiodic temporal fluctuations were observed, consistent with end-over-end tumbling of the molecule. Maier and Rädler (1999) used fluorescence videomicroscopy for investigating the conformation of DNA molecules electrostatically bound to fluid cationic lipid bilayers, and Maier et al. (2002) looked at the elastic response of surface tethered DNA to an external electric field. Both studies found good agreement between observed conformational behavior and theoretical modeling. Melnikov et al. (1999) studied the compaction of single large double strand DNA chains in aqueous solution in the presence of primary alcohols, acetone, and ethylene glycol. It was found that in the presence of all studied organic solvents single DNA molecules exhibited a discrete phase transition from an elongated coiled to a compacted globular conformation. Melnikov et al. (1998) studied the interaction of T4 DNA and liposomes, and Melnikov and Lindman (1999) studied the solubilization of such complexes in low-polar organic solvents. The obtained data implied that T4 DNA-lipid complexes in hydrophobic liquids have a highly compacted globular conformation and consist of a single T4 DNA macromolecule. Babcock et al. (2003) investigated the phase transition between coiled and elongated conformation of linear polymer molecules. Similarly, Crut et al. (2003) studied transverse fluctuations of long DNA strands that were fastened at both ends to a glass surface.

Kang et al. (2001) used TIRF microscopy to study individual DNA molecules undergoing adsorption/desorption at a glass interface. They found a similar adsorption/desorption dynamics as can be deduced from elution peaks found in capillary liquid chromatography and capillary electrophoresis. Hydrophobic interaction rather than electrostatic interaction was the major driving force for adsorption of individual DNA molecules. Osborne et al. (2001) studied the attachment of singly fluorescently labeled DNA to a silica surface by either a streptavidin-biotin or a covalent linkage. Hybridization experiments using a complementary strand of DNA labeled with a different fluorophore gave a low level of colocalized fluorescence, indicating a significant fraction of the surface attached

DNA was not available for hybridization, a result consistent with the idea that the surface attached DNA spends significant time in a collapsed state. In (Kang and Yeung, 2002), they used the same technique for studying the adsorption/desorption of R-phycoerythrin, finding again similarity between the observed adsorption/desorption dynamics and the elution peaks of the proteins in capillary electrophoresis and capillary liquid chromatography. Le Goff et al. (2002) used fluorescence videomicroscopy for studying the end-to-end fluctuations of semiflexible actin filaments. In order to specifically measure the position of the polymer's ends, they developed a novel noninvasive method that consists of annealing short end tags to the filaments. This allowed them to probe polymer fluctuations to a very high accuracy. They found excellent agreement between measured fluctuation dynamics and theoretical predictions.

Deniz et al. (2000) used spFRET for studying a small, single-domain protein, chymotrypsin inhibitor 2 (CI2). Conformationally assisted ligation methodology was used to synthesize the proteins and site-specifically label them with donor and acceptor dyes. Simultaneous coexistence of folded and unfolded subpopulations of the protein could be observed, with changing partition as a function of the added guanidinium chloride concentration. The same technique, spFRET, was used by this group for investigating the opening and closing events of surface-immobilized DNA hairpins (Grunwell et al., 2001). Ha et al. (2002) used spFRET with the donor/acceptor pair Cy3/Cy5 for investigating the action of *E. coli* Rep helicase, a protein that couples conformational changes induced by ATP binding and hydrolysis with unwinding of duplex nucleic acid. They found indication that a Rep monomer uses ATP hydrolysis to move toward the junction between single-stranded and double-stranded DNA but then displays conformational fluctuations that do not lead to DNA unwinding. DNA unwinding initiates only if a functional helicase is formed via additional protein binding. Jia et al. (1999) measured the folding and unfolding fluctuations of single bichromophorically conjugated GCN4-Pf molecules in a two-channel confocal microscope with which donor and acceptor fluorescence trajectories are measured simultaneously. The measurement results indicated that single molecule GCN4-Pf is in dynamic folding equilibrium with the position of the equilibrium being altered by the concentration of urea. Katiliene et al. (2003) studied the DNA looping by NgoMIV restriction endonuclease with spFRET using a linear double-stranded DNA molecule labeled with a fluorescence donor molecule, Cy3, and fluorescence acceptor molecule, Cy5, and by varying the concentration of NgoMIV endonuclease. Schuler et al. (2002) also employed spFRET for studying folding of a small cold-shock protein. Their measurements exposed equilibrium collapse of the unfolded polypeptide and allowed them to calculate limits on the polypeptide reconfiguration time. From these results, limits on the height of the free energy barrier to folding are obtained that are consistent with a simple statistical mechanical model, but not with the barriers derived from simulations using molecular dynamics. Unlike the activation energy, the free-energy barrier includes the activation entropy and thus has been elusive to experimental determination for any kinetic process in solution.

Ha (2001) has given a thorough overview of spFRET, discussing practical considerations for its implementation including experimental apparatus, fluorescent probe selection, surface immobilization, spFRET analysis schemes, and interpretation.

Edman et al. (1996, 1998) used fluorescence quenching of TMR by guanosine-rich stretches of DNA for studying conformational dynamics of DNA conjugated with a single dye molecule. Similarly, Kästner et al. (2003) used quenching of the

Alexa dye AF546 when bound to the Na^+-dependent citrate carrier (CitS) from *Klebsiella pneumoniae* to study the conformational changes of CitS upon citrate addition.

Jia et al. (1997) used lifetime measurements for elucidating conformational changes in specifically TMR-labeled $tRNA^{Phe}$ molecules. The measurement results indicated that the $tRNA^{Phe}$-probe adduct fluctuates between two states, one of which provides conditions that quench the probe fluorescence. Similarly, Wennmalm et al. (1999a) also used guanosine mediated quenching of TMR fluorescence for investigating the switching between two conformational states (quenching and non-quenching state) in a small TMR-labeled oligonucleotide that was attached to a streptavidin-coated glass surface. For different molecules, they found different rates for the conformational switching, which however remained unchanged for a given molecule during the measurement time. Zhuang et al. (2000b) used fluorescence quenching also as a tool to study protein folding. Multiply labeled titin molecules were used as a model system and shown to be able to fold to the native state. In the native folded state, the fluorescence from dye molecules was quenched due to the close proximity between the dye molecules. Unfolding of the titin did lead to a dramatic increase in the fluorescence intensity. Such a change made the folded and unfolded states of a single titin molecule clearly distinguishable and allowed the measurement of the folding dynamics of individual titin molecules in real time. Neuweiler et al (2003) used quenching of a rhodamine derivative by tryptophan for investigating fast conformation dynamics of small peptides on a sub-microsecond timescale. This quenching mechanism was used by Neuweiler et al. (2002) for designing a peptidic molecular beacon, which changes its fluorescence intensity upon binding to a p53-autoantibody. This allowed for a single-molecule sensitive detection of p53-autoantibodies in blood serum in early-stage cancer diagnosis.

Ladoux et al. (2000) combined force spectroscopy with fluorescence videomicroscopy to study chromatin assembly on individual DNA molecules. They found that DNA compaction is a sequential process with at least three steps, involving DNA wrapping as the final event.

Talaga et al. (2000) reported SM measurements on the folding and unfolding conformational equilibrium distributions and dynamics of a disulfide crosslinked version of a short stretch of a two-stranded coiled coil from the yeast transcription factor, GCN4. They used spFRET for monitoring the end-to-end-distance of the peptide. Their studies provided information concerning the distributions of conformational states in the folded, unfolded, and dynamically interconverting states.

Rhoades et al. (2003) performed SMS on individual fluorophore-labeled molecules of the protein adenylate kinase which was trapped within surface-tethered lipid vesicles, thereby allowing spatial restriction without inducing any spurious interactions with the environment. The conformational fluctuations of these protein molecules, prepared at the thermodynamic midtransition point, were studied by using fluorescence resonance energy transfer between two specifically attached labels. By analyzing the temporal changes and fluctuations in observed FRET efficiency, they found direct evidence for heterogeneous folding pathways of the protein.

3.3. Chemical analysis and screening

3.3.1. DNA Sequencing

Historically, the driving force behind the SMS efforts of Richard Keller's group in the early nineties was to develop a fast DNA sequencing technique, capable of sequencing uninterrupted long DNA-sequences – see the historical overview presented by Enderlein and Zander in (Zander, 2002). The core idea of these efforts was to label every nucleotide of a DNA molecule with a fluorescent marker, four different markers for the four different bases, and subsequently detecting the labeled nucleotides, one by one, after having them cleaved off by an exonuclease. Although in the beginning of these efforts, the technical challenges of handling and detecting single molecules seemed to be the biggest problem, it finally occurred that the biochemistry of fluorescently labeling every base of a DNA sequence, and of finding a suitable exonuclease for efficiently cleaving such a sequence, are the most serious obstacles towards a successful realization of SM sequencing. Nonetheless, there has been recent progress towards this goal, and in 2001, a special issue of the *Journal of Biotechnology*, vol. 86, issue 3, was devoted to this topic. We will not discuss the papers presented in that issue but refer the reader directly to it.

Machara et al. (1998) introduced the optical trap for handling and holding single DNA molecules in a fluid stream for subsequent exonucleatic cleavage and sequencing. Sauer et al. (1999, 2000) reported on detection and identification of single dye labeled mononucleotide molecules released from an optical fiber in a microcapillary. Brakmann and Nieckchen (2001) report on enzymatic synthesis of completely labeled DNA, which could be a cornerstone for the successful implementation of SM sequencing. A different approach to SM DNA sequencing is based not on the sequential cleavage of completely labeled DNA strands, but on the observation of sequential incorporation of labeled nucleotides upon DNA polymerization. Braslavsky et al. reported the observation of polymerase catalyzed incorporation of single fluorescently labeled nucleotides into DNA, allowing the determination of sequence fingerprints up to 5 base-pairs in length. An interesting step towards observation of single nucleotides without labeling is the work by Ishikawa et al. (2002a, 2002b) and Ye et al. (2000) where the authors observed single complexes composed of a fluorescent nucleotide analogue and the Klenow fragment of DNA polymerase I using violet excitation.

3.3.2. DNA Fagment Sizing and DNA Mapping

Fragment sizing of DNA double strand fragments is one of the most successful applications of SMD in bioanalysis. The core idea is to measure the fluorescence brightness of single DNA fragments that were fluorescently labeled with an intercalating dye. Assuming that the label density is growing linearly with fragment length, the measured brightness values will be linearly dependent on fragment length.

Schins et al. (1998) describe a simple epifluorescence based setup for DNA fragment sizing. They employed this system for observing the orientation of the intercalating dye TOTO in TOTO-DNA complexes (Schins et al., 1999). Huang et al. (1999) applied DNA fragment sizing in the SM flow cytometer setup for fingerprinting the bacterial genome of *Staphylococcus aureus*. Van Orden et al. (1999) demonstrated that it is even possible to perform DNA fragment sizing by employing two-photon excitation of the intercalating dyes. Van Orden et al. (2000)

used a CCD for parallel detection of many DNA fragments simultaneously, thus significantly increasing the throughput of the technique. Werner et al. (2000) performed a systematic study of the relation between the fluorescence intensity measured from individual stained DNA fragments and the lengths of the fragments. For optimized excitation geometries, they found that linearity of the relation holds for DNA samples that exhibit a wide range of conformations. Chou et al. (1999) realized DNA fragment sizing on a micro-fluidic device. Fouquet et al. (2002) report on fragment sizing in submicrometer-size closed fluidic channels. The small dimensions facilitate single molecule detection and minimize events of simultaneous passage of more than one molecule through the measurement volume. Filippova et al. (2003) used SM fragment sizing for investigating the frequency of double-strand breaks induced by various treatments such as ionizing radiation.

Related to fragment sizing is the technique of molecular combing, the stretching and aligning of single DNA molecules bound to a surface. Herrick and Bensimon (1999a) reviewed the application of molecular combing for the high-resolution mapping of the human genome, the detection and quantification of subtle genomic imbalances and the positional cloning of disease-related genes. In (Herrick and Bensimon, 1999b), they use molecular combing for analyzing DNA replication.

Jing et al. (1998) describe a highly automated system for mapping fragments from fluorescence microscope images of individual, endonuclease-digested DNA molecules. The optical mapping system is based on fluid flows developed within tiny, evaporating droplets to elongate and fix DNA molecules onto derivatized surfaces. Such evaporation-driven molecular fixation produces well elongated molecules accessible to restriction endonucleases, and notably, DNA polymerase I. As in fluid-flow fragment sizing, measurements of relative fluorescence intensity and apparent length of the fluorescently labeled fragments determine the sizes of restriction fragments.

3.3.3. Specific Site Detection in DNA

Another important application of SMD is the detection of specific nucleotide sequences within a given DNA target. Trabesinger et al. (1999) employed SM imaging for observing the hybridization of single DNA probes against surface bound DNA targets. Coincident determination of the positions of both the target and the probe oligonucleotides using dual-wavelength fluorescence labeling allowed for highly reliable discrimination of specifically bound probe molecules from those being physiosorbed. The figures of merit of the assay were characterized by the low probability for false positive (10^{-4}) events and the high speed for detection of up to hundreds of different DNA fragments per second. The probability for false negative events was limited by the biochemical binding probability of short oligonucleotides. Trabesinger et al. (2001) extended their work on SM DNA hybridization assays by presenting a thorough statistical analysis of colocalization measurements for distinguishing between true interaction of two molecules and their accidental colocalization.

Since several years, Castro et al. have pioneered specific site detection in DNA with SM coincidence measurements (Castro and Okinaka, 2000). The method is based on detecting the simultaneous binding of two different probes onto the target DNA and using coincidence detection for discriminating successful double binding against accidental binding of only one probe or any other background signals. This is an archetypical SMS application for analytical purposes – in a bulk measurement, it would be impossible to distinguish between a mixture of target molecules with only a single probe bound to each target molecule and a sample of target molecules where both probes are binding simultaneously to the target. In a similar paper, Li et al. (2003) used coincidence detection of dual-labeled DNA molecules in the presence of a 1000-fold excess of single-fluorophore-labeled DNA.

Knemeyer et al. (2000) applied SMD for counting single DNA molecules with fluorescently labeled nucleotides. They called the fluorescently labeled oligonucleotides smart probes because they report the presence of complementary target sequences by a strong increase in fluorescence intensity. The smart probes consisted of a fluorescent dye attached at the terminus of a hairpin oligonucleotide. The presented technique took advantage of the fact that the used oxazine dye JA242 was efficiently quenched by complementary guanosine residues. Upon specific hybridization to the target DNA, the smart probe underwent a conformational change that increased the fluorescence intensity about six fold.

Li et al. (2001) demonstrated the ability to monitor polymerase chain-reaction on a SM level, opening the possibility of highly selective and sensitive disease diagnosis at a very early stage by using selective primer design.

3.3.4. Screening

Until now, there are only few publications describing the application of SMD (besides FCS or FIDA) in screening. Shortreed et al. (2000) demonstrated a high-throughput imaging approach that allowed the determination of the individual electrophoretic mobilities of many molecules at a time. Electrophoretic mobility is an important parameter widely used for identifying DNA or proteins based on charge and hydrodynamic radius. Many protein and DNA assays relevant to disease diagnosis are based on determining this parameter. Hesse et al. (2002) reviewed state-of-the-art SM microscopy with respect to its applicability to ultrasensitive screening.

4. THEORETICAL ASCPECTS

Few papers are dealing with the theoretical aspects of SMS. A brief look on SMS experiments and their possible scientific contribution from a theorist's point of view is given by Bai et al. (1999). Schenter et al. (1999) dealt with the statistical analysis of single-molecule fluorescence time traces. They show how to derive from these time traces information about fluctuations in the rate of the activation step of enzymatic reactions. They discuss models of the dynamical disorder behavior and relate them to observables of SM experiments. Yang and Xie (2002a) presented the theoretical rationales for data analysis protocols that afford an efficient extraction of conformational dynamics on a broad range of time scales from SM fluorescence lifetime trajectories. They derive and discuss analytical expressions relating fluorescence lifetime fluctuation correlations to a Brownian diffusion model and to an anomalous diffusion model. They extended their analysis in (Yang and Xie, 2002b) by introducing and discussing a three-time correlation analysis, which is based on time series analyses, and the Kullback-Liebler distance, which is based on information theory principles. Jung et al. (2002) presented a theoretical analysis of SMS experiments, discussing among others the following aspects: photon counting statistics for time-dependent fluctuations in SM spectroscopy, fluorescence intensity fluctuations for non-ergodic systems, time-resolved SM fluorescence for conformational dynamics of single biomolecules, and SM reaction dynamics at room temperature. Gopich and Szabo (2003) theoretically studied the potential dynamical and structural information that can be extracted from spFRET measurements, taking into account the finite measurement time and thus finite number of detectable photons as well as the temporal dynamics of the FRET-pair distance.

ACKNOWLEDGMENTS

I thank Thomas Gensch, Ingo Gregor (IBI-1, FZ Jülich) and Jörg Fitter (IBI-2, FZ Jülich) for many helpful discussions and comments during preparation of the manuscript.

REFERENCES

Adachi K., Kinosita K., Ando T., 1999, Single-fluorophore imaging with an unmodified epifluorescence microscope and conventional video camera, *J. Microsc.* **195**:125.

Ambrose W. P., Goodwin P. M., Jett J. H., Van Orden A., Werner J. H., Keller R. A., 1999, Single molecule fluorescence spectroscopy at ambient temperature, *Chem. Rev.* **99**:2929.

Amediek A., Haustein E., Scherfeld D., Schwille P., 2002, Scanning dual-color cross-correlation analysis for dynamic co-localization studies of immobile molecules, *Single Molecules* **3**:201.

Arden-Jacob J., Frantzeskos J., Kemnitzer N. U., Zilles A., Drexhage K. H., 2001, New fluorescent markers for the red region, *Spectrochimica Acta A* **57**:2271.

Babcock H. P., Teixeira R. E., Hur J. S., Shaqfeh E. S. G., Chu S., 2003, Visualization of molecular fluctuations near the critical point of the coil-stretch transition in polymer elongation, *Macromolecules* **36**:4544.

Bacia K., Majoul I. V., Schwille P., 2002, Probing the endocytic pathway in live cells using dual-color fluorescence cross-correlation analysis, *Biophys. J.* **83**:1184.

Bagshaw C. R., Conibear P. B., 2000, Single molecule enzyme kinetics: Critical aspects exemplified by myosin ATPase activity, *Single Molecules* **1**:271.

Bai C., Wang C., Xie X. S., Wolynes P. G., 1999, Single molecule physics and chemistry, *Proc. Nat. Acad. Sci. USA* **96**:11075.

Balakrishnan J., 2000, Spatial curvature effects on molecular transport by diffusion, *Phys. Rev. E* **61**:4648.

Bartko A. P., Dickson R. M., 1999a, Imaging three-dimensional singe molecule orientations, *J. Phys. Chem. B* **103**:11237.

Bartko A. P., Dickson R. M., 1999b, Three-dimensional orientations of polymer-bound single molecules, *J. Phys. Chem. B* **103**:3053.

Bartko A. P., Xu K., Dickson R. M., 2002, Three-dimensional single molecule rotational diffusion in glassy state polymer films, *Phys. Rev. Lett.* **89**:026101.

Basché T., 1998, Fluorescence intensity fluctuations of single atoms, molecules and nanoparticles, *J. Luminescence* **76**:263.

Becker W., Hickl H., Zander C., Drexhage K. H., Sauer M., Siebert S., Wolfrum J., 1999, Time-resolved detection and identification of single analyte molecules in microcapillaries by time-correlated single-photon counting (TCSPC), *Rev. Sci. Instrum.* **70**:1835.

Bennink M. L., Schärer O. D., Kanaar R., Sakata-Sogawa K., Schins J. M., Kanger J. S., De Grooth B. G., Greve J., 1999, Single-molecule manipulation of double-stranded DNA using optical tweezers: Interaction studies of DNA with RecA and YOYO-1, *Cytometry* **36**:200.

Blum C., Stracke F., Becker S., Müllen K., Meixner A. J., 2001, Discrimination and interpretation of spectral phenomena by room-temperature single-molecule spectroscopy, *J. Phys. Chem. A* **105**:6983.

Blum C., Subramaniam V., Schleifenbaum F., Stracke F., Angres B., Terskikh A., Meixner A. J., 2002, Single molecule fluorescence spectroscopy of mutants of the discosoma red fluorescent protein DsRed, *Chem. Phys. Lett.* **362**:355.

Böhmer M., Enderlein J., 2003, Orientation imaging of single molecules by wide-field epi-fluorescence microscopy, *J. Opt. Soc. B* **20**:554.

Böhmer M., Pampaloni F., Wahl M., Rahn H. J., Erdmann R., Enderlein J., 2001, Advanced time-resolved confocal scanning device for ultrasensitive fluorescence detection, *Rev. Sci. Instrum.* **72**:4145.

Böhmer M., Wahl M., Rahn H. J., Erdmann R., Enderlein J., 2002, Time-resolved fluorescence correlation spectroscopy, *Chem. Phys. Lett.* **353**:439.

Bopp M. A., Sytnik A., Howard T. D., Cogdell R. J., Hochstrasser R. M., 1999, The dynamics of structural deformations of immobilized single light-harvesting complexes, *Proc. Nat. Acad. Sci. USA* **96**:11271.

Borisenko V., Lougheed T., Hesse J., Füreder-Kitzmüller E., Fertig N., Behrends J. C., Woolley G. A., Schütz G. J., 2003, Simultaneous optical and electrical recording of single gramicidin channels, *Biophys. J.* **84**:612.

Börsch M., Turina P:, Eggeling C., Fries J. R., Seidel C. A. M., Labahn A., Graber P., 1998, Conformational changes of the H^+-ATPase from *Escherichia coli* upon nucleotide binding detected by single molecule fluorescence, *FEBS Lett.* **437**:251.

Brakmann S., Nieckchen P., 2001, The large fragment of *Escherichia coli* DNA polymerase I can synthesize DNA exclusively from fluorescently labeled nucleotides, *ChemBioChem.* **2**:773.

Braslavsky I., Hebert B., Kartalov E., Quake S. R., 2003, Sequence information can be obtained from single DNA molecules, *Proc. Nat. Acad. Sci. USA* **100**:3960.

Brasselet S., Moerner W. E., 2000, Fluorescence behavior of single-molecule pH-sensors, *Single Molecules* **1**:17.

Brasselet S., Peterman E. J. G., Miyawaki A., Moerner W. E., 2000, Single-molecule fluorescence resonant energy transfer in calcium concentration dependent cameleon, *J. Phys. Chem. B* **104**:3676.

Burden D. L., Kasianowicz J. J., 2000, Diffusion bias and photophysical dynamics of single molecules in unsupported lipid bilayer membranes probed with confocal microscopy, *J. Phys. Chem. B* **104**:6103.

Buschmann V., Weston K. D., Sauer M., 2003, Spectroscopic study and evaluation of red-absorbing fluorescent dyes, *Bioconj. Chem.* **14**:195.

Byassee T. A., Chan W. C. W., Nie S., 2000, Probing single molecules in single living cells, *Anal. Chem.* **72**:5606.

Castro A., Okinaka R. T., 2000, Ultrasensitive direct detection of a specific DNA sequence of *Bacillus anthracis* in solution, *Analyst* **125**:9.

Chan W. C. W., Nie S., 1998, Quantum dot bioconjugates for ultrasensitive nonisotopic detection, *Science* **281**:2016.

Chen Y., Müller J. D., Berland K. M., Gratton E., 1999a, Fluorescence fluctuation spectroscopy, *Methods: A Companion to Methods in Enzymology* **19**:234.

Chen Y., Müller J. D., So P. T. C., Gratton E., 1999b, The photon counting histogram in fluorescence fluctuation spectroscopy, *Biophys. J.* **77**:553.

Chen Y., Müller J. D., Tetin S. Y., Tyner J. D., Gratton E., 2000, Probing ligand protein binding equilibria with fluorescence fluctuation spectroscopy, *Biophys. J.* **79**:1074.

Chen Y., Müller J. D., QiaoQiao Ruan, Gratton E., 2002, Molecular brightness characterization of EGFP in vivo by fluorescence fluctuation spectroscopy, *Biophys. J.* **82**:133.

Cherepanov D. A., Junge W., 2001, Viscoelastic dynamics of actin filaments coupled to rotary F-ATPase: curvature as an indicator of the torque, *Biophys. J.* **81**:1234.

Chirico G., Fumagalli C., Baldini G., 2002, Trapped Brownian motion in single- and two-photon excitation fluorescence correlation experiments, *J. Phys. Chem. B* **106**:2508.

Chou H. P., Spence C., Scherer A., Quake S., 1999, A microfabricated device for sizing and sorting DNA molecules, *Proc. Nat. Acad. Sci. USA* **96**:11.

Cognet L., Harms G. S., Blab G. A., Lommerse P. H. M., Schmidt T., 2000, Simultaneous dual-color and dual-polarization imaging of single molecules, *Appl. Phys. Lett.* **77**:4052.

Cognet L., Coussen F., Choquet D., Lounis B., 2002, Fluorescence microscopy of single autofluorescent proteins for cellular biology, *C. R. Acad. Sci.* **3 Ser. IV**:645.

Conibear P. B., Kuhlman P. A., Bagshaw C. R., 1998, Measurement of ATPase activities of myosin at the level of tracks and single molecules, *Adv. Exp. Medicine. Biol.* **453**:15.

Corrie J. E. T., Brandmeier B. D., Ferguson R. E., Trentham D. R., Kendrick-Jones J., Hopkins S. C., Van Der Heide U. A., Goldman Y. E., Sabido-David C., Dale R. E., Criddle S., Irving M., 1999, Dynamic measurement of myosin light-chain-domain tilt and twist in muscle contraction, *Nature* **400**:425.

Cotlet M., Hofkens J., Habuchi S., Dirix G., Van Guyse M., Michiels J., Vanderleyden J., De Schryver F. C., 2001a, Identification of different emitting species in the red fluorescent protein DsRed by means of ensemble and single-molecule spectroscopy, *Proc. Nat. Acad. Sci. USA* **98**:14398.

Cotlet M., Hofkens J., Kohn F., Michiels J., Dirix G., Van Guyse M., Vanderleyden J., De Schryver F. C., 2001b, Collective effects in individual oligomers of the red fluorescent coral protein DsRed, *Chem. Phys. Lett.* **336**:415.

Craig D. B., Dovichi N. J., 1998, *Escherichia Coli* β-galactosidase is heterogeneous with respect to the activity of individual molecules, *Can. J. Chem.* **76**:623.

Crut A., Lasne D., Allemand J. F., Dahan M., Desbiolles P., 2003, Transverse fluctuations of single DNA molecules attached at both extremities to a surface, *Phys. Rev. E* **67**:051910.

Dahan M., Deniz A. A., Ha T., Chemla D. S., 1999, Ratiometric measurement and identification of single diffusing molecules, *Chem. Phys.* **247**:85.

Dahan M., Laurence T., Pinaud F., Chemla D. S., Alivisatos A. P., Sauer M., Weiss S., 2001, Time-gated biological imaging by use of colloidal quantum dots, *Opt. Lett.* **26**:825.

Daniel D. C., Thompson M., Woodbury N. W., 2002, DNA-binding interactions and conformational fluctuations of Tc3 transposase DNA binding domain examined with single molecule fluorescence spectroscopy, *Biophys. J.* **82**:1654.

Demas J. N., Wu M., Goodwin P. M., Affleck R.L., Keller R. A., 1998, Fluorescence detection in hydrodynamically focused sample streams : Reduction of diffusional defocusing by association of analyte with high-molecular weight species, *Appl. Spectrosc.* **52**:755.

Deniz A. A., Dahan M., Grunwell J. R., Ha T., Faulhaber A. E., Chemla D. S., Weiss S., Schultz P. G., 1999, Single-pair fluorescence resonance energy transfer on freely diffusing molecules: Observation of Förster distance dependence and subpopulations, *Proc. Nat. Acad. Sci. USA* **96**:3670.

Deniz A. A., Laurence T. A., Beligere G. S., Dahan M., Martin A. B., Chemla D. S., Dawson P. E., Schultz P. G., Weiss S., 2000, Single-molecule protein folding: Diffusion fluorescence resonance energy transfer studies of the denaturation of chymotrypsin inhibitor 2, *Proc. Nat. Acad. Sci. USA* **97**:5179.

Deniz A. A., Laurence T. A., Dahan M., Chemla D. S., Schultz P. G., Weiss S., 2001, Ratiometric single-molecule studies of freely diffusing molecules, *Annu. Rev. Phys. Chem.* **52**:233.

Dennis J. R., Howard J., Vogel V., 1999, Molecular shuttles: directed motion of microtubules along nanoscale kinesin tracks, *Nanotechnol.* **10**:232.

Deschenes L. A., Vanden Bout D. A., 2002, Single molecule photobleaching: increasing photon yield and survival time through suppression of two-step photolysis, *Chem. Phys. Lett.* **365**:387.

Dickson R. M., Norris D. J., Moerner W. E., 1998, Simultaneous imaging of individual molecules aligned both parallel and perpendicular to the optic axis, *Phys. Rev. Lett.* **81**:5322.

Dietrich A., Buschmann V., Müller C., Sauer M., 2002, Fluorescence resonance energy transfer (FRET) and competing processes in donor-acceptor substituted DNA strands: a comparative study of ensemble and single-molecule data, *Rev. Mol. Biotechnol.* **82**:211.

Dittrich P. S., Schwille P., 2001, Photobleaching and stabilization of fluorophores used for single-molecule analysis with one- and two-photon excitation, *Appl. Phys. B* **73**:829.

Dörre K., Stephan J., Lapczyna M., Stuke M., Dunkel H., Eigen M., 2001, Highly efficient single molecule detection in microstructures, *J. Biotechnol.* **86**:225.

Dovichi N.J., Martin J.C., Jett J.H., Keller R.A. 1983, Attogram detection limit for aqueous dye samples by laser-induced fluorescence, *Science* **219**:845.

Edman L., Mets Ü., Rigler R., 1996, Conformal transitions monitored for single molecules in solution, *Proc. Nat. Acad. Sci. USA* **93**:6710.

Edman L., Wennmalm S., Tamsen F., Rigler R., 1998, Heterogeneity in single DNA conformational fluctuations *Chem. Phys. Lett.* **292:**15.

Edman L., Földes-Papp Z., Wennmalm S., Rigler R., 1999, The fluctuating enzyme: a single molecule approach, *Chem. Phys.* **247**:11.

Edman L., 2000, Theory of fluorescence correlation spectroscopy on single molecules, *J. Phys. Chem. A* **104**:6165.

Eggeling C., Brand L., Seidel C. A. M., 1997, Laser-induced fluorescence of coumarin derivatives in aqueous solution: photochemical aspects for single molecule detection, *Bioimaging* **5**:105.

Eggeling C., Fries J. R., Brand L., Günther R., Seidel C. A. M., 1998a, Monitoring conformational dynamics of a single molecule by selective fluorescence spectroscopy, *Proc. Nat. Acad. Sci. USA* **95**:1556.

Eggeling C., Widengren J., Rigler R., Seidel C. A. M., 1998b, Photobleaching of fluorescent dyes under conditions used for single-molecule detection: Evidence of two-step photolysis, *Anal. Chem.* **70**:2651.

Eggeling C., Widengren J., Rigler R., Seidel C. A. M., 1999, Photostability of fluorescent dyes for single-molecule spectroscopy: Mechanisms and experimental methods for estimating photobleaching in aqueous solution, in: *Applications of Fluorescence in Chemisty, Biology, and Medicine*, W. Rettig, ed., Springer, New York/Berlin, pp. 193-240.

Eggeling C., Berger S., Brand L., Fries J. R., Schaffer J., Volkmer A., Seidel C. A. M., 2001, Data registration and selective single-molecule analysis using multi-parameter fluorescence detection, *J. Biotechnol.* **86**:163.

Egner A., Schrader M., Hell S. W., 1998, Refractive index mismatch induced intensity and phase variations in fluorescence confocal, multiphoton and 4Pi-microscopy, *Opt. Commun.* **153**:211.

Enderlein J., 2000a, Tracking of fluorescent molecules diffusing within membranes, *Appl. Phys. B* **71**:773.

Enderlein J., 2000b, Positional and temporal accuracy of single molecule tracking, *Single Molecules* **1**:225.

Enderlein J., 2000c, A theoretical investigation of single molecule fluorescence detection on thin metallic layers, *Biophys. J.* **78**:2151.

Enderlein J., 2002a, Spectral properties of a fluorescing molecule within a spherical metallic nanocavity, *Phys. Chem. Chem. Phys.* **4**:2780.

Enderlein J., 2002b, Theoretical study of single molecule fluorescence in a metallic nanocavity, *Appl. Phys. Lett.* **80**:315.

Enderlein J., Sauer M., 2001, Optimal algorithm for single molecule identification with time-correlated single photon counting, *J. Phys. Chem. A* **105**:48.

Filippova E. M., Monteleone D. C., Trunk J. G., Sutherland B. M., Quake S. R., Sutherland J. C., 2003, Quantifying double-strand breaks and clustered damages in DNA by single-molecule laser fluorescence sizing, *Biophys. J.* **84**:1281.

Fister J. C., Jacobson S. C., Davis L. M., Ramsey J. M., 1998, Counting single chromophore molecules for ultrasensitive analysis and separations on microchip devices, *Anal. Chem.* **70**:431.
Fister J. C., Jacobson S. C., Ramsey J. M., 1999, Ultrasensitive cross-correlation electrophoresis on microchip devices, *Anal. Chem.* **71**:4460.
Foquet M., Korlach J., Zipfel W., Webb W. W., Craighead H. G., 2002, DNA fragment sizing by single molecule detection in submicrometer-sized closed fluidic channels, *Anal. Chem.* **74**:1415.
Forkey J. N., Quinlan M. E., Goldman Y. E., 2001, Protein structural dynamics by single-molecule fluorescence polarization, *Prog. Biophys. Mol. Biol.* **74**:1.
Forkey J. N., Quinlan M. E., Shaw M. A., Corrie J. E. T., Goldman Y. E., 2003, Three-dimensional structural dynamics of myosin V by single-molecule fluorescence polarization, *Nature* **422**:399.
Fradin C., Abu-Arish A., Granek R., Elbaum M., 2003, Fluorescence correlation spectroscopy close to a fluctuating membrane, *Biophys. J.* **84**:2005.
Friedman D. S., Vale R. D., 1999, Single-molecule analysis of kinesin motility reveals regulation by the cargo-binding tail domain, *Nature Cell. Biol.* **1**:293.
Fries J. R., Brand L., Eggeling C., Köllner M., Seidel C. A. M., 1998, Quantitative identification of different single molecules by selective time-resolved confocal fluorescence spectroscopy, *J. Phys. Chem. A* **102**:6601.
Funatsu T., Harada Y., Tokunaga M., Saito K., Yanagida T., Imaging of single fluorescent molecules and individual ATP turnovers by single myosin molecules in aqueous-solution, 1995, *Nature* **374:**555.
Funatsu T., Harada Y., Higuchi H., Tokunaga M., Saito K., Ishii Y., Vale R., Yanagida T., 1997, Imaging and nano-manipulation of single biomolecules, *Biophys. Chem.* **68:**63.
Ganic D., Gan X., Gu M., 2000, Reduced effects of spherical aberration on penetration depth under two-photon excitation, *Appl. Opt.* **39**:3945.
Gao X., Chan W. C. W., Nie S., 2002, Quantum-dot nanocrystals for ultrasensitive biological labeling and multicolor optical encoding, *J. Biomed. Opt.* **7**:532.
Gennerich A., Schild D., 2000, Fluorescence correlation spectroscopy in small cytosolic compartments depends critically on the diffusion model used, *Biophys. J.* **79**:3294.
Gensch T., Hofkens J., Heirmann A:, Tsuda K., Verheijen W., Vosch T., Christ T., Basché T., Müllen K., De Schryver F. C., 1999, Fluorescence detection from single dendrimers with multiple chromophores, *Angew. Chem. Int. Ed.* **38**:3752.
Gensch T., Hofkens J., Kohn J., Vosch T., Herrmann A., Müllen K., De Schryver F. C., 2001, Polarisation sensitive single molecule fluorescence detection with linear polarised excitation light and modulated polarisation direction applied to multichromophoric entities, *Single Molecules* **2**:35.
Gerken U., Wolf-Klein H., Huschenbett C., Gotze B., Schuler S., Jelezko F., Tietz C., Wrachtrup J., Paulsen H., 2002, Single molecule spectroscopy of oriented recombinant trimeric light harvesting complexes of higher plants, *Single Molecules* **3**:183.
Gopich I. V., Szabo A., 2003, Single-macromolecule fluorescence resonance energy transfer and free-energy profiles, *J. Phys. Chem. B* **107**:5058.
Goulian M., Simon S. M., 2000, Tracking single proteins within cells, *Biophys. J.* **79**:2188.
Grama L., Somogyi B., Kellermayer M. S. Z., 2001, Direct visualization of surface-adsorbed single fluorescently labeled titin molecules, *Single Molecules* **2**:79.
Grunwell J. R., Glass J. L., Lacoste T. D., Deniz A. A., Chemla D. S., Schultz P. G., 2001, Monitoring the conformational fluctuations of DNA hairpins using single-pair fluorescence resonance energy transfer, *J. Am. Chem. Soc.* **123**:4295.
Ha T., Enderle T., Ogletree D. F., Chemla D. S., Selvin P. R., Weiss S., 1996, Probing the interaction between two single molecules: Fluorescence resonance energy transfer between a single donor and a single acceptor, *Proc. Natl. Acad. Sci. USA* **93**:6264.
Ha T., Ting A. Y., Liang J., Deniz A. A., Chemla D. S., Schultz P. G., Weiss S., 1999a, Temporal fluctuations of fluorescence resonance energy transfer between two dyes conjugated to a single protein, *Chem. Phys.* **247**:107.
Ha T., Zhuang X:, Kim H. D., Orr J. W., Williamson J. R., Chu S., 1999b, Ligand-induced conformational changes observed in single RNA molecules, *Proc. Nat. Acad. Sci. USA* **96**:9077.
Ha T. J., Ting A. Y., Liang J., Caldwell W. B., Deniz A. A., Chemla D. S., Schultz P. G., Weiss S., 1999c, Single-molecule fluorescence spectroscopy of enzyme conformational dynamics and cleavage mechanism, *Proc. Nat. Acad. Sci. USA* **96**:893.
Ha T., 2001, Single-molecule fluorescence resonance energy transfer, *Methods* **25**:78.
Ha T., Rasnik I., Cheng W., Babcock H. P., Gauss G. H., Lohman T. M., Chu S., 2002, Initiation and re-initiation of DNA unwinding by the *Escherichia coli* Rep helicase, *Nature* **419**:638.
Han M., Gao X., Su J. Z., Nie S., 2001, Quantum-dot-tagged microbeads for multiplexed optical coding of biomolecules, *Nature Biotech.* **19**:631.
Harada Y., Funatsu T., Murakami K., Nonoyama Y., Ishihama A., Yanagida T., 1999, Single-molecule imaging of RNA polymerase-DNA interactions in real time, *Biophys. J.* **76**:709.
Harms G. S., Sonnleitner M., Schütz G. J., Gruber H. J., Schmidt T., 1999, Single-molecule anisotropy imaging, *Biophys. J.* **77:**2864.

Harms G., Cognet L., Lommerse P. H. M., Blab G. A., Schmidt T., 2001a, Autofluorescent proteins in single-molecule research: Applications to live cell imaging microscopy, *Biophys. J.* **80**:2396.

Harms G. S., Cognet L., Lommerse P. H. M., Blab G. A., Kahr H., Gamsjäger R., Spaink H. P., Soldatov N. M., Romanin C., Schmidt T., 2001b, Single-molecule imaging of L-type Ca^{2+} channels in live cells, *Biophys. J.* **81**:2639.

Hashimoto F., Tsukahara S., Watarai H., 2003, Lateral diffusion dynamics for single molecules of fluorescent cyanine dye at the free and surfactant-modified dodecane-water interface, *Langmuir* **19**:4197.

Heilemann M., Herten D. P., Heintzmann R., Cremer C., Müller C., Tinnefeld P., Weston K. D., Wolfrum J., Sauer M., 2002, High-resolution colocalization of single dye molecules by fluorescence lifetime imaging microscopy, *Anal. Chem.* **74**:3511.

Heinze K., Koltermann A., Schwille P., 2000, Simultaneous two-photon excitation of distinct labels for dual-color fluorescence crosscorrelation analysis, *Proc. Nat. Acad. Sci. USA* **97**:10377.

Heinze K. G., Rarbach M., Jahnz M., Schwille P., 2002, Two-photon fluorescence coincidence analysis: Rapid measurements of enzyme kinetics, *Biophys. J.* **83**:1671.

Herrick J., Bensimon A., 1999a, Imaging of single DNA molecule: Applications to high-resolution genomic studies, *Chromosome Res.* **7**:409.

Herrick J., Bensimon A., 1999b, Single molecule analysis of DNA replication, *Biochimie Paris* **81**:859.

Herten D. P., Tinnefeld P., Sauer M., 2000, Identification of single fluorescently labeled mononucleotide molecules in solution by spectrally resolved time-correlated single photon counting, *Appl. Phys. B* **71**:765.

Hess S. T., Huang S., Heikal A. A., Webb W. W., 2002, Biological and chemical applications of fluorescence correlation spectroscopy: A review, *Biochem.* **41**:697.

Hesse J., Wechselberger C., Sonnleitner M., Schindler H., Schütz G. J., 2002, Single-molecule reader for proteomics and genomics, *J. Chromatography B* **782**:127.

Hirono-Hara Y., Noji H., Nishiura M., Muneyuki E., Hara K. Y., Yasuda R., Kinosita K., Yoshida M., 2001, Pause and rotation of F_1-ATPase during catalysis, *Proc. Nat. Acad. Sci. USA* **20**:13649.

Hisabori T., Kondoh A., Yoshida M., 1999, The γ subunit in chloroplast F_1-ATPase can rotate in a unidirectional and counter-clockwise manner, *FEBS Lett.* **463**:35.

Hofkens J., Verheijen W., Shulka R., De Haen W., De Schryver F. C., 1998, Detection of a single dendrimer macromolecule with a fluorescent dihydropyrrolopurroledione (DPP) core embedded in a thin polystyrene film, *Macromolecules* **32:**4493.

Hofkens J., Maus M., Gensch T., Vosch T., Cotlet M., Köhn F., Herrmann A., Müllen K., De Schryver F. C., 2000, Probing photophysical processes in individual multichromophoric dendrimers by single-molecule spectroscopy, *J. Am. Chem. Soc.* **122**:9278.

Hofkens J., Schroeyers W., Loos D., Cotlet M., Köhn F., Vosch T., Maus M., Herrmann A., Müllen K., Gensch T., De Schryver F. C., 2001a, Triplet state as non-radiative traps in multichromophoric entities: single molecule spectroscopy of an artificial and natural antnna system, *Spectrochimica Acta A* **57**:2093.

Hofkens J., Vosch T., Maus M., Kohn F., Cotlet M., Weil T., Herrmann A., Müllen K., De Schryver F. C., 2001b, Conformational rearrangements in and twisting of a single molecule, *Chem. Phys. Lett.* **333**:255.

Hou Y., Bardo A. M., Martinez C., Higgins D. A., 2000, Characterization of molecular scale environments in polymer films by single molecule spectroscopy, *J. Phys. Chem. B* **104**:212.

Hou Y., Higgins D. A., 2002, Single molecule studies of dynamics in polymer thin films and at surfaces: effect of ambient relative humidity, *J. Phys. Chem. B* **106**:10306.

Hu D., Lu H. P., 2003, Single-molecule nanosecond anisotropy dynamics of tethered protein motions, *J. Phys. Chem. B* **107**:618.

Huang Z., Jett J. H., Keller R. A., 1999, Bacteria genome fingerprinting by flow cytometry, *Cytometry* **35**:169.

Hübner C. G., Renn A., Renge I., Wild U. P, 2001, Direct observation of the triplet lifetime quenching of single dye molecules by molecular oxygen, *J. Chem. Phys.* **115**:9619-22.

Ide T., Takeuchi Y., Yanagida T., 2002, Development of an experimental apparatus for simultaneous observation of optical and electrical signals from single ion channels, *Single Molecules* **3**:33.

Iino R., Koyama I., Kusumi A., 2001, Single molecule imaging of green fluorescent proteins in living cells: E-Cadherin forms oligomers on the free cell surface, *Biophys. J.* **80**:2667.

Inoue Y., Iwane A. H., Miyai T., Muto E., Yanagida T., 2001, Motility of single one-headed kinesin molecules along microtubules, *Biophys. J.* **81**:2838.

Ishii Y., Yoshida T., Funatsu T., Wazawa T., Yanagida T., 1999, Fluorescence resonance energy transfer between single fluorophores attached to a coiled-coil protein in aqueous solution, *Chem. Phys.* **247**:163.

Ishii Y., Yanagida T., 2000, Single molecule detection in life science, *Single Molecules* **1**:5.

Ishikawa M., Maruyama Y., Jing Yong Ye, Futamata M., 2002a, Single-molecule imaging and spectroscopy of adenine and an analog of adenine using surface-enhanced Raman scattering and fluorescence, *J. Luminesc.* **98**:81.

Ishikawa M., Maruyama Y., Ye J.-Y., Futamata M., 2002b, Single-molecule imaging and spectroscopy using fluorescence and surface-enhanced Raman scattering, *J. Biol. Phys.* **28**:573.

Jelezko F., Tietz C., Gruber A., Popa I., Nizovtsev A., Kilin S., Wrachtrup J., 2001, Spectroscopy of single N-V centers in diamond, *Single Molecules* **2**:255.

Jia Y. W., Sytnik A., Li L. Q., Vladimirov S., Cooperman B. S., Hochstrasser R. M., 1997, Nonexponential kinetics of a single $tRNA^{Phe}$ molecule under physiological conditions, *Proc. Nat. Acad. Sci. USA* **94**:7932.

Jia Y., Talaga D. S., Lau W. L., Lu H. S. M., DeGrado W. F., Hochstrasser R. M., 1999, Folding dynamics of single GCN4 peptides by fluorescence resonant energy transfer confocal microscopy, *Chem. Phys.* **247**:69.

Jing J. P., Reed J., Huang J., Hu X. H., Clarke V., Edington J., Housman D., Anantharaman T. S., Huff E. J., Mishra B., Porter B., Shenker A., Wolfson E., Hiort C., Kantor R., Aston C., Schwartz D., 1998, Automated high-resolution optical mapping using arrayed, fluid-fixed DNA molecules, *Proc. Nat. Acad. Sci. USA* **95**:8046.

Jung Y.J., Barkai E., Silbey R. J., 2002, Current status of single-molecule spectroscopy: theoretical aspects, *J. Chem. Phys.* **117**:10980.

Kaim G., Prummer M., Sich B., Zumofen G., Renn A., Wild U. P., Dimroth P., 2002, Coupled rotation within single F_0F_1 enzyme complexes during ATP synthesis or hydrolysis, *FEBS Lett.* **525**:156.

Kang S. H., Gong X., Yeung E. S., 2001, Real-time dynamics of single-DNA molecules undergoing adsorption and desorption at liquid-solid interfaces, *Anal. Chem.* **73**:1091.

Kang S. H., Yeung E. S., 2002, Dynamics of single-protein molecules at a liquid/solid interface: Implications in capillary electrophoresis and chromatography, *Anal. Chem.* **74**:6334.

Kapanidis A. N., Weiss S., 2002, Fluorescent probes and bioconjugation chemistries for single-molecule fluorescence analysis of biomolecules, *J. Chem. Phys.* **117**:10953.

Kaseda K., Yokota H., Ishii Y., Yanagida T., Inoue T., Fukui K., Kodama T., 2000, Single-molecule imaging of interaction between dextran and glucosyltransferase from *Streptococcus sobrinus*, *J. Bacteriol.* **182**:1162.

Kask P., Palo K., Ullmann D., Gall K., 1999, Fluorescence-intensity distribution analysis and its application in biomolecular detection technology, *Proc. Nat. Acad. Sci. USA* **96**:13756.

Kask P., Palo K., Fay N., Brand L., Mets Ü., Ullmann D., Jungmann J., Pschorr J., Gall K., 2000, Two-dimensional fluorescence intensity distribution analysis: theory and applications, *Biophys. J.* **78**:1703.

Kästner C. N., Prummer M., Sick B., Renn A., Wild U. P., Dimroth P., 2003, The citrate carrier CitS probed by single-molecule fluorescence spectroscopy, *Biophys. J.* **84**:1651.

Katiliene Z., Katilius E., Woodbury N. W., 2003, Single molecule detection of DNA looping by NgoMIV restriction endonuclease, *Biophys. J.* **84**:4053.

Ke P. C., Naumann C. A., 2001, Single molecule fluorescence imaging of phospholipid monolayers at the air-water interface, *Langmuir* **17**:3727.

Keller R. A., Emory S. R., Ambrose W. P., Goodwin P. M., Arias A. A., Jett J. J., Cai H., 2002, Analytical applications of single-molecule detection, *Anal. Chem.* **74**:316-24A.

Kettling U., Koltermann A., Schwille P., Eigen M., 1998, Real-time enzyme kinetics monitored by dual-color fluorescence cross-correlation spectroscopy, *Proc. Nat. Acad. Sci. USA* **95**:1416.

Kitamura K., Tokunaga M., Iwane A. H., Yanagida T., 1999, A single myosin head moves along an actin filament with regular steps of 5.3 nanometers, *Nature* **397**:129.

Knemeyer J. P., Marmé N., Sauer M., 2000, Probes for detection of specific DNA sequences at the single-molecule level, *Anal. Chem.* **72**:3717.

Knemeyer J.-P., Herten D.-P., Sauer M., 2003, Detection and identification of single molecules in living cells using spectrally resolved fluorescence lifetime imaging microscopy, *Anal. Chem.* **75**:2147.

Köhn F., Hofkens J., De Schryver F. C., 2000, Emission of the contact ion pair of rhodamine dyes observed by single molecule spectroscopy, *Chem. Phys. Lett.* **321**:372.

Köhn F., Hofkens J., Wiesler U. M., Cotlet M., Van der Auweraer M., Müllen K., C. De Schryver F., 2001, Single-molecule spectroscopy of a dendrimer-based host-guest system, *Chem. Eur. J.* **7**:4126.

Köhn F., Hofkens J., Gronheid R., Cotlet M., Müllen K., Van der Auweraer M., De Schryver F. C., 2002a, Excitation energy transfer in dendritic host-guest donor-acceptor systems, *ChemPhysChem* **3**:1005.

Köhn F., Hofkens J., Gronheid R., Van der Auweraer M., De Schryver F. C., 2002b, Parameters influencing the on- and off-times in the fluorescence intensity traces of single cyanine dye molecules, *J. Phys. Chem. A* **106**:4808.

Koltermann A., Kettling U., Bieschke J., Winkler T., Eigen M., 1998, Rapid assay processing by integration of dual-color fluorescence cross-correlation spectroscopy: High throughput screening for enzyme activity, *Proc. Nat. Acad. Sci. USA* **95**:1421.

Kubitscheck U., Kuckmann O., Kues T., Peters R., 2000, Imaging and tracking of single GFP molecules in solution, *Biophys. J.* **78**:2170.

Kubitschek U., 2002, Single protein molecules visualized and tracked in the interior of eukaryotic cells, *Single Molecules* **3**:267.

Kues T., Peters R., Kubitscheck U., 2001a, Visualization and tracking of single protein molecules in the cell nucleus, *Biophys. J.* **80**:2954.

Kues T., Dickmanns A., Lührmann R., Peters R., Kubitscheck U., 2001b, High intranuclear mobility and dynamic clustering of the splicing factor U1 snRNP observed by single particle tracking, *Proc. Nat. Acad. Sci. USA* **98**:12021.

Kues T., Kubitschek U., 2002, Single molecule motion perpendicular to the focal plane of a microscope: Application to splicing factor dynamics within the cell Nucleus, *Single Molecules* **3**:218.

Kühnemuth R., Seidel C. A. M., 2001, Principles of single molecule multiparameter fluorescence spectroscopy, *Single Molecules* **2**:251.

Lacoste T. D., Michalet X., Pinaud F., Chemla D. S., Alivisatos A. P., Weiss S., 2000, Ultrahigh-resolution multicolor colocalization of single fluorescent probes, *Proc. Nat. Acad. Sci. USA* **97**:9461.

Ladoux B., Quivy J. P., Doyle P., Du Roure O., Almouzni G., Viovy J. L., 2000, Fast kinetics of chromatin assembly revealed by single-molecule videomicroscopy and scanning force microscopy, *Proc. Nat. Acad. Sci. USA* **97**:14251.

Lakowicz J. R., 2001, Radiative decay engineering: biophysical and biomedical applications, *Anal. Biochem.* **298**:1.

Le Goff L., Hallatschek O., Frey E., Amblard F., 2002, Tracer studies on F-actin fluctuations, *Phys. Rev. Lett.* **89**:258101.

Lee M., Tang J., Hochstrasser R. M., 2001, Fluorescence lifetime distribution of single molecules undergoing Förster energy transfer, *Chem. Phys. Lett.* **344**:501.

Lee T. H., Gonzalez J. I., Dickson R. M., 2002, Strongly enhanced field-dependent single-molecule electroluminescence, *Proc. Nat. Acad. Sci. USA* **99**:10272.

Levene M. J., Korlach J., Turner S. W., Foquet M., Craighead H. G., Webb W. W., 2003, Zero-mode waveguides for single-molecule analysis at high concentrations, *Science* **299**:682.

Li H., Xue G., Yeung E. S., 2001, Selective detection of individual DNA molecules by capillary polymerase chain reaction, *Anal. Chem.* **73**:1537.

Li H., Ying L., Green J. J., Balasubramanian S., Klenerman D., 2003, Ultrasensitive coincidence fluorescence detection of single DNA molecules, *Anal. Chem.* **75**:1664.

Lounis B., Deich J., Rosell F. I., Boxer S. G., Moerner W. E., 2001, Photophysics of DsRed, a red fluorescent protein, from the ensemble to the single-molecule level, *J. Phys. Chem. B* **105**:5048.

Lounis B., Moerner W. E., 2000, Single photons on demand from a single molecule at room temperature, *Nature* **407**:491.

Lu H. P., Xun L., Xie X. S., 1998, Single-molecule enzymatic dynamics, *Science* **282**:1877.

Ludes M. D., Wirth M. J., 2002, Single-molecule resolution and fluorescence imaging of mixed-mode sorption of a dye at the interface of C18 and acetonitrile/water, *Anal. Chem.* **74**:386.

Lyon W. A., Fang M. M., Haskins W. E., Nie S. M., 1998, A dual beam optical microscope for observation and cleavage of single DNA molecules, *Anal. Chem.* **70**:1743.

Ma Y., Shortreed M. R., Yeung E. S., 2000, High-throughput single-molecule spectroscopy in free solution, *Anal. Chem.* **72**:4640.

Machara N. P., Goodwin P. M., Enderlein J., Semin D. J., Keller R. A., 1998, Efficient detection of single molecules eluting off an optically trapped microsphere, *Bioimaging* **6**:33.

Maier B., Rädler J.O., 1999, Conformation and self-diffusion of single DNA molecules confined to two dimensions, *Phys. Rev. Lett.* **82:**1911.

Maier B., Seifert U., Rädler J. O., 2002, Elastic response of DNA to external electric fields in two dimensions, *Europhys. Lett.* **60**:622.

Masaike T., Mitome N., Noji H., Muneyuki E., Yasuda R., Kinosita K., Yoshida M., 2000, Rotation of F_1-ATPase and the hinge residues of the beta subunit, *J. Exp. Biol.* **203**:1.

Maus M., Cotlet M., Hofkens J., Gensch T., De Schryver F. C., Schaffer J., Seidel C. A. M., 2001, An experimental comparison of the maximum likelihood estimation and nonlinear least-squares fluorescence lifetime analysis of single molecules, *Anal. Chem.* **73**:2078.

Mei E., Tang J., Vanderkooi J. M., Hochstrasser R. M., 2003, Motions of single molecules and proteins in trehalose glass, *J. Am. Chem. Soc.* **125**:2730.

Meixner A. J., 1998, Optical single-molecule detection at room temperature, *Adv. Photochem.* **24**:1.

Melnikov S. M., Melnikova Y. S., Lofroth J. E., 1998, Single-molecule visualization of interaction between DNA and oppositely charged mixed liposomes, *J. Phys. Chem. B* **102**:9367.

Melnikov S.M., Khan M. O., Lindman B., Jonsson B., 1999, Phase behavior of single DNA in mixed solvents,
J. Am. Chem. Soc. **121**:1130.

Melnikov S.M., Lindman B., 1999, Cationic lipid complexes in hydrophobic solvents. A single-molecule visualization by fluorescence microscopy, *Langmuir* **15:**1923.

Michalet X., Weiss S., 2002, Single-molecule spectroscopy and microscopy, *C. R. Acad. Sci. Ser. IV.* **3**:619.

Miyamoto Y., Muto E., Mashimo T., Iwane A. H., Yoshiya I., Yanagida T., 2000, Direct inhibition of microtubule-based kinesin motility by local anesthetics, *Biophys. J.* **78**:940.

Moerner W. E., 2002, A dozen years of single-molecule spectroscopy in physics, chemistry, and biophysics, *J. Phys. Chem. B* **106**:910.
Moerner W. E., Peterman E. J. G., Brasselet S., Kummer S., Dickson R. M., 1999, Optical methods for exploring dynamics of single copies of green fluorescent protein, *Cytometry* **36**:232.
Moerner W. E., 2002, Single-molecule optical spectroscopy of autofluorescent proteins, *J. Chem. Phys.* **117**:10925.
Mörtelmaier M., Kögler E. J., Hesse J., Sonnleitner M., Huber L. A., Schütz G. J., 2002, Single molecule microscopy in living cells: Subtraction of autofluorescence based on two color recording, *Single Molecules* **3**:225.
Müller J. D., Chen Y., Gratton E., 2000, Resolving heterogeneity on the single molecular level with the photon-counting histogram, *Biophys. J.* **78**:474.
Neuweiler H., Schulz A., Böhmer M., Enderlein J., Sauer M., 2003, Measurement of submicrosecond intramolecular contact formation in peptides at the single-molecule level, *J. Am. Chem. Soc.* **125**:5324.
Neuweiler H., Schulz A., Vaiana A. C., Smith J. C., Kaul S., Wolfrum J., Sauer M., 2002, Detection of individual p53-autoantibodies by using quenched peptide-based molecular probes, *Angew. Chem. Int. Ed. Engl.* **41**:4769.
Noji H., Yasuda R., Yoshida M., Kinosita K., 1997, Direct observation of the rotation of F_1-ATPase, *Nature* **386:**299.
Oiwa K., Eccleston J. F., Anson M., Kikumoto M., Davis C. T., Reid G. P., Ferenczi M. A., Corrie J. E. T., Yamada A., Nakayama H., Trentham D. R., 2000, Comparative single-molecule and ensemble myosin enzymology: Sulfoindocyanine ATP and ADP derivatives, *Biophys. J.* **78**:3048.
Okada Y., Hirokawa N., 1999, A processive single-headed motor: Kinesin superfamily protein KIF1A, *Science* **283**:1152.
Orrit M., Bernard J., Brown R., Lounis B., 1996, Optical Spectroscopy of Single Molecule in Solids, in: *Progress in Optics vol. XXXV*, E. Wolf, ed., Elsevier, Amsterdam, pp. 63-144.
Osborne M. A., Balasubramanian S., Furey W. S., Klenerman D., 1998, Optically biased diffusion of single molecules studied by confocal fluorescence microscopy, *J. Phys. B* **102**:3160.
Osborne M. A., Barnes C. L., Balasubramanian S., Klenerman D., 2001, Probing DNA surface attachment and local environment using single molecule spectroscopy, *J. Phys. Chem. B* **105**:3120.
Paige M. F., Bierneld E. J., Moerner W. E., 2001, A comparison of through-the-objective total internal reflection microscopy and epifluorescence microscopy for single-molecule fluorescence imaging, *Single Molecules* **2**:191.
Palo K., Mets U., Jäger S., Kask P., Gall K., 2000, Fluorescence intensity multiple distributions analysis: concurrent determination of diffusion times and molecular brightness, *Biophys. J.* **79**:2858.
Palo K., Brand L., Eggeling C., Jäger S., Kask P., Gall K., 2002, Fluorescence intensity and lifetime distribution analysis: Toward higher accuracy in fluorescence fluctuation spectroscopy, *Biophys. J.* **83**:605.
Pänke O., Cherepanov D. A., Gumbiowski K., Engelbrecht S., Junge W., 2001, Viscoelastic dynamics of actin filaments coupled to rotary F-ATPase: Angular torque profile of the enzyme, *Biophys. J.* **81**:1220.
Peleg G., Ghanouni P., Kobilka B. K., Zare R. N., 2001, Single-molecule spectroscopy of the β_2 adrenergic receptor: Observation of conformational substates in a membrane protein, *Proc. Nat. Acad. Sci. USA* **98**:8469.
Peterman E. J. G., Brasselet S., Moerner W. E., 1999, The fluorescence dynamics of single molecules of green fluorescent protein, *J. Phys. Chem. A* **103**:10553.
Peterman E. J. G., Sosa H., Goldstein L. S. B., Moerner W. E., 2001, Polarized fluorescence microscopy of individual and many kinesin motors bound to axonemal microtubules, *Biophys. J.* **81**:2851.
Pierce D. W., Hom-Booher N., Otsuka A. J., Vale R. D., 1999, Single-molecule behavior of monomeric and heteromeric kinesins, *Biochem.* **38**:5412.
Polakowski R., Craig D. B., Skelley A., Dovichi N. J., 2000, Single molecules of highly purified bacterial alkaline phosphatase have identical activity, *J. Am. Chem. Soc.* **122**:4853.
Qiu X. H., Nazin G. V., Ho W., 2003, Vibrationally resolved fluorescence excited with submolecular precision, *Science* **299**:542.
Rarbach M., Kettling U., Koltermann A., Eigen M., 2001, Dual-color fluorescence cross-correlation spectroscopy for monitoring the kinetics of enzyme-catalyzed reactions, *Methods* **24**:104.
Rhoades E., Gussakovsky E., Haran G., 2003, Watching proteins fold one molecule at a time, *Proc. Nat. Acad. Sci. USA* **100**:3197.
Rigler R., Orrit M., Basché T., eds., 2001, *Single Molecule Spectroscopy*, Springer, New York/Berlin.
Rigler R., Elson E., eds., 2001, *Fluorescence Correlation Spectroscopy*, Springer, New York/Berlin.
Rock R. S., Rief M., Mehta A. D., Spudich J. A., 2000, In vitro assays of processive myosin motors, *Methods* **22**:373.
Rock R. S., Rice S. E., Wells A. L., Purcell T. J., Spudich J. A., Sweeney H. L., 2001, Myosin VI is a processive motor with a large step size, *Proc. Nat. Acad. Soc. USA* **98**:13655.

Romberg L., Pierce D. W., Vale R. D., 1998, Role of the kinesin neck region in processive microtubule-based motility, *J. Cell. Biol.* **140**:1407.

Rothwell P. J., Berger S., Kensch O., Felekyan S., Antonik M., Wöhrl B. M., Restle T., Goody R. S., Seidel C. A. M., 2003, Multiparameter single-molecule fluorescence spectroscopy reveals heterogeneity of HIV-1 reverse transcriptase: primer/template complexes, *Proc. Nat. Acad. Sci. USA* **100**:1655.

Saga Y., Wazawa T., Nakada T., Ishii Y., Yanagida T., Tamiaki H., 2002, Fluorescence emission spectroscopy of single light-harvesting complex from green filamentous photosynthetic bacteria, *J. Phys. Chem. B* **106**:1430.

Sakmann B., Neher E., 1984, Patch clamp techniques for studying ionic channels in excitable membranes, *Annu. Rev. Physiology* **46**:455.

Sako Y., Hibino K., Miyauchi T., Miyamoto Y., Ueda M., Yanagida T., 2000a, Single-molecule imaging of signaling molecules in living cells, *Single Molecules* **1**:159.

Sako Y., Minoguchi S., Yanagida T., 2000b, Single-molecule imaging of EGFR signalling on the surface of living cells, *Nature Cell Biol.* **2**:168.

Sauer M., Angerer B., Han K.-T., Zander C., 1999, Detection and identification of single dye labeled mononucleotide molecules released from an optical fiber in a microcapillary: First steps towards a new single molecule DNA sequencing technique, *Phys. Chem. Chem. Phys.* **1**:2471.

Sauer M., Göbel F., Han K. T., Zander C., 2000, Single molecule DNA sequencing in microcapillaries, *Trends in Optics and Photonics Laser - Applications to Chemical and Environmental Analysis* **36**:18.

Schaffer J., Volkmer A., Eggeling C., Subramaniam V., Striker G., Seidel C. A. M., 1999, Identification of single molecules in aqueous solution by time-resolved fluorescence anisotropy, *Phys. Chem. A* **103**:331.

Schenter G. K., Lu H. P., Xie X. S., 1999, Statistical analyses and theoretical models of single-molecule enzymatic dynamics, *J. Phys. Chem. A* **103**:10477.

Schins J. M., Agronskaya A., De Grooth B. G., Greve J., 1998, New technique for high resolution DNA sizing in epi-illumination, *Cytometry* **32**:132.

Schins J. M., Agronskaia A., De Grooth B. G., Greve J., 1999, Orientation of the chromophore dipoles in the TOTO-DNA system, *Cytometry* **37**:230.

Schmidt T., Kubitscheck U., Rohler D., Nienhaus U., 2002, Photostability data for fluorescent dyes: An update., *Single Molecules* **3**:327.

Schuler B., Lipman E. A., Eaton W. A., 2002, Probing the free-energy surface for protein folding with single-molecule fluorescence spectroscopy, *Nature* **419**:743.

Schuster J., Cichos F, Wrachtrup J., Von Borczyskowski C., 2000, Diffusion of single molecules close to interfaces, *Single Molecules* **1**:299.

Schütz G. J., Trabesinger W., Schmidt T., 1998, Direct observation of ligand colocalization on individual receptor molecules, *Biophys. J.* **74**:2223.

Schütz G. J., Hesse J., Freudenthaler G., Pastushenkoo V. P., Knaus H. G., Pragl B., Schindler H., 2000a, 3D-mapping of individual ion channels on living cells, *Single Molecules* **1**:153.

Schütz G. J., Pastushenko V. P., Gruber H. J., Knaus H. G., Pragl B., Schindler H., 2000b, 3D imaging of individual ion channels in live cells at 40 nm resolution, *Single Molecules* **1**:25.

Schütz G. J., Kada G., Pastushenko V. P., Schindler H. G., 2000c, Properties of lipid microdomains in a muscle cell membrane visualized by single molecule microscopy, *EMBO J.* **19**:892.

Schütz G. J., Axmann M., Schindler H., 2001, Imaging single molecules in three dimensions, *Single Molecules* **2**:69.

Schwille P., 2001, Fluorescene correlation spectroscopy and its potential for intracellular applications, *Cell. Biochem. Biophys.* **34**:383.

Seebacher C., Hellriegel C., Deeg F.-W., Bräuchle C., Altmaier S., Behrens P., Müllen K., 2002, Observation of translational diffusion of single terrylenediimide molecules in a mesostructured molecular sieve, *J. Phys. Chem. B* **106**:5591.

Seisenberger G., Ried M. U., Endreß T., Büning H., Hallek M., Bräuchle C., 2001, Real-time single-molecule imaging of the infection pathway of an adeno-associated virus, *Science* **294**:1929.

Shelby J. P., Chiu D. T., 2003, Mapping fast flows over micrometer-length scales using flow-tagging velocimetry and single-molecule detection, *Anal. Chem.* **75**:1387.

Shera E.B., Seitzinger N.K., Davis L.M., Keller R.A., Soper S.A., 1990, Detection of single fluorescent molecules, *Chem. Phys. Lett.* **174**:553.

Shimizu T., Thorn K. S., Ruby A., Vale R. D., 2000, ATPase kinetic characterization and single molecule behavior of mutant human kinesin motors defective in microtubule-based motility, *Biochem.* **39**:5265.

Shoemaker G. K., Juers D. H., Coombs J. M. L., Matthews B. W., Craig D. B., 2003, Crystallization of β-galactosidase does not reduce the range of activity of individual molecules, *Biochem.* **42**:1707.

Shortreed M. R., Li H., Huang W. H., Yeung E. S., 2000, High-throughput single-molecule DNA screening based on electrophoresis, *Anal. Chem.* **72**:2879.

Smith D. E., Chu S., 1998, Response of flexible polymers to a sudden elongational flow, *Science* **281**:1335.

Smith D. E., Babcock H. P., Chu S., 1999, Single-polymer dynamics in steady shear flow, *Science* **283**:1724.
Sonnleitner M., Schütz G. J., Schmidt T., 1999a, Free brownian motion of individual lipid molecules in biomembranes, *Biophys. J.* **77**:2638.
Sonnleitner M., Schütz G. J., Schmidt T., 1999b, Imaging individual molecules by two-photon excitation, *Chem. Phys. Lett.* **300**:221.
Sytnik A., Vladimirov S., Jia Y. W., Li L. Q., Cooperman B. S., Hochstrasser R. M., 1998, Peptidyl transferase center activity observed in single ribosomes, *J. Mol. Biol.* **285**:49.
Taguchi H., Ueno T., Tadakuma H., Yoshida M., Funatsu T., 2001, Single-molecule observation of protein-protein interactions in the chaperonin system, *Nature Biotech.* **19**:861.
Talaga D. S., Lau W. L., Roder H., Tang J., Jia Y., DeGrado W. F., Hochstrasser R. M., 2000, Dynamics and folding of single two-stranded coiled-coil peptides studied by fluorescent energy transfer confocal microscopy, *Proc. Nat. Acad. Sci. USA* **97**:13021.
Talley C. E., Dunn R. C., 1999, Single molecule as probes of lipid membrane microenvironments, *Phys. Chem. B* **103**:10214.
Tamarat P., Maali A., Lounis B., Orrit M., 2000, Ten years of single-molecule spectroscopy, *J. Phys. Chem. A* **104**:1.
Tietz C., Jelezko F., Gerken U., Schuler S., Schubert A., Rogl H., Wrachtrup J., 2001, Single molecule spectroscopy on the light-harvesting complex II of higher plants, *Biophys. J.* **81**:556.
Tinnefeld P., Buschmann V., Herten D. P., Han K. T., Sauer M., 2000, Confocal fluorescence lifetime imaging microscopy (FLIM) at the single molecule level, *Single Molecules* **1**:215.
Tinnefeld P., Herten D. P., Sauer M., 2001a, Photophysical dynamics of single molecules studied by spectrally-resolved fluorescence lifetime imaging microscopy (SFLIM), *J. Phys. Chem. A* **105**:7989.
Tinnefeld P., Müller C., Sauer M., 2001b, Time-varying photon probability distribution of individual molecules at room temperature, *Chem. Phys. Lett.* **345**:252.
Tinnefeld P., Weston K. D., Vosch T., Cotlet M., Weil T., Hofkens J., Müllen K., De Schryver F. C., Sauer M., 2002, Antibunching in the emission of a single tetrachromophoric dendritic system, *J. Am. Chem. Soc.* **124**:14310.
Tinnefeld P., Buschmann V., Weston K. D., Sauer M., 2003, Direct observation of collective blinking and energy transfer in a bichromophoric system, *J. Phys. Chem. A* **107**:323.
Trabesinger W., Schütz G. J., Gruber H. J., Schindler H., Schmidt T., 1999, Detection of individual oligonucleotide pairing by single-molecule microscopy, *Anal. Chem.* **71**:279.
Trabesinger W., Renn A., Hecht B., Wild U. P., Montali A., Smith P., Weder C., 2000, Single-molecule imaging revealing the deformation-induced formation of a molecular polymer blend, *J. Phys. Chem. B* **104**:5221.
Trabesinger W., Hecht B., Wild U. P., Schütz G. J., Schindler H., Schmidt T., 2001, Statistical analysis of single-molecule colocalization assays, *Anal. Chem.* **73**:1100.
Ueda M., Sako Y., Tanaka T., Devreotes P., Yanagida T., 2001, Single-molecule analysis of chemotactic signaling in *Dictyostelium* cells, *Science* **294**:864.
Vacha M., Kotani M., 2003, Three-dimensional orientation of single molecules observed by far- and near-field fluorescence microscopy, *J. Chem. Phys.* **118**:5279.
Van Orden A., Machara N. P., Goodwin P. M., Keller R. A., 1998, Single molecule identification in flowing sample streams by fluorescence burst size and intraburst fluorescence decay rate, *Anal. Chem.* **70**:1444.
Van Orden A., Cai H., Goodwin P. M., Keller R. A., 1999, Efficient detection of single DNA-fragments in flowing sample streams by two-photon fluorescence excitation, *Anal. Chem.* **71**:2108.
Van Orden A., Keller R. A., Ambrose W. P., 2000, High-throughput flow cytometric DNA fragment sizing, *Anal. Chem.* **72:**37.
Van Sark W. G. J. H. M., Frederix P. L. T. M., Van den Heuvel D. J., Asselbergs M. A. H., Senf I., Gerritsen H. C., 2000, Fast imaging of single molecules and nanoparticles by wide-field microscopy and spectrally resolved confocal microscopy, *Single Molecules* **1**:291.
Vargas F., Hollricher O., Marti O., De Schaetzen G., Tarrach G., 2002, Influence of protective layers on the blinking of fluorescent single molecules observed by confocal microscopy and scanning near field optical microscopy, *J. Chem. Phys.* **117**:866.
Viteri C. R., Gilliland J. W., Yip W. T., 2003, Probing the dynamic guest-host interactions in sol-gel films using single molecule spectroscopy, *J. Am. Chem. Soc.* **125**:1980.
Vosch T., Hofkens J., Cotlet M., Kohn F., Fujiwara H., Gronheid R., Van Der Biest K., Weil T., Herrmann A., Müllen K., Mukamel S., Van der Auweraer M., De Schryver F. C., 2001, Influence of structural and rotational isomerism on the triplet blinking of individual dendrimer molecules, *Ang. Chem.* **40**:4643.
Vrljic M., Nishimura S. Y., Brasselet S., Moerner W. E., McConnell H. M., 2002, Translational diffusion of individual class II MHC membrane proteins in cells, *Biophys. J.* **83**:2681.
Wang H. M., Bardo A. M., Collinson M. M., Higgins D. A., 1998, Microheterogeneity in dye-doped silicate and polymer films, *J. Phys. Chem. B* **102**:2731.

Warshaw D. M., Hayes E., Gaffney D., Lauzon A. M., Wu J. R., Kennedy G., Trybus K., Lowey S., Berger C., 1998, Myosin conformational states determined by single fluorophore polarization, *Proc. Nat. Acad. Sci. USA* **95**:8034.

Watanabe N., Mitchison T. J., 2002, Single-molecule speckle analysis of actin filament turnover in lamellipodia, *Science* **295**:1083.

Wazawa T., Ishii Y., Funatsu T., Yanagida T., 2000, Spectral fluctuation of a single fluorophore conjugated to a protein molecule, *Biophys. J.* **78**:1561.

Webb W. W., 2001, Fluorescence correlation spectroscopy: inception, biophysical experimentations, and prospectus, *Appl. Opt.* **40**:3969.

Weber M. A., Stracke F., Meixner A. J., 1999, Dynamics of single dye molecules observed by confocal Imaging and spectroscopy, *Cytometry* **36**:217.

Weiss S., 1999, Fluorescence spectroscopy of single biomolecules, *Science* **283**:1676.

Weiss S., 2000, Measuring conformational dynamics of biomolecules by single molecule fluorescence spectroscopy, *Nature Struct. Biol.* **7**:724.

Wennmalm S., Edman L., Rigler R., 1999a, Non-ergodic behaviour in conformational transitions of single DNA molecules, *Chem. Phys.* **247**:61.

Wennmalm S., Rigler R., 1999b, On death numbers and survival times of single molecules, *J. Phys. Chem. B* **103**:2516.

Werner J. H., Larson E. J., Goodwin P. V., Ambrose W. P., Keller R. A., 2000, Effects of fluorescence excitation geometry on the accuracy of DNA fragment sizing by flow cytometry, *Appl. Opt.* **39**:2831.

Weston K. D., Buratto S. K., 1998, Millisecond intensity fluctuations of single molecules at room temperature, *J. Phys. Chem. A* **102**:3635.

Weston K. D., Carson P. J., Metiu H., Buratto S. K., 1998, Room temperature fluorescence characteristics of single dye molecules adsorbed on a glass surface, *J. Chem. Phys.* **109**:7474.

Weston K. D., Carson P. J., DeAro J. A., Buratto S. K., 1999, Single-molecule detection fluorescence of surface-bound species in vacuum, *Chem. Phys. Lett.* **308**:58.

Weston K. D., Goldner L. S., 2001, Orientation imaging and reorientation dynamics of single dye molecules, *J. Phys. Chem. B* **105**:3453.

Weston K. D., Dyck M., Tinnefeld P., Müller C., Herten D. P., Sauer M., 2002, Measuring the number of independent emitters in single-molecule fluorescence images and trajectories using coincident photons, *Anal. Chem.* **74**:5342.

Willets K. A., Ostroverkhova O., He M., Twieg R. J., Moerner W. E., 2003, Novel fluorophores for single-molecule imaging, *J. Am. Chem. Soc.* **125**:1174.

Winkler T., Kettling U., Koltermann A., Eigen M., 1999, Confocal fluorescence coincidence analysis: An approach to ultra high-throughput screening, *Proc. Nat. Acad. Sci. USA* **96**:1375.

Wirth M. J., Swinton D. J., 1998, Single molecule probing of mixed mode adsorption at a chromatographic interface, *Anal. Chem.* **70**:5264.

Wirth M. J., Swinton D. J., 2001, Single-molecule study of an adsorbed oligonucleotide undergoing both lateral diffusion and strong adsorption, *J. Phys. Chem. B* **105**:1472.

Wirth M. J., Swinton D. J., Ludes M. D., 2003, Adsorption and diffusion of single molecules at chromatographic interfaces, *J. Phys. Chem. B* **107**:6258.

Xie S., 2001, Single-molecule approach to enzymology, *Single Molecules* **2**:229.

Xie X. S., 2002, Single-molecule approach to dispersed kinetics and dynamic disorder: Probing conformational fluctuation and enzymatic dynamics, *J. Chem. Phys.* **117**:11024.

Xie X. S., Lu P. H., 1999, Single-Molecule Enzymology, *J. Biol. Chem.* **274**:15967.

Xie X. S., Trautman J. K., 1998, Optical studies of single molecules at room temperature, *Annu. Rev. Phys. Chem.* **49**:441.

Yanagida T., Esaki S., Iwane A. H., Inoue Y., Ishijima A., Kitamura K., Tanaka H., Tokunaga M., 2000a, Single-motor mechanics and models of the myosin motor, *Philos. Trans. R. Soc. Lond. B Biol. Sci.* **355**:441.

Yanagida T., Kitamura K., Tanaka H., Iwane A. H., Esaki S., 2000b, Single molecule analysis of the actomyosin motor, *Curr. Opinion Cell. Biol.* **12**:20.

Yang H., Xie X. S., 2002a, Probing single-molecule dynamics photon by photon, *J. Chem. Phys.* **117**:10965.

Yang H., Xie X. S., 2002b, Statistical approaches for probing single-molecule dynamics photon-by-photon, *Chem. Phys.* **284**:423.

Yasuda R., Noji H., Kinosita K., Yoshida M., 1998, F_1-ATPase is a higly efficient molecular motor that rotates with discrete 120-degree steps, *Cell* **93**:1117.

Yasuda R., Noji H., Yoshida M., Kinosita K. J., Itoh H., 2001, Resolution of distinct rotational substeps by submillisecond kinetic analysis of F_1-ATPase, *Nature* **410**:898.

Ye J. Y., Ishikawa M., Yogi O., Okada T., Maruyama Y., 1998, Bimodal site distribution of a polymer film revealed by flexible single molecule probes, *Chem. Phys. Lett.* **288**:885.

Ye J. Y., Yamane Y., Yamauchi M., Nakatsuka H., Ishikawa M., 2000, Direct observation of the interaction of single fluorescent nucleotide analogue molecules with DNA polymerase I, *Chem. Phys. Lett.* **320**:607.

Yildiz A., Forkey J. N., McKinney S. A., Ha T., Goldman Y. E., Selvin P. R., 2003, Myosin V walks hand-over-hand: Single fluorophore imaging with 1.5-nm localization, *Science* **300**:2061.

Ying L., Wallace M. I., Balasubramanian S., Klenerman D., 2000, Ratiometric analysis of single-molecule fluorescence resonance energy transfer using logical combinations of threshold criteria: A study of 12-mer DNA, *J. Phys. Chem. B* **104**:5171.

Ying L. M., Xie X. S., 1998, Fluorescence spectroscopy, exciton dynamics, and photochemistry of single allophycocyanin trimers, *J. Phys. Chem. B* **102**:10399.

Yip W. T., Hu D. H., Yu J., Vanden Bout D. A., Barbara P. F., 1998, Classifying the photophysical dynamics of single-chromophoric and multiple-chromophoric molecules by single molecule spectroscopy, *J. Phys. Chem. A* **102**:7564.

Yokota H., Saito K., Yanagida T., 1998, Single molecule imaging of fluorescently labeled proteins on metal by surface plasmons in aqueous solution, *Phys. Rev. Lett.* **80**:4606.

Yu J., Hu D., Barbara P. F., 2000, Unmasking electronic energy transfer of conjugated polymers by suppression of O_2 quenching, *Science* **289**:1327.

Zander C., Enderlein J., Keller R. A., eds., 2002, *Single-Molecule Detection in Solution - Methods and Applications*, VCH-Wiley, Berlin/New York.

Zang L., Liu R., Holman M. W., Nguyen K. T., Adams D. M., 2002, A single-molecule probe based on intramolecular electron transfer, *J. Am. Chem. Soc.* **124**:10640.

Zehetmayer P., Hellerer T., Parbel A., Scheer H., Zumbusch A., 2002, Spectroscopy of single phycoerythrocyanin monomers: dark state identification and observation of energy transfer heterogeneities, *Biophys. J.* **83**:407.

Zheng J., Dickson R. M., 2002, Individual water-soluble dendrimer-encapsulated silver nanodot fluorescence, *J. Am. Chem. Soc.* **124**:13982.

Zheng J., Yeung E. S., 2002, Anomalous radial migration of single DNA molecules in capillary electrophoresis, *Anal. Chem.* **74**:4533.

Zhuang X., Bartley L. E., Babcock H. P., Russell R., Ha T., Herschlag D., Chu S., 2000a, A single-molecule study of RNA catalysis and folding, *Science* **288**:2048.

Zhuang X., Ha T., Kim H. D., Centner T., Labeit S., Chu S., 2000b, Fluorescence quenching: A tool for single-molecule protein-folding study, *Proc. Nat. Acad. Sci. USA* **97**:14241.

Zicha D., Dobbie I. M., Holt M. R., Monypenny J., Soong D. Y. H., Gray C., Dunn G. A., 2003, Rapid actin transport during cell protrusion, *Science* **300**:142.

Zumbusch A., Jung G., 2000, Single molecule spectroscopy of the green fluorescent protein: A critical assessment, *Single Molecules* **1**:261.

APPLICATIONS OF DISTRIBUTED OPTICAL FIBER SENSING: FLUORESCENT ASSAYS OF LINEAR COMBINATORIAL ARRAYS

Peter Geissinger* and Alan W. Schwabacher*

1. INTRODUCTION

Distributed optical fiber sensing has been referred to as "the highest state of the art in optical sensing."[1] Measurements of desired parameters can take place spatially resolved along a continuous section of an optical fiber (distributed sensing) or on a number of discrete regions along a fiber (quasi-distributed sensing), allowing for multi-parameter sensing on a single fiber.

Spatial resolution is achieved using the Optical Time-Domain Reflectometry (OTDR) technique, which was first introduced to locate faults in optical fibers.[2, 3] In this application, a brief laser pulse is coupled into the fiber and the intensity of light scattered back to the same end of the fiber is monitored. A crack or a refractive index inhomogeneity in the fiber changes the backscattered intensity, which is detected with a characteristic time delay that encodes the location of the fault. The OTDR method has since come into widespread use for distributed fiber-optic sensing (see e.g. Refs.[1, 4-8]), employing various physical and chemical effects that can lead to a modulation of the signal pulse.[9-17]

In recent years, fiber-optic chemosensors have found increasing use for monitoring environmental conditions (pH, temperature, pressure, radiation, etc.) or for detecting the presence of substances in chemical or biological analytes (pollutants, toxins, antibodies, etc.). Microarrays of sensors[18, 19] with multiple sensor regions of different type arrayed on an optical fiber offer highly specific responses, and discrimination between related analytes. Combined with the OTDR method, each sensor region of the microarray can be independently addressed and its signal change assigned to a specific location.

Distributed sensing using Rayleigh,[14] Raman,[9, 10, 15] and Brillouin[11, 13] scattering has been demonstrated over the past years. The use of fluorescent species as sensor molecules has the potential to provide much larger signals, since the quantum efficiencies for fluorescence may be orders of magnitude higher than those for scattering.[1, 20, 21]

* Department of Chemistry, University of Wisconsin-Milwaukee, Milwaukee, WI, 53211, U.S.A.

Placing fluorophores into the fiber core permits efficient collection of the emitted fluorescence; however, it also attenuates the exciting light in the fiber available for subsequent sensor regions. While for short fibers the increase in signal strength is useful,[21] these losses limit the number of distinct sensor regions that can be placed on the fiber.

Placing the fluorescent sensor molecules into the cladding of the fiber can alleviate this problem. In practice, the existing cladding of the fiber is often replaced with a cladding that is (1) a suitable host material for the fluorescent sensors, (2) allows for the diffusion of the analyte to the sensor molecules, and (3) preserves the guiding condition of the fiber. The latter condition implies that there is no refractive coupling of light from the fiber core to the cladding, meaning that fluorophores in the cladding can only be excited through the evanescent field of the core modes (more detail below). Since evanescent coupling is weak, the attenuation of the light in the core is significantly less than for refractive coupling, making long arrays of chemosensors feasible. Moreover, since the evanescent-field intensity drops off exponentially with distance from the core/cladding interface, only fluorophores close to this interface can be excited. The emitted fluorescence may be coupled back evanescently into the fiber core, which allows it to propagate under guided conditions back along the fiber to a detector that is placed at the start of the fiber. The response of a fluorescent sensor molecule can be a change of fluorescence intensity, fluorescence wavelength, and/or fluorescence lifetime.[22] Evanescent excitation of fluorophores located in the fiber cladding and subsequent evanescent capture of the fluorescence by core modes were first combined with the OTDR method by Kvasnik et al.[12] and Chronister et al.[16] As fluorescence output is not reflection, the method has been referred to as Optical Time of Flight (OTOF) detection.[23, 24] It is this sensing mode that is of interest in our research.

The intent of this chapter is twofold. First, we introduce a new application area for (quasi)-distributed fiber-optic sensing techniques in the field of combinatorial chemistry. This branch of chemistry is concerned with the mass synthesis of compounds and their systematic evaluation for a desired property and/or function. Schwabacher et al.[25, 26] introduced a novel combinatorial chemistry that allows for synthesizing large libraries of compounds on linear supports. The evaluation of the libraries for a desired property and/or function is performed using fluorescent assays, where the fluorescence characteristics of a library compound change upon positive identification. We show that optical fibers present a highly suitable and flexible linear support for combinatorial synthesis: reactions leading the compound library can be carried out in the fiber cladding (or in suitable materials replacing the original fiber cladding), whereas the fluorescence characteristics of the compound library can be probed spatially resolved using the OTDR method.[27-29]

Second, we present a new readout scheme for fiber-optic sensor arrays that allows significantly reduced spacing of sensor regions on a fiber, overcoming limitations given by the fluorescence lifetimes of the sensor molecules. This scheme, which employs a second optical fiber to provide an optical delay, is not limited to applications in combinatorial chemistry. It is generally applicable to many sensing tasks that require dense packing of compounds onto an optical fiber support.

The chapter is structured as follows. After briefly reviewing fiber-optic sensing basics and stating equations relevant for the interpretation of our experimental data, we will summarize the basic ideas of combinatorial chemistry to show that the use of linear supports can be superior to other methods. Next we show how the two fields combine to form our "Fiber-Optic Combinatorial Chemistry" technique. An important aspect of this technique is our unique approach to library analysis. Fourier analysis of fluorescent sensor output provides useful generalizations and compact summary of library structural variation on activity. For a fabricated array of sensors, such analysis also can be useful for interpretation of complex output. This is followed by a description of the two-fiber detection scheme and by experimental data verifying the feasibility of this scheme. The two-fiber scheme is also tested in a sensing application, where a gel-film doped with fluorescent groups replaces the original fiber cladding.

2. DISTRIBUTED OPTICAL FIBER SENSING

Optical fibers guide light using the phenomenon of total internal reflection. For this to occur, the fiber core, in which the light propagates, has to have a refractive index n_{co} that is larger than the refractive index of the surrounding fiber cladding, n_{cl}. In step-index fibers, which are considered here exclusively, the core refractive index is constant over the core cross section, dropping to the cladding refractive index value at the core-cladding interface. These refractive indices determine the critical angle Θ_z (measured with respect to the surface normal at the core-cladding interface) according to

$$\Theta_z = \arcsin\left\{n_{cl}/n_{co}\right\} \tag{1}$$

as well as the numerical aperture $N.A.$ (or equivalently the half-angle α_c of the acceptance cone) of the fiber

$$N.A. = n_{air}\sin\alpha_c = \left(n_{co}^2 - n_{cl}^2\right)^{1/2} \tag{2}$$

Interpreting Eq. (1) in terms of geometrical (or ray) optics† could lead to the conclusion that once the critical incident angle is exceeded, an infinite number of guided light modes (each characterized by an incident angle Θ with respect to the normal on the core-cladding interface[38]) is possible. This is not the case, as a proper treatment requires solving Maxwell's equations using the appropriate boundary conditions for time-varying electric and magnetic fields in a waveguide.[38-40] Alternatively, Marcuse[41] showed that ray optics can be extended to obtain a discrete number of light paths (i.e. incident angles) by demanding that the phase fronts (which are represented by light rays) interfere constructively after each subsequent internal reflection at the core-cladding interface. The number of optical modes N that a step-profile fiber can sustain is given by[42]

$$N \cong V^2/2 \tag{3}$$

† Note that Snell's Law is applicable only to planar interfaces. Therefore, in optical fibers it is strictly applicable only to rays whose path crosses the center axis (i.e. meridional rays), but not the helical rays (i.e. skew rays)[30-37]. Snell's law remains a useful tool, though, for the first interpretation of our results.

V is termed the fiber parameter[37] or the normalized frequency parameter.[43, 44] It is determined by the radius r of the fiber core, the wavelength λ of the light in free space, and the critical angle Θ_z as follows

$$V = (2\pi r/\lambda)\, n_{co} \cos\Theta_z \tag{4}$$

Equation (3) is valid for $V \gg 1$.

Even for light propagating in the fiber core under guided conditions, the electric-field amplitude of a light mode does not equal zero at the core-cladding interface. In fact, the electric field $E(z)$ extends normal to the core-cladding interface into the fiber cladding as[45, 46]

$$E(z) = E_0 \exp(-z/d_p) \tag{5}$$

where z is the distance outside of the core and E_0 is the amplitude at the interface. These evanescent fields are characterized by their penetration depths d_p, which are given by[45, 46]

$$d_p = \frac{\lambda}{2\pi n_{co}\left(\sin^2\Theta - \sin^2\Theta_z\right)^{1/2}} \tag{6}$$

Equation (6) shows that each core mode has its own characteristic penetration depth and that modes closer to the critical angle Θ_z, larger critical angles, and larger wavelengths of the light in the fiber core (λ/n_{co}) all lead to increasing penetration depths. While the penetration depth is infinite for light incident at the critical angle, it quickly drops to small values when the incident angles exceed the critical angle. For example, the penetration depth for a light mode incident at angle that is 10% larger than the critical angle is approximately half of the wavelength of the light (for a fiber with $n_{co} = 1.457$ and $n_{cl} = 1.404$ - see Figure 1)

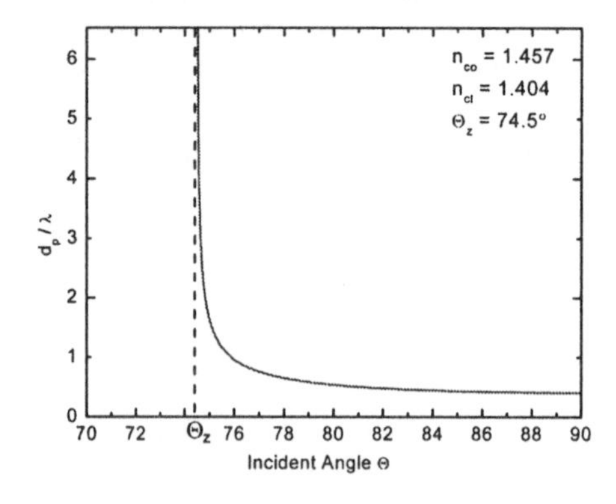

Figure 1. Penetration depth of evanescent fields with respect to the free space wavelength versus incident angle in degrees.

The fact that the electric fields extend into the cladding does not imply that there are losses of light intensity from the fiber core. On the contrary, a calculation of the time average of the component of the Poynting vector (i.e. the power flow density vector) normal to the interface yields zero, showing that no intensity is lost from the core. A calculation of the time average of the Poynting vector component parallel to the fiber axis at the interface, however, yields finite values,[47] demonstrating that there is power flow parallel to the interface in the cladding close to the interface.

The situation is different if absorbing species are present within the range of the evanescent field. Chromophores present in the cladding within the range of the evanescent field can be optically excited, leading to an overall power loss in the fiber.[48, 49] For a multimode fiber with homogeneously distributed absorbers characterized by a cladding absorption coefficient α along a length l of the fiber, the overall ratio of output and input powers P_{out}/P_{in} is approximately given by[42, 50]

$$P_{out}/P_{in} \cong V/\alpha l \tag{7}$$

Equation (7) states that for increasing V, a smaller proportion of the total light admitted into the fiber is available for interaction with absorbers in the cladding, which was pointed out in Ref. [44]

The light emitted by evanescently excited fluorophores in the fiber cladding can couple back into the fiber core and propagate under guided conditions to the fiber ends: light sources in the fiber cladding can interact with the evanescent tails of the guided core modes, transferring power to them.[43, 51] Fluorophores located closer to the core/cladding interface couple light more efficiently into the core and light modes with a larger penetration depths d_p of their evanescent tails are able to collect more light than those with smaller values of d_p. Marcuse[43] showed that the overall light collection efficiency of a fiber from sources distributed homogeneously in the cladding depends approximately linearly on V and on the length l of the region. However, as Eqs. (5) and (6) show, each core mode will interact quite differently with the light sources in the cladding. This is exemplified by the fact that for light sources located directly at the core/cladding interface, the collection efficiency is proportional to V^2, since all modes have reasonably high intensities at the interface.[43]

As indicated in the Introduction, spatial information regarding the location of an emitting fluorophore along the fiber can be obtained using the OTDR method[2, 3, 52-54] (for a review see Ref. [1]). The basic scheme is shown in Figure 2. Laser pulses fed into the front end of the fiber cause pulsed evanescent excitation of the fluorophores.[16] The returning fluorescence 'pulses' arrive at front of the fiber with a time characteristic time delay τ_d that allows for the calculation of the location L of the emitting fluorophore on the fiber according to

$$L = \left(c/2n_{co}\right)\tau_d \tag{8}$$

c is the speed of light in vacuum. Of course, many different sensor regions can be prepared along the fiber. In this case, the data consist of a train of pulses, encoding in the time domain the spatial location of the emitting molecules revealing where a sensing event takes place.

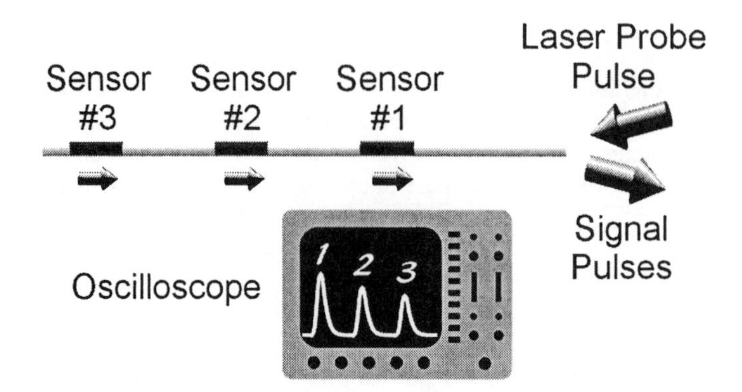

Figure 2. Optical Time Domain Reflectometry (OTDR): Laser pulses are coupled into the fiber and excite fluorophores contained in the sensor regions. The fluorescence pulses from each region arrive with a characteristic time delay at the front end of the fiber, encoding the locations of the sensors along the fiber.

The spatial resolution, that is, the minimum separation of adjacent sensor regions, depends on the fluorescence lifetimes of the fluorophores. This is particularly critical if the sensor response to a triggering event is a change in the fluorescence lifetime of the fluorophore. For an accurate determination of fluorescence lifetime changes, the fluorescence intensity decay curves must be sampled over sufficiently long time intervals to allow for reliable fitting of the decay curves. Using Eq. (8), we obtain for the minimum separation ΔL_{i+1} of adjacent fluorescent regions "i + 1" and "i"

$$\Delta L_{i+1} = L_{i+1} - L_{i} = \left(c/2n_{co}\right) w_{i+1}\, \tau_{fl}(i) \tag{9}$$

where $\tau_{fl}(i)$ is the fluorescence lifetime of the ith fluorophore and w_{i+1} specifies the desired length of the of the detection time window as a multiple of the fluorescence lifetime. For a silica-core fiber with $n_{co} = 1.457$ and for $w_{i+1}\, \tau_{fl} = 30$ ns, Eq. (9) yields for the minimum separation $\Delta L_{i+1} = 3.09$ m. While this may not be critical for some applications, for large, high-density sensor arrays or for combinatorial libraries subjected to fluorescent assays, the minimum separation requirement is a limiting factor. Section 6 will describe a scheme to overcome this limitation.

3. COMBINATORIAL CHEMISTRY

New chemical substances are of value in many areas; Combinatorial Chemistry provides methods to prepare and evaluate large numbers of compounds more easily and more rapidly than was previously possible.[55-61] Since its inception through the innovative work of Geysen et al.,[62] combinatorial chemistry has experienced dramatic growth. Much of the initial work focused on medicinal chemistry.[57, 63-68] Increasingly, however, combinatorial synthesis has found applications in materials science, catalysis, and electrochemistry[69-72] as well as in chemical genetics.[73]

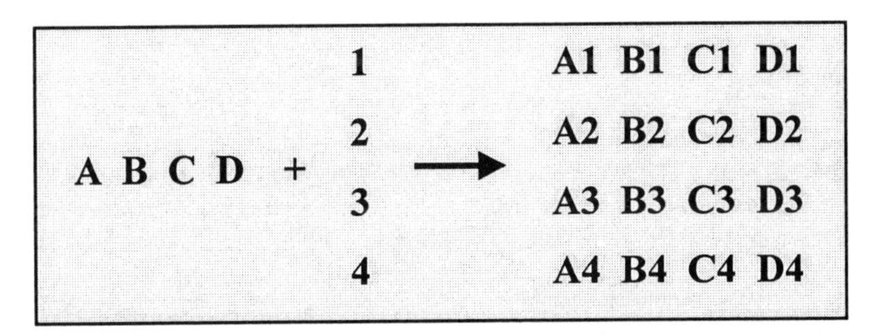

Figure 3. The basic idea of combinatorial synthesis is to create from a set of initial reactants and for a given number of reaction steps a compound library that contains all possible combinations of reactants, thus producing large numbers of substances with few steps.

As shown in Figure 3, the goal is to create from a basic set of reactants and for a given number of reaction steps, a library of all possible compounds that result from the sequential combination of the initial set of reactants. Although computational methods for the calculation of the structure and the properties of molecules have achieved a high degree of accuracy, for the purposes of designing new drugs, however, the degree of uncertainty associated with quantum chemical calculations for complex biological systems often renders this approach less effective. Structure-based drug design, while being a powerful approach, does not afford the reliability needed for targeted synthesis of molecules having the desired properties, particularly since in addition to the structure of a binding site, the dynamics of the whole system will also play a role. Not only are the properties of relevance to drug design calculable only approximately, but it is simply unknown in most cases which properties are desired. A combination of approaches for drug development is most robust, including the synthesis of large numbers of compounds and their systematic evaluation for some set of properties that can be correlated with pharmaceutical efficacy. This is the domain of Combinatorial Chemistry.

A useful method must address three basic questions:

(1) How can one make all these compounds systematically and efficiently?
(2) How can one identify *whether* a compound has the desired properties?
(3) How can one determine *which* compound has the desired properties?

Much attention has focused on the combinatorial synthetic schemes that address the first question, as the compounds must be prepared in order to be studied. The second question is addressed by devising an assay scheme suitable to the compounds prepared, in the amount and format produced by the synthetic method. For example, if the goal of the library is to find members that bind to a target molecule, a common assay is to expose the entire compound library to the target molecule. If the target molecule is fluorescent or tagged with a fluorescent group, the positively identified library member will acquire the fluorescence characteristics of the target molecule and can, thus, be identified. It may at first seem redundant to distinguish between questions two and three, but scenarios can be envisioned, where an assay proves only that there is a useful member in the library, but fails to identify this member in the library (the library members can be of the order of millions[74]). Thus, the "decoding" of the library is required to isolate the desired product compound.

Most reaction sequences that lead to combinatorial libraries are now carried out on solid support materials, which were first introduced to chemical synthesis by Merrifield.[75] The support assumes the role of the "reaction vessel." In the following, three combinatorial methods, all of which require a solid support, will be discussed. For our purposes, we classify these methods according to the dimensionality of the supports that are used:

(1) the "split-mix" method uses zero-dimensional supports such as polymer beads;
(2) the spatial array uses two-dimensional supports such as glass slides;
(3) the linear array method use one-dimensional supports such as cotton threads.

3.1. Zero-Dimensional Combinatorial Chemistry

The "split-mix" technique (also known as split-pool technique or proportioning-mixing method[58, 76-79]) uses quasi-zero-dimensional beads as supports for the reactions leading the compound library. Typical bead diameters are of the order of 100 μm. In the "One-Bead-One-Compound" concept[77] each bead carries just one type of compound, although there may be many copies of this compound on one bead.

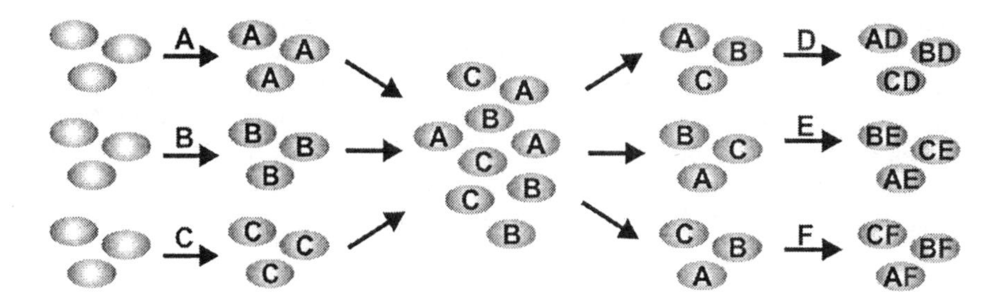

Figure 4. Split-Mix combinatorial method. The reactions are carried out in beads made from porous materials. Repeated diving and mixing of the beads after exposure to the reactants ensures that all combinations of reactants are created. Due to the mixing process, however, the synthetic history of each compound is lost.

The procedure starts by subdividing the beads into as many groups as there are reactants in the first step. Each group is exposed to a different reactant (reactants A, B, and C in Figure 4). Subsequently, the beads are mixed. After resubdivision into the desired number of groups for the second reaction step, each group now carries representatives of the reactions from the first step. Now each group is exposed to one of the reactants in the second step (reactants D, E, and F in Figure 4), after which the beads are again mixed, etc. In this way all possible combinations of reactants are formed. This is a parallel synthesis: reaction step *m* is carried out for each library member at the same time. But the scheme is particularly advantageous in a way we describe as "fully parallel." Note that reagent E in Figure 4 may react with many different species, forming a large number of different products. With this scheme, reagent E (and every other reagent) need be handled only once, and each of the many reactions will result! The major disadvantage is that the synthetic history of all compounds is lost due to the mixing process, that is, a positively identified compound at the end of the procedure cannot be immediately identified, as information of the reaction sequence that led to its creation is no longer available. This necessitates extensive library encoding methodologies or library deconvolution procedures[68, 80, 81] such as spectroscopic bar-coding of the beads.[82]

3.2. Two-Dimensional Combinatorial Chemistry

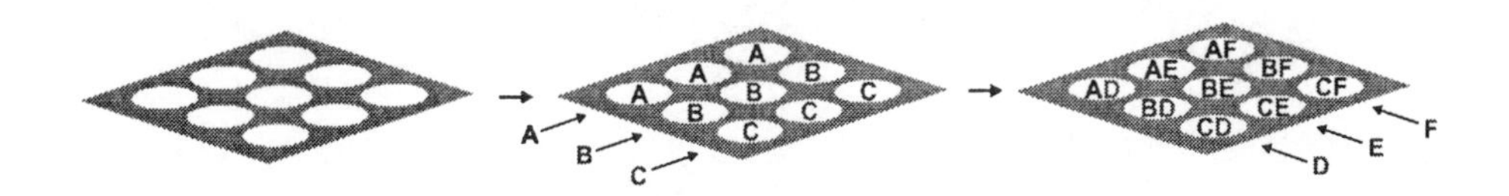

Figure 5. Two-dimensional array method. The synthetic history of each compound is always known by specifying the (x,y)-coordinates of the compound on the two-dimensional support. The process of synthesis itself, however, is a slow, serial procedure.

A second group of methods employs a two-dimensional layout of compounds, such that the position of a molecule on the support identifies its composition.[83] An example is given in Figure 5, where a supporting glass slide is divided into individual regions, each of which carries just one type of compound. All combinations of reactants are created by sequential application of the reactants to these spots. With this scheme, the full synthetic history of each compound is available at any reaction step, however, the synthesis proceeds in a slow, serial fashion. The synthesis is, however, amenable to automation using robotic systems.

3.3. One-Dimensional Combinatorial Chemistry

The one-dimensional combinatorial chemistry method introduced by Schwabacher et al.[25, 26, 84] relies on a support of linear morphology. In its first manifestation, a cotton thread was chosen as support for the combinatorial synthesis.[26] We will show below that the use of linear supports provides fully parallel synthesis *and* access to the complete synthetic history of each library compound through knowledge of the position on the support. The procedure is depicted in Figure 6.

The linear support is wrapped around a cylinder of precisely determined circumference in a single spiral layer. First, the cylinder surface is divided lengthwise in surface regions by laying down divisions (using, e.g., paraffin wax) on the cylinder surface parallel to cylinder axis (black lines in Figure 6). The space between a pair of divisions – hereafter referred to as a "segment" – is now considered to be a reaction vessel. As many segments are created as there are reactants in the first step. In first reaction step, a separate reactant is added to each segment. If three reagents were used for the first reaction step (as in Figure 6), the sequence of compounds attached to the thread would be A B C A B C A B C A.... The identity of each species is specified by its distance from the end of the thread.

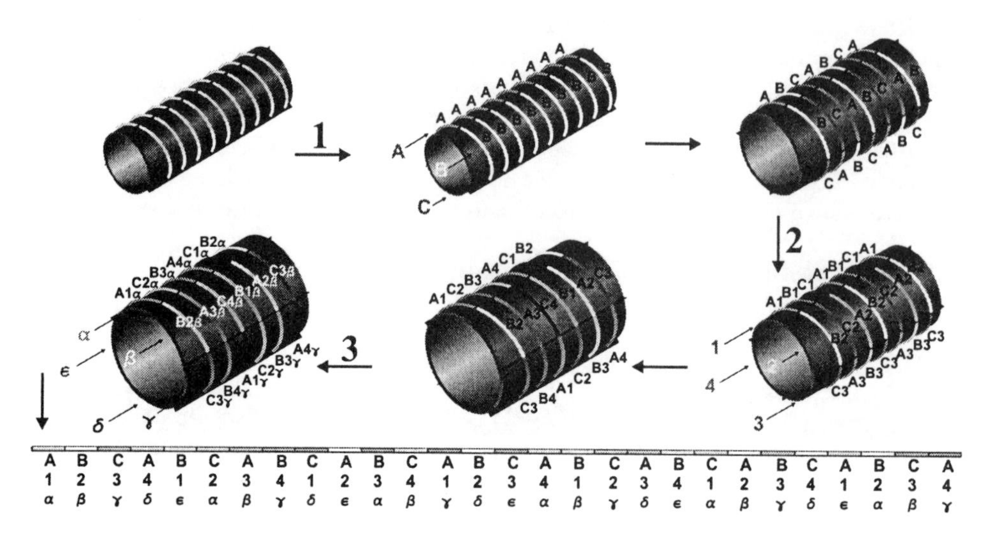

Figure 6. One-dimensional combinatorial synthesis. A three-step reaction sequence is shown, with three, four, and five reactants in steps one, two, and three, respectively. Details see text.

Subsequently, the thread is wrapped around another cylinder of a different, well-defined circumference, depending on the number of reactants in the second reaction step. In the example shown in Figure 6, the circumference is increased by one segment width. Therefore, wrapping the thread spirally onto this cylinder results in each of the newly created segments containing multiple regions of all compounds used in the previous step. Exposure of each segment to the next set of reactants (Step 2 in Figure 6, using compounds 1, 2, 3, and 4) creates all possible combinations of the first step and second reactants. The net result is the sequence A1 B2 C3 A4 B1 C2 A3 B4 C1 A2 B3 C4, etc. Repetition of this process allows use of as many steps as desired, each set of reagents on a cylinder of distinct diameter. The identity of each functional group installed by a given reaction varies periodically as a function of distance along the thread, with each cylinder providing a distinct period for the corresponding portion of the molecule. Thus, after each reaction step, *the full synthetic history of every compound is known from its position on the support!* Furthermore, the synthesis is fully parallel, as each reactant is used only once in each reaction step.‡

Figure 6 also shows the unwrapped linear support after three-step reaction using three, four, and five reactants in steps one, two, and three, respectively. After 60 distinct compounds, which are created in this three-step process, the sequence of compounds will repeat. This periodicity can be exploited to create fitness profiles for each reactant. This issue will be explored further in Section 5 of this chapter.

Having answered the question of synthesis (i.e. Question 1), we now turn to the question of library assay and library decoding. In our one-dimensional combinatorial chemistry scheme, these issues can be addressed simultaneously by an assay that

‡ Other research groups have described procedures that use a thread to organize and identify library members during synthesis, though these involved substantially more manipulation during synthesis.[85, 86]

modifies the fluorescence characteristics of a domain in case of a positive identification. For example, the simple assay described in the beginning of this section, would result in the target molecule – tagged with a fluorescent group – binding to any suitable compound on the thread. In these locations, the thread would become fluorescent. Pulling the thread at a constant speed through a fluorimeter[26] allows recording a fluorescence time trace, from with the locations of fluorescent domains can be extracted.

4. COMBINATORIAL CHEMISTRY ON OPTICAL FIBERS

The one-dimensional combinatorial chemistry method is not limited to cotton thread supports: optical fibers constitute promising and in many ways more versatile linear supports. The reactions leading to the compounds libraries can be carried in the original cladding of the fiber or in a suitable substance replacing the original cladding. From the point of view of chemical synthesis, using silica core fibers with polymeric cladding of the same type used in other combinatorial methods (e.g. in the split-mix method) allows for a wider range of chemistries compared to a cotton thread support. From the point of view of property/function evaluation, employing fluorescent assays for compound library allows for reading out the fluorescent properties of the library spatially resolved with the OTDR method. At this point it is important to recall that it is an inherent virtue of our one-dimensional combinatorial chemistry method that the location of a fluorophore along the fiber encodes the complete synthetic history of an assayed compound. Since compound libraries built on fibers and assayed with fluorophores constitute fluorescent optical fiber sensor arrays, the optical evaluation such libraries are identical to the readout of the quasi-distributed fluorescent chemosensor arrays described above!

The one-dimensional combinatorial chemistry method readily allows for the creation compounds libraries with tens of thousands of members. Any optical assay method has to provide for sufficient light intensity to be able to excite fluorescent groups along the entire length of the optical fiber. This is why it has to be ensured that the interaction of the light in the fiber core with fluorophores in the cladding has to be only through the evanescent fields, keeping light losses from the core minimal. Clearly, the most beneficial assay scenario is one where at the chosen wavelength of the readout pulse, no absorption takes places in the fiber cladding (neglecting losses due to other mechanisms like scattering). As described in section 2, the evanescent fields then probe the cladding at no "cost" in light intensity. Only if a successful binding event positively identifies a compound – causing it to become fluorescent – will excitation and subsequent fluorescence emission take place.

In addition to the generation of lead structures for new drugs, our fiber-optic combinatorial chemistry method offers a variety of other applications. For example, one significant application of arrays of substances prepared in a combinatorial manner is as a sensor array. We envision the preparation of large arrays of fluorescent chemosensors to be quite useful for the selective detection of a variety of analytes. The method provides great flexibility for the fabrication of sensors for a multitude of sensing tasks (e.g. immunosensors, gas and vapor sensors, sensors for toxins, sensors characterizing the chemical and physical environment including pH, metal ions, pCO, pCO_2, pO_2, solvent polarity, temperature, pressure, presence of explosives and other organics, etc.). Furthermore, for the readout of these sensors no imaging is required: the entire array output is obtained as a single time-varying signal. The method offers great flexibility as to the layout of the compounds in the library. For example, the library could consist of only a few different compounds, with the sequence repeating hundreds of times. Conversely, the library could consist of hundreds of different compounds.

The use of fiber claddings as supports for chemical reactions places special demands on them:

- The cladding must be optically transparent.
- The cladding refractive index must be less than that of the fiber core in order to maintain total internal reflection conditions.
- The cladding has to exhibit good swelling properties in a range of solvents.
- The cladding has to be porous to take up reactant and analytes.
- The cladding has to be stable to many reaction conditions.

Unfortunately, claddings of commercially available optical fibers meet none or only some of these requirements. Therefore, for most applications, the original fiber cladding has to be replaced. For example, Blyler, Lieberman et al.[87, 88] used polydimethyl siloxane claddings as hosts for fluorescent dyes that are sensitive to the presence of oxygen. In our work we are focusing on functionalized, crosslinked polyethylene glycol (PEG) gels for use as fiber claddings and as hosts for fluorescent sensor molecules. These polymers were first introduced to split-mix combinatorial chemistry by Meldal et al.[89, 90] under the name SPOCC-gels (super permeable organic combinatorial chemistry gels). In order to achieve a more flexible functionalization of the gels, we developed a new synthetic route that allows for a wider range of concentration of functional groups.[91] A first sensor application of this gel – doped with covalently attached fluorescein – is described below.

5. FOURIER ANALYSIS OF PERIODIC ARRAY OUTPUT DATA

The one-dimensional combinatorial chemistry method leads to a cyclic sequence of compounds along the fiber as each reactant cycles with a period determined by the circumference of the cylinder used to append this particular reactant. If the activity of the compound library (e.g. the affinity for binding the fluorescently tagged target molecule) depends to a significant degree upon a particular portion of the molecule, the activity data (in our example the fluorescence trace) will be periodic. The period will specify the portion of the molecule important for activity. The fact that the output data is periodic suggests an analysis using the Fourier transform.[26]

Assume that the fluorescence trace of the compound library is sampled with a sampling interval Δ and that there are a total number of P consecutively samples h_k with $k = 0, 1, 2, \ldots, P - 1$. A Fourier transform of P input values yields P output values in the frequency domain within the range $-\nu_c < \nu < \nu_c$, where ν_c is the Nyquist frequency $\nu_c = 1/(2\Delta)$. The discrete Fourier transform is given by[92]

$$H(\nu_n) = \Delta \sum_{k=0}^{P-1} h_k \exp\left[2\pi i (kn/P)\right] \tag{10}$$

with $0 \le n \le P - 1$. In this convention,[92] the DC-component of the Fourier transform corresponds to $n = 0$, positive frequencies ($0 < \nu < \nu_c$) correspond to $1 \le n \le (P/2) - 1$, negative frequencies ($-\nu_c < \nu < 0$) to $(P/2) + 1 \le n \le P - 1$. $n = P/2$ corresponds to both $\nu = \nu_c$ and $\nu = -\nu_c$. The discrete frequencies νn are given by $\nu_n = n/(P\Delta)$.

To illustrate the information that can be obtained by applying the Fourier transform to the array output data, we model output data based on the linear combinatorial array pictured in Figure 6. We assign amplitudes to each reactant in each step as well as to combinations of reactants in separate steps. The output signal amplitude is obtained under the assumption that all contributions are additive. The amplitude assignments are given in Table 1.

Table 1. Rules for simulating the periodic output signal of a compound library. Given are the amplitudes *A* for each compound in each step of the three-step combinatorial synthesis depicted in Figure 6. A favorable combination of compounds in steps 1 and 3 is also included.

Step 1	Step 2	Step 3	Steps 1+3
$A(A) = 0$	$A(1) = 0$	$A(\alpha) = 0$	$A(C \wedge \varepsilon) = 1.0$
$A(B) = 0.5$	$A(2) = 0$	$A(\beta) = 0$	
$A(A) = 0.3$	$A(3) = 0$	$A(\gamma) = 0$	
	$A(4) = 0$	$A(\delta) = 0$	
		$A(\varepsilon) = 0.5$	

Figure 7a shows the simulated output based on these rules (assuming $\Delta = 1$). The length of one full sequence is 60, however, the data shows periodicities of three (two "good" reactants in step 1), five (one good reactant in step 3) and 15 (favorable combination of one step-1 reactant and one step-3 reactant).

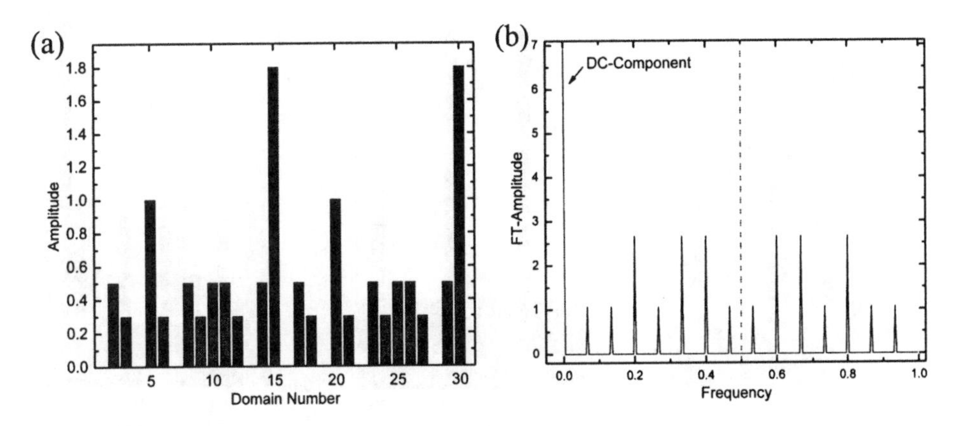

Figure 7. (a) Simulated array output data according to the rules given in Table 1. (b) Fourier transform of the simulated array output. The frequencies 1/3, 1/5, and 1/15 and some of their harmonics are present. No frequency larger than 1/2 (which is the Nyquist frequency in this case) is present. The dashed line at the Nyquist frequency separates positive and negative frequencies.

The data set was extended by repeating this sequence to obtain 255 points. This data set was Fourier-transformed using Mathematica.[93] The frequency spectrum is shown in Figure 7b. It consists of peaks representing the cylinder frequencies and their harmonics. Furthermore, the combination frequency 1/15 corresponding to the product periodicity of steps 1 and 3 is also present. The magnitude of a given frequency domain signal indicates the significance to activity, averaged over the library, of variation of the molecule in the region appended on the corresponding cylinder. Thus, trends in the entire library data are immediately apparent from the Fourier spectrum of the library.

More information can be gained by selective inverse Fourier transform. An inverse transform of the frequency data in Figure 7b will, of course, reproduce exactly the array output data. Having identified the frequencies in the FT-spectrum, though, we can establish "fitness profiles" to summarize how well each specific functional group added at a certain step at a given molecular position contributes to activity. This library-averaged trend is easily extracted from the Fourier spectrum as follows. Only the frequency domain peak corresponding to that molecular region, its harmonics and the DC-component are retained in the Fourier data set – all other values are set to zero. An inverse transform on this reduced frequency data set yields the fitness for that position. The results are shown in Figure 8.

Figure 8a shows the inverse transform for periodicity three. The repeating set of three peaks corresponds to the three different compounds A, B, and C used in step one. The amplitude for each compound can be interpreted a "fitness" number for the contribution of the corresponding compound to the overall activity. As assumed in the rules for the simulation rules for the assay data, compounds B and C are clearly better suited than compound A. Contrary to the rules, however, the fitness numbers for B and C are equal and the number for A is not zero. One reason for this behavior is that the fifth harmonic of the frequency (1/15) interferes with the frequency (1/3). Therefore, the filtering in the frequency domain was imperfect.

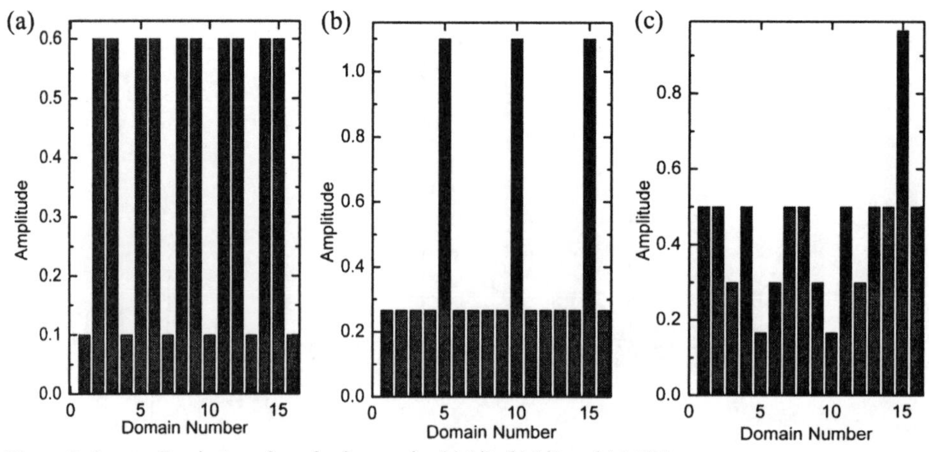

Figure 8. Inverse Fourier transform for frequencies (a) 1/3, (b) 1/5, and (c) 1/15.

Figure 8b shows the fitness numbers for the five compounds used in this step three (i.e. α, β, γ, δ, and ε). The fifth compound ε is recognized to have a significantly higher fitness than the other compounds in this step. This agrees with the amplitude assigned to this compound in the simulation rules. Again, however, the amplitudes for the remaining compounds are non-zero, in contrast to the simulation rules in Table 1. The reason is again interference for harmonics of frequency (1/15).

Figure 8c shows the fitness numbers for the combination frequency (1/15). Shown is one complete period of the inversely transformed data. The fifteenth compound in the sequence is correctly identified as having a large fitness than other combinations of compounds. The numbers for other combinations are again influenced by harmonic interference. Clearly, the Fourier spectrum of a library is convenient for identifying general features important for binding, and may also provide a useful measure of the extent to which activity variation in a library is describable by trends. The reduced reverse transform allows – within the constraints described above – assigning fitness numbers to each compound in each step. A better filtering scheme in the frequency domain could improve these numbers.

6. SPATIAL RESOLUTION INCREASE: FIBER COUPLING

The one-dimensional combinatorial method readily allows for synthesis of compound libraries on optical fibers with tens of thousands of members. As discussed in section 2, if the library is assayed using the changes of fluorescence properties, a minimum spacing must be maintained between adjacent domains in order to be able to fully resolve fluorescence pulses emitted from these domains. Maintaining the minimum separation for large numbers of compounds leads to total lengths of fiber that may be impractical to use in one-dimensional combinatorial synthesis. Moreover, absorption in the fiber-core, starts to become an issue,[94-96] necessitating pulse amplification schemes. Of course, combinatorial libraries can be designed in ways to minimize the probability that two adjacent regions both become fluorescent under the same assay. Moreover, those fiber regions showing non-resolved responses to fluorescent assays can be reproduced quickly with a higher spatial resolution.

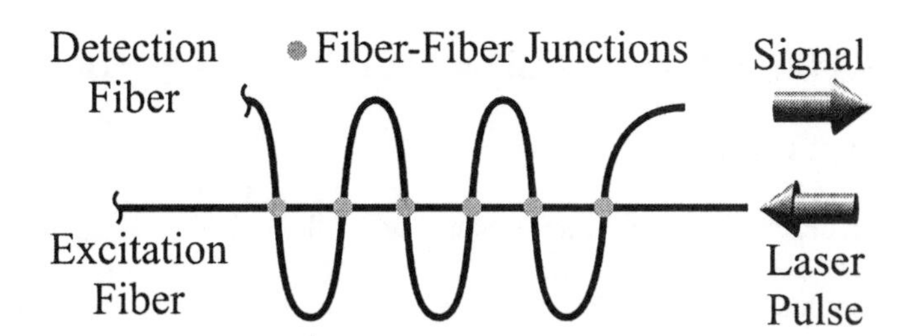

Figure 9. Readout scheme with two optical fibers. On one fiber, the fluorescent regions are closely spaced. Fiber-fiber junctions are created at the fluorescent regions. The fluorophores are excited evanescently through one fiber, whereas the other fiber evanescently captures the fluorescence. The longer fiber provides an optical delay sufficient for temporal resolution of signals from adjacent fiber-fiber junctions.

Accepting restrictions on the spatial resolution implies foregoing a major benefit of our combinatorial synthetic scheme, which is the possibility of efficient fabrication of multi-parameter, high-density arrays of fluorescent chemosensors. To overcome the spatial resolution limitations, we have developed a readout scheme that employs two optical fibers (see Figure 9). One fiber contains the sensor regions that are space much closer than the fluorescence-lifetime limit, while the other fiber provides that optical delay needed to temporally separate the returning signal pulses. The second fiber periodically contacts the first fiber orthogonally at the sensor regions. Each fiber can act as the excitation or the detection fiber, meaning that either fiber can provide evanescent excitation of the fluorophores or collect evanescently the emitted fluorescence.

For this scheme to work, the separation of two adjacent contact points along the delaying fiber has to be sufficiently large to create the required temporal separation. With this scheme, even for large numbers of sensor regions, one of the fibers can be very compact, allowing easy handling and high spatial resolution. In the two-fiber readout scheme the location of a fluorescent sensor is given by

$$L_i = \frac{c}{n_{co}} \tau_i - d_i \tag{11}$$

where L_i is the position of the ith sensor region along the fiber that provides the optical delay between sensor regions with respect to a fluorescent reference region.[27] τ_i is the time delay between the arrival of the signal from region i and that of the reference region, d_i is the distance between sensor region i and the reference region along the fiber on which the regions are closely spaced, and n_{co} is the core refractive index at the wavelength of the emitted fluorescence. The spacing of adjacent sensor regions on the short fiber can be less than 1 mm. For such compact sensor arrays $L_i \gg d_i$, meaning that d_i in Eq. (11) can be neglected. The factor 1/2 appearing in Eq. (8) is absent here.

In addition to reducing the spatial resolution from meters in the one-fiber scheme to less than millimeters, the two-fiber scheme offers the additional advantage of separate optimization of the two fibers for the very different roles they play. In the following sections, we will present data demonstrating the feasibility of the two-fiber scheme for the readout of large, closely packet arrays of fluorescent chemosensors.

7. PROPERTIES OF TWO-FIBER DETECTION SCHEMES

7.1 Experimental Details

The results described below were obtained using three different experimental setups. Figure 10 shows details of an experimental setup consisting of two fibers with multiple fiber-fiber junctions. The "excitation fiber" carried the 337-nm light pulses from a Photon Technology International PL2300 nitrogen laser (typically 0.6-ns pulse width and 1.4-mJ pulse energy with repetition rates of 3 – 10 Hz) to the sensor regions, and the "detection fiber" delivered the resulting fluorescence pulses to a Burle C31034 (2.5-ns rise-time) photomultiplier tube (PMT). The resulting signals were collected on either a Hewlett Packard HP54505B or a LeCroy LC564DL digital storage oscilloscope (DSO) with bandwidths of 300 MHz and 1 GHz, and sampling rates of 500 MSa/s and 4 GSa/s, respectively. The signal from a second PMT (RCA 1P28, 1.6-ns rise-time), which collected light scattered from the front of the excitation fiber, provided a trigger pulse for the DSO. For the spectral decomposition of the fluorescence signals – if used – colored glass filters, a Jobin Yvon H10 monochromator (0.1-m focal length, 8-nm/mm linear dispersion), or a Jarrell Ash 82-000 monochromator (0.5-m focal length, 1.6-nm/mm linear dispersion) were employed.

The sensor regions at the fiber junctions were prepared by removing the fiber jacket and cladding at the junction(s) so that the two fiber cores were just touching. The contact pressure was minimal to avoid losses from microbending. The junctions were immersed in a solution of a fluorescent dye in a solvent with a refractive index that preserved the guiding condition of the fiber (more detail on the refractive index dependence below). In Figure 10, it is the detection fiber that provides an optical delay between detection of the fluorescence from adjacent regions. In the setups that contain multiple fiber junctions, the separation between adjacent regions along the fiber providing the optical delay was typically two orders of magnitude greater than their separation with respect to the other fiber. All setups used multimode silica fibers with a TECS™ (trademark of 3M Corp.) cladding (Thorlabs/3M FT-200-UMT and FT-400-UMT with core diameters of 200 μm and 400 μm, respectively). All fibers had core and cladding refractive indices of n_{co} = 1.457 and n_{cl} = 1.404, respectively. Although the TECS cladding shows some absorption in the UV region leading to transmission losses in the fiber, its use is advantageous since the cladding can be removed with acetone, thereby preserving the core surface. This avoids the HF treatment often applied for the removal of silicone claddings, which may etch the core surface and lead to refractive losses.

The first fiber setup was used for measuring the sensitivity of evanescent coupling between the two fibers using a single junction that was located approximately 1 m along the excitation fiber: a 400-μm fiber for excitation and 200-μm fiber for detection. A single junction was located approximately 1 m along the excitation fiber. The evanescently captured fluorescence signal had to travel through a length of 500 m of the detection fiber to reach the PMT.

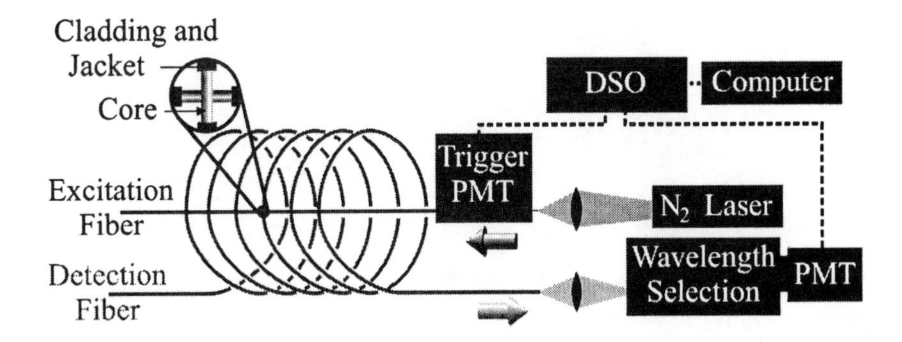

Figure 10. Details of a two-fiber experimental setup with multiple fiber-fiber junctions. The enlargement shows one of the fiber junctions, where the core of both fibers was exposed and immersed into solutions of fluorescent dyes. PMT = photomultiplier tube, DSO = digital storage oscilloscope. Copyright 2002 Association for Laboratory Automation.

The second setup was used to test the attenuation of the exciting light pulse when multiple sensor regions are present on the fiber. For these experiments a 200-μm fiber was coiled up on a 20-cm-diameter cylinder. Every 9 windings (corresponding to a separation of 5.84 m or 28 ns), 1 cm of the fiber core was exposed. A 400-μm fiber with cladding and jacket removed touched each of these exposed regions. The crossing points were approximately 1 cm apart along the 400-μm fiber. Six fiber-fiber coupling regions were created in this way and submerged in a 2×10^{-5}-mol/L aqueous solution of Rhodamine 590 (R6G). The refractive index of this solution was approximately that of water, which ensured that the guiding conditions of both fibers were retained throughout the exposed regions.

This setup was also used to measure the dependence of the intensity of the signal pulses on the refractive index of the solution, which was in direct contact with both fibers at the fiber-fiber junctions. In these experiments, the junctions were submerged in glycerol/water mixtures of varying composition, maintaining a constant R6G concentration of 2.5×10^{-5} mol/L. The refractive indices ranged from 1.3797 to 1.4712 (measured with an Abbé-type refractometer – AO Scientific Instruments).

The third setup was prepared to model multi-parameter sensor systems with different sensor dyes. The goal was to see how a number of closely-packed sensor regions containing different dyes (Stilbene 420 (S420), Coumarin 500 (C500), R6G, LDS698 – all Exciton) can be spatially and spectrally resolved. Ten fluorescent regions were prepared, separated by only 3 windings (which corresponds to a temporal separation of approximately 9.5 ns) of the detection fiber. Subsequently, the regions were immersed in a drop of agarose gel swelled with water containing the dye molecules of interest. These sensor regions were spaced by approximately 5 mm along the 400-μm-diameter excitation fiber. The dyes in each sensor region, from the end of the detection fiber closest to the detector, were: S420; LDS698; R6G; C500; S420; LDS698; C500; S420; LDS698; C500.

7.2. Sensitivity of Fiber-Fiber Coupling

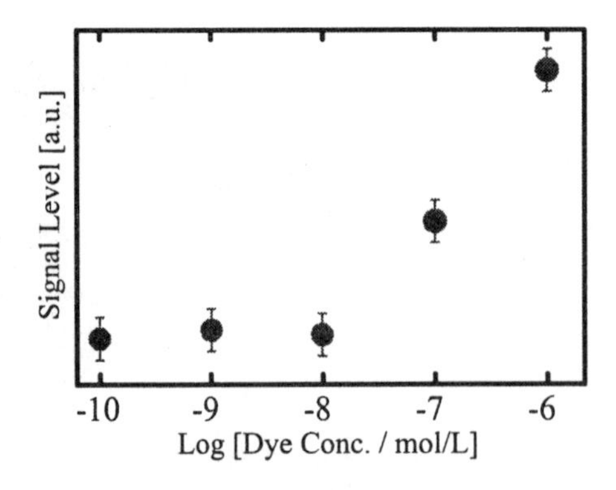

Figure 11. Single fiber-fiber junction immersed in aqueous solutions of R6G with increasing concentrations. A concentration of 10^{-7} mol/L can readily be detected. Copyright 2002 Association for Laboratory Automation.

Setup 1 with a single fiber-fiber junction was used to determine the sensitivity of the measurements in the evanescent fiber-fiber coupling scheme. Aqueous solutions of RG6 were created, with the R6G concentration increasing by factors of ten starting from a solution of 10^{-10} mol/L until an R6G fluorescence signal could be discerned above the level of the background signal. The average of 1000 time-traces obtained for each concentration was integrated and divided by the integrated signal time-trace from the reference PMT. The results in Figure 11 show that an R6G concentration of 10^{-7} mol/L can be easily detected above the background signal using an effective sample volume in the sub-nanoliter region.

7.3 Attenuation in the Two-Fiber Scheme

As described above, setup 2 consists of six fiber-fiber junctions, each immersed in a 2×10^{-5}-mol/L aqueous solution of R6G. Figures 12a and 12b show signal pulse trains obtained by averaging 64 fluorescence time-traces. Even though there are equal R6G concentrations at each fiber-fiber junction, the resulting intensities vary. This is due to the exponential dependence of the evanescent fields on distance (see Eq. (5)): a small variation in the separation of the fibers there has a large effect on the signals. Note that no particular attention was paid with regard to maintaining equal fiber spacings in all regions when setup 2 was constructed.

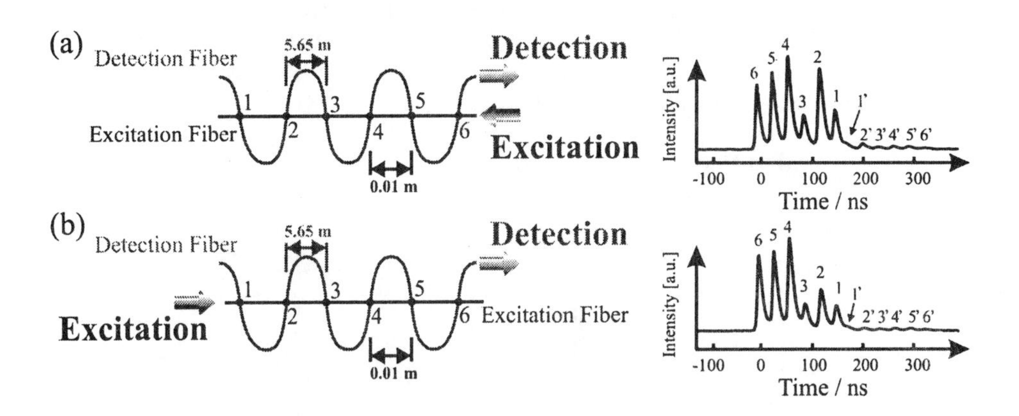

Figure 12. Signal traces obtain with a six-junction array using R6G fluorophores. Varying the excitation end provides information about losses in the excitation fiber. For strong losses, the signal intensity should be skewed in favor of junction "6" in (a) and "1" in (b). This is not the case, showing that attenuation is minimal.

Since the one-dimensional combinatorial chemistry method allows for the construction of large arrays of fiber-optic fluorescent chemosensors, it was important to the test the suitability of the fiber-fiber coupling scheme for such arrays. A primary concern is the attenuation of the exciting and signal pulses. This issue was investigated by switching excitation and detection ends of the of the six-junction array.

Figure 12 shows one such test in which the PMT was kept at the same end of the detection fiber; however, excitation took place at the different ends of the excitation fiber. Since the spacing of the fiber-fiber junctions along the excitation fiber on 1cm and the extent of a laser pulse inside the fiber is approximately 12.4 cm, the dye molecules in each junction are essentially excited simultaneously and their fluorescence is coupled simultaneously into the detection fiber. Irrespective of the excitation-fiber end at which the pulse was coupled into the fiber, the subsequent fluorescence pulses travel the same path in the detection fiber. Therefore, in both cases the pulse from fiber junction six arrives first at the PMT, followed by the pulse from junction five, etc. Comparing the relative peak intensities of traces shown in Figures 12a and 12b provides information about the losses within the excitation fiber: with zero attenuation in this fiber these traces will be exact duplicates. It can be seen that spite of slight variations in the relative peak intensities, the overall pattern remains, suggesting minimal attenuation.

A similar comparison can be made by keeping the excitation end fixed and varying the detection end. This case provides information on the losses in the detection fiber, as the fluorescence pulse now travels different paths in that fiber – in the ideal case the traces monitored at the different ends will exactly mirror one another. Experimentally we found only minor variations,[27] showing again that the attenuation is minimal.

These observations are corroborated by the presence of a second set of peaks, which can be seen in Figures 12a and 12b starting at approximately 200 ns after all fluorescence signals of the six junctions reached the PMT (labeled 1'-6'). These arise from reflections of the sensor fluorescence from the far-end (i.e. the non-detection end) of the fiber, which can be verified by placing a mirror at this end of the fiber. The second set of peaks, which arrives in inverse order at the PMT compared to the first set, is amplified accordingly.[28] The fact that, for example, the signal peak arising from the fluorescence of junction "6" but arriving in the second set, travels the length of the fiber twice and passes by 11 sensor regions before reaching the detector, is detectable again indicates low loss and weak evanescent coupling in the fiber junctions.

7.4. Refractive Index Dependence

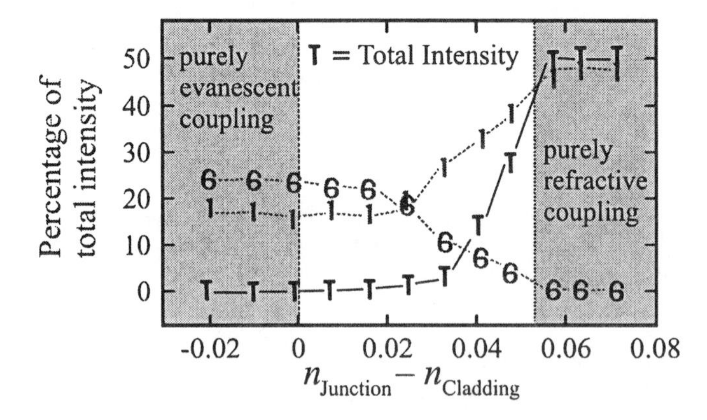

Figure 13. Integrated signal intensities originating from all six fiber junction (symbols "T") and from junctions "1" and "6," respectively. The abscissa is the difference of the refractive index at the fiber junctions and that of the original fiber cladding (which is retained outside of the junctions). The ordinate values are the percent intensities of peaks "1" and "6" with respect to the integrated intensity from all junctions. Copyright 2002 Association for Laboratory Automation.

As describe above, for most applications in fiber-optic sensing and/or combinatorial chemistry, the cladding on most commercially available optical fibers is not suitable as host for sensor molecules and/or chemical reactions. Section 4 describes the general requirements that a substance replacing the fiber cladding has to meet. This is particularly true in the two-fiber scheme, where the coupling efficiency of the two fibers depends critically on the refractive index. This is demonstrated using setup 2, which contains six fiber-fiber junctions, by varying the refractive index n_r the solution that surrounds the fiber junctions. Since the cladding of both fibers was removed at the sensor regions, the solution was in direct contact with the fibers' cores. Thus, the refractive index of the solution n_r determined the value of the critical angle Θ_z and therefore, the extent of guidance of the light modes in the fiber core at the sensor regions.

Figure 13 shows the refractive-index dependence for a series of 12 solutions. The solid line connecting the "T" symbols represents the integrated fluorescence intensity from dye molecules in all six fiber junctions. The refractive indices are given with respect to the index of the original fiber cladding, which is retained outside of the fiber-fiber junctions. The relative refractive index range $\Delta n_r := n_r - n_{cl}$ can be divided in three ranges:

(1) $\Delta n_r < 0$: All guided modes in regions where the original cladding is present are still guided in the junction region. Coupling between the fibers is purely by the evanescent mechanism. Lowering the solvent refractive index within this range makes the fiber more strongly guiding at the sensor regions. While coupling is still evanescent, it is also weaker since those modes that lie close to the critical angle (and, therefore, have the largest evanescent penetration depth – see Eqs. (5) and (6)) when launched into the fiber, lie further from the critical angle of the fiber at the sensor regions and consequently, have a smaller penetration depth. The result is that the relative intensities of the individual peaks are preserved, but the absolute intensities drop.

(2) $0 < \Delta n_r < n_{co} - n_{cl}$: The solvent refractive index for n_r is above that of the original cladding, but still below the core refractive index. With increasing Δn_r more modes that propagate close to the critical angle when launched into the fiber, propagate at angles greater than the critical angle at the junctions, and, therefore, couple refractively out of the fiber. They can also enter the detection fiber via refraction and be guided to end of the fiber when the original is encountered outside of the junction. The total intensity increases.

(3) $n_{co} - n_{cl} < \Delta n_r$: All modes that are guided in regions where the original cladding is present are now able to refract out of the core of the excitation fiber and refract into the detection fiber at the junction. After coupling by refraction into the detection fiber, the light will encounter the original cladding in regions outside of the junctions, making guided propagation to the PMT possible. The integrated intensity plateaus for refractive index values greater than that of the fiber core due to the fact that most of the light is lost from the fiber via a refractive coupling mechanism at the first few junctions.

The dashed lines in Figure 13 show the integrated intensities of the signal peaks "1" (the first peak to arrive at the detector in this particular readout configuration) and "6" (the last peak to arrive at the PMT) as a percentage of the total intensity of all peaks. For solvent refractive indices with $\Delta n_r < 0$, the intensities of all peaks remain largely unchanged. This is true for all six peaks.[28] In range (3) with $\Delta n_r > n_{cl} - n_{co}$, the intensities are very different from those of range (1), but constant relative to each other. In the intermediate range (2), the intensity of the first peak increases with increasing refractive index. The same behavior is observed for the second peak (not shown in Figure 13 – see Ref.[28]), although the increase is not as pronounced. All other peaks decrease in intensity with increasing refractive index, although the magnitude of the decrease consistently rises from peak 3 to peak 6. Overall, there is a redistribution of relative intensity towards those sensor regions that lie closer to the input ends of the fibers. As a matter of fact, once the solution refractive index reaches the core refractive index, the signal peaks 3-6 disappear completely as most of the light intensity is quickly lost from the excitation fiber at the first two junctions. The numbers "6" in Figure 13 show the behavior of the intensity of the last signal peak reaching the PMT.

For the readout of large arrays of fluorescent regions it is important to strike a careful balance between preserving light intensity in the fiber core, yet allow for sufficient interaction at the fiber junctions to generate a measurable signal. Favorable conditions to meet these requirements are found in range (1) where the refractive index in the junction is less than that of the original fiber cladding – provided that the original cladding is retained outside of the junction region.

7.5 Arrays with Multiple Dyes

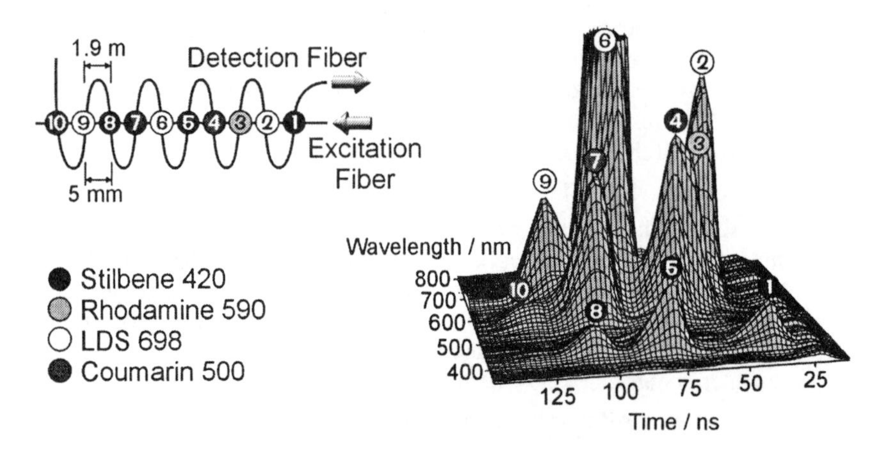

Figure 14. Fluorescence intensity vs. time and wavelength for an array with ten fiber-fiber junctions containing multiple fluorescent dyes. The spacing of the dyes along the delaying fiber is insufficient for full temporal resolution; however, the signals are sufficiently strong for spectral decomposition with a 0.5 m monochromator.

The third setup, in which four different dyes were distributed among ten sensor regions, was prepared to model multi-parameter sensors. As described in section 7.1., the spacing of adjacent fiber regions along the detection fiber, which provides the optical delay in this case, corresponds to only ~ 9.5 ns. This figure is comparable to the fluorescence lifetimes of these dyes, meaning that there is insufficient spacing for full temporal resolution of the array signal output.[27] Using the 0.5-m monochromator for the spectral decomposition of the light exiting the detection fiber, the signals from the individual regions can still be distinguished. Figure 14 shows a surface plot, in which the emission intensity is plotted versus wavelength and time of detection. For each wavelength, 128 time traces were averaged. While the coupling efficiency varies significantly for the different regions, all 10 regions are accounted for.

These results show that even when the sensor regions are closely spaced along the fiber that provides the optical delay, they may still be spectrally resolved. It should be noted that the fluorescent signal from one region could be reabsorbed by a different dye in a subsequent region (for example, the S420 from regions "5" and "8" passes two regions containing LDS698, i.e. "2" and "6", which can be excited in this way). The significance of reabsorption processes for the design of large fiber arrays requires further study.

8. APPLICATION: A CROSSED FIBER PH-SENSOR

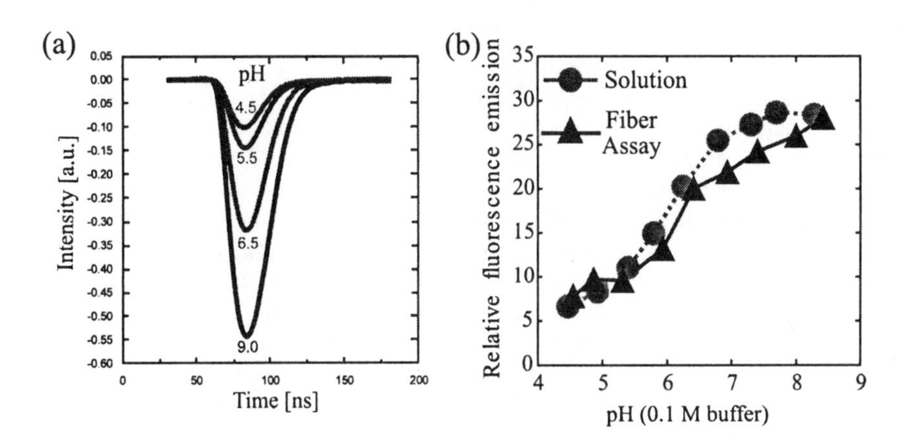

Figure 15. Fluorescein doped into a polyethylene glycol (PEG) gel as a pH sensor. PEG acts as replacement cladding and immobilizes fluorescein by covalent attachment. (a) Dependence of signal peak on pH; (b) comparison of integrated intensity in fiber assay and solution measurement in a fluorimeter.

Setup 1 is now employed to evaluate the two-fiber scheme in a real sensing application, using as a replacement cladding and as host for the fluorescent sensor molecules the polyethylene glycol gel described in section 4. Sol-gel films have been used earlier to host dye molecules for oxygen concentration[97] and for pH measurements.[16, 98] In our experiment, we use as pH-sensor the fluorescent dye fluorescein, which was covalently attached to the PEG gel matrix. The emission intensity of fluorescein changes with pH.[99] A thin film of the gel was placed between the two fibers at the junction, with the effect that the sensor molecules were immobilized within the range of the evanescent fields.

Figure 15a shows the signal pulse originating from the single fiber junction for various pH-values. The signal decreases with decreasing pH as expected. This demonstrates that the solvent mixture used to change the pH can permeate the gel and that there is full access to the sites in the gel to which the fluorescein is attached. Figure 15b plots the integrated intensities of the signal peaks (triangles) versus pH and compares it data obtained from fluorescein solutions in a fluorimeter (circles). The agreement is quite satisfactory, which verifies that fluorescein in the gel shows the expected response to pH changes

9. DISCUSSION AND CONCLUSIONS

The results presented in the previous sections demonstrate that the two-fiber scheme overcomes restrictions on the spatial resolution found in applications involving fluorescent regions on optical fibers. In our studies, these regions were separated by 5–10 mm; however, a further reduction is possible. A practical limit is given by the diameter of the fiber providing the optical delay, whereas the penetration depth of the evanescent waves imposes a theoretical limit. As long as the fiber junctions are separated by approximately 5 λ, the probability that the fluorescence from one such region is picked up by two regions on the detection fiber, or that one sensor region is excited repeatedly from two regions on the excitation fiber is negligible.

As in the one-fiber fluorescent OTDR scheme, the two-fiber scheme relies on two evanescent coupling steps: transfer of exciting light from core to cladding and capture of emitted light into guided core modes. The two-fiber scheme, however, offers more flexibility since fiber parameters (like core diameter, refractive indices, numerical aperture) can be optimized separately for excitation and detection fibers. This flexibility is particularly important for large arrays of fluorescent regions on fibers. The magnitude of light intensity available for excitation at the fiber junctions can be controlled to some extent by the choice of refractive indices of the excitation fiber. Likewise, the detection fiber can be optimized in this way for maximum capture efficiency. Furthermore, the values of E_0 (Eqn. (5)) and d_p (Eqn. (6)) depend on the fiber parameters and on the particular core mode.[45, 46] The range of core modes can be controlled to some extent by the cone angle of the light focused into the end of the fiber. The total amount of light that a fiber accepts increases with the number of guiding modes it can support,[100] which again is directly related to the basic fiber parameters (see Eqs. (1)-(4)). For large arrays of fluorescent regions, the optical transmission characteristics of the delaying fiber have to be considered. The spectral position of the transmission maximum of the fiber and the excitation and emission wavelengths in a particular experiment will determine whether the delaying fiber is best used as the excitation or detection fiber.

The excitation fiber ideally has to provide equal intensities to each sensor region. Even if there is only evanescent coupling, a gradual decrease in the amount of light in the fiber, and therefore, the extent of sensor-region excitation, is expected along the fiber. To minimize this decline, the percentage loss of light from the excitation fiber at each sensor region must be kept small. Based on the results presented above, there are a number of ways that may be used to approach the characteristics of an ideal excitation fiber. It was shown[51, 53] that the percentage of the total power within a fiber that is carried by the collective evanescent fields decreases as the fiber becomes more strongly guiding (see Eq. (7)). As the number of modes that a fiber can support increases, the *percentage* of light available for coupling through the evanescent fields decreases. Thus, by choosing very strongly-guiding fibers for excitation, only a small proportion of the modes (those propagating close to the critical angle) would be available for evanescent coupling at each sensor region.

The efficiency of evanescent coupling from the excitation fiber could also be controlled by selectively populating a subset of its modes.[101] For example, if only modes propagating at angles far from the critical angle are populated in a strongly guiding fiber,[42, 100] their evanescent penetration depths will be small, and, consequently, there will be small losses from the fiber at each sensor region. Alternatively, if the refractive index of sensor regions n_r is smaller than that of the excitation fiber cladding – a case that was identified in section 7 as the ideal case – then modes propagating close to the critical angle when launched into the fiber will be far from the critical angle in the sensor regions. Consequently, these modes will experience smaller losses at each sensor region. In all cases, the more light that the excitation fiber can initially carry, the better, so large-diameter fibers are preferred for the excitation fiber.

For the detection fiber a different picture arises. The light collection efficiency of a fiber is proportional to V for homogeneously distributed light sources in the cladding and proportional to V for sources located directly at the core/cladding interface.[43] The fiber parameter V, which is defined in Eq. (4) , is related to the number of modes that a fiber can sustain (Eq. (3)). Therefore, maximizing the fiber parameter leads to increased collection efficiency and, therefore, to larger signals in the detection fiber.

The sensitivity of the evanescent fiber-fiber coupling scheme is remarkable considering that the cladding of the fibers used in our experiments produced a background fluorescent signal. R6G concentrations of 10^{-7} mol/L could easily be detected against this background. Using a longer-wavelength light source for excitation, this background luminescence effectively eliminates leading to a significant improvement in the signal to noise ratio (S/N) and, therefore, an enhancement of the sensitivity. It is also worthy of note that the previous experiments of Kvasnik et al.[12] and Chronister et al.,[16] which utilized a single fiber for both excitation and detection, probed sensor regions which spanned ~25 cm and ~4 cm, respectively. In the two-fiber scheme, the sensor regions probed are orders of magnitude smaller, since the sensor regions encompass only the volume that lies within ~ 5λ of both fibers at the fiber junction. It is thus, all the more remarkable that this technique, in which such small volumes are probed, is as sensitive as we have determined here. For the data collected using setup 1, all detectable contributions to the noise were stochastic in nature. Thus, although the points plotted in Figure 11 were the result of averaging 10^3 traces, the LeCroy DSO has the capability of averaging 10^6 traces, providing the potential for a further 30-fold increase in S/N. At the present time such an experiment is not feasible due to the low repetition rate of the laser (recording 10^6 traces at 10 Hz would take ~28 hrs). An excitation source with higher pulse repetition rates will further enhance the sensitivity of this readout scheme.

Our experiments show that changing the refractive index of the substance replacing the cladding has a profound influence on the signals. Not only are the overall intensities affected, but also the relative intensities of the signals of the individual regions. Clearly, the losses that occur at the sensor regions under refractive conditions will prohibit the creation of long arrays of sensors. For minimal attenuation the refractive index n_r of the substance replacing the cladding (the role played by the solution in the experiments described in section 7.4.) has to be equal or less than the refractive index of the original cladding, provided that the original fiber cladding is retained outside of the fiber junctions. The results shown in Figure 12 were obtained for such a case: the attenuation is minimal.

One of the most important fields of application of our techniques is in the synthesis of biologically relevant molecules and in the area of biosensors (e.g. immunoassays). In these cases, the chemistry takes places in aqueous solutions at low concentrations. This means that the refractive index is determined by that of water ($n_W \sim 1.34$). These solutions permeate the porous, low-density replacement claddings. The refractive index of the replacement cladding including the solution is still close to n_W. A typical core refractive index for silica fibers is $n_{co} = 1.457$, which implies that for guided propagation the cladding index has to be smaller than n_{co}. Substituting a replacement cladding containing an aqueous solution with a refractive index of approximately n_W, means that biologically relevant chemistries can be carried out in the most beneficial refractive index range, i.e. range (1).

10. SUMMARY AND OUTLOOK

The one-dimensional combinatorial chemistry method described in this chapter allows fabricating large libraries of compounds on linear supports. Depending on the selected reactants, a multitude of libraries for many different purposes can be created. Whether fluorescent chemosensors are made this way or whether a library of potential drugs is subjected to a fluorescent assay, common to both is that the evaluation of the library is performed optically using the fluorescence originating from a certain region along the linear support. When optical fibers are used as linear supports, spatially resolved optical evaluation can take place by sending laser pulses down the fiber core (OTDR/OTOF method). We show that with a two-fiber readout scheme the minimum spacing between adjacent fluorophores can be much less than dictated by the fluorescence lifetimes in a one-fiber scheme. This was verified with a single-component six-region array and a multi-component array containing ten sensor regions. Signal attenuation in two-fiber arrays is lower than that reported in single-fiber arrays.[16] Using the multi-component array, we showed that spectral discrimination of the different fluorophores is possible, in spite of the fact that the two fibers are only weakly coupled through their evanescent fields.

The impact of our fiber-optic combinatorial chemistry technique is the ease with which diverse multi-parameter sensor arrays can be fabricated and their output obtained and evaluated. The combinatorial approach provides a particularly simple method of chemosensor preparation and array fabrication. A distinctive feature of the readout method is the multiplexed use of a single light source and a single detector to provide real-time output from an entire array of sensors. A focus here has been close packing of quasi-distributed sensor regions. We also note that sparse packing can provide spatially resolved measurements of interest for some applications. We anticipate exciting new developments in the near future.

ACKNOWLEDGEMENTS

We would like to acknowledge valuable contributions from the members of our research groups. In the Geissinger group these are postdocs Dr. Barry Prince and Dr. Nadejda Kaltcheva; in the Schwabacher group graduate students Yixing Shen, Christopher Johnson, and Anna Benko and postdoc Dr. Maureen Prince. This material is based upon work supported by the National Science Foundation under Grants CHE-9874241 (AWS) and CHE-0078895 (PG). PG would also like to acknowledge support from the University of Wisconsin-Milwaukee Campus Opportunity Fund.

REFERENCES

1. J. P. Dakin, *Distributed Optical Fiber Sensors*; Proceedings of the Conference on Distributed and Multiplexed Fiber Optic Sensors II, Boston, MA, U.S.A., SPIE-The International Society for Optical Engineering, Vol. 1797, pp. 76-108, 1992.
2. Y. Ueno, and M. Shimizu, An Optical Fiber Fault Location Method, *IEEE J. Quantum Electron.* **QE-11**(9), 77D-78D (1975).
3. Y. Ueno, and M. Shimizu, Optical Fiber Fault Location Method, *Appl. Opt.* **15**(6), 1385-1388 (1976).
4. J. P. Dakin, and B. Culshaw, Editors, *Optical Fiber Sensors: Principles and Components* (Artech House, Inc., Norwood, MA, 1988) Artech House Optoelectronics Library Vol. 1.
5. B. Culshaw, and J. P. Dakin, Editors, *Optical Fiber Sensors: Systems and Applications* (Artech House, Inc., Norwood, MA, 1989) Artech House Optoelectronics Library Vol. 2.
6. B. Culshaw, and J. P. Dakin, Editors, *Optical Fiber Sensors: Components and Subsystems* (Artech House, Inc., Norwood, MA, 1996) Artech House Optoelectronics Library Vol. 3.
7. J. P. Dakin, and B. Culshaw, Editors, *Optical Fiber Sensors: Applications, Analysis, and Future Trends* (Artech House, Inc., Norwood, MA, 1997) Artech House Optoelectronics Library Vol. 4.
8. R. A. Potyrailo, S. E. Hobbs, and G. M. Hieftje, Optical Waveguide Sensors in Analytical Chemistry: Today's Instrumentation, Applications and Trends for Future Development, *Fresenius J. Anal. Chem.* **17**(10), 593-604 (1998).
9. J. P. Dakin, D. J. Pratt, G. W. Bibby, and J. N. Ross, Distributed Optical Fibre Raman Temperature Sensor Using a Semiconductor Light Source and Detector, *Electron. Lett.* **21**(13), 569-570 (1985).
10. A. H. Hartog, A. Leach, and M. P. Gold, Distributed Temperature Sensing in Solid-Core Fibres, *Electron. Lett.* **21**(23), 1061-1062 (1985).
11. D. Culverhouse, F. Farahi, C. N. Pannell, and D. A. Jackson, *Exploitation of Stimulated Brillouin Scattering as a Sensing Mechanism for Distributed Temperature Sensors and as a Means of Realising a Tunable Microwave Generator*; Proceedings of the OFS '89, Paris, France, Springer Verlag, pp. 552-559, 1989.
12. F. Kvasnik, and A. D. McGrath, *Distributed Chemical Sensing Utilising Evanescent Wave Interactions*; Proceedings of the Conference on Chemical, Biochemical, and Environmental Fiber Sensors, Boston, MA, SPIE-The International Society for Optical Engineering, Vol. 1172, pp. 75-82, 1989.
13. G. J. Cowle, J. P. Dakin, P. R. Morkel, T. P. Newson, C. N. Pannell, D. N. Payne, and J. E. Townsend, *Optical Fibre Sources, Amplifiers and Special Fibres for Application in Multiplexed and Distributed Sensor Systems*; Proceedings of the O/E Fibers 91, Boston, SPIE-The International Society for Optical Engineering, Vol. 1586, pp. 130-145, 1991.
14. A. A. Boiarski, and V. D. McGinniss, Electric Power Research Institute, Inc., U.S. Patent 5,191,206 (1993).
15. M. Höbel, J. Ricka, M. Wuthrich, and T. Binkert, High-Resolution Distributed Temperature Sensing with the Multiphoton-Timing Technique, *Appl. Opt.* **34**(16), 2955-2967 (1995).
16. C. A. Browne, D. H. Tarrant, M. S. Olteanu, J. W. Mullens, and E. L. Chronister, Intrinsic Sol-Gel Clad Fiber-Optic Sensors with Time-Resolved Detection, *Anal. Chem.* **68**(14), 2289-2295 (1996).
17. K. Nakamura, N. Uchino, Y. Matsuda, and T. Yoshino, Distributed Oil Sensors by Eccentric Core Fibers, *IEICE Trans. Commun.* **E80-B**, 528-534 (1997).
18. A. D. Kersey, *Multiplexed Fiber Optic Sensors*; Proceedings of the Conference on Distributed and Multilplexed Fiber Optic Sensors II, Boston, MA, U.S.A., SPIE-The International Society for Optical Engineering, Vol. 1797, pp. 161-185, 1992.
19. K. J. Albert, N. S. Lewis, C. L. Schauer, G. A. Sotzing, S. E. Stitzel, T. P. Vaid, and D. R. Walt, Cross-Reactive Chemical Sensor Arrays, *Chem. Rev.* **100**(7), 2595-2626 (2000).

20. J. P. Dakin, The Plessey Company, U.K. Patent GB 2156513A (1984).
21. J. P. Dakin, and D. J. Pratt, *Fibre-Optic Distributed Temperature Measurement-A Comparative Study of Techniques*; Proceedings of the IEE Colloquium on 'Distributed Optical Fibre Sensors', London, U.K., Vol. Digest No.74, pp. 10/11-16, 1986.
22. M. E. Lippitsch, S. Draxler, and M. J. P. Leiner, *Time-Domain Fluorescence Methods as Applied to pH Sensing*; Proceedings of the Conference on Chemical, Biochemical, and Environmental Fiber Sensors IV, Boston, MA, SPIE-The International Society for Optical Engineering, Vol. 1796, pp. 202-209, 1992.
23. R. A. Potyrailo, and G. M. Hieftje, Optical Time-of-Flight Chemical Detection: Absorption-Modulated Fluorescence for Spatially Resolved Analyte Mapping in a Bidirectional Distributed Fiber-Optic Sensor, *Anal. Chem.* **70**(16), 3407-3412 (1998).
24. R. A. Potyrailo, and G. M. Hieftje, Optical Time-of-Flight Chemical Detection: Spatially Resolved Analyte Mapping with Extended-Length Continuous Chemically Modified Optical Fibers, *Anal. Chem.* **70**(8), 1453-1461 (1998).
25. A. W. Schwabacher, WiSys Technology Foundation, U.S. Patent 09,253,153 (1999), Pending.
26. A. W. Schwabacher, Y. Shen, and C. W. Johnson, Fourier Transform Combinatorial Chemistry, *J. Am. Chem. Soc.* **121**(37), 8669-8670 (1999).
27. B. J. Prince, A. W. Schwabacher, and P. Geissinger, A Readout Scheme for Closely Packed Fluorescent Chemosensors on Optical Fibers, *Anal. Chem.* **73**(5), 1007-1015 (2001).
28. B. J. Prince, A. W. Schwabacher, and P. Geissinger, Fluorescent Fiber-Optic Sensor Arrays Probed Utilizing Evanescent Fiber-Fiber Coupling, *Appl. Spectrosc.* **55**(8), 1018-1024 (2001).
29. B. J. Prince, A. W. Schwabacher, and P. Geissinger, An Optical Readout Scheme Providing High Spatial Resolution for the Evaluation of Combinatorial Libraries on Optical Fibers, *J. Assoc. Laboratory Automation* **7**(1), 66-73 (2002).
30. A. W. Snyder, and D. J. Mitchell, Leaky Rays Cause Failure of Geometric Optics on Optical Fibres, *Electron. Lett.* **9**(19), 437-438 (1973).
31. A. W. Snyder, D. J. Mitchell, and C. Pask, Failure of Geometric Optics for Analysis of Circular Optical Fibers, *J. Opt. Soc. Am.* **64**(5), 608-614 (1974).
32. M. J. Adams, D. N. Payne, and F. M. E. Sladen, Leaky rays on Optical Fibres of Arbitrary (Circularly Symmetric) Index Profiles, *Electron. Lett.* **11**(11), 238-240 (1975).
33. J. D. Love, and A. W. Snyder, Generalized Fresnel's Laws for a Curved Absorbing Interface, *J. Opt. Soc. Am.* **65**(9), 1072-1074 (1975).
34. J. D. Love, and A. W. Snyder, Fresnel's and Snell's Laws for the Multimode Optical Waveguide of Circular Cross Section, *J. Opt. Soc. Am.* **65**(11), 1241-1247 (1975).
35. A. W. Snyder, and J. D. Love, Reflection of a Curved Dielectric Interface-Electromagnetic Tunnelling, *IEEE Trans. Microwave Theor. Tech.* **MTT-23**(1), 134-141 (1975).
36. A. Ankiewicz, and C. Pask, Geometric Optics Approach to Light Acceptance and Propagation in Graded Index Fibres, *Opt. Quant. Electr.* **9**(2), 87-109 (1977).
37. A. W. Snyder, and J. D. Love, *Optical Waveguide Theory* (Chapman and Hall, London, UK, 1983).
38. D. Marcuse, *Light Transmission Optics*, 2nd ed. (Van Nordstrand Reinhold Company, Inc., New York, 1982).
39. J. D. Jackson, *Classical Electrodynamics*, 3rd ed. (John Wiley, New York, 1999).
40. D. Marcuse, Cutoff Condition of Optical Fibers, *J. Opt. Soc. Am.* **63**(11), 1369-1371 (1973).
41. D. Marcuse, *Theory of Dielectric Optical Waveguides*, 2nd ed. (Academic Press, Boston, 1991).
42. F. P. Payne, and Z. M. Hale, Deviation from Beer's Law in Multimode Optical Fibre Evanescent Field Sensors, *Int. J. Optoelectron.* **8**(5/6), 743-748 (1993).
43. D. Marcuse, Launching Light into Fiber Cores from Sources Located in the Cladding, *J. Lightwave Technol.* **6**(8), 1273-1279 (1988).
44. M. D. DeGrandpre, and L. W. Burgess, Long Path Fiber-Optic Sensor for Evanescent Field Absorbance Measurements, *Anal. Chem.* **60**(23), 2582-2586 (1988).
45. N. J. Harrick, Electric Field Strengths at Totally Reflecting Interfaces, *J. Opt. Soc. Am.* **55**(7), 851-857 (1965).
46. N. J. Harrick, *Internal Reflection Spectroscopy* (Interscience Publishers, New York, 1967).
47. M. Born, *Optik: Ein Lehrbuch der elektromagnetischen Lichttheorie*, 3rd ed. (Springer Verlag, Berlin, 1981).
48. A. W. Snyder, and J. D. Love, Attenuation Coefficient for Rays in Graded Fibres with Absorbing Cladding, *Electron. Lett.* **12**(10), 255-257 (1976).
49. A. Messica, A. Greenstein, and A. Katzir, Theory of Fiber-Optic, Evanescent-Wave Spectroscopy and Sensors, *Appl. Opt.* **35**(13), 2274-2284 (1996).
50. C. O. Egalon, E. A. Mendoza, A. N. Khalil, and R. A. Lieberman, Modeling an Evanescent Field Absorption Optical Fiber Sensor, *Opt. Eng.* **34**(12), 3583-3586 (1995).

51. D. Christensen, J. Andrade, J. Wang, J. Ives, and D. Yoshida, *Evanescent-Wave Coupling of Fluorescence into Guided Modes: FDTD Analysis*; Proceedings of the Conference on Chemical, Biochemical, and Environmental Fiber Sensors, Boston, MA, SPIE-The International Society for Optical Engineering, Vol. 1172, pp. 70-74, 1989.
52. M. K. Barnoski, and S. M. Jensen, Fiber Waveguides: A Novel Technique for Investigating Attenuation, *Appl. Opt.* **15**(9), 2112-2115 (1976).
53. M. K. Barnoski, M. D. Rourke, S. M. Jensen, and R. T. Melville, Optical Time Domain Reflectometer, *Appl. Opt.* **16**(9), 2375-2379 (1977).
54. S. D. Personick, Photon Probe-An Optical-Fiber Time-Domain Reflectometer, *Bell Syst. Tech. J.* **56**(3), 355-366 (1977).
55. L. A. Thompson, and J. A. Ellman, Synthesis and Applications of Small Molecule Libraries, *Chem. Rev.* **96**(1), 555-600 (1996).
56. G. Jung, Editor, *Combinatorial Peptide and Non-Peptide Libraries. A Handbook* (VCH, Weinheim, Germany, 1996).
57. Combinatorial Chemistry, *Acc. Chem. Res.* **29**(3), (Special Issue) (1996).
58. K. S. Lam, M. Lebl, and V. Krchnák, The "One-Bead-One-Compound" Combinatorial Library Method, *Chem. Rev.* **97**(2), 411-448 (1997).
59. A. Nefzi, J. M. Ostresh, and R. A. Houghten, The Current Status of Heterocyclic Combinatorial Libraries, *Chem. Rev.* **97**(2), 449-472 (1997).
60. C. Gennari, H. P. Nestler, U. Piarulli, and B. Salom, Combinatorial Libraries: Studies in Molecular Recognition and the Quest for New Catalysts, *Liebigs Ann./Recueil* (4), 637-647 (1997).
61. D. J. Gravert, and K. D. Janda, Organic Synthesis on Soluble Polymer Supports: Liquid-Phase Methodologies, *Chem. Rev.* **97**(2), 489-509 (1997).
62. H. M. Geysen, R. H. Meloen, and S. J. Barteling, Use of Peptide Synthesis to Probe Viral Antigens for Epitopes to a Resolution of a Single Amino Acid, *Proc. Natl. Acad. Sci. U.S.A.* **81**(13), 3998-4002 (1984).
63. M. A. Gallop, R. W. Barrett, W. J. Dower, S. P. A. Fodor, and E. M. Gordon, Applications of Combinatorial Technologies to Drug Discovery. 1. Background and Peptide Combinatorial Libraries, *J. Med. Chem.* **37**(9), 1233-1251 (1994).
64. E. M. Gordon, R. W. Barrett, W. J. Dower, S. P. A. Fodor, and M. A. Gallop, Applications of Combinatorial Technologies to Drug Discovery. 2. Combinatorial Organic Synthesis, Library Screening Strategies, and Future Directions, *J. Med. Chem.* **37**(10), 1385-1401 (1994).
65. G. Lowe, Combinatorial Chemistry, *Chem. Soc. Rev.* **24**, 329-341 (1995).
66. J. C. Hogan, Jr., Combinatorial Chemistry in Drug Discovery, *Nature Biotechnol.* **15**(4), 328 (1997).
67. S. R. Wilson, and A. W. Czarnik, *Combinatorial Chemistry -- Synthesis and Application* (John Wiley, New York, 1997).
68. Combinatorial Chemistry, *Chem. Rev.* **97**(2), 347-509 (1997).
69. G. Briceño, H. Chang, X. Sun, P. G. Schultz, and X.-D. Xiang, A Class of Cobalt Oxide Magnetoresistance Materials Discovered with Combinatorial Synthesis, *Science* **270**(5234), 273-275 (1995).
70. J. Wang, Y. Yoo, C. Gao, I. Takeuchi, X. Sun, H. Chang, X.-D. Xiang, and P. G. Schultz, Identification of a Blue Photoluminescent Composite Material from a Combinatorial Library, *Science* **279**(5357), 1712-1714 (1998).
71. E. Danielson, M. Devenney, D. M. Giaquinta, J. H. Golden, R. C. Haushalter, E. W. McFarland, D. M. Poojary, C. M. Reaves, W. H. Weinberg, and X. D. Wu, A Rare-Earth Phosphor Containing One-Dimensional Chains Identified Through Combinatorial Methods, **279**(5352), 837-839 (1998).
72. E. Reddington, A. Sapienza, B. Gurau, R. Viswanathan, S. Sarangapani, E. S. Smotkin, and T. E. Mallouk, Combinatorial Electrochemistry: A Highly Parallel, Optical Screening Method for Discovery of Better Electrocatalysts, *Science* **280**(5370), 1735-1737 (1998).
73. G. R. Rosania, Y.-T. Chang, O. Perez, D. Sutherlin, H. Dong, D. J. Lockhart, and P. G. Schultz, Myoseverin, a Microtubule-Binding Molecule with Novel Cellular Effects, *Nature Biotechnol.* **18**(3), 304-308 (2000).
74. J. C. Hogan, Jr., Directed Combinatorial Chemistry, *Nature* **384 (Suppl.)**(6604), 17-19 (1997).
75. R. B. Merrifield, Solid Phase Peptide Synthesis. I. The Synthesis of a Tetrapeptide, *J. Am. Chem. Soc.* **85**, 2149-2153 (1963).
76. Á. Furka, F. Sebestyén, M. Asgedom, and G. Dibó, General Method for Rapid Synthesis of Multicomponent Peptide Mixtures, *Int. J. Pept. Protein Res.* **37**(6), 487-493 (1991).
77. K. S. Lam, S. E. Salmon, E. M. Hersh, V. J. Hruby, W. M. Kazmierski, and R. J. Knapp, A New Type of Synthetic Peptide Library for Identifying Ligand-Binding Activity, *Nature* **354**(6348), 82-84 (1991).
78. R. A. Houghten, C. Pinilla, S. E. Blondelle, J. R. Appel, C. T. Dooley, and J. H. Cuervo, Generation and Use of Synthetic Peptide Combinatorial Libraries for Basic Research and Drug Discovery, *Nature* **354**(6348), 84-86 (1991).

79. Á. Furka, L. K. Hamaker, and M. L. Peterson, *Synthesis of Combinatorial Libraries Using the Proportioning-Mixing Procedure* in: *Combinatorial Chemistry. A Practical Approach*, edited by H. Fenniri, (Oxford University Press, Oxford, UK, 2000).
80. H. Fenniri, Editor, *Combinatorial Chemistry. A Practical Approach* (Oxford University Press, Oxford, UK, 2000) The Practical Approach Series Vol. 233.
81. H. Fenniri, H. G. Hedderich, K. Haber, J. Achkar, B. Taylor, and D. Ben-Amotz, Towards the Dual Recursive Deconvolution (DRED) of Resin-Supported Combinatorial Libraries: A Non-invasive Methodology Based on Bead Self-Encoding and Multispectral Imaging, *Angew. Chem. Int. Ed. Engl.* **39**, 4483-4485 (2000).
82. H. Fenniri, L. Ding, A. E. Ribbe, and Y. Zyrianov, Barcoded Resins: A New Concept for Resin-Supported Combinatorial Library Self-Deconvolution, *J. Am. Chem. Soc.* **123**, 8151-8152 (2001).
83. M. C. Pirrung, Spatially Addressable Combinatorial Libraries, *Chem. Rev.* **97**(2), 473-488 (1997).
84. A. W. Schwabacher, and P. Geissinger, *One-Dimensional Spatial Encoding: Split/Mix Synthetic Parallelism with Tag-Free Identification and Assays at the Speed of Light* in: *Peptides: The Wave of the Future*, edited by M. Lebl, and R. A. Houghten, (American Peptide Society, San Diego, CA, 2001).
85. J. M. Smith, J. Gard, W. Cummings, A. Kanizsai, and V. Krchnák, Necklace-Coded Polymer-Supported Combinatorial Synthesis of 2-Arylaminobenzimidazoles, *J. Comb. Chem.* **1**(5), 368-370 (1999).
86. Á. Furka, J. W. Christensen, E. Eric Healy, H. R. Tanner, and H. Saneii, String Synthesis. A Spatially Addressable Split Procedure, *J. Comb. Chem.* **2**(3), 220 -223 (2000).
87. L. L. Blyler, R. A. Lieberman, L. G. Cohen, J. A. Ferrara, and J. B. MacChesney, Optical Fiber Chemical Sensors Utilizing Dye-Doped Silicone Polymer Claddings, *Polymer Eng. Sci.* **29**(17), 1215-1218 (1989).
88. R. A. Lieberman, L. L. Blyler, and L. G. Cohen, A Distributed Fiber Optic Sensor Based on Cladding Fluorescence, *J. Lightwave Technol.* **8**(2), 212-220 (1990).
89. J. Rademann, M. Grøtli, M. Meldal, and K. Bock, SPOCC: A Resin for Solid-Phase Organic Chemistry and Enzymatic Reactions on Solid Phase, *J. Am. Chem. Soc.* **121**(23), 5459-5466 (1999).
90. M. Grøtli, C. H. Gotfredsen, J. Rademann, J. Buchardt, A. J. Clark, J. O. Duus, and M. Meldal, Physical Properties of Poly(ethylene glycol) (PEG)-Based Resins for Combinatorial Solid Phase Organic Chemistry: A Comparison of PEG-Cross-Linked and PEG-Grafted Resins, *J. Comb. Chem.* **2**(2), 108-119 (2000).
91. M. J. Prince, A. Benko, and A. W. Schwabacher, in preparation, (2003).
92. W. H. Press, S. A. Teukolsky, W. T. Vetterling, and B. P. Flannery, *Numerical Recipes in C: The Art of Scientific Computing*, 2nd ed. (Cambridge University Press, Cambridge, UK, 1992).
93. Mathematica, Vers. 4.2.0.0, (2002).
94. D. B. Keck, R. D. Maurer, and P. C. Schultz, On the Ultimate Lower Limit of Attenuation in Glass Optical Waveguides, *Appl. Phys. Lett.* **22**(7), 307-309 (1973).
95. J. B. MacChesney, and D. J. DiGiovanni, Materials Development of Optical Fiber, *J. Am. Ceram. Soc.* **73**(12), 3537-3556 (1990).
96. G. A. Thomas, B. I. Shraiman, P. F. Glodis, and M. J. Stephen, Towards the Clarity Limit in Optical Fibres, *Nature* **404**(6775), 262-264 (2000).
97. B. D. MacCraith, G. O'Keeffe, A. K. McEvoy, and C. McDonagh, *Development of a LED-Based Fibre Optic Oxygen Sensor Using a Sol-Gel-Derived Coating*; Proceedings of the Conference on Chemical, Biochemical, and Environmental Fiber Sensors VI, San Diego, CA, SPIE-The International Society for Optical Engineering, Vol. 2293, pp. 110-120, 1994.
98. R. Blue, and G. Stewart, Fibre Optic Evanescent Wave pH Sensing with Dye-Doped Sol-Gel Films, *Int. J. Optoelectron.* **10**(3), 211-222 (1996).
99. J. R. Lakowicz, *Principles of Fluorescence* (Kluwer Academic/Plenum Publishers, New York, 1999).
100. D. Gloge, Weakly Guiding Fibers, *Appl. Opt.* **10**(10), 2252-2258 (1971).
101. T. B. Colin, K.-H. Yang, and W. C. Stwalley, The Effect of Mode Distribution on Evanescent Field Intensity: Applications in Optical Fiber Sensors, *Appl. Spectrosc.* **45**(8), 1291-1295 (1991).

FLUORESCENCE TECHNIQUES IN BIOMEDICINE: FROM THE MONITORING OF CELL METABOLISM TO IMAGE PROCESSING IN CANCER DETECTION

Olaf Minet[1], Jürgen Beuthan[1], Vida Mildažiene[2], Rasa Baniene[2]

ABSTRACT

The state of the art in fluorescence imaging for biomedical applications is demonstrated. 2-dimensional profiles of fluorophores gained with non-contact techniques show the quantitative distribution of endogenous NADH in the UV and synthetic markers in the NIR spectral range. The biomedical use extends from basic research investigations on the metabolism in mitochondria to clinical applications when differentiating the tumor border zone.

One of the outstanding advantages of near infrared so-called Optical Molecular Imaging (OMI) is bright fluorescence of the markers by specific molecular interaction with tumor specific enzymes. For the *in vivo* tests of the dyes an experimental NIR imager was used. NIR fluorescence of the entire body of small animals can be imaged.

The analysis of fluorescences from the interior of the probes shows strong intensity distortion due to tissue optics. Rescaling as the physical basis for image processing taking into account biochemical and biooptical methods result in the real concentration of the fluorophore under consideration. For example, the diameter of the fluorescent volume is apparently larger without rescaling. This new interpretation of fluorescence pictures has useful applications in biomedicine now and in the future.

---

[1] Olaf Minet, Jürgen Beuthan: Freie Universität Berlin, University Hospital Benjamin Franklin, Institute for Medical Physics, Fabeckstr. 60-62, 14195 Berlin, Germany, e-mail: minet@zedat.fu-berlin.de

[2] Vida Mildažiene, Rasa Baniene: Kaunas University of Medicine, Institute for Biomedical Research, Eiveniu 4, LT-3007 Kaunas, Lithuania

1. INTRODUCTION

By means of optical measuring methods it is possible to gain information on the structural and metabolic state of biological tissues. Such information may prove useful when various tissue states have to be clearly differentiated.

The optical measuring methods used either individually or in a combined form are remission spectroscopy, fluorescence spectroscopy (time integral, time resolved), Raman spectroscopy, FTIR spectroscopy, ellipsometry, ultrafast spectroscopy and photon density wave technique[1].

Thanks to their several remarkable properties lasers are finding widespread applications in medicine. The use of lasers in surgery, which started the very year they had been invented, is now well established. Lasers have made minimally invasive ultra-precise surgery possible, thus reducing patient trauma and hospitalization time. The use of lasers for medical diagnosis is also attracting considerable interest and significant advancements are being made. This work is motivated by the fact that the onset and the progression of a disease is often accompanied by biochemical and/or morphological changes, which can be sensitively monitored by laser spectroscopic techniques. These techniques can therefore lead to disease diagnosis at an early stage before the disease becomes difficult to manage. The other important advantages offered by laser spectroscopic techniques are their potential for in-situ, near real time diagnosis and the use of non-ionizing radiation which makes them particularly suited for mass screening because they can be repeatedly used without any adverse effects and monitoring of surgical interventions in minimal invasive medicine.

In case of imaging procedures, however, it is well possible to replace the laser by other light sources. Depending on the fact whether endogenous or exogenous matters are optically excited and their fluorescences detected, the procedures based on fluorescence measurements are subdivided into autofluorescent (optical biopsy) and xenofluorescent ones (e.g. Optical Molecular Imaging – OMI). Bearing this in mind, optical biopsy is deemed to complement and improve surgical biopsy, too.

The potentials of these two methods for observing carcinogenic processes are different. Cancer is a common term for tumors (neoplasms) characterized by uncontrolled growth of cells. Unlike normal cells, cancer cells are atypical in structure and do not have specialized functions. They compete with normal cells for nutrients, eventually killing normal tissue. Cancerous, or malignant, tissue can remain localized, invading only neighbouring tissue, or can spread to other tissues or organs via the lymphatic system or in the blood (i.e. metastasize). Virtually all tissues and organs are susceptible. Tumors detected early (before metastasis) have the best cure rates. Known methods of detection include visual observation, palpation, X-ray study, endoscopy, CT scans, ultrasound, and surgical biopsy. Contrary to these established methods for clinical daily routine optical methods are still in the beginning.

One of the reasons for that is that cancer has many faces. At the molecular level, cancer is characterized specific reactions that are frequently very different. On the other hand, tumor-specific time profiles can often be recorded over several days, which indicate the proliferation of the events. This may lead to organ-specific macroscopic fluorescence profiles (e.g. a squamous cell carcinoma with hot edge and necrotic centre).

Indeed, a central issue of oncological therapy is the correct evaluation of malignant tumor extension and local metastasis spread, if any. Based on the macroscopic evaluation of the tumor extension the surgeon will resect the tumor with a certain safety margin in the healthy tissue. That safety margin often lets the surgery become a mutilating intervention, which refers e.g. to larynx or brain tumors. The fluorescence technology may be a valuable tool to optimize the safety margin, i.e. to make it as small as possible and as large as necessary.

Potentials	**Restrictions**
• Minimally invasive	• Surface diagnosis only
• Surface diagnosis	• Low penetration depths in the UV range
• Early and *in situ* diagnosis	• Low specificity statements
• Evaluation of radicalism / dignity	• Metabolism-dependent
• Simultaneous photodynamic therapy is possible	• Individual variations
• Cost-favourable	• Work in progress for quantitative assessment

2. NADH AS ENDOGENOUS CHROMOPHOR: BACKGROUND AND OBSERVATIONS

2.1. Chemistry and Spectroscopy

For several years laser-induced fluorescence spectroscopy has been a valuable tool in medical diagnosis detecting the relative distribution of endogenous chromophors. In the case of the reduced form of NADH, the fluorescence signal reveals the activity status of the observed cells, because NADH takes on some essential features of cellular energy metabolism.

The nicotinamide adenine dinucleotides are natural compounds having the most obvious potential for being used as the fluorescent markers of the metabolic state of the tissue [2-24].

These rather complex substances are named dinucleotides because their molecule consists of two nucleotide moieties – nicotinamide mononucleotide and adenosine monophosphate (Figure 1). The main metabolic function of nicotinamide adenine nucleotide cofactors (called also pyridine nucleotides) is participation in numerous hydrogen transfer reactions that stems from the ability of the nicotinamide ring to undergo reversible oxidation/reduction.

Concentration of $NADH^{+}$ as well as its close analog nicotinamide adenine dinucleotide phosphate NADP (phosphorylated at the position indicated by arrow in Figure 1) is easily detectable by optical methods. The spectral properties of the molecule are determined both by adenine ring (absorption maximum at 260 nm) and by nicotinamide ring (Figure 2).

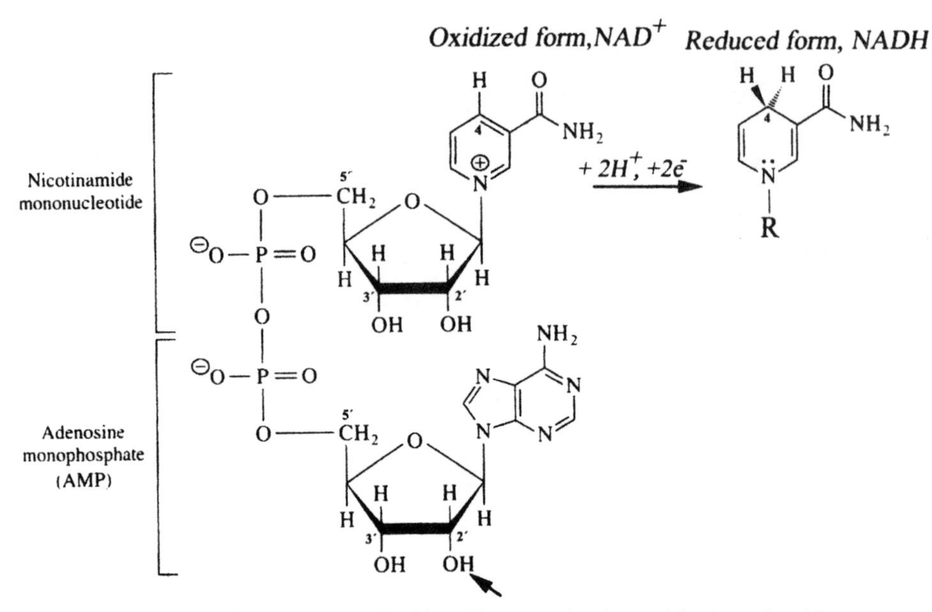

Figure 1. Structure of nicotinamide adenine dinucleotide and its conversion from oxidized to reduced form

The absorption as well as fluorescence spectrum of the reduced forms of nicotinamide nucleotides differs from that of their oxidized counterparts – the reduced forms absorb light at 340 nm; the oxidized forms do not (Figure 2). The fluorimetric method for estimation of concentration NADH (excitation wavelength 340 nm, emission wavelength 465 nm) is about 10 times more sensitive than spectrophotometry.

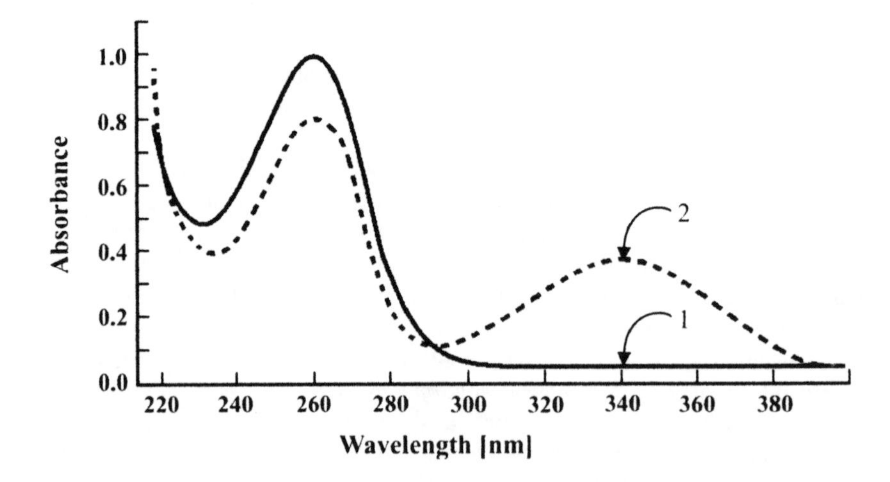

Figure 2. Absorption spectra of NAD^+ (curve 1) and NADH (curve 2)

2.2. Biological Importance of NAD(P)H.

The water soluble and freely diffusable pyridine nucleotides are referred to as central metabolites because they participate in many metabolic pathways and play important roles in the integration of the metabolism.

The NAD^+ is the cofactor for most enzymes that act in the direction of substrate oxidation. The reduction of NAD^+ in catabolic processes results in the conservation of energy released by oxidation of various substrates. NADPH usually functions as a cofactor for enzymes that catalyse substrate reduction in biosynthetic pathways. It is also essential for cellular antioxidant defence since it is substrate for regeneration of the reduced glutathione.

More than 200 enzymes are known to catalyse reactions in which NAD^+ or $NADP^+$ are involved. The ability of nicotine adenine dinucleotides to modulate allosterically the activity of key regulatory enzymes determines the integrating function of pyridine nucleotides in the metabolism. The total concentration of NAD^+ + NADH in undamaged tissues is about 400-500 μM [25], and that of $NADP^+$ + NADPH is about 10 times lower. In many cells and tissues the ratio of NAD^+ to NADH is high, favoring hydride transfer from substrate to NADH. By contrast, NADPH is normally present in larger amounts than its oxidized form $NADP^+$. The ratio of their reduced to oxidized forms – NADH/NAD^+ (NADPH/$NADP^+$) is therefore a sensitive measure of the balance between catabolic and anabolic processes and thus, the energy state of the cell.

It is well known that any disturbance of the cell metabolism leads to very fast changes in the ratio of NADH/NAD^+. In the early stages of metabolic response to different injuries the change in that ratio occurs without significant changes in the total pyridine nucleotide pool, however the decrease in their total pool might be also expected under the more pronounced or prolonged deleterious treatment.

Under metabolic stress condition, particulary under oxidative stress, massive NAD^+ catabolism is caused by activation of poly(ADP-ribose) polymerase (PARP), an enzyme that perform the most severe modification of proteins by attaching long (reaching up to 200 monomers) and branched ADP-ribosyl polymer chains with the simultaneous release of nicotinamide (for reviews, see [26,27]). The majority of the acceptor proteins are implicated in the DNA metabolism: chromatin remodelling, DNA repair, transcription, apoptosis, neoplastic transformation, etc. The activity of the waste array of important target proteins is regulated upon ADP rybosylation. NAD^+ is a precursor and immediate substrate for the synthesis of the large ADP-ribose polymers. Therefore by extensive activating PARP, cellular NAD^+ resources may be almost completely depleted. That occurs within 5-15 min after severe genotoxic insult[28]. In addition, NAD^+ depletion blocks glycolysis and results in ATP depletion that finally may lead to irreversible cell damage and death[29,30].

Another important group of reactions of $NAD(P)^+$ degradation is generation of calcium mobilizing derivatives catalysed by enzymes called NAD^+-glycohydrolases (NADases) (for reviews see [31,32]). $NAD(P)^+$ is a precursor of secondary signal messengers cyclic ADP-ribose (cADPR) and nicotinic acid adenine nucleotide phosphate (NAADP) that both release Ca^{2+} from different cellular calcium stories and therefore are important in modulation of Ca^{2+} signal transduction under stress condition. The activation of NADases hardly may cause significant depletion of total pyridine nucleotide pool, however even small amounts of Ca^{2+}mobilising second messengers (both effective at nanomolar concentration range) could play essential roles in cellular responses to insult or survival.

NAD(P)H/NAD(P)$^+$ ratio in the cell is in homeostasis under normal condition, but NAD(P)H is rapidly exhausted upon response to stress. Therefore the detection of NAD(P)H amount in the tissue by optical sampling seems a very promising development providing a powerful tool for non-invasive fundamental and clinical investigations directed to the waste scope of problems. It may be used both for evaluation of metabolic state by changes in ratio of NAD(P)H/NAD(P)$^+$ as well as for evaluation of viability or survival status by following the depletion of the total pyridine nucleotide pool.

In contrast to flavine nucleotides that are tightly attached to proteins by covalent bonding, pyridine coenzymes NAD(P)H are generally supposed to be "free" or only temporarily bound to proteins by noncovalent interaction forces. However, NADH molecule undergoes significant changes from folded to extended conformation upon binding to proteins[33,34]. The conformational change is reflected by essential changes in spectral properties. Free NADH in folded conformation has emission maximum around 465 nm, whereas emission maximum of bound NADH is around 440 nm; binding results in 2-fold increase in fluorescence intensity and larger[35] lifetime (400 ps in comparison to 1 ns). In the cell, certain equilibrium exist between bound and free NAD(P)H, implying that measurements of total NAD(P)H fluorescence in tissues might be not easily interpreted due to enhanced fluorescence of bound NADH, and differential binding of NAD(P)H/ NAD(P)$^+$ to proteins. On the other hand, the differences in fluorescence may provide important information for investigation of compartmentation of pyridine nucleotides *in vivo*.

The NAD(P)H concentration is lower in cancer tissue in comparison to normal tissue. It is slightly higher at the edge of the tumor [9,20,21] corresponding to the proliferation and the induced new growth of vessels. It still remains to be determined if lower NAD(P)H amount in the tumor may be explainable by the increased energy demand for intensive biosynthetic reactions and for the rapid division. On the other hand, tumor cells show perturbations in poly(ADP-ribose) metabolism, in most cases - PARP activation that may also result in lower NAD(P)H. PARP inhibitors are adjuvants for certain anticancer treatment (mutilating agents, topoisomerase inhibitors and ionizing radiation) since they delay DNA repair and divert cells to apoptosis orchestrated by p53 (for the reviews see [36,37]).

Another example is heart tissue undergoing ischemic stress. During heart perfusion ischemia induces very fast initial changes in NADH since oxidative reactions are stopped due to the lack of oxygen [38]. Sudden initial rise NADH after 15 min of hypoxic perfusion is followed by the continuos decay that is not yet clearly defined, but might possibly be explained by NADH catabolism upon PARP activation [39].

Nevertheless, the study of NAD(P)H transients and oscillations in response to metabolite flux provides a unique insight into the regulatory and deregulatory processes within the microenvironment of the living cell [40].

2.3 Experimental Observations [41]

The experimental setup (Figure 3) consists mainly of a Lumatec SUV-DC light source with a lens Computar (f=25mm, 1:1,8), a liquid core wave guide (LC), and an intensified CCD-camera CPL-22B by Canadian Photonics Labs. The excitation light has a wavelength of 347 $\pm$ 4 nm and the fluorescence light 467 + 3 nm after passing the band pass filters respectively. NADH was dissolved in an aqueous suspension of intralipid Fresenius Kabi AB, Upsala, of a concentration of 2% in incubation medium (buffer pH 7.2) to give a phantom with approximated optical scattering properties of human skin.

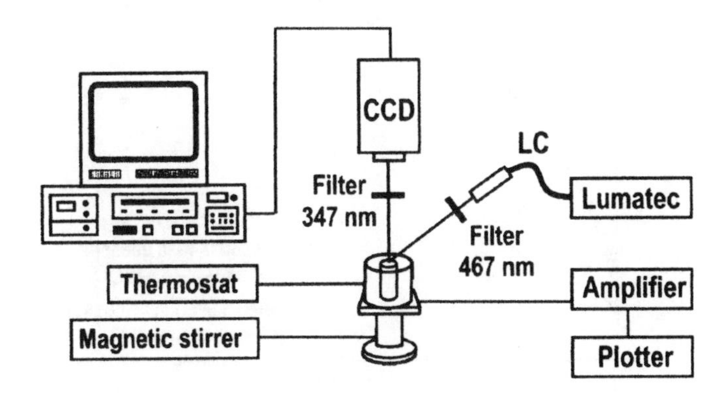

Figure 3. Experimental setup showing the laboratory monitoring of cell metabolism

In an aqueous solution the NADH is free (ref. chapter 2.2.). Figure 4 is an example of a picture for a concentration of 300 µM/l. After frame grabbing the mean intensity in a region of interest in the middle of the cuvette was determined by an appropriate software tool.

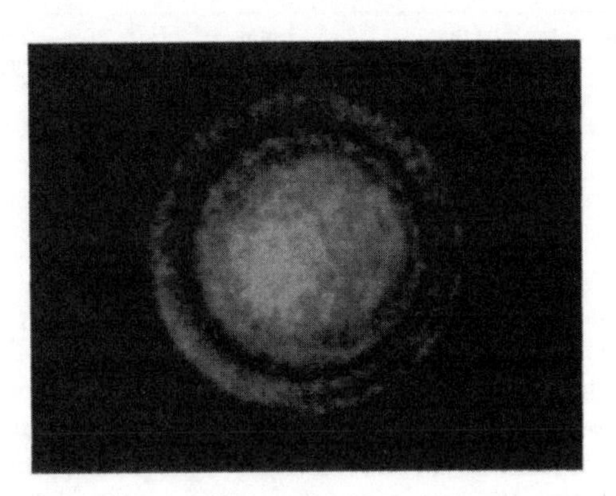

Figure 4. Example of a fluorescence picture in cuvette (sample volume: 2 ml, NADH concentration: 300 µM/l)

Concentrations of NADH from 0 up to 500 μM/l were measured and the results for two sample volumes 1 ml and 2ml, resp. plotted in Figure 5. Using the gaussian method of minimizing the sum of the squares of the deviations the dashed lines show the fit acc. to equation (1) from an spectrometric model explained in chapter 3.1 (see below).

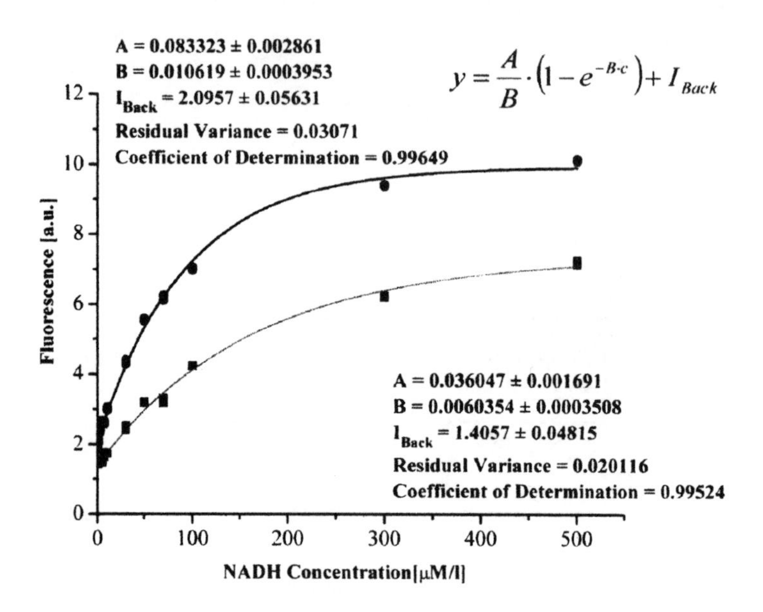

Figure 5. Calibration curves for two different diameters of the cuvette (solid lines: interpolated experimental results, dashed lines: spectrometric model)

Table 1. Values for the parameters A and B for different diameters of the cuvette

Diameter of the cuvette	1 mm	2 mm
A [l/μM]	$36.05 \cdot 10^{-3}$	$83.33 \cdot 10^{-3}$
B [l/μM]	$6.04 \cdot 10^{-3}$	$10.62 \cdot 10^{-3}$
I_{Back} [a.u.]	1.41	2.10

The results of the numerical analysis are shown in Table 1. The values for A and B for diameter of 2 mm are clearly larger in comparison for 1 mm. This is in accordance to the spectrometric model because A is the parameter between chromophore concentration and fluorescence intensity which growth up with the volume of the cuvette. B is the parameter for the absorption of the fluorescence light passing the turbid suspension. It grows up, too, according to the increase of the mean path length for the fluorescence light between the location the fluorescence is developing in the suspension and the surface, since this path length is proportional to the absorption probability.

The offset I_{Back} is nearly constant. Thus, the calibration has all the important features predicted by a general ansatz from tissue optics and together with an experimental setup as shown Figure 3 a non-destructive determination of the NADH concentration in scattering media is possible.

In a second experiment mitochondria from the heart of male Wistar rats were isolated by differential centrifugation. The animals were killed according to the rules defined by the European Convention for the Protection of Vertebrate Animals Used for Experimental and Other Scientific Purposes (Licence No.0006). The tissue of two hearts was cut into small pieces and homogenized by a glass-teflon homogenizer in a buffer containing 160 mM KCl, 10 mM NaCl, 20 mM Tris-HCl, 5 mM EGTA and 1 mg/ml BSA (pH 7.7, 2-4 °C). The homogenate was centrifuged at 750 x g, and the obtained supernatant was centrifuged at 7000 x g. The obtained pellet of heart mitochondria was suspended in a buffer containing 160 mM KCl, 20 mM Tris-HCl and 3 mM EGTA (pH 7.35). The preparations of isolated mitochondria were stored on ice. The concentration of mitochondria in final suspension was approximately 50 mg/ml protein. The protein concentration was determined by the biuret method [42].

Respiration of heart mitochondria and NADH fluorescence was measured in the stirred and thermostated (25°C) 1.0 ml vessel fitted both a Clark type oxygen electrode (Rank Brothers LTD, Cambridge, UK). Respiration was measured in a closed chamber, and NADH fluorescence was determined from the top of the open chamber following exactly the same experimental protocol with an experimental setup based on Figure 3. Values of the concentration of free Ca^{2+} and Mg^{2+} in the incubation media were stabilized by EGTA buffers and calculated using the apparent stability constants for EGTA [43]. The run was started by adding mitochondria (2 mg protein) to the incubation medium containing 30 mM Tris-HCl, 5 mM KH_2PO_4, 125 mM KCl, 10 mM NaCl, 3 mM EGTA, 1.5 mM $MgCl_2$ (1 mM free Mg^{2+}) and 0.09 mM $CaCl_2$ (5 nM free Ca^{2+}), pH 7.2. The final concentrations of further additions were: oxidizable substrate, 5 mM pyruvate + 5 mM malate; uncoupler, carbonyl cyanide 4-trifluoromethoxyphenylhydrazone 1.5 μM, inhibitor of Complex I, rotenone 2 μM; hydrogen peroxide, 0,3%.

The results of NADH fluorescence measurement in the respiring heart mitochondria are presented in Figure 6. Figure 7 demonstrates what changes in the rate of oxygen consumption are induced by addition of the same substances that were used for monitoring changes in the amount of NADH. After each addition the NADH changes almost similar to the rate of mitochondrial respiration.

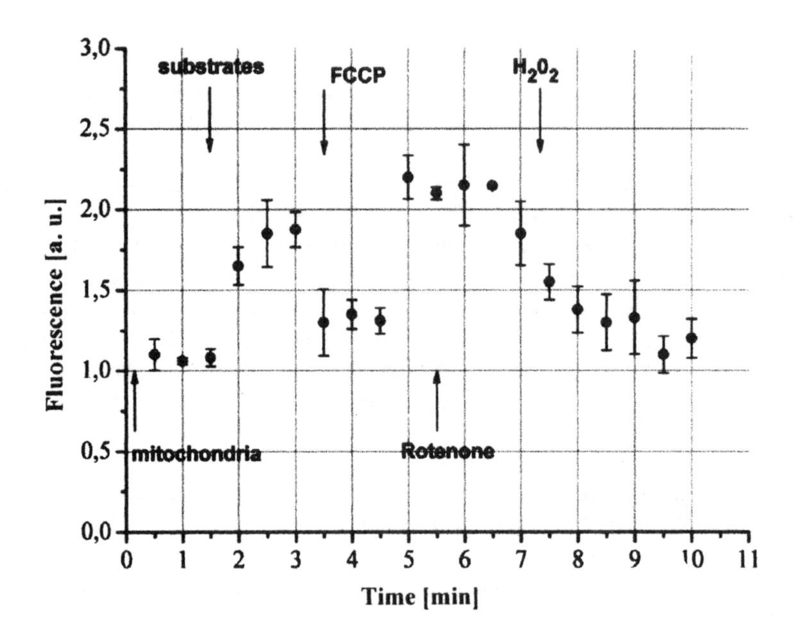

Figure 6. Result of NADH fluorescence measurements in the respiring heart mitochondria (substrates pyruvate and malate, uncoupler FCCP, inhibitor Rotenone, and hydrogen peroxide H_2O_2)

The scheme of the oxidative phosphorylation system in mitochondria is presented in Figure 8. Electrons from oxidizable substrates may enter the respiratory chain via different routes: electrons to Complex I (NADH dehydrogenase) are donated by the substrates reducing NAD^+ (in our experiment by pyruvate), while electrons originating from succinate are passed to ubiquinone (UQ) via Complex II (succinate dehydrogenase).

Further, the electrons are transferred to oxygen via Complex III (cytochrome bc_1 complex), cytochrome c and Complex IV (cytochrome oxidase). Electron flow through Complexes I, III and IV is coupled with the outward pumping of protons producing both chemical (ΔpH) and electrical ($\Delta\Psi$) gradient because the inner mitochondrial membrane has very low natural permeability to protons (the proton leak). The ATP synthase makes use of the electrochemical gradient by coupling the inward proton flux with the ADP phosphorylation reaction. Upon increase on the proton permeability of the inner membrane by uncouplers of oxidative phosphorylation (substances like FCCP) the membrane potential collapses, therefore electron transport in the respiratory chain (substrate oxidation) attains the maximal rate however the ADP phosphorylation is not operating as it lacks the driving force.

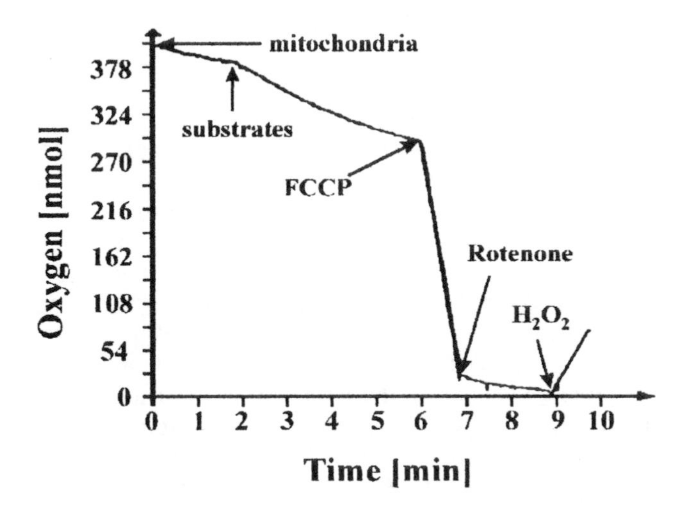

Figure 7. Polarographic trace of respiration of heart mitochondria plotted acc. to the setup in Figure 3

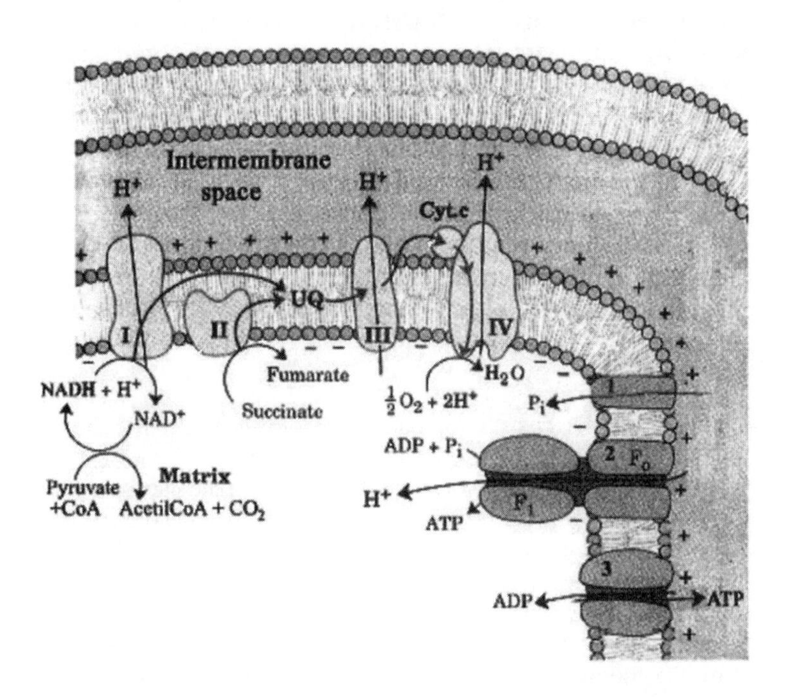

Figure 8. Components of the oxidative phosphorylation system. Complexes of the respiratory chain are denoted by Roman numbers I, II, III, IV, respectively, the natural membrane permeability to protons (the membrane leak) by dotted arrow, and the components responsible for phosphorylation by Latin numbers: 1 - phosphate carrie, 2 - ATP synthase, 3 - ATP/ADP carrier

This scheme explains the simultaneous changes in the respiratory rate and NADH fluorescence, observed after addition of certain substances to the incubation medium containing mitochondrial suspension (Figures 6 and 7). Both the respiration rate and NADH fluorescence is low after addition of mitochondria to the medium, because of limited amount of endogenous substrates in mitochondrial matrix. After addition of external substrates (pyruvate and malate) to the incubation medium, they are quickly transported by specific carriers inside of mitochondria and oxidized by NAD dependent dehydrogenases (e.g., by pyruvate dehydrogenase) with a concomitant reduction of NAD^+ to NADH, so that both the respiratory rate and NADH fluorescence increases. In the presence of oxygen and substrate mitochondria generate high electrochemical proton potential (approaching 180-200 mV) since due to the lack of ADP the membrane potential consumers (ATP synthase or ATP/ADP carrier) do not operate. Under these conditions further protein pumping to the positively charged intermembrane space (as well as simultaneous electron transport through the respiratory chain) is prevented by charge repulsion and proton concentration gradient, therefore the rate of mitochondrial respiration is slow (Fig 8). Further addition of uncoupler FCCP leads to the dissipation of the membrane potential, therefore the electron transport from NADH to oxygen greatly increases (respiratory rate is maximal). At the same time NADH fluorescence decreases because of very fast NADH oxidation. Rotenone specifically inhibits the respiratory complex I that transfers electrons from NADH to the mobile electron carrier ubichinone or coenzyme Q (UQ). Therefore rotenone blocks oxidation of NADH downstream of Complex I and leads to inhibition of the rate of oxidation of pyruvate + malate (Figure 8) and to accumulation of NADH (Figure 7). The following addition of H_2O_2 leads to complete and unspecific oxiditation of NADH (the obtained fluorescence level might be referred to as a zero point).

3. FROM RESCALING TO IMAGE PROCESSING I

3.1. Rescaling: Fluorescence and Tissue Optics

For the differential diagnosis of metabolic disturbance and for the quantitative assessment of intraoperative sites or transplants, in most cases, it is indispensable to know the quantitative values of the indicating fluorophore in the particular region. Since the first use of fluorescence in biomedical applications [4-7] the quantitative interpretation of the signal becomes more and more important [7, 10-12, 14, 15, 20-22, 24] and the possibilities of the diagnostics in medicine grow rapidly with such tools.

The general problem arising therefrom is fluorescence measuring in a strongly scattering medium. The fluorophores are embedded in this medium which is characterized by the following optical parameters: absorption coefficient μ_a, scattering coefficient μ_s and anisotropy factor g [44-46]. When detecting intrinsic fluorescence on the surface of the tissue, the intensity values measured will change with different optical parameters. From this physical point of view, the quantitative evaluation of laser induced fluorescence in strongly scattering media remains a challenge.

A fashion honoured in time is based on the following two principles [7, 11, 12] (Figure 9):

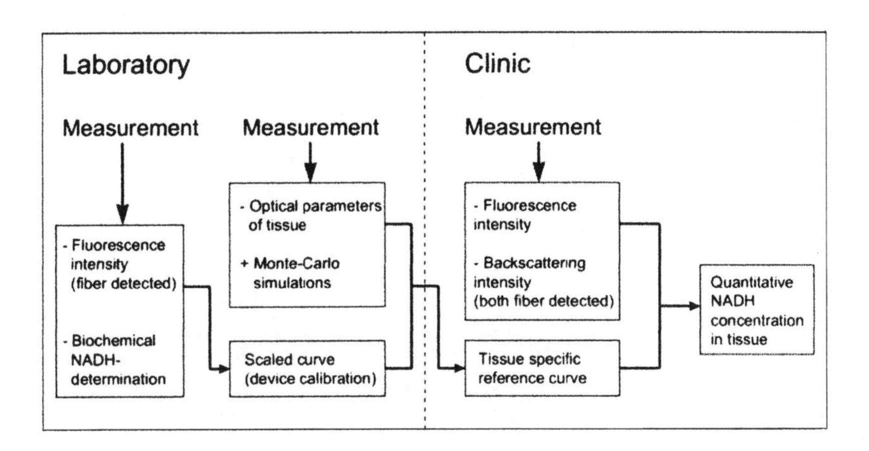

Figure 9. Scheme of the rescaling procedure

1. The calibration function of the experimental setup is determined by biochemical reference measurements which fit the parameters of a mathematical model [11]. To get the analytical expression we make the assumption that the fluorescence intensity I is proportional to the concentration c of NADH and the absorption of the fluorescence radiation is proportional to the intensity itself:

$$I(c) = \frac{A}{B} \cdot (1 - e^{-B \cdot c}) + I_{Back} \qquad (1)$$

   whereas A is the constant linking the fluorescence intensity I to the concentration c without absorption and B describes the overall absorption of the fluorescence light within the tissue geometry under consideration. This equation is valid only for spatial constant conditions and must be supplied for different regions of optical parameters separately.

2. The physical basis of rescaling remains the light transport in turbid media. Clinical fluorescence results often differ from this calibration curve because of the complex relationship of scattering, absorption and intrinsic fluorescence within the tissue. Especially, the optical parameters μ_a, μ_s and the anisotropy factor g vary individually and locally in the target tissues. For example, the natural variation of the absorption and scattering coefficients for the excitation wavelength in human liver [21] is shown in Figure 6. The separate variations are remarkably high, up to 70%.

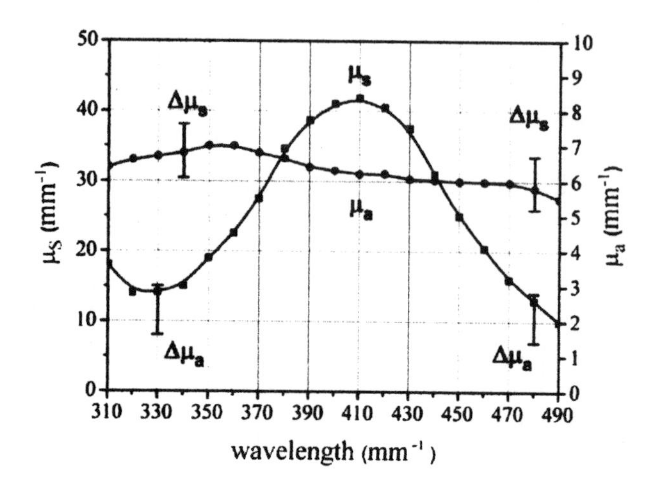

Figure 10. Variation of the absorption and scattering coefficient for human liver tumor

In principle, the main optical parameters, i.e. the absorption and scattering coefficient μ_a and μ_s must be known at first. Then, these parameters of the tissue into consideration are used to determine the deviation from Eq. (1).

One possible approach for an underlying model is the diffusion approximation within the context of the transport theory for light in turbid media. Such analytical methods [22, 24] suffer by different disadvantages like failing the necessary precession using common approximation arguments or breaking down even for less complicated geometries. Therefore, these optical parameters of the tissue into consideration were used for Monte Carlo simulations of the fluorescence intensity and backscattering intensity. The Monte Carlo simulation permits to separate the influence of all parameters as follows. All but one of the involved parameters are fixed to their mean values. Although the dependence of fluorescence and backscattering intensity from these parameters may vary in a magnitude of one order (for example Figure 10), the Monte Carlo simulations advise a nearly linear functional relation between the optical tissue parameters [12, 21] . Therefore, an approximation with clear meaning like Equ. (2) is appropriate in general,

$$\Delta c = \frac{\partial c}{\partial I} \sum_{i=x,f} \left(\frac{\partial I}{\partial \mu_{ai}} \cdot \Delta\mu_{ai} + \frac{\partial I}{\partial \mu_{si}} \cdot \Delta\mu_{si} + \frac{\partial I}{\partial g_i} \cdot \Delta g_i \right), \tag{2}$$

whereas x denotes the excitation wavelength and f the fluorescence wavelength. The first partial derivation in (2) is due to Eq. (1) and the other result from Monte Carlo simulations. According to these Monte Carlo simulations the backscattering intensity also depends essentially on the same parameters. A simple simultaneous measurement of these parameters provides the data for rescaling of the fluorescence intensity with respect to the deviations of the given values from their mean values.

3.2. Image Processing: Optical Biopsy in the UV Range

Optical Biopsy is a valuable tool in biomedical research and medical diagnosis detecting the relative distribution of endogenous chromophores (ref. chap. 2).

The experimental setup for two-dimensional fluorescence imaging (Figure 11) is based on early investigations taking line scans using an optical fiber.

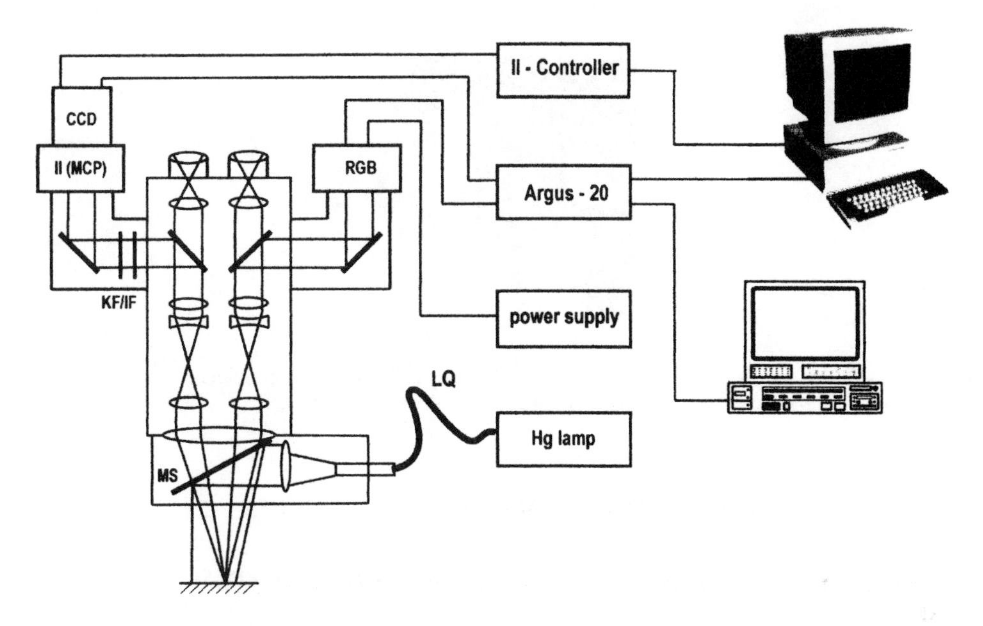

Figure 11. Experimental setup for two-dimensional fluorescence imaging [47] . II: Image intensifier (Micro channel plate), LC - Liquid core wave guide, MM - Monochromatic mirror (HR 365 nm), EF - Edge filter, IF - Interference filter (bandpass 460 nm)

As target tissue we used human tissue from of a tumor of the base of the tongue (squamous cell carcinoma) *in vivo*. As an example for a 2-dimensional fluorescence mapping a native picture of a tumor of the base of a human tongue is shown (Figure 12). Right next to it is the appendant fluorescence and the rescaled image. The rescaling is made by using the optical parameters in relationship to the tissue region of interest as discussed in chapter 4.

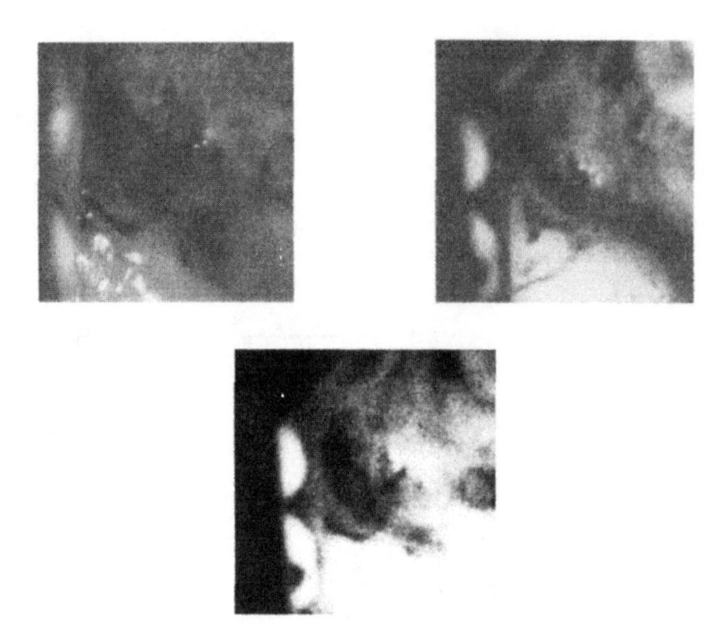

Figure 12. Example for the rescaling method in Optical Biopsy of a base tumor of a human tongue *in vivo* , upper left native image, upper right fluorescence image, middle down rescaled image

4. FROM RESCALING TO IMAGE PROCESSING II

4. 1. Exogenous Contrast agents for Optical Molecular Imaging (OMI)

Most of the known, not only optical, imaging techniques are based on physical properties of the object under consideration like absorption, scattering and physiologic features. Molecular Imaging is based on such techniques with the aim of using specific molecules as the source of image contrast [48]. Key elements are special imaging probes with high specificity, amplification strategies, and systems for imaging with high resolution. Our task consists in the last point by establishing a procedure for image processing which is based on the modelling of the light transport in biological tissue.

Approaches using autofluorescence are increasingly replaced by contrast-enhanced optical imaging methods using drugs such as indocyanine green (ICG) or 5-ALA in order to enhance the fluorescence intensity within tissues and improve the diagnostic specificity. A variety of novel dyes has recently been synthesized and characterized regarding its potential to detect tumors and other diseases. As mentioned above, such dyes have to be non-toxic and absorb in the near infrared (NIR) to ensure deep photon penetration in general. The class of cyanine dyes has shown greatest potential since many derivatives out of this structural family are already known and have proven to be efficient fluorescent labels. For *in vivo* diagnostic applications, different pharmacological and molecular principles to achieve tumor-specific signal enhancement can be followed. For example, fluorescence-tagged receptor-specific peptides have been used to impart molecular specificity for the detection of tumor-specific receptor expression [49, 50] .

One of the outstanding advantages of near infrared OMI using injectable fluorescence probes is the fact that the fluorescence signal of an organic dye molecule can potentially be affected by its local environment. For instance, energy transfer, fluorescence quenching or dequenching processes can be employed to report molecular or physiological conditions. Weissleder et al.[51, 52] have introduced a novel approach which employs so-called fluorescence-quenched, enzyme-activatable polymers which carry cyanine dye molecules in close distance to each other so that an efficient quenching of fluorescence emission results. Upon enzymatic cleavage and liberation of dye molecules by tumor-specific enzymes, a bright fluorescence recovers in areas where enzymes are active, thus permitting the optical detection of specific protein expression, e. g. proteolytic enzymes, *in vivo*. Later a prostate-specific adenovirus vector was used to identify metastases in a human-prostate cancer model in living mice by transcriptional targeting[53].

The different design approaches for optical contrast agents have in common that those resulting images have to be analyzed regarding its diagnostic accuracy, whether or not image processing is used.

One dye is presented here as a more simple approach involving the synthesis of indotricarbocyanines with hydrophilic chemical substitution. As described in [54, 55] different structures comprising two different amino-sugar derivatives, D-glucamine and D-glucosamine, were synthesized and characterized for their physicochemical and pharmacological properties. The plasma protein bond decreases in agreement with increasing hydrophilicity. The most hydrophilic compound NIR96010 (bis-1,1'-(4-sulfobutyl)indotricarbocyanine-5,5'-dicarboxylic acid diglucamide, monosodium salt, chemical structure see Figure 13) was thoroughly studied in different animal models[56].

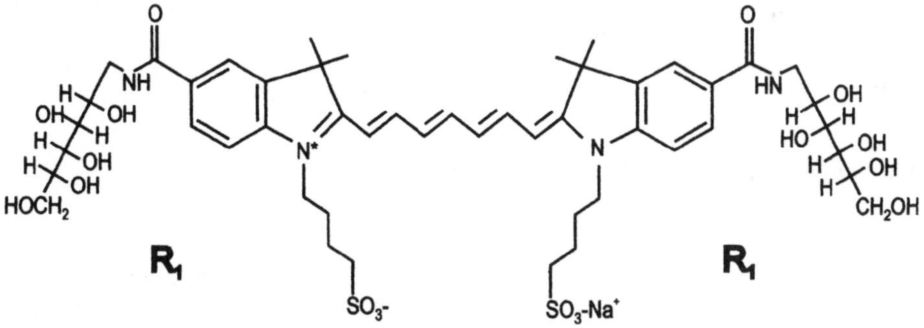

Figure 13. Chemical structure of NIR96010 (bis-1,1'-(4-sulfobutyl)indotricarbocyanine-5,5'-dicarboxylic acid diglucamide, monosodium salt)

NIR96010 has an absorption maximum at 754 nm, a fluorescence maximum at 782 nm and a fluorescence quantum yield of 7- 12 % in physiological environment. In Figure 14 the absorption and fluorescence emission spectrum is depicted.

At doses in the range of 0.5 – 2 μmol/kg it shows improved properties regarding its ability to generate increased tumor-to-normal tissue concentration ratios compared to the known drug indocyanine green (ICG) [54].

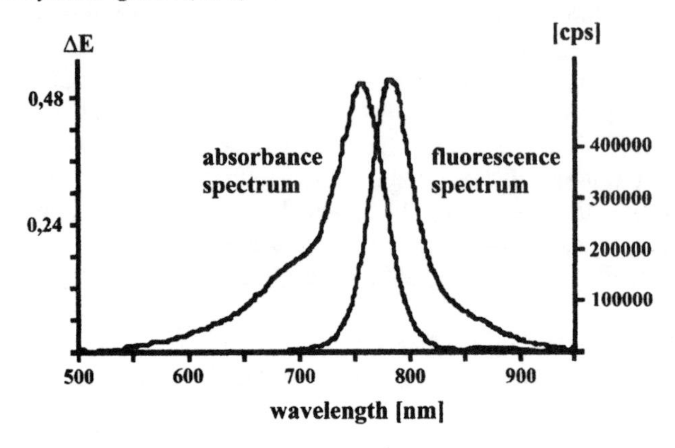

Figure 14. Absorbance and fluorescence spectrum of NIR96010 in bovine plasma (concentration 2 μM)

An additional asset was observed in fluorescence imaging studies when looking at late times after intravenous injection. While ICG is strongly protein-bound, thus distributed in the intravascular space and rapidly cleared from the intravascular space by the liver, NIR96010 extravagates into the extra cellular compartment. As a consequence, a significantly decreased tissue clearance was apparently leading to a preferential retention in tumor tissues and thus an subsequently improving contrast which shows highest values at 24 h after injection of the compound [57,58]. This behaviour makes the compound suitable as fluorescent probe for studying imaging methods and rescaling principles (ref. chap. 4) throughout a variety of *in vivo* models.

4.2. Image Processing: OMI in the NIR Range

For the case of Optical Molecular Imaging, the rescaling procedure described in chapter 3 changes slightly according to Figure 15. At first, the signal/noise ratio of the CCD camera is set off in the black and white bitmap mainly by cutting off those signals, which are below the noise limit compared to the maximum intensity. In a second step, tissue optics knowledge is used for rescaling fluorescence intensities by means of Monte Carlo simulations as described above.

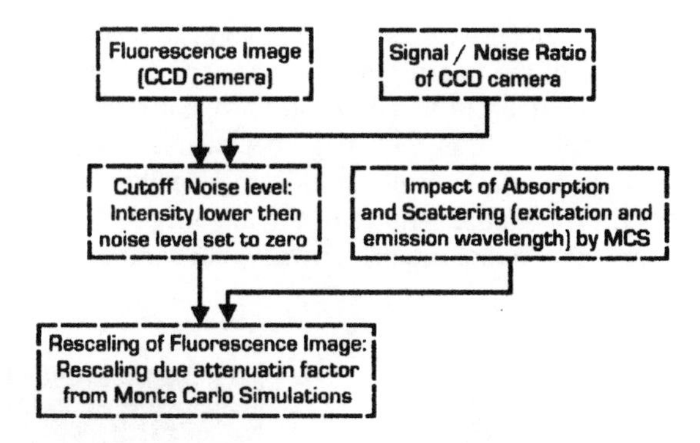

Figure 15. Scheme of the rescaling procedure for Optical Molecular Imaging

For the *in vivo* tests of the new dyes [56] a NIR Imager was designed (Figure 16). The imager consists of two laser systems and a cooled CCD-camera (Hamamatsu). The first laser system is a 742 nm cw diode laser (1.5 W, CeramOptec) with an interference filter 1x for 740 nm. The NIR fluorescence at 800 nm of an entire mouse body can be imaged with a specific interference filter 1f acc. to Figure 14. The second laser can be used for different purposes. It is possible, e.g., to excite EGFP (CLONTECH) with an argon laser that is adjusted and filtered at 488 nm (filter 2x). The fluorescence light of this laser excitation can be imaged with another interference filter for 509 nm (filter 2f).

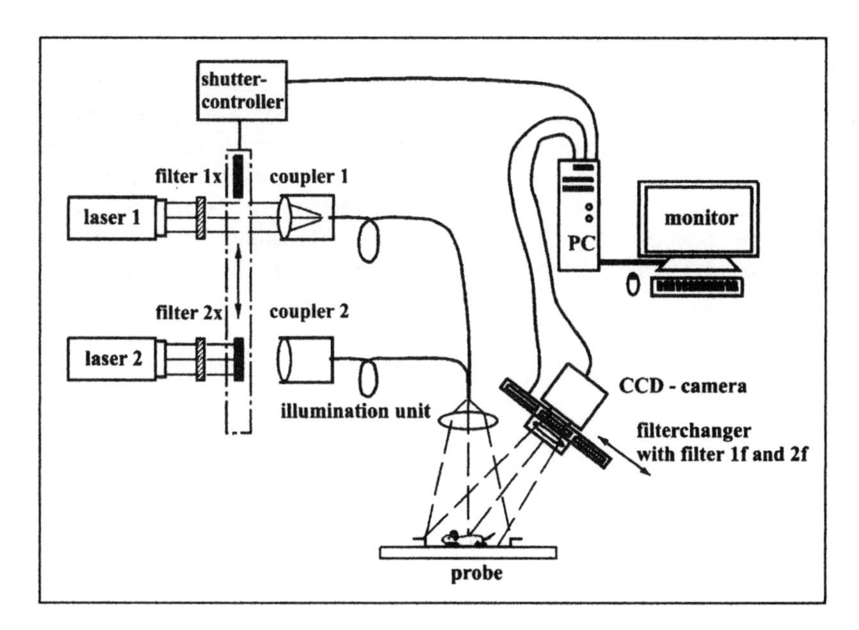

Figure 16. Experimental setup for OMI; Laser 1: excitation wavelength λ=742nm, filter 1: transmission wavelength λ=742nm, filter 1a: fluorescence wavelength λ=800nm; Laser 2 (example): excitation wavelength λ=488nm, filter 2: transmisson wavelength λ=488nm, filter 2a: fluorescence wavelength λ=509nm

The *in vivo* imaging potential was studied in nude mice bearing F9 teratocarcinoma. After intravenous injection of the dye, a typical contrast performance was observed with highest contrast at approximately 20 h. The contrast signal ratio (tumor /normal tissue) was determined to be about 10 from the image depicted according to Figure 17b. As described above, this high contrast results from tumor retention of a small amount of NIR96010 accompanied by a renal clearance of the majority of injected dose. Compared to previous studies using other tumor models [57, 58] a higher contrast was observed.

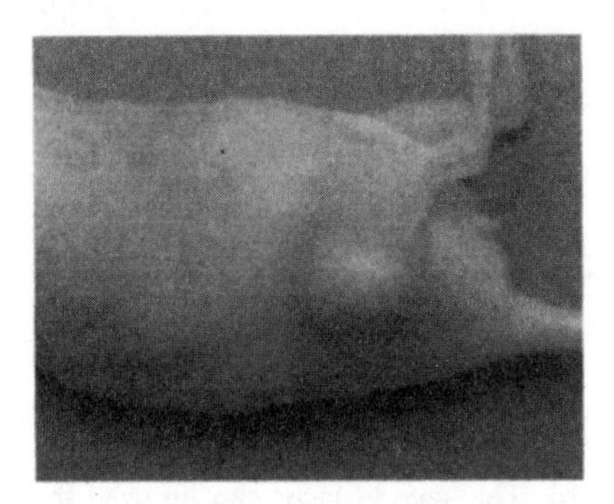

Figure 17. Native (left) and fluorescence image with line scan (right) of superficial tumor of a mouse thigh (nude mouse, F9 teratocarcinoma of approx. 8 mm x 8 mm in size) at 20 h after intravenous injection of NIR96010 at a dose of 2μmol dye / kg body weight

To improve the image of the concentration of the NIR-fluorophore in the mouse the rescaling principle according to chapter 4 was applied. For tissue-optics related corrections it is to be considered that the light scattering in the NIR range is stronger compared to the UV fluorescence used in Optical Biopsy. The diameter of the fluorescent volume is apparently larger for the same reason. Furthermore, the signal/noise ratio can be improved by suppressing parasitic fluorescences. Rescaling the above fluorescence image of the mouse thigh results in a rather sharp intensity profile (Figure18).

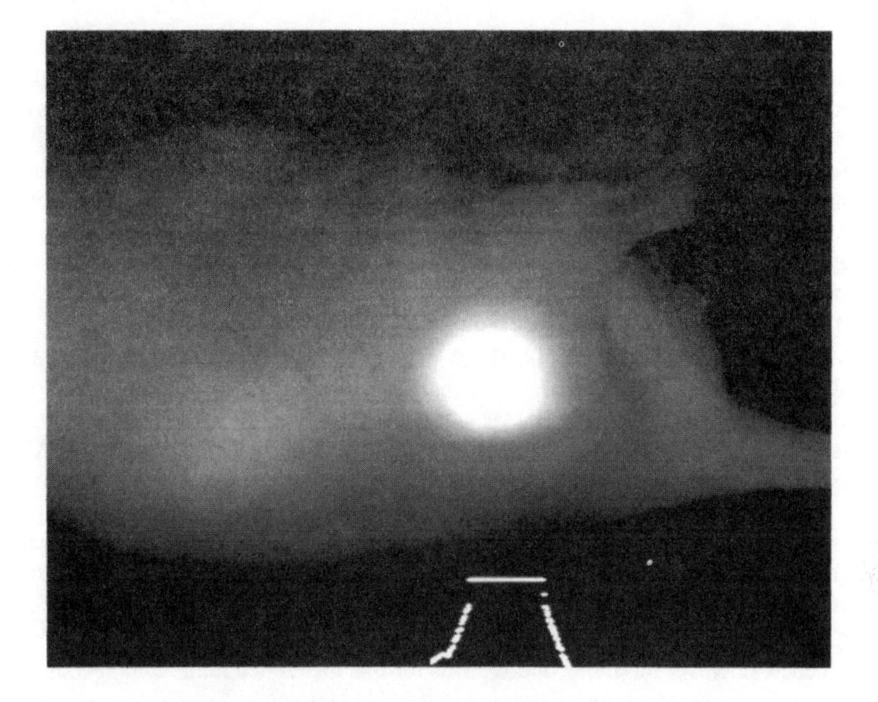

Figure 18. Fluorescence image from figure 17 after rescaling procedure

5. DISCUSSION

The presented investigations and findings show exemplarily the complexity of using auto- and xenofluorescences for medical diagnostics and intraoperative fluorescence monitoring. Even an approximate qualitative evaluation of the auto- or xenofluorescences by a medical doctor require prior fulfillment of at least the following four steps:

- Medical and biochemical investigation of the relevant actual biodynamic fluorescence processes in relation to the disease
- Determination of the tissue-optical environment, into which these processes are embedded
- Estimation of optical methods for realizing an appropriate biomedical technique
- Hard- and software for rescaling the detected fluorescence signals aiming at an evaluation adapted to the attending doctor's requirements

The biochemical fluorescence reactions of NADH in mitochrondria, similar to those that – in a modified form, of course, may occur during fluorescence monitoring at the open heart, have been chosen as an example for the first step. In a suspension of respiring mitochondria very fast and reasonable response of NADH fluorescence signal to various perturbations of intrinsic biochemical reactions was obtained. This signal could be transformed to the concentration dimension. Nevertheless, it still remains to elucidate what impact on this transformation might have changes in bound versus free NADH upon transition between the metabolic states of mitochondria. On one side, model experiments with solution of NADH and proteins having different affinity for NADH (or following NAD-dependent enzyme kinetics) could help in finding solution for this question. Another alternative is rapid extraction and chromatographic analysis of samples followed by analysis of correlation between the monitored level of fluorescence and determined concentration of NADH. However, low NADH stability is a highly expected problem on this way and it implies a need to develop more appropriate methods for the extraction and analysis of pyridine nucleotides from biological samples.

Moreover, the analysis of NADH fluorescence signal in tissues will need more extended biochemical background not only because of more complicated optical environment in the tissue. A doctor should be provided by some guidelines for understanding the NAD(P)H dynamics. Biomedical interpretation of NAD(P)H fluorescence changes should approach the state of the art, since in various tissues and under different disease condition these changes are not going to follow the same pattern. The understanding of NADH fluorescence signal is complicated because the direction of fast changes in actual amount of NAD(P)H in the tissue may be caused by multiple reasons.

Although it is already known that tumor area in many tissues show lower NAD(P)H fluorescence, however the molecular reasons for this difference remain unclear as well as their dependence on the type of cancer. Lower NADH content in cancer cell might be explained either by fast energy need for intensive biosynthesis or by lower NADH production in glycolytic pathway to compare to normal cell metabolism, or by extensive use of NADH for lactate production, or by all mentioned (and other) reasons at once.

The image blurred by photon migration is to be rescaled according to the respective tissue optical parameters. Various methods are suitable for rescaling. The suggested methodical combination of a cut off according to the signal/noise ratio and the intensity scaling for setting off the partial elimination of the signal in the tissue is an applicable first approximation. The described rescaling is part of the imaging system and remains unchanged in terms of its physical and mathematical principles switching from Optical Biopsy to Optical Molecular Imaging. Based on NADH monitoring, it was possible to demonstrate the rescaling process by means of an exemplary clinical case.

The tremendous efforts required for determining the tissue optics in the diseased partial volume cannot be avoided. Such efforts include the preparation of a comprehensive statistic for characterizing the tissue optics of the specific disease process by prototype. Even then, it must always be determined whether or not tissue-optical variances would render the diagnosis of the specific case of disease impossible. The investigations related to fluorescence dynamics and tissue optics in different situations, such as the laboratory investigations on mitochondria of the myocardium, the intraoperative monitoring of squamous cell carcinoma in the upper aerodigestive tract and of superficial tumors in living nude mice impressively show that this problem can be solved.

For OMI the concentration profile of NIR-fluophores in the border region of tumors can be achieved just like the NADH concentration profile in Optical Biopsy.

An instantaneous and non-destructive determination of the NADH concentration in scattering media is possible by these techniques. Steps forward to a 3D reconstruction are described elsewhere [59]. This aspect (along with the tumor specific quenching of the fluorophore for OMI) enhances the border and the local resolution thus offering a chance to analyse even early-stage tumors.

In a total of 30 measurements in subcutaneous tumors, a rescaling error of 12 % was determined. At a tumor process of 10 mm in diameter this corresponds to a resulting surgical unreliability of approx. 1 mm. These statements, of course, have to be considered strictly conditioned. In the case of OMI, further investigations on the influences of the marker concentration, the saturation behaviour and the respective quantum yield will follow. These findings may serve to improve the presented technique.

7. ACKNOWLEDGEMENTS

We thank the German Ministry BMBF for grant LTU 02/004 for scientific exchange between Germany and Lithuania under the title “Combination of methods for heart tissue viability monitoring” and chief coordinator Dr. Algimantas Krisciukaitis from Kaunas University of Medicine. The equipment for laboratory cell monitoring was realized by ELINTA Corp. Kaunas. Prof. Weissleder from the Harvard Medical School was helpful to discuss the principles of OMI.

7. REFERENCES

1. Paul H (ed.): *Lexikon der Optik.* Spektrum Akademischer Verlag, Heidelberg, Berlin, 1999
2. Chance B, Legallais V, Schoener B: Metabolically linked changes in fluorescence emission spectra of cortex of rat brain, kidney and adrenal. *Nature* 195 (1962) 1073-1075
3. Chance B, Schoener B, Oshino R, Itshak F, Nakase Y: Oxidation-Reduction Ratio studies of mitichondria in freeze-trapped samples. *J. Biol. Chem.* 254 (1979) 4766-4771
4. Renault G, Raynal E, Sinet M, Berthier JP, Godard B, Cornillault J: A laser fluorimeter for direct cardiac metabolism monitoring. *Opt. Laser Technol.* 14 (1982) 143-148
5. Renault G, Sinet M, Muffat-Joly M, Fourati T, Polianski J, Meric P, Weiser M, Pocidalo J: Evaluation *in situ* du métabolisme tissulaire par fluorimétrie laser. *La Presse Médicale* 13 (1984) 2381-2385
6. Richards-Kortum, R, Rava R, Fitzmaurice M, Tong L, Ratliff N, Kramer J: A one-layer model of laser-induced fluorescence for diagnosis in human tissue: Applications to artherosclerois. *IEEE Trans. Biomed. Eng.* 36 (1989) 1222-1232
7. Beuthan J, Zur C, Hofmann H: Quantitative (*in vivo*) NADH-Messung - ein methodisch klinischer Ansatz zur biologischen Äquivalentdosimetrie. *Adv. Laser Medicine* 5 (1990) 253-260
8. Beuthan J, Minet O, Müller G: Observations of the fluorescence response of the coenzyme NADH in biological samples, *Opt. Lett.* 18 (1993) 1098-1099
9. Lohmann W, Schill WB, Bucher D, Peters T, Nilles M, Schulz A, Bohle R, Schramm W: Tissue diagnosis using autofluorescence. *Proc. SPIE* 2081 (1993) 10-24
10. Wu J, Field S, Rava RP: Analytical model for extracting intrinsic fluorescence in turbid media. *Appl. Opt.* 32 (1993) 3285-3295
11. Beuthan J, Weber A, Minet O, Hagemann R, Roggan A, Schmitt I, Müller G, Germer C, Albrecht D, Bocher T: Untersuchungen zur NADH-Konzentrationsbestimmung mittels optischer Biopsie. *Lasermedizin* 10 (1994) 57-63
12. Beuthan J, Bocher T, Minet O, Roggan A, Schmitt I, Weber A, Müller G: Investigations concerning the determination of NADH-concentrations using optical biopsy. *Proc. SPIE* 2135 (1994) 147-156

13. Chung YG, Schwartz J, Gardner C, Sawaya R, Jacques SL: Fluorescence of normal and cancerous brain tissues: the excitation/emission matrix. *Proc. SPIE* 2135 (1994) 66-75
14. Gandjbakhche A, Gannot I: Quantitative fluorescent imaging of specific markers of diseased tissue. *IEEE Sel. Topics QE* 2 (1996) 914-921
15. Pogue BW, Hasan T: Fluorophore quantition in tissue-simulating media with confocal detection. *IEEE Sel. Topics QE* 2 (1996) 959-964
16. Yova D, Atlamazoglou V, Davaris P, Kavantzas N, Loukas S: Colon cancer diagnosis using fluorescence spectroscopy and fluorescence imaging technique. *Proc. SPIE* 3197 (1997) 4-15
17. af Klinteberg C, Wang I, Lindquist C, Vaitkuviene A, Svanberg K: Laser-induced fluorescence studies of premalignant and benign lesions in the female genital tract. *Proc. SPIE* 3197 (1997) 34-40
18. Padilla-Ybarra JJ, Bourg-Heckly G, A'Amar O, Blais J, Etienne J, Guillemin F: UV induced autofluorescence spectroscopy in Barrett's esophagus. *Proc. SPIE* 3197 (1997) 54-59
19. Lohmann W, Schill WB, Bohle RM, Dreyer T: Autofluorescence of seborrheic keratosis (warts) and of tissue surrounding malignant tumors. *Proc. SPIE* 3197 (1997) 140-150
20. Beuthan J, Minet O, Müller G: Optical Biopsy of Cytokeratin and NADH in the Tumor Border Zone. *Ann. N.Y. Acad. Sci.* 838 (1998) 150-170
21. Beuthan J, Minet O: Fluorescence Diagnosis in the Border Zone of Liver Tumors. In: Rettig W. et al (eds.): *Applied Fluorescence in Chemistry, Biology and Medicine.* Springer, Berlin 1999, pp.537-551
22. Shehada REN, Marmarelis VZ, Mansour HN, Grundfest WS: Laser Induced Fluorescence Attenuation Spectroscopy: Detection of Hypoxia. *IEEE Trans. Biomed. Eng.* 47 (2000) 301-312
23. Schuchmann S, Kovacs R, Kann O, Heinemann U, Buchheim K: Monitoring NAD(P)H autofluorescence to assess mitochondrial metabolic functions in rat hippocampal-entorhinal cortex slices. *Brain Res. Prot.* 7 (2001) 267-276
24. Zellweger M, Goujon D, Conde R, Forrer M, van den Bergh, H, Wagnières G: Absolute autofluorescence spectra of human healthy, metaplastic, and early cancerous bronchial tissue *in vivo*. *Appl. Opt.* 40 (2001) 3784-3791
25. Loetscher P, Alvarez-Gonsalez R, Althaus FR: Poly(ADP-ribose) may signal changing metabolic conditions to the chromatin of mammalian cells. *Proc. Natl. Acad. Sci. U.S.A.* 84 (1987), 1286-1289
26. D'Amours D, Desnoyers S, D'Silva I, Poirier GG: Poly(ADP-ribosyl)ation reactions in the regulation of nuclear functions. *Biochem. J.* 324 (1999) 249-268
27. Ziegler M: New functions of a long-known molecule. Emerging roles of NAD in cellular signalling. Eur. J. Biochem. 267 (2000) 1550-1564
28. Skidmore CJ, Davies MI, Goodwin PM, Halldorsson H, Lewis PJ, Shall S, Zia'ee AA: The involvement of poly(ADP-ribose) polymerase in the degradation of NAD caused by gamma-radiation and *N*-methyl-*N*-nitrosourea. *Eur. J. Biochem.* 101 (1979) 135-142
29. Berger NA: Poly(ADP-ribose) in the cellular response to DNA damage. Radiat. Res. 101 (1985) 4-15
30. Ha HC, Snyder SH: Poly(ADP-ribose) polymerase is a mediator of necrotic cell death by ATP depletion. *Proc. Natl. Acad. Sci. U.S.A.* 96 (1999) 13978-13982
31. Lee HC: Physiological functions of cyclic ADP-ribose and NAADP as calcium messengers. *Ann. Rev. Pharmacol. Toxicol.* 41 (2001) 317-345
32. Guse AH: Cyclic ADP-ribose: a novel Ca2+-mobilising second messenger. *Cell. Signal.* 11 (1999) 309-316
33. Salman JM, Kohen E, Viallet P, Hirschberg JG, Wouters AW, Kohen C, Thorell B: Microspectrofluometric approach to the study of free/bound NAD(P)H ratio as metabolic indicator in various cell types. *Photochem. Photobiol.* 36 (1982) 585-593
34. Galeotti T, van Rossum GDV, Mayer DH, Chance B: On the fluorescence of NAD(P)H in Whole-Cell Preparations of Tumours and Normal Tissues. *Eur. J. Biochem.* 17 (1970) 485-496
35. Lakowicz JR, Szmacinski H, Nowaczyk K, Johnson ML: Fluorescence Lifetime Imaging of Free and Protein-Bound NADH. *Proc. Natl. Acad. Sci. USA.* 89 (1992) 1271–1275
36. Virag L, Szabo C: The therapeutic potential of poly(ADP-ribose) polymerase inhibitors. *Pharmacol. Rev.* 54 (2002) 375-429
37. Tentori L, Portarena I, Graziani G: Potential clinical applications of poly(ADP-ribose) polymerase (PARP) inhibitors. *Pharmacol Res.* 45 (2002) 73-85

38. Varadarajan SG, An J, Novalija E, Smart SC, Stowe DF: Changes in $[Na^+]_i$, compartmental $[Ca^{2+}]$, and NADH with dysfunction after global ischemia in intact hearts. *Am. J. Physiol. Heart Circ. Physiol.* 280 (2001) H280-293

39. Halmosi R, Berente Z, Osz E, Toth K, Literati-Nagy P, Sumegi B: Effect of Poly(ADP-Ribose) Polymerase Inhibitors on the Ischemia-Reperfusion-Induced Oxidative Cell Damage and Mitochondrial Metabolism in Langendorff Heart Perfusion System. Mol. Pharmacol. 59 (2001) 1479-1505 Kohen E, Santus R, Hirschberg JG: *Fluorescence Probes in Oncology.* Imperial College Press, London 2002

41. Mildažiene V, Baniene R: private communication July 2003

42. Gornal AG, Bardavill GJ, David MM: Determination of serum proteins by means of the biuret reaction. *J. Biol. Chem.* 177 (1949) 751-766

43. Fabiato A, Fabiato F: Calculator programs for computing the composition of the solutions containing multiple metals and ligands used for experiments in skinned muscle cells. *J. Physiol.*(Paris), 75 (1979) 463-505

44. Roggan A, Minet O, Schröder C, Müller G: Measurements of optical tissue properties using integrating sphere technique. In: *Medical Optical Tomography.* G. Müller et al. (eds.), *SPIE IS* 11, Bellingham 1993, pp. 149-165

45. Ishimaru I: *Wave Propagation and Scattering in Random Media.* Academic Press, San Diego, New York 1978

46. Tuchin V: Selected Papers on Tissue Optics. *SPIE MS* 102, Bellingham 1994

47. Bocher T, Luhmann T, Baier S, Dierolf M, Naumann M, Beuthan J, Berlien HP, Müller GJ: Multispectral fluorescence imaging device for malignancy detection. *Proc. SPIE* 3197 (1997) 60-67

48. Weissleder R: Molecular Imaging: Exploring the next frontier. Radiology 212 (1999) 609-614

49. Becker A, Hessenius C, Licha K, Ebert B, Sukowski U, Semmler W, Wiedenmann B, Grötzinger C: Receptor-targeted optical imaging of tumors with near-infrared fluorescent ligands. *Nature Biotech.* 19 (2001) 327-331

50. Bugaj JE, Achilefu S, Dorshow RB, Rajagopalan R: Novel fluorescent contrast agents for optical imaging of *in vivo* tumors based on a receptor-targeted dye-peptide conjugate platform. J. Biomed. Opt. 6 (2001) 122-133

51. Weissleder R, Tung C, Mahmood U, Bognanov A: *In vivo* imaging of tumors with protease-activated near-infrared fluorescent probes. *Nature Biotech.* 17 (1999) 375-378

52. Weissleder R: Scaling down Imaging: Molecular Mapping of Cancer in Mice. *Nature Rev.* 2 (2002) 1-7

53. Adams JY, Johnson M, Sato M, Berger F, Gambhie SS, Carey M, Iruela-Arispe ML, Wu L: Visualization of advanced human prostate cancer lesions in mice by a targeted gene transfer vector and optical imaging. *Nature Med.* 8 (2002) 891-896

54. Licha K, Riefke B, Ntziachristos V, Becker A, Chance B, Semmler W: Hydrophilic Cyanine Dyes as Contrast Agents for Near-Infrared Tumor Imaging: Synthesis, Photophysical Properties and Spectroscopic *In Vivo* Characterization. *Photochem. Photobiol.* 72 (2000) 392-398

55. Licha K: Contrast Agents for Optical imaging. In: Krause W (ed.): *Contrast Agents II (Topics in Current Chemistry*, Vol. 222). Springer, Berlin, Heidelberg, NY. 2002, pp. 1-29

56. Minet O, Beuthan J, Licha K, Mahnke C: The biomedical use of rescaling procedures in Optical Biopsy and Optical Molecular Imaging. In R. Kraayenhof, A.J.W.G. Visser, H.C. Gerritsen. (ed): Fluorescence Spectroscopy, Imaging and Probes. Springer, Berlin, Heidelberg, NY. 2002, pp. 349-360

57. Riefke, B, Licha, K, Nolte, D, Ebert, B, Rinneberg, H, Semmler, W: *In vivo* characterization of cyanine dyes as contrast agents for near-infrared imaging. *Proc. SPIE* 2927 (1996) 199-208

58. Ebert B, Sukowski U, Grosenick D, Wabnitz H, Moesta KT, Licha K, Becker A, Semmler W, Schlag PM, Rinneberg H: Near-infrared fluorescent dyes for enhanced contrast in optical mammography: phantom experiments. *J. Biomed. Opt.* 6 (2001) 134-40

59. Minet O: Zur Bestimmung der räumlichen Verteilung von Fluoreszenzzentren in streuenden Medien. *Fortschritte in der Lasermedizin* 12 (1995) 98

DETECTION OF GENOMIC ABNORMALITIES BY FLUORESCENCE *IN SITU* HYBRIDIZATION

Larry E. Morrison*

1. INTRODUCTION

Fluorescence has been used to great advantage in detecting specific nucleotide sequences within cell-free nucleic acids (Morrison, 1999). These cell-free assays frequently take advantage of various nucleic acid amplification methods, such as the polymerase chain reaction (PCR), to obtain detection limits that can be as low as several molecules. However, the separation of nucleic acids from their cells and tissues of origin results in the loss of valuable information that resides in the structural features, such as cell and nuclear morphology, cellular organization, and relationships between cells and structural components of tissue. In the medical community it is well known that histopathology, which relies upon examination of intact tissues, is significantly more accurate in diagnosing cancer than cytopathology, which examines cells that have detached from their original tissues. In addition, only a very few cell types within a specimen may be of importance, and lysing all of the cells to release their nucleic acids will result in diluting the nucleic acid of interest, for example tumor DNA, with the nucleic acid from irrelevant cells, such as the normal cell background, making analysis difficult or impossible.

In situ hybridization is a method that detects specific nucleic acid sequences within the framework of intact cells and tissues, thereby retaining the information contained within the cellular and tissue structure. This permits a cell-by-cell analysis of nucleic acids, and prevents dilution of nucleic acid from one cell by that of another. Even without amplification, *in situ* hybridization can detect single genetic loci, such as single genes. *In situ* hybridization has many features in common with immunohistochemistry, in which labeled antibody reagents bind to their haptens within a slide-mounted specimen, and their locations are subsequently visualized via microscopy. To perform an *in situ* hybridization the specimen is adhered to a microscope slide, the nucleic acid within the specimen is denatured to provide single-stranded regions (if not already single-stranded), and labeled nucleic acid probes are allowed to hybridize to their complementary sequences within the specimen nucleic acid. After washing to remove

* Larry E. Morrison, Vysis/Abbott Laboratories, Downers Grove, IL 60515.

unbound probe, the slide is viewed under a microscope to observe the location of the targeted nucleic acid sequence within the specimen.

In fluorescence *in situ* hybridization, abbreviated as FISH, the labels ultimately detected are fluorophores. The fluorophores are either directly conjugated to the nucleic acid probes, or associate with the probes post-hybridization, for example via antibody binding to hapten-conjugated probes. For the most part FISH takes advantage of the ability of fluorescence to provide bright staining of small features against a dark background, and the ability to distinguish different fluorophores by their different absorbance and emission spectra. Fluorescence properties such as emission polarization, emissive lifetime, quenching, and energy transfer have been little utilized in FISH. Generating sufficient intensity for visual detection has been the dominating goal, requiring many fluorophores per targeted region, and resulting in targets that are on the order of thousands to hundreds of thousands of nucleotides long. With labels spanning such large sequences, inter-label distances are much larger than those necessary for efficient static or dynamic quenching, or for non-radiative energy transfer. The elucidation of microenvironments within specimens has not been of interest so that polarization and environmental dependence of fluorescence have not been applied. However, spectral separation of fluorophores has been used to great advantage, permitting the simultaneous detection of a number of different nucleic acid targets within single cells, which is necessary to conserve specimen and to establish links between multiple genetic abnormalities.

This chapter will discuss how FISH is applied to cellular analysis, first describing the FISH reagents and methodologies, including descriptions of nucleic acid probe types, methods of probe preparation, protocols for performing FISH, and multiple target detection methods. This will be followed by various examples of FISH being used to detect genomic abnormalities. Although FISH can be used to detect both RNA and DNA sequences within specimens, the applications here will be confined to DNA targets and the use of FISH to detect genomic aberrations relevant to both peri-natal analyses and the characterization of cancers.

2. FISH REAGENTS AND METHODS

2.1. Types of FISH Probes

Several different types of FISH probes can be defined with respect to the general characteristics of the chromosomal target sequence. The first type of probe is the whole chromosome painting probe. This is a high complexity probe that has sequence complementarity across an entire chromosome, such that when hybridized the entirety of a specific chromosome is uniformly stained. Whole chromosome painting probes are of practical use only on metaphase chromosomes because the decondensed chromosomes of interphase nuclei are spread over too large an area to be useful for structural analysis or chromosome enumeration. Primarily whole chromosome painting probes are used for identifying structural rearrangements between chromosomes or for determining chromosome copy number.

The second probe type is the repetitive sequence probe. This is the lowest complexity probe and is comprised of sequences that repeat hundreds and thousands of times within a relatively small region, for example 5 to 10 megabases (Mb). A common repetitive sequence is the alpha satellite sequence which occurs in the vicinity of chromosome centromeres. The alpha satellite sequences are sufficiently different from one another that for most chromosomes a specific alpha satellite probe can be prepared that stains only the area near the centromere of a single chromosome (a pair of autosomes in diploid nuclei). The alpha satellite repeat sequences for chromosomes 13 and 21, 14 and 22, and 5 and 19 are too similar to separately distinguish each of these chromosomes. Repeat probes usually provide bright and rapid hybridizations because of their lower complexity, and are most useful for determining the copy number of chromosomes. Because they hybridize to a relatively small region, most repetitive sequence probes appear as tight spots that are easily counted to provide chromosome copy number, even in interphase nuclei in which the chromosomes are decondensed. This is particularly important in solid tumor specimens in which metaphase chromosome spreads are rare or nonexistent.

The third probe type is the locus specific probe. Locus specific probes are unique sequences and typically target between 40 and 500 kilobases (kb) of DNA. Probes smaller than this can be difficult to detect routinely, especially on formalin-fixed paraffin-embedded specimens, although it is possible to detect some probes as small as several kilobases (Rogan et al., 2001). Probes larger than 500 kb can become diffuse in interphase nuclei, and therefore difficult to distinguish as single spots. The locus specific probes are useful for determining the status of a single genetic locus, such as a single gene or group of genes. When co-hybridized with a probe of a different fluorescent color controlling for the chromosome copy number, such as a centromeric repeat probe, or other locus specific probe distant from the first probe, the ratio provides a measure of locus amplification or deletion. The MYC and ERBB2 genes are examples of loci that are commonly amplified in cancers.

Another way to classify probes is by the method in which the fluorescent label is associated with the probe. Indirectly labeled probes are associated with their fluorescent labels following the hybridization step. The probe itself is conjugated with one member of an affinity binding pair, such as a hapten or biotin. Following *in situ* hybridization the specimen is incubated with fluorophore-labeled antibody or avidin. A second layer of reagents can be added in the form of labeled species-specific antibody directed to the first antibody. If avidin is used in the first layer, biotin or fluorophore labeled anti-avidin is used to form the second layer. More layers can be added, however, building up layers of proteins to increase the fluorescence signal also increases backgrounds, and care must be taken to minimize non-specific adsorption of antibodies and avidins to the slide surface and specimen.

Directly labeled FISH probes have fluorophores chemically linked to the probes prior to hybridization. Protocols are generally shorter when using directly labeled probes because the post-hybridization antibody or avidin incubations are not necessary. Backgrounds are generally lower because of the absence of the secondary binding proteins. With directly labeled probes, though, the fluorescence intensity is not increased through multiple rounds of antibody or avidin binding. However, locus specific probes of 70 kb and greater can be easily visualized with directly labeled probes, and the option exists for using anti-fluorophore antibodies for signal enhancement, if necessary.

2.2. Sources of FISH Probes

FISH probes are composed of nucleic acids with appended fluorophores, haptens, or biotins. For genomic analysis the nucleic acid is usually DNA. The DNA for FISH probes can come from a variety of sources. Whole chromosome painting probes start with flow sorted (Carter et al., 1992; Telenius et al., 1993) or microdissected (Guan et al., 1994) chromosomes. Originally the chromosomes were cleaved with restriction enzymes and the fragments combined in vectors to form phage or bacterial libraries (Collins et al., 1991; Deaven et al., 1986). However, with time the libraries tended to loose complexity, and coverage was not uniform for many chromosomes, especially near the chromosome ends. Whole chromosome probes are now more commonly prepared by degenerate oligonucleotide primed polymerase chain reaction (DOP-PCR) of flow sorted or microdissected chromosomes (Telenius et al., 1992a; Telenius et al., 1992b). Locus specific probes are generally prepared from DNA cloned into vectors that support large inserts, such as cosmids, P1s, PACs, BACs or YACs. With the success of the human genome project, BACs covering the entire genome are readily available. Repetitive sequence DNA in the range of 300 to 10,000 bp can be inserted into single bacterial plasmids. Once cloned the probe DNA is generated through culturing the bacteria or yeast, harvesting the cells, lysing the cells, and extracting the DNA.[†] Bacterial DNA can be isolated by common methods that included alkaline lysis or boiling (Sambrook et al., 1989).[‡] Alternatively, extraction kits are available from several commercial suppliers (e.g. Qiagen, Inc., Chatsworth, CA, Stratagene, La Jolla, CA, and Gentra Systems, Inc., Minneapolis, MN).

2.3. FISH Probe Labeling

The attachment of fluorophores, haptens, or biotins to nucleic acids can be performed either chemically or ezymatically (Morrison et al., 2002). Early fluorescent labeling was performed using periodate oxidation of a 3'-terminal ribonucleotide, followed by reaction with a hydrazine derivative of fluorescein (Bauman and van Duijn, 1981). Early labeling of FISH probes was also accomplished through conjugation with aminoacetylfluorene (Cremer et al., 1988; Landegent et al., 1984) and mercuration of DNA (Dale and Ward, 1975). Following mercuration, the DNA was further reacted with a thiol-containing label (Hopman et al., 1986; Hopman et al., 1987). Other chemistries used to label FISH probes include bisulfite mediated transamination of cytosine (Draper, 1984; Reisfeld et al., 1987), bromination of thymine, guanine and cytosine with N-bromosuccinimide, followed by reaction with amine-containing labels (Keller et al., 1988), photochemical reaction with photobiotin (Keller et al., 1989), and condensation of terminal phosphate groups with diamines, followed by coupling with amine reactive labels (Chu et al., 1983; Morrison et al., 1989). Platinum complexes containing label as a ligand have also been used to form coordinate covalent bonds with nucleic acids, primarily through the guanine bases (van Belkum et al., 1994).

[†] see various volumes in the series: Current Protocols in Molecular Biology, K. Janssen., ed., John Wiley and Sons, New York.

[‡] see various volumes in the series: Current Protocols in Human Genetics, A. L. Boyle., ed., John Wiley and Sons, New York.

The most common forms of probe labeling utilize enzymatic reactions, particularly reactions using DNA polymerases. DNA polymerases catalyze the addition of nucleoside triphosphates to the 3'-OH terminus of a DNA primer strand that is hybridized to a strand of DNA serving as the template, in this case the probe DNA. Each nucleotide added by the polymerase is complementary to the next base in the template strand. For labeling purposes, one of the four nucleoside triphosphates is appended with a label, which becomes incorporated in the newly synthesized strand with the nucleotide. Incorporation of biotinylated nucleoside triphosphate is the oldest and most frequently used method (Langer et al., 1981; Langer-Safer et al., 1982; Manuelidis et al., 1982). Haptens incorporated by this method include dinitrophenol (Ried et al., 1992b), digoxigenin (Arnoldus et al., 1990), and fluorescein (Wiegant et al., 1991). Fluorescein can serve the dual purpose of being either a hapten or a directly incorporated fluorescent label. A number of fluorophores are commercially available attached to different nucleoside triphosphates (for example, from Molecular Probes Inc., Eugene, OR; New England Nuclear, Boston, MA; and Vysis/Abbott Laboratories, Downers Grove, IL).

Different labeling reactions utilizing DNA polymerase include nick translation (Cherif et al., 1989; Kelly et al., 1970), PCR (Ried et al., 1992a), DOP-PCR (Telenius et al., 1992b), and random priming (Feinberg and Vogelstein, 1983, 1984). In the most popular polymerase reaction, nick translation, double-stranded probe DNA is nicked (cleavage of the phosphodiester bond of one strand) in multiple places, and DNA polymerase adds nucleotides to the 3'-OH terminus formed at the nick. A DNA polymerase is used that also contains a 5'-to-3' exonuclease activity such that the strand ahead of the nick is digested to make way for the growing labeled strand. No net gain in the amount of probe DNA occurs during nick translation, however, in the other polymerase methods of PCR, DOP-PCR, and random-priming, the amount of probe DNA is increased. Detailed protocols for these and other DNA and RNA probe labeling methods can be found elsewhere (Morrison et al., 2002; Ramakrishnan and Morrison, 2002).

2.4. *In Situ* Hybridization

In situ hybridization involves the steps of: 1) adhering the specimen to the surface of a microscope slide, 2) treating the specimen to improve the accessibility of the target nucleic acid to the probe, 3) denaturing the specimen nucleic acid to create regions of single strands, 4) contacting the denatured specimen with single-stranded labeled nucleic acid probe, 5) allowing the probe to hybridize to complementary regions of the specimen nucleic acid, 6) washing the slide to remove all unbound probe, 7) developing the probe label, if using indirect detection methods, and 8) viewing or imaging the hybridization pattern created by the probe on the specimen.

Specimens for genomic *in situ* hybridizations include cultured and uncultured cells, and range from cells that are dispersed in a fluid singly or in small clusters, to cells grown as monolayers, to tissues. Special cell culturing methods can be used to increase the proportion of metaphase nuclei, such as using stimulators of mitosis together with agents that halt the cell cycle after condensation of the chromosomes. The cells are typically 'fixed' by some method that improves the structural integrity of the cells, thereby preserving the cell morphology that is critical for cytological or histological interpretation. Common fixatives include ethanol, methanol, methanol with acetic acid, and formaldehyde. Tissue specimens include thin sections cut from frozen tissue, touch preparations of fresh tissue, or sections cut from tissue that has been strongly fixed and embedded in paraffin. Methods for preparing specimens and adhering specimens to microscope slides are varied and can be found in detail elsewhere (Barch et al., 1997).

Slide-mounted specimens can be further prepared for *in situ* hybridization by various treatments that include incubation in buffer at elevated temperature, incubation with chaotropes, digestion with proteinase and/or ribonuclease, and further fixation. Specimens such as lymphocytes may need no treatment prior to hybridization, or only mild treatment such as incubation in 2xSSC (0.3 N NaCl, 30 mM sodium citrate, pH 7.0-7.5) at 72° C for 2 minutes. Incubation in 0.005% pepsin/10 mN HCl for 5 minutes at 37° C followed by fixation in 1% formaldehyde/50 mM $MgCl_2$ for 2 minutes can further improve performance in poorer lymphocyte preparations. Cytology specimens generally require stronger treatment with pepsin, for example, 0.05% pepsin for 5 minutes, followed by 5 minutes of formaldehyde fixation. Formaldehyde-fixed paraffin-embedded specimens generally require the strongest treatment. Concentrated chaotropes at elevated temperature generally precede a pepsin digestion and fixation. Proteinase K can be required in place of pepsin for some paraffin-embedded specimens. Variations between different types of tissues can be large, and the literature should be consulted for protocols optimized for specific specimen types (Jacobson et al., 2000).

Specimens may be denatured chemically by means of strong chaotropes or extremes in pH but are most commonly denatured by heating in a buffered formamide solution, such as 70% formamide/2xSSC at 72° C for 5 minutes. Following denaturation, the specimen is dehydrated in a series of ethanol solutions and dried, denatured probe is added in hybridization solution (typically 50-60% formamide, 10% dextran sulfate, 2xSSC), and hybridization is allowed to proceed at an optimal temperature for the stringency of the hybridization solution employed (e.g. 37°-42° C). Hybridization can require minutes for repetitive sequence probes to several hours or overnight for more complex probes. Alternatively, the specimen and probe can be denatured simultaneously by adding the probe in hybridization solution to the specimen, heating to the denaturation temperature (e.g. 70°-85° C) for several minutes, and cooling to the hybridization temperature.

Unhybridized probe is removed by washing the specimen after the hybridization step. For many specimen types, such as lymphocytes and cytology specimens, washing can be as simple as a 2 minute incubation in 0.4X SSC/ 0.3% NP-40, pH 7.0-7.5 at 72° C. Paraffin-embedded specimens tend to require lower wash stringency, such as 2 minutes in 2xSSC/0.3% NP-40 at 72° C.

If the probes used were directly labeled a mounting medium and cover slip are placed over the specimen, and the specimen is ready for viewing on a fluorescence microscope. If the probes were indirectly labeled, one or more incubations with labeled antibody and/or avidin reagents are required for visualization (Dauwerse et al., 1992; Kearns and Pearson, 1994). Antibody solutions should be titrated to provide optimal signal with minimal background. An agent to reduce non-specific protein adsorption, such as 5% non-fat dry milk, is generally included in each antibody/avidin solution, and is preferably incubated with the specimen prior to incubation with the affinity reagents. Incubations with each affinity reagent are typically 30 minutes long followed by washing steps to remove unbound reagent. After the incubations are complete the mounting medium and coverslip are added for viewing with the fluorescence microscope. Mounting media typically contain a DNA counterstain such as 4,6-diamidino-2-

phenylindole (DAPI) or propidium iodide, and an antifade compound such as 1,4-diazabicyclo(2,2,2)-octane (DABCO) or p-phenylenediamine in buffered glycerol. The counterstain is a fluorescent molecule that binds to all DNA, through intercalation or binding to the major or minor groove of the double helix, thereby highlighting the nuclei and metaphase chromosomes. Good counterstains have high quantum efficiency when bound to DNA, and low efficiency when free in solution, so that background fluorescence is minimized. Antifades are compounds that protect the fluorescent labels from photodestruction.

The hybridized specimens are viewed using single or multiple bandpass filter sets appropriate for the counterstain and probe labels used in the hybridization. Each filter set consists of an excitation filter through which the lamp excitation light first passes, an emission filter placed before the eyepiece and imaging device, and a beam splitter. In the epifluorescence microscope format, the beamsplitter is designed to reflect the filtered light from the excitation lamp onto the specimen, while transmitting the light emitted by the specimen to the emission filter. Hybridization results are recorded for further analysis using either conventional photography, or by digital imaging, typically with the aid of a cooled CCD camera. A number of software packages are commercially available for general image analysis or more specific applications such as comparative genomic hybridization (CGH), M-FISH, SKY, and automated spot counting (e.g. Scanalytics Inc., Fairfax, VA, Applied Imaging Corp., Santa Clara CA, MetaSystems, Altlussheim, Germany, Applied Spectral Imaging, Migdal Ha'Emek, Israel).

3. MULTI-COLOR FISH

Perhaps the most important fluorescence characteristic utilized by FISH is the ability to distinguish multiple fluorescent labels based on their excitation and emission spectra (Morrison and Legator, 1999). *In situ* hybridization based on chromagen deposition has been limited to distinguishing two (Hopman et al., 1998) to three (Speel et al., 1994) different targets simultaneously, while 24 or more targets can be detected on a routine basis by FISH (Schrock et al., 1996; Speicher et al., 1996), and, in fact, commercial kits are available for detecting all 24 human chromosomes simultaneously (e.g. Vysis/Abbott Laboratories and Applied Spectral Imaging).

3.1. Distinguishing Multiple FISH Labels

The simplest approach to detecting multiple chromosome targets is to use a different spectrally distinct fluorescent label for each FISH probe. The various FISH probes are mixed together in the hybridization solution and allowed to hybridize to the specimen simultaneously. The specimen can then be viewed with various single bandpass filter sets optimized for viewing each fluorescent label to the exclusion of the other labels present. The fluorescence excitation and emission spectra for eight FISH labels commonly used in our laboratory to span the visible spectrum are displayed in Figure 1.

With single bandpass filter sets five[§] and even 7 (Morrison and Legator, 1999) different fluorescent labels can be viewed or imaged with minimal spectral crosstalk between

[§] Morrison L. E. and Legator M. S., 1991, Multi-color *in situ* hybridization using direct-labeled fluorescent probes, *Cytometry* supplement 5, Abstract 712B. Presented at the XV Congress of the International Society for Analytical Cytology, August 25-30, Bergen, Norway.

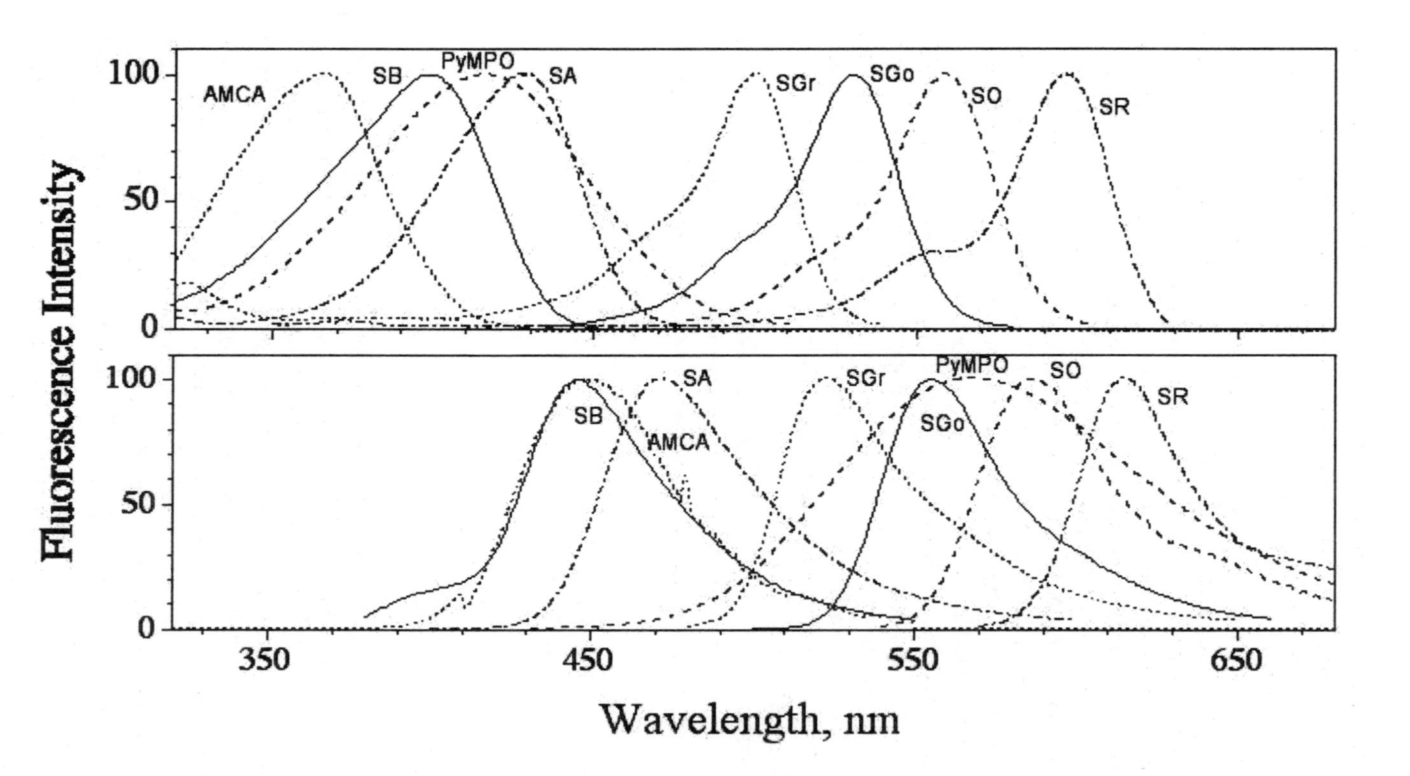

Figure 1. Corrected fluorescence excitation spectra (top) and emission spectra (bottom) for eight FISH labels with visible emission. AMCA = 7-amino-4-methylcoumarin-3-acetic acid, SB = SpectrumBlue™, PyMPO = 5-(4-methoxyphenyl)oxazol-2-yl)pyridinium bromide, SA = SpectrumAqua™, SGr = SpectrumGreen™, SGo = SpectrumGold™, SO = SpectrumOrange™, SR = SpectrumRed™ (indicated trademarks are for Vysis/Abbott Laboratories).

labels. Alternatively, one multiple bandpass filter set can be used to view several fluorescent labels simultaneously. Dual, triple, and quadruple bandpass filter sets are commercially available for this common application (e.g. Chroma Technology Corp., Brattleboro, VT; Vysis/Abbott Laboratories.).

Spectral crosstalk is the detection of additional fluorophores when using a filter set intended for isolating the fluorescence of a different fluorophore. This usually results from using fluorophores which have too closely spaced excitation and emission maxima for the width of the transmission bands of the fluorescence filter sets being utilized. Narrower filter spectra and better placement of the filter transmission bands can reduce spectral crosstalk. However, as greater numbers of targets are analyzed simultaneously by FISH, greater numbers of fluorescence labels are required, and overlap in their excitation and emission spectra become more severe. This is particularly a problem when fluorophores are confined to only the visible part of the spectrum for the convenience of manual focusing, viewing, and human interpretation. A method of correcting for spectral overlaps has been described and demonstrated for three fluorophores (Castleman, 1993). Application of this method requires knowledge of the spectral response of the imaging system, the main components of which include the microscope lamp spectrum, the individual filter spectra, the spectral response of the imaging camera, and the spectral characteristics of the optical elements and their coatings. Practically, these are all measured in conglomerate as described below.

Images are corrected for spectral crosstalk according to the solutions of n simultaneous linear equations that describe the amount each of the n different fluorophores contributes to the fluorescence measured using each of the n different single bandpass filter sets:

$$
\begin{aligned}
I'_1 &= a_{11}\, I^o_1 + a_{12}\, I^o_2 + \ldots + a_{1n}\, I^o_n \\
I'_2 &= a_{21}\, I^o_1 + a_{22}\, I^o_2 + \ldots + a_{2n}\, I^o_n \\
&\vdots \\
I'_n &= a_{n1}\, I^o_1 + a_{n2}\, I^o_2 + \ldots + a_{nn}\, I^o_n
\end{aligned}
$$

where I^o_j are the true (crosstalk corrected) images, I'_j are the observed images, a_{ij} are the crosstalk coefficients, and the subscripts 1,2,3... represent different fluorophores. Each coefficient, a_{ij}, is the amount of fluorescence measured for the jth fluorophore using the ith filter set, relative to the fluorescence measured for the jth fluorophore using the jth filter set (the filter combination intended to isolate the jth fluorophore). The values of a_{ij} are determined when calibrating the imaging system by measuring a specimen stained only with the jth fluorophore using the $i=1^{st}$ through n^{th} filter combinations. Backgrounds are subtracted from corresponding regions of specific staining in each image and the net average intensity (I_{ij}) is used to calculate $a_{ij}=I_{ij}/I_{jj}$. The values of a_{ij} form the determinant of the crosstalk which is used to solve the simultaneous equations for the set of n equations that specify what fraction of each of the n recorded images are to be added or subtracted from each other to provide the n corrected images.

We previously applied this method to the use of eight different fluorophores simultaneously, labeling eight whole chromosome painting probes, all emitting in the

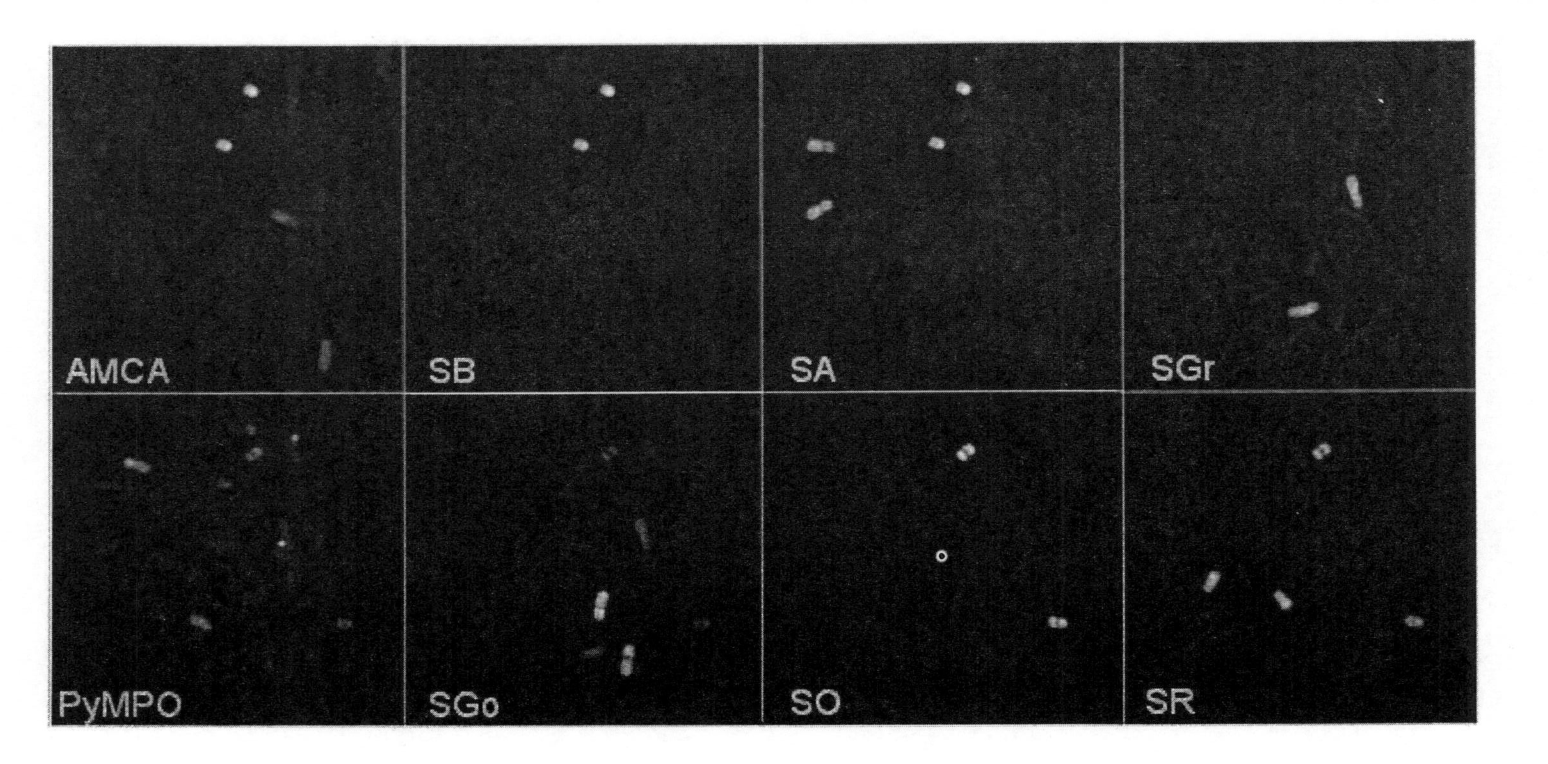

Figure 2. Eight monochrome images (uncorrected) recorded on a single metaphase chromosome spread hybridized simultaneously with whole chromosome painting probes to eight different chromosomes. The images were recorded with a cooled CCD camera using eight single bandpass filter sets optimized for each of the fluorescent labels. Label abbreviations marking each image are defined in the Figure 1 legend. Probes for chromosomes 3, 4, 6, 7, 8, 12, 16, and 22 were labeled with SGr, AMCA, SGo, SA, PyMPO, SR, SO, and SB, respectively.

visible region of the spectrum.[**] The images of a single metaphase chromosome spread are displayed in Figure 2; each recorded with a single bandpass filter set optimized for one the fluorophores. Significant crosstalk can be seen in several of the images as the staining of more than the expected chromosome pair target by each chromosome-specific painting probe.

Using crosstalk coefficients determined for these eight fluorophores, the following equations for correcting each individual fluorophore image were derived (ignoring coefficients smaller than 0.01; see Figure 1 legend for label abbreviations):

$$I^o_{AMCA} = 1.06I'_{AMCA} - 0.47I'_{SB} + 0.080I'_{SA}$$
$$I^o_{SB} = 1.15I'_{SB} - 0.16I'_{AMCA} - 0.20I'_{SA}$$
$$I^o_{SA} = 1.08I'_{SA} - 0.47I'_{SB}$$
$$I^o_{SGr} = 1.06I'_{SGr} + 0.010I'_{SB} - 0.024I'_{SA} - 0.37I'_{SGo}$$
$$I^o_{PyMPO} = I'_{PyMPO} + 0.017I'_{AMCA} - 0.069I'_{SB} - 0.12I'_{SA} + 0.16I'_{SGo} - 0.044I'_{SO}$$
$$I^o_{SGo} = 1.17I'_{SGo} - 0.18I'_{SGr} - 0.024I'_{SO}$$
$$I^o_{SO} = 1.14I'_{SO} - 0.019I'_{SA} + 0.85I'_{SGr} - 5.71I'_{SGo} - 0.18I'_{SR}$$
$$I^o_{SR} = 1.03I'_{SR} - 0.13I'_{SGr} + 0.84I'_{SGo} - 0.17I'_{SO}$$

After adding and subtracting the recorded images according to these equations, the eight images in Figure 3 were obtained, showing removal of the crosstalk, and leaving only a single pair of chromosomes stained in each image, thereby removing any ambiguity in the identity of the eight different chromosome pairs. More recently we have successfully applied this to FISH using 11 fluorophores simultaneously, adding three far red dyes (Cy5™, Cy5.5™, and Cy7™; trademarks of Amersham Biosciences, Piscataway, NJ) to the eight fluorophores described above.[††]

3.2. Coded Multi-Color FISH Formats

Since the potential for complexity in chromosomal changes associated with cancer or other genetic diseases is great, it is desirable to be able to monitor more than the eight to eleven labels described above on a cell-by-cell basis. In order to achieve this, coding methods can be used to increase the number of distinguishable targets beyond the number of fluorescent labels used in an *in situ* hybridization. Combinatorial coding and ratio coding are two such methods. Combinatorial coding is the use of fluorescent label combinations to represent different targets. For example, if the probe for one chromosomal target is labeled with a red emitting fluorophore, and the probe for a second chromosomal target is labeled with a green emitting fluorophore, then a third target can be identified with a probe labeled with both the red and the green emitting fluorophores. The number of targets that can be uniquely identified with combinatorial coding is equal to 2^N-1, where N is the number of different fluorescent labels utilized in the hybridization. With five labels, all 24 of the human chromosomes have been simultaneously

[**] Morrison, L., 1998, *Cytometry*, Supplement 9:140, Abstract CT 108. Presented at the XIX Congress of the International Society for Analytical Cytology, February 28-March 5, Colorado Springs, CO.

[††] Morrison, L. and Legator, M., 2002, *Cytometry*, Supplement 11:93-94, Abstract 43766. Presented at the XXI Congress of the International Society for Analytical Cytology, May 4-9, San Diego, CA.

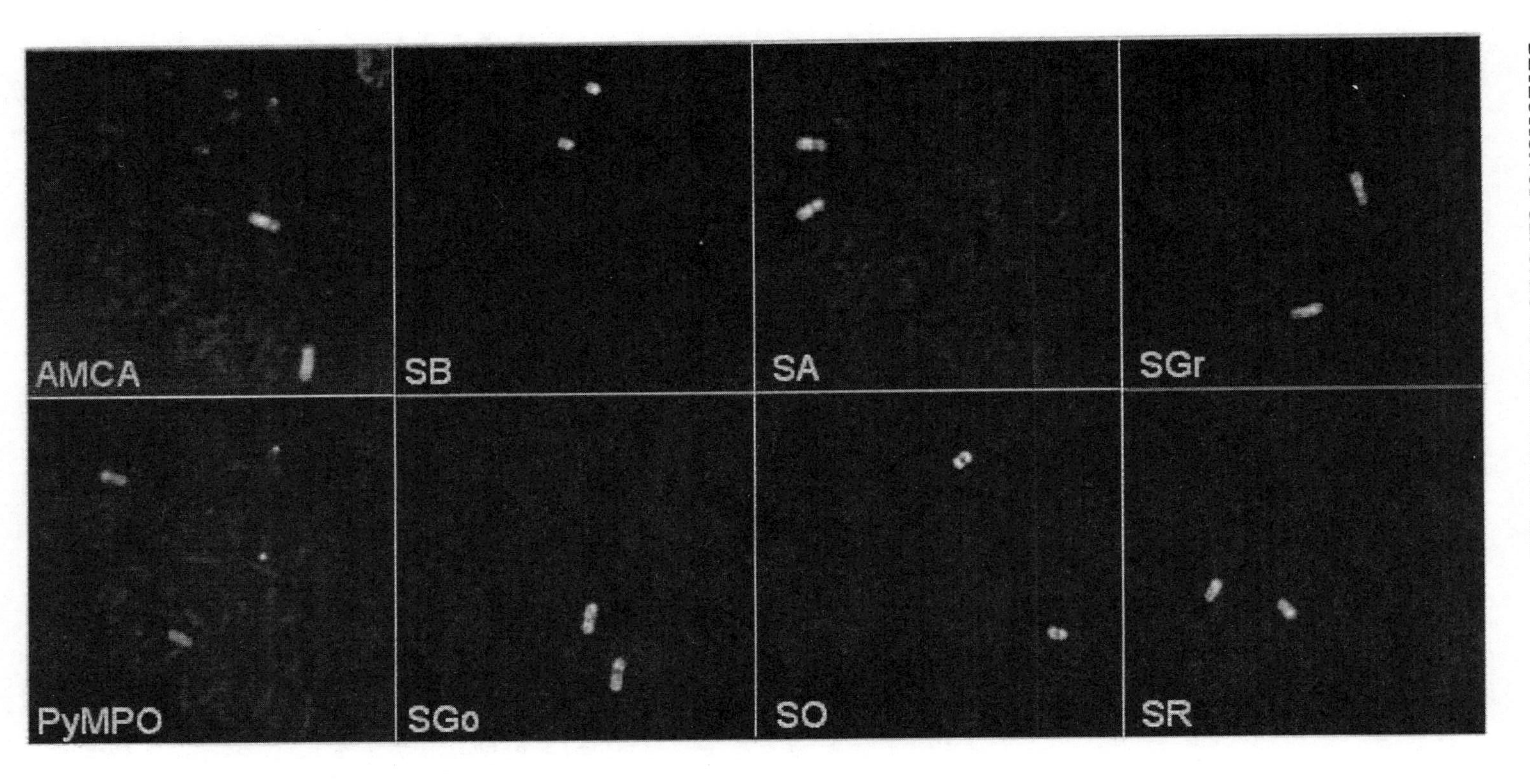

Figure 3. The eight monochrome images shown in Figure 2 corrected for spectral crosstalk. See the Figure 2 legend for FISH probes and labels used in this hybridization.

Table 1. Coding patterns in the SpectraVysion™ probe set for detecting all 24 human chromosomes in an M-FISH format. Solid circles indicate which labels are attached to probes for the corresponding chromosome.

Chrom.	SA	SGr	SGo	SR	SFR	Chrom.	SA	SGr	SGo	SR	SFR
1			●			13	●	●			
2				●		14		●	●	●	
3	●					15	●		●	●	
4		●		●		16		●			●
5			●		●	17		●		●	●
6		●				18			●	●	●
7					●	19		●	●		●
8				●	●	20	●			●	●
9			●	●		21	●	●	●		
10	●		●		●	22	●	●		●	
11	●			●		X	●				●
12		●	●			Y	●		●		

distinguished. In combinatorial coding the amount of each label present in the combination is not important – only the presence or absence of each label.

Ratio coding also relies upon combinations of labels, however, the relative amount of each label present is also important. Ratio coding requires not only detecting the presence or absence of each label, but also requires a quantitative measurement of the fluorescence intensity of each label at each target location. The number of different targets that can be distinguished is only limited by the precision of the intensity measurements. Using only two fluorescent labels, eight different targets have been simultaneously identified (Morrison and Legator, 1997). With four labels, all 24 human chromosomes have been distinguished (Tanke et al., 1999).

Because quantitative measurements are not required for combinatorial coding, this form of coding has become the most commonly used coding method. Commercial five-label probe sets are available for staining the full complement of human chromosomes in a single metaphase chromosome spread, and commercial software is available for analyzing the staining patterns and assigning the chromosome identities based on the combinations of labels detected at each target. While the combinatorial probe labeling approach is the same, two variations on the methods have been developed based on the method of imaging the fluorescent staining patterns.

The Multiplex FISH method (M-FISH) relies upon a more straightforward approach to imaging, using a different single bandpass filter set to image the staining pattern of each fluorophore separately (Speicher et al., 1996). Each image is analyzed to determine which fluorophores are present at each image pixel location, and the identity of each target is determined from the combination of fluorophores at that location. An example of one such coding pattern, the coding table for the SpectraVysion™ (Vysis/Abbott Laboratories) commercial M-FISH probe set, is listed in Table 1. The spectra of four of the probe labels utilized can be found in Figure 1. The fifth label, abbreviated SFR (SpectrumFRed) in Table 1, is a far red emitting dye.

The Spectral Karyotyping method (SKY) uses a multi-bandpass filter set to excite and transmit the emission from all five probe labels simultaneously, using an interferometer to subsequently interpret the individual fluorophore staining patterns (Schrock et al., 1996). For the interferometry approach, the path of the collected emission from the specimen is split into two paths, one path being variable in length. The paths are then recombined, resulting in an interference pattern dependent upon the wavelengths of the fluorophores present and the difference in the path lengths. A number of images are recorded at various path length differences, from which the spectrum of the emission falling on each image pixel can be computed.

It should be noted that while the probe coding approach is the same in M-FISH and SKY, the labels must be optimized differently. In M-FISH it is preferable to space the labels spectrally far apart from one another to reduce spectral crosstalk among the separate single bandpass filter sets. In SKY the probe labels must be selected to have appreciable absorbance and emission within the several available bandpasses of the multi-bandpass filter set (typically three excitation bands and three emission bands). This requires selecting some of the fluorophores to have excitation and emission bands relatively close to one another.

Figure 4A shows an example of the SpectraVysion probe set hybridized to a spread of highly abnormal chromosomes. The 5 images recorded with different single bandpass fluorescence filter sets were processed by Quips SpectraVysion software (Applied Imaging Corp.) to determine the fluorophore combination present at each image pixel, and a distinct pseudo-color was assigned to represent each combination in the final composite image. This hybridization was part of a study intended to identify genomic changes responsible for reversing the effects of a mutation in the mitochondrial DNA of an osteosarcoma-derived cell line (Hao et al., 1999). Due to the complexity of the DNA alterations, the exact genomic change associated with the reversion could not be identified. However, this hybridization serves to demonstrate the power of M-FISH. The frequent rearrangements are easily identified by the changes in color down the length of the various abnormal chromosomes.

Although coding methods provide an important economy of label usage, the non-coded hybridization method provides the least ambiguity of target assignments. When a different fluorophore is used to label each FISH probe, overlapping chromosome targets can be clearly identified by the presence of the fluorescence from each of the two overlapping targets. In coded hybridizations, however, it is not clear if the combination

---

Figure 4 (See color insert section). Color composite images of metaphase chromosome spreads and interphase nuclei hybridized with panels of multi-color FISH probes. A: Pseudo-colored image of a metaphase spread hybridized with the SpectraVysion™ M-FISH probe set. Each color represents a different label combination, and therefore a different chromosome. B: Image of a single blastomere (XY, -13,+21) hybridized at Reproductive Genetics Institute, Chicago, IL, with the MultiVysion™ PGT probe set. C: Image of metaphase chromosomes from a patient with an abnormal chromosome 4 hybridized with SO-WCP® 8 and SGr WCP® 4. D: Image of a bladder cell isolated from the urine of a patient monitored for recurrence, and hybridized with the 4-color UroVysion™ probe set. E: Image of a cell from a bronchial washing of a patient with lung cancer hybridized with SR-LSI®EGFR, SG-LSI® 5p15, SGo-LSI® MYC, and SA-CEP® 1. F: Image of a cell from ductal lavage fluid taken from a patient with breast cancer and hybridized with the Vysis Breast Cancer Aneusomy Probe Set. G: Image of cells in a formaldehyde-fixed paraffin-embedded breast tumor specimen hybridized with SO-LSI® TOP2A, SGr-LSI® HER2, and SA-CEP® 17. H: Image of cells in a formaldehyde-fixed paraffin-embedded larynx tumor specimen hybridized with SO-LSI® EGFR, and SGr-CEP® 7. All probe sets are from Vysis/Abbott Laboratories. Metaphase chromosomes and interphase nuclei in parts C through H are stained blue with DAPI counterstain.

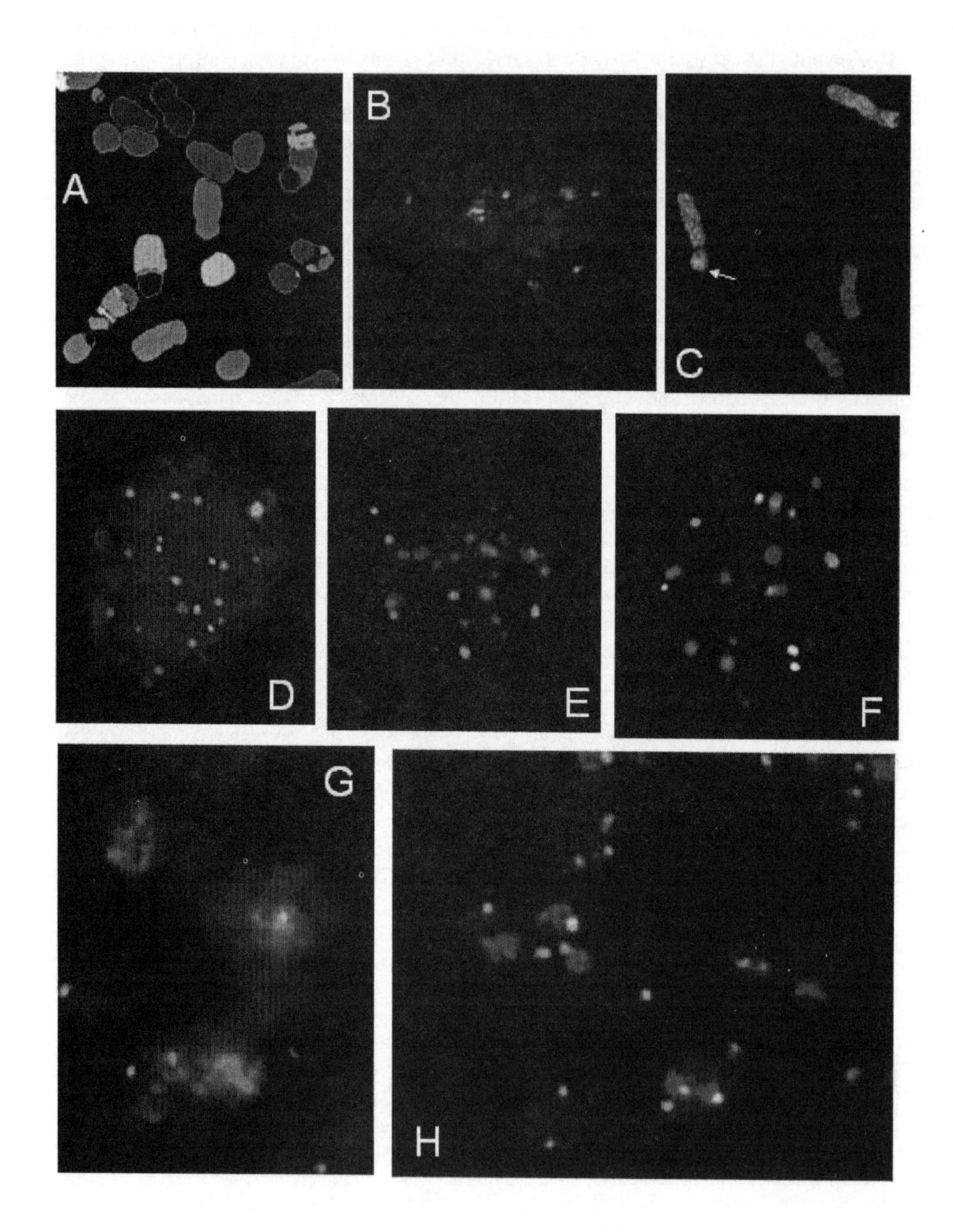
A
B
C
D
E
F
G
H

of targets overlapping is actually two overlapping targets, or a single target associated with the same label combination. A more advanced combinatorial approach has been demonstrated that can relieve much of this ambiguity (Azofeifa et al., 2000). In one implementation of this approach, all targets are identified by combinations of two labels. This permits the easy identification of certain small translocations in metaphase chromosome spreads by guaranteeing that each chromosome piece will have at least one fluorescent label that is different from the label combination of the adjacent translocated region. Using seven different fluorophores nearly all of the 24 human chromosomes could be assigned two-label combinations to improve the ability to detect translocations (Azofeifa et al., 2000). The use of eight labels to provide two-label combinations for all 24 chromosomes has also been demonstrated, as well as five-label two-combination hybridizations to detect ten small chromosomal targets in interphase cells[‡‡]. Distinguishing overlapping targets can be more important in interphase nuclei where chromosome morphology cannot be utilized to distinguish individual chromosomes. Although this labeling method requires more fluorescent labels than the simpler combinatorial coding method to identify the same number of targets, the improved ability to recognize chromosome rearrangements and the reduction of ambiguity in target assignments can be worth the increase in fluorescent labels.

4. EXAMPLES OF FISH IN HUMAN GENOME ANALYSIS

The ability of FISH to detect chromosome rearrangements, gains and losses of whole chromosomes or chromosome arms, and deletion or amplification of genetic loci allows it to play a major role in the diagnosis of inherited diseases, birth defects, and mental retardation as well as the diagnosis, prognosis, and prediction of therapeutic response in cancers. To demonstrate the breadth of FISH applications examples from both the perinatal and cancer fields are presented below.

4.1. Application of FISH to Perinatal Analysis

Although many inherited defects are single base or several base alterations that are too small to be detectable by FISH, many important genetic conditions are easily distinguished by FISH. These include a number of translocations that result in mental retardation, including cryptic translocations at chromosome telomeres, that are too small to be detected by conventional cytogenetics but can be visualized with locus specific telomeric probes hybridized to metaphase chromosome spreads (Knight et al., 1999). Deletions of genetic loci lead to such conditions as Prader-Willi, Angelman, and Smith-Magenis syndromes, and some deletions that are too small to be detected by conventional karyotyping, including deletions associated with Williams and DiGeorge syndromes (Shapira, 1998). Each of these deletions can be identified using FISH with locus specific probes applied to metaphase chromosomes.

Chromosome aneusomies (gain or loss of whole chromosomes) result in a sizable number of serious birth defects, and abnormal copy numbers of chromosomes 21, 13, 18,

[‡‡] Legator, M., Piper, J., and Morrison, L., 2002, *Cytometry* Supplement 11:44, Abstract 43959. Presented at the XXI Congress of the International Society for Analytical Cytology, May 4 -9, San Diego, CA.

X, and Y account for 95% of chromosome abnormalities accompanied by birth defects.[§§] FISH probe sets to these five chromosomes have been used in prenatal testing, providing highly accurate results much sooner than conventional karyotyping (Eiben et al., 1999). Another important application of these aneusomy probe sets is in preimplantation genetic diagnosis (PGD). Chromosomally abnormal nuclei are not uncommon among *in vitro* fertilized embryos, and these defective embryos are not likely to produce full term pregnancies. Prior to implantation, FISH testing can be performed on one or two cells that have been removed from a blastomere without impacting its development. Those blastomeres shown to be aneusomic will not be implanted, reducing the number of embryos needed to produce the same rate of live births (Munne et al., 1998b).

Figure 4B shows a single blastomere that has been hybridized with the five probe panel containing two locus specific probes for chromosomes 13 and 21, and three centromeric repetitive sequence probes for chromosomes X, Y, and 18, each labeled with a different fluorophore. The blastomere in this example is shown to have a normal copy number of chromosomes X (blue), Y (yellow), and 18 (aqua), but has an abnormal gain of chromosome 21 (green) and loss of chromosome 13 (red). One of the three chromosome 21 signals is overlapping the X signal in this cell, showing the value of having a different label on each FISH probe. When viewed through the individual single bandpass filter sets, each signal is seen clearly and is correctly identified. If color combinations were used to identify each chromosome, overlapping signals potentially could be misidentified as the chromosome corresponding to the sum of the label colors present in the signal. Despite the potential problems of overlapping FISH signals, coded hybridizations have been used to advantage in PGD by increasing the amount of information that can be obtained from blastomeres (Munne et al., 1998c). Another way to maximize the information gathered from a single cell is to analyze the results of the initial probe set, repeat the denaturing procedure to remove those signals, and re-hybridize using a different probe set. Multiple hybridizations have permitted the enumeration of nine chromosomes in a single blastomere (Munne et al., 1998a).

Balanced translocations can exist in adults without causing noticeable abnormalities because the full genetic complement is present, but their offspring are prone to having unbalanced translocation that do result in abnormalities. While conventional cytogenetics can identify many of these imbalances the rearranged chromosome is often difficult to interpret by banding patterns alone, and FISH with whole chromosome painting probes provides a definitive answer. M-FISH and SKY probe sets can provide results on all of the chromosomes simultaneously, or pairs of chromosomes labeled with two different fluorophores can be used if the identity of the rearranged chromosomes is suspected. Whole chromosome probe sets in two or more fluorescent colors can also be used for confirmation of the M-FISH and SKY results. Figure 4C shows metaphase chromosomes hybridized with whole chromosome painting probes for chromosomes 4 (green) and 8 (red). An unbalanced rearrangement is easily identified as a small portion of chromosome 8 (arrow in Figure 4C) relocated to the derivative chromosome 4.

[§§] Whiteman, D. A. H. and Klinger, K. W., 1991, Am J Hum Genet supplement, 49, Abstract 1279. Presented at the 8th International Congress of Human Genetics, October 6-11, Washington D.C.

4.2. Application of FISH to Cancer

Genetic alterations are common to all cancers, and are required to alter key cell cycle and signaling pathway proteins that permit the cancer cell to proliferate and evade the many checks and balances that normally prevent this disease. Chromosome aneusomy is a common and frequently early event in cancer progression, and can be identified using repetitive sequence or locus specific probes, since metaphase spreads are rarely found in the uncultured tumor specimens used for diagnosis. Rearrangements of chromosomes can result in suppression or activation of an enzymatic activity by placing the relevant gene behind an inactive or an activated promoter region, respectively. Rearrangements can also produce fusion proteins, such as the BCR-ABL fusion product found in the Philadelphia chromosome. Translocations such as this can be identified in interphase cells by using locus specific probes that target loci bordering the break point region on each chromosome. These loci are well separated in normal cells, being on different chromosomes, but are brought next to one another in the cells containing the translocation. If a different color is used for each probe, a fusion of the two colors is seen in abnormal nuclei as evidence of the rearrangement. More complex schemes for identifying rearrangements have been implemented that permit the detection of lower numbers of abnormal cells (DeWald et al., 2002). Inactivation of tumor suppressor genes is another method of circumventing normal cell control. Tumor suppressor genes typically code for cell cycle components that serve to suppress cell proliferation. One gene is often silenced by single base mutation, while the remaining normal copy becomes deleted, either by deletion of the gene locus, detectable with small locus specific FISH probes, or by loss of the entire chromosome, detectable with locus specific or repetitive sequence probes.

Figure 4D shows an example of detecting cancer cells in the urine of a patient being monitored for the recurrence of a previously diagnosed and treated bladder cancer. The bladder cell shown in this figure was isolated from a urine specimen by centrifugation, and was hybridized with a probe set containing three repetitive sequence probes for aneusomy detection and a locus specific probe targeting the p16 tumor suppressor gene located on chromosome 9 at 9p21. Repetitive sequence probes for chromosomes 3, 7, and 17 were labeled with red, green, and aqua emitting fluorophores, respectively, while the p16 probe was labeled with a yellow emitting fluorophore. The cell pictured in Figure 4D clearly shows abnormal gains of chromosome 3 (nine copies), chromosome 7 (seven copies), and chromosome 17 (six copies), as well as deletion of both of the p16 alleles (no yellow signals visible in the nucleus). In a study of over 200 patient specimens, this 4-probe FISH test was found to provide significantly improved sensitivity (81%) over conventional cytology (58%; $p=0.001$) (Halling et al., 2000). Overall specificities for the two testing methods were essentially the same.

An example of detecting chromosome abnormalities associated with lung cancer is shown in Figure 4E. A bronchial washing specimen was taken from a patient with lung cancer, smeared on a slide, treated, and hybridized with a 4-color probe set. This probe set was composed of locus specific probes targeting the EGFR gene at 7p12, the MYC gene at 8q24, the 5p15 locus, and an alpha satellite repetitive sequence on chromosome 1, labeled with red, yellow, green, and aqua emitting fluorophores, respectively. Chromosomal abnormalities are clearly present in the cell pictured in Figure 4E as greater than two signals are present for each of the four chromosome targets. In a study of 74 bronchial washing specimens, 48 from patients with confirmed lung cancer and 26 from patients with normal clinical diagnoses, the 4-color probe set provided 82% sensitivity for detecting lung cancer, as compared to 54% sensitivity for conventional cytology (statistically significant at $p<0.007$) (Sokolova et al., 2002). The difference between FISH and cytology specificities was not statistically significant.

Figure 4F shows an example of FISH applied to the detection of chromosomal abnormalities in breast cancer. The cell pictured in this figure is from a ductal lavage specimen obtained by injecting saline into the active ducts of the nipple. FISH was performed with a 4-probe set containing three repetitive sequence probes for chromosomes 8, 11, and 17, labeled respectively with red, green, and aqua emitting fluorophores, and a locus specific probe for chromosome 1 at 1p12 labeled with a yellow emitting fluorophore. Again abnormal gain of all four targets is observed in this cell. In a study of 39 cases with ductal lavage performed prior to surgical excision of lesions, FISH achieved a sensitivity and specificity of 71% and 89%, respectively, which compared favorably to conventional cytology with respective sensitivity and specificity of 47% and 79% (King et al., 2002). In the future this assay may prove important for the early detection of breast cancer or for improved risk assessment.

An example of FISH applied to cancer prognosis and prediction of therapy is shown in Figure 4G. In this image a thin section from a formaldehyde-fixed paraffin-embedded breast tumor specimen has been hybridized with a 3-probe set containing two locus specific probes targeting the ERBB2 (HER2) and TOP2A genes, both on the q-arm of chromosome 17, labeled with green and orange emitting fluorophores, respectively. The third probe of the set is an alpha satellite repetitive sequence probe to chromosome 17 labeled with an aqua emitting fluorophore that serves as a measure of chromosome 17 aneusomy. The control probe allows amplification or deletion of the ERBB2 and TOP2A loci to be differentiated from chromosome 17 copy number changes. In this example the ERBB2 gene is highly amplified as evidenced by the large region staining green within each nucleus. The individual ERBB2 targets are too close together to be individually distinguished. In contrast, the neighboring TOP2A gene is present at a normal number of 2 (note that fewer signals are seen in some cells due to the imaging depth of field being smaller than the nuclear diameter). ERBB2 amplification has been correlated with poor patient outcome (Press et al., 1997; Slamon et al., 1987). Therapeutic agents are available that target the ERBB2 and TOP2A gene products, and commercial FISH assays are available currently for predicting patient response to one anti-ERBB2 therapy.

As a final prognostic example, Figure 4H shows an *in situ* hybridization to a thin section of a formaldehyde-fixed paraffin-embedded larynx tumor specimen. The 2-color probe set used in this hybridization contained a locus specific probe targeting the EGFR gene at 7p12 and an alpha satellite repetitive sequence probe to chromosome 7, labeled with orange and green emitting fluorophores, respectively. Amplification of the EGFR locus relative to the chromosome 7 is evidenced by the large regions of red staining in the nuclei compared to the two to three green signals per nucleus (note that orange fluorescence can appear red depending upon the filter set used for visualization or the method by which the color composite image was formed). In a study of 59 larynx cancer patients, survival curves indicated that EGFR gene status and chromosome 7 copy number provided considerably improved prediction of patient outcome compared to expression of EGFR by immunohistochemistry.***

5. SUMMARY

FISH has made great contributions to studying the human genome, from mapping genes along chromosomes to illuminating the mechanisms of chromosome rearrangements. In medicine FISH provides a valuable means to diagnose inherited diseases and cancer, as well as to predict the outcome of cancer patients and to predict the most effective anti-cancer therapy. FISH probes can be made to stain and identify entire chromosomes, or to stain sub-chromosomal regions such as individual genes, groups of genes, or satellite repetitive sequences. A large number of different fluorophores have been linked to FISH probes, either directly or indirectly, and as many as eleven fluorophores have been employed simultaneously in FISH experiments. Complex multi-label hybridizations using fluorescent coding can provide information on all 24 human chromosomes in a single hybridization, and the reagents for these complex hybridizations are sufficiently robust to be available in commercial kits. FISH kits for diagnosis and prognosis of cancer are not only commercially available, but a number are now approved by the US Food and Drug Administration for *in vitro* diagnostic use. The future should see considerably more expansion of FISH to provide greater diagnostic, prognostic, and predictive information in all cancers, as well as to provide diagnosis for an increasing number of inherited diseases and disease susceptibilities.

REFERENCES

Arnoldus, E. P., Wiegant, J., Noordermeer, I. A., Wessels, J. W., Beverstock, G. C., Grosveld, G. C., van der Ploeg, M., and Raap, A. K., 1990, Detection of the Philadelphia chromosome in interphase nuclei, *Cytogenet Cell Genet* **54**(3-4):108-11.

Azofeifa, J., Fauth, C., Kraus, J., Maierhofer, C., Langer, S., Bolzer, A., Reichman, J., Schuffenhauer, S., and Speicher, M. R., 2000, An optimized probe set for the detection of small interchromosomal aberrations by use of 24-color FISH, *Am J Hum Genet* **66**:1684-1688.

Barch, M. J., Knutsen, T., and Spurbeck, J. L., 1997, *The AGT Cytogenetics Laboratory Manual,* Lippincott-Raven, Philadelphia.

Bauman, J. G., and van Duijn, P., 1981, Hybrido-cytochemical localization of specific DNA sequences by fluorescence microscopy, *Histochem J* **13**(5):723-33.

Carter, N. P., Ferguson-Smith, M. A., Perryman, M. T., Telenius, H., Pelmear, A. H., Leversha, M. A., Glancy, M. T., Wood, S. L., Cook, K., Dyson, H. M., and et al., 1992, Reverse chromosome painting: a method for the rapid analysis of aberrant chromosomes in clinical cytogenetics, *J Med Genet* **29**(5):299-307.

Castleman, K. R., 1993, Color compensation for digitized FISH images, *Bioimaging* **1**:159-165.

Cherif, D., Bernard, O., and Berger, R., 1989, Detection of single-copy genes by nonisotopic in situ hybridization on human chromosomes, *Hum Genet* **81**(4):358-62.

Chu, B. C., Wahl, G. M., and Orgel, L. E., 1983, Derivatization of unprotected polynucleotides, *Nucleic Acids Res* **11**(18):6513-29.

*** Jacobson, K., Morrison, L., Schroeder, J., Friedman, M., Henderson, B., Blondin, B., Seelig, S., Coon, 2003, AACR 94th Annual Meeting Proceedings, 44, Abstract 3541. Presented at the 94th Annual Meeting of the American Association for Cancer Research, July11-14, Washington D.C.

Collins, C., Kuo, W. L., Segraves, R., Fuscoe, J., Pinkel, D., and Gray, J. W., 1991, Construction and characterization of plasmid libraries enriched in sequences from single human chromosomes, *Genomics* **11**(4):997-1006.

Cremer, T., Tesin, D., Hopman, A. H., and Manuelidis, L., 1988, Rapid interphase and metaphase assessment of specific chromosomal changes in neuroectodermal tumor cells by in situ hybridization with chemically modified DNA probes, *Exp Cell Res* **176**(2):199-220.

Dale, R. M., and Ward, D. C., 1975, Mercurated polynucleotides: new probes for hybridization and selective polymer fractionation, *Biochemistry* **14**(11):2458-69.

Dauwerse, J. G., Wiegant, J., Raap, A. K., Breuning, M. H., and van Ommen, G. J. B., 1992, Multiple colors by fluorescence *in situ* hybridization using ratio-labelled DNA probes create a molecular karyotype, *Hum Mol Genet* **1**:593-598.

Deaven, L. L., Van Dilla, M. A., Bartholdi, M. F., Carrano, A. V., Cram, L. S., Fuscoe, J. C., Gray, J. W., Hildebrand, C. E., Moyzis, R. K., and Perlman, J., 1986, Construction of human chromosome-specific DNA libraries from flow- sorted chromosomes, *Cold Spring Harb Symp Quant Biol* **51**(Pt 1):159-67.

DeWald, G. W., Ketterling, R. P., Wyatt, W. A., and Stupca, P. J., 2002, Cytogenetic studies in neoplastic hematologic disorders, in: *Clinical Laboratory Medicine,* K. D. McClatchey, ed., Lippincott Williams & Wilkins, Philadelphia.

Draper, D. E., 1984, Attachment of reporter groups to specific, selected cytidine residues in RNA using a bisulfite-catalyzed transamination reaction, *Nucleic Acids Res* **12**(2):989-1002.

Eiben, B., Trawicki, W., Hammans, W., Goebel, R., Pruggmayer, M., and Epplen, J. T., 1999, Rapid prenatal diagnosis of aneuploidies in uncultured amniocytes by fluorescence in situ hybridization. Evaluation of >3,000 cases, *Fetal Diagn Ther* **14**(4):193-7.

Feinberg, A. P., and Vogelstein, B., 1983, A technique for radiolabeling DNA restriction endonuclease fragments to high specific activity, *Anal Biochem* **132**(1):6-13.

Feinberg, A. P., and Vogelstein, B., 1984, "A technique for radiolabeling DNA restriction endonuclease fragments to high specific activity". Addendum, *Anal Biochem* **137**(1):266-7.

Guan, X. Y., Meltzer, P. S., and Trent, J. M., 1994, Rapid generation of whole chromosome painting probes (WCPs) by chromosome microdissection, *Genomics* **22**(1):101-7.

Halling, K. C., King, W., Sokolova, I. A., Meyer, R. G., Burkhardt, H., Halling, A. C., Cheville, J. C., Sebo, T. J., Ramakumar, S., Stewart, C. S., Pankratz, S., O'Kane, D. J., Seelig, S. A., Lieber, M. M., and Jenkins, R. B., 2000, A comparison of cytology and fluorescence in situ hybridization for the detection of bladder cancer, *J Urol* **164**.

Hao, H., Morrison, L. E., and Moraes, C. T., 1999, Suppression of a mitochondrial tRNA gene mutation phenotype associated with changes in the nuclear background [In Process Citation], *Hum Mol Genet* **8**(6):1117-24.

Hopman, A. H., Wiegant, J., Tesser, G. I., and Van Duijn, P., 1986, A non-radioactive in situ hybridization method based on mercurated nucleic acid probes and sulfhydryl-hapten ligands, *Nucleic Acids Res* **14**(16):6471-88.

Hopman, A. H., Wiegant, J., and van Duijn, P., 1987, Mercurated nucleic acid probes, a new principle for non-radioactive in situ hybridization, *Exp Cell Res* **169**(2):357-68.

Hopman, A. H. N., Ramaekers, F. C. S., and Speel, E. J. M., 1998, Rapid synthesis of biotin-, digoxigenin-, trinitrophenyl-, and fluorochrome-labeled tyramides and their application for *in situ* hybridization using CARD amplification, *J Histochem Cytochem* **46**(6):771-777.

Jacobson, K., Thompson, A., Browne, G., Shasserre, C., Seelig, S. A., and King, W., 2000, Automation of fluorescence in situ hybridization pretreatment: a comparative study of different sample types, *Mol Diagn* **5**(3):209-20.

Kearns, W. G., and Pearson, P. L., 1994, Fluorescent *in situ* hybridization using chromosome-specific DNA libraries, in: *Methods in Molecular Biology, Vol. 33: In Situ Hybridization Protocols,* K. H. A. Choo, ed., Humana Press, Totowa, pp. 15-22.

Keller, G. H., Cumming, C. U., Huang, D. P., Manak, M. M., and Ting, R., 1988, A chemical method for introducing haptens onto DNA probes, *Anal Biochem* **170**(2):441-50.

Keller, G. H., Huang, D. P., and Manak, M. M., 1989, Labeling of DNA probes with a photoactivatable hapten, *Anal Biochem* **177**(2):392-5.

Kelly, R. B., Cozzarelli, N. R., Deutscher, M. P., Lehman, I. R., and Kornberg, A., 1970, Enzymatic synthesis of deoxyribonucleic acid. XXXII. Replication of duplex deoxyribonucleic acid by polymerase at a single strand break, *J Biol Chem* **245**(1):39-45.

King, B. L., Tsai, S. C., Gryga, M. E., D'Aquila, T. G., Seelig, S. A., Morrison, L. E., Jacobson, K. K. B., Legator, M., Ward, D. C., Rimm, D. L., and Phillips, R. F., 2002, Detection of chromosomal instability in paired breast surgery and ductal lavage specimens by interphase fluorescence *in situ* hybridization, *Clin Cancer Res* **9**:1509-1516.

Knight, S. J., Regan, R., Nicod, A., Horsley, S. W., Kearney, L., Homfray, T., Winter, R. M., Bolton, P., and Flint, J., 1999, Subtle chromosomal rearrangements in children with unexplained mental retardation, *Lancet* **354**(9191):1676-1681.

Landegent, J. E., Jasen in de Wal, N., Baan, R. A., Hoeijmakers, J. H., and Van der Ploeg, M., 1984, 2-Acetylaminofluorene-modified probes for the indirect hybridocytochemical detection of specific nucleic acid sequences, *Exp Cell Res* **153**(1):61-72.

Langer, P. R., Waldrop, A. A., and Ward, D. C., 1981, Enzymatic synthesis of biotin-labeled polynucleotides: novel nucleic acid affinity probes, *Proc Natl Acad Sci U S A* **78**(11):6633-7.

Langer-Safer, P. R., Levine, M., and Ward, D. C., 1982, Immunological method for mapping genes on Drosophila polytene chromosomes, *Proc Natl Acad Sci U S A* **79**(14):4381-5.

Manuelidis, L., Langer-Safer, P. R., and Ward, D. C., 1982, High-resolution mapping of satellite DNA using biotin-labeled DNA probes, *J Cell Biol* **95**(2 Pt 1):619-25.

Morrison, L. E., 1999, Homogeneous detection of specific DNA sequences by fluorescence quenching and energy transfer, *J Fluoresc* **9**(3):187-196.

Morrison, L. E., Halder, T. C., and Stols, L. M., 1989, Solution-phase detection of polynucleotides using interacting fluorescent labels and competitive hybridization, *Anal Biochem* **183**(2):231-44.

Morrison, L. E., and Legator, M. S., 1997, Two-color ratio-coding of chromosome targets in fluorescence in situ hybridization: quantitative analysis and reproducibility, *Cytometry* **27**(4):314-26.

Morrison, L. E., and Legator, M. S., 1999, Multi-color fluorescence in situ hybridizations techniques, in: *An Introduction to Fluorescence in situ Hybridization: Principles and Clinical Applications,* M. Andreeff, and D. Pinkel, eds., Wiley-Liss, New York, pp. 77-118.

Morrison, L. E., Ramakrishnan, R., Ruffalo, T. M., and Wilber, K. A., 2002, Labeling fluorescence *in situ* hybridization probes for genoimc targets, in: *Molecular Cytogenetics: Protocols and Applications (Methods of Molecular Medicine Series),* Y.-S. Fan, ed., Humana Press, Totowa, pp. 21-40.

Munne, S., Magli, C., Bahce, M., Fung, J., Legator, M., Morrison, L., Cohen, J., and Gianaroli, L., 1998a, Preimplantation diagnosis of the aneuploidies most commonly found in spontaneous abortions and live births: XY, 13, 14, 15, 16, 18, 21, 22, *Prenatal Diag* **18**:1459-1466.

Munne, S., Magli, C., Bahce, M., Fung, J., Legator, M., Morrison, L., Cohert, J., and Gianaroli, L., 1998b, Preimplantation diagnosis of the aneuploidies most commonly found in spontaneous abortions and live births: XY, 13, 14, 15, 16, 18, 21, 22, *Prenatal Diag* **18**(13):1459-66.

Munne, S., Marquez, C., Magli, C., Morton, P., and Morrison, L., 1998c, Scoring criteria for preimplantation genetic diagnosis of numerical abnormalities for chromosomes X, Y, 13, 16, 18 and 21, *Mol Hum Reprod* **4**(9):863-70.

Press, M. F., Bernstein, L., Thomas, P. A., Meisner, L. F., Zhou, J. Y., Ma, Y., Hung, G., Robinson, R. A., Harris, C., El-Naggar, A., Slamon, D. J., Phillips, R. N., Ross, J. S., Wolman, S. R., and Flom, K. J., 1997, HER-2/neu gene amplification characterized by fluorescence in situ hybridization: poor prognosis in node-negative breast carcinomas, *J Clin Oncol* **15**(8):2894-904.

Ramakrishnan, R., and Morrison, L. E., 2002, Labeling fluorescence *in situ* hybridization probes for RNA targets, in: *Molecular Cytogenetics: Protocols and Applications (Methods of Molecular Medicine Series),* Y.-S. Fan, ed., Humana Press, Totowa, pp. 41-49.

Reisfeld, A., Rothenberg, J. M., Bayer, E. A., and Wilchek, M., 1987, Nonradioactive hybridization probes prepared by the reaction of biotin hydrazide with DNA, *Biochem Bioph Res Co* **142**(2):519-26.

Ried, T., Baldini, A., Rand, T. C., and Ward, D. C., 1992a, Simultaneous visualization of seven different DNA probes by in situ hybridization using combinatorial fluorescence and digital imaging microscopy, *Proc Natl Acad Sci U S A* **89**(4):1388-92.

Ried, T., Landes, G., Dackowski, W., Klinger, K., and Ward, D. C., 1992b, Multicolor fluorescence in situ hybridization for the simultaneous detection of probe sets for chromosomes 13, 18, 21, X and Y in uncultured amniotic fluid cells, *Hum Mol Genet* **1**(5):307-13.

Rogan, P. K., Cazcarro, P. M., and Knoll, J. H. M., 2001, Sequence-based design of single-copy genomic DNA probes for fluorescence *in situ* hybridization, *Genome Res* **11**:1086-1094.

Sambrook, J., Fritsch, E. F., and Maniatis, T., 1989, *Molecular Cloning, A Laboratory Manual,* Cold Spring Harbor Laboratory Press, Cold Spring Harbor.

Schrock, E., du Manoir, S., Veldman, T., Schoell, B., Wienberg, J., Ferguson-Smith, M. A., Ning, Y., Ledbetter, D. H., Bar-Am, I., Soenksen, D., Garini, Y., and Ried, T., 1996, Multicolor spectral karyotyping of human chromosomes, *Science* **273**(5274):494-7.

Shapira, S. K., 1998, An update on chromosome deletion and microdeletion syndromes, *Curr Opin Pediatr* **10**(6):622-627.

Slamon, D. J., Clark, G. M., Wong, S. G., Levin, W. J., Ullrich, A., and McGuire, W. L., 1987, Human breast cancer: correlation of relapse and survival with amplification of the HER-2/neu oncogene, *Science* **235**(4785):177-82.

Sokolova, I. A., Bubendorf, L., O'Hare, A., Legator, L., Jacobson, K., Grilli, B., Dalquen, P., Halling, K., Tamm, M., Seelig, S., and Morrison, L. E., 2002, A FISH-based assay for effective detection of lung cancer cells in bronchial washing specimens, *Cancer Cytopathol* **96**(5):306-315.

Speel, E. J. M., Jansen, M. P. H. N., Ramaekers, F. C. S., and Hopman, A. H. N., 1994, A novel triple-color detection procedure for brightfield microscopy, combining *in situ* hybridization with immunocytchemistry, *J Histochem Cytochem* **42**:1299-1307.

Speicher, M. R., Gwyn Ballard, S., and Ward, D. C., 1996, Karyotyping human chromosomes by combinatorial multi-fluor FISH, *Nat Genet* **12**(4):368-75.

Tanke, H. J., Wiegant, J., van Gijlswijk, R. P., Bezrukoove, V., Pattenier, H., Heetebrij, R. J., Talman, E. G., Raap, A. K., and Vrolijk, J., 1999, New strategy for multi-colour fluorescence *in situ* hybridisation: COBRA: COmbined Ratio Binary labelling, *Eur J Hum Genet* **7**:2-11.

Telenius, H., Carter, N. P., Bebb, C. E., Nordenskjold, M., Ponder, B. A., and Tunnacliffe, A., 1992a, Degenerate oligonucleotide-primed PCR: general amplification of target DNA by a single degenerate primer, *Genomics* **13**(3):718-25.

Telenius, H., de Vos, D., Blennow, E., Willat, L. R., Ponder, B. A., and Carter, N. P., 1993, Chromatid contamination can impair the purity of flow-sorted metaphase chromosomes, *Cytometry* **14**(1):97-101.

Telenius, H., Pelmear, A. H., Tunnacliffe, A., Carter, N. P., Behmel, A., Ferguson-Smith, M. A., Nordenskjold, M., Pfragner, R., and Ponder, B. A., 1992b, Cytogenetic analysis by chromosome painting using DOP-PCR amplified flow-sorted chromosomes, *Gene Chromosome Canc* **4**(3):257-63.

van Belkum, A., Linkels, E., Jelsma, T., van den Berg, F. M., and Quint, W., 1994, Non-isotopic labeling of DNA by newly developed hapten-containing platinum compounds, *Biotechniques* **16**(1):148-53.

Wiegant, J., Ried, T., Nederlof, P. M., van der Ploeg, M., Tanke, H. J., and Raap, A. K., 1991, In situ hybridization with fluoresceinated DNA, *Nucleic Acids Res* **19**(12):3237-41.

[1]LUMINESCENT SEMICONDUCTOR QUANTUM DOTS NANOASSEMBLIES FOR BIOANALYSIS

Yongfen Chen[1] and Zeev Rosenzweig[1]*

1. INTRODUCTION

CdS and CdSe luminescent quantum dots (QDs) have been widely investigated for their luminescence properties [1-20]. Luminescent QDs show a number of advantages compared to organic fluorophores commonly used in biological applications. QDs exhibit higher photostability than organic dyes. Their emission band is size dependent due to quantum confinement effects. For example, 3 nm CdSe QDs emit green light while 6 nm CdSe QDs emit red light. It is therefore possible to prepare a series of solutions showing different emission colors by using only one type of semiconductor material. The emission spectra of semiconductor QDs are symmetric and sharp with a full width at half maximum (FWHM) as narrow as 30 nm. On the other hand, emission spectra of organic dyes are asymmetric and broad. This precludes the simultaneous use of several organic dyes to analyze multi-analyte samples due to overlap between their broad emission peaks. Luminescent semiconductor QDs have a wide excitation spectrum which enable the excitation of QDs of different size with a single excitation wavelength. In contrast, multiple excitation wavelengths are needed to simultaneously excite several organic fluorophores. Recognizing their potential bioanalytical researchers have recently applied luminescent QDs as biological labels [21-26], selective ions probes [27] and luminescent gas sensors [28].

Another research direction in the Area of luminescent QDs has focused on the development of composite nanomaterials and thin films that contain luminescent QDs. These nanocomposite materials could find use in photonic applications. Various methods were developed to form nanocomposites of QDs with other materials, like polystyrene and silica. Han et al encapsulated CdSe-ZnS QDs of different emission colors into 1.2 μm porous polystyrene beads with different ratios and formed barcodes for DNA hybridization assays [29]. Gaponik et al developed a method to pump charged species into

[1] Department of Chemistry and the Advanced Material Research Institute, University of New Orleans, New Orleans, LA 70148

specially designed polymer spheres whose interior composed the solution of an oppositely charged polyelectrolyte. Using this method, they created single-color and multicolored tagged beads with controlled emission intensity ratios [30]. Chang et al developed a method to fabricate silica coated CdS QDs with different nanoscale complex morphologies by using microemulsion based synthetic techniques. The CdS QDs were homogeneously dispersed in silica particles, were incorporated as a large inclusion in the sphere, or coated the silica spheres [31-32]. Kotov et al used mercaptopropyltrimethylsilane derived CdTe QDs as seeds to synthesize CdTe QDs doped silica nanospheres that were about 100nm in diameter. The CdTe QDs-doped uniform silica nanospheres formed a 3-D colloid crystal suitable for photonic crystal applications [33].

Research in our laboratory has focused in recent years on the formation of nanoparticle assemblies that contain CdSe QDs and their use in bioassays. While singleluminescent QDs exhibit high emission quantum yield it is difficult to observe them using ordinary fluorescence microscopy due to limited emission signals. A laser is required as a light source and a high performance, intensified or very sensitive, expensive charge coupled device (CCD) camera is needed to obtain quantitative images of the emission of individual QDs. Nanoparticle assemblies of QDs with an average diameter of around 100 nm (20 folds larger than single QDs) could be used successfully in bioassays and could still provide spatial resolution higher than the limit of diffraction. Throughout our studies we reasoned that the expected signal increase due to the encapsulation of multiple QDs in the larger but still nanometric particle assemblies would enable their use in quantitative cellular and bioassays using conventional microscopic techniques. This review summarizes the development of three highly luminescent nanoparticle assemblies containing CdSe-ZnS QDs. In the first system hydrophobic CdSe-ZnS QDs were encapsulated in siloxane surfactant micelles. In the second and third systems hydrophilic CdSe-ZnS QDs were encapsulated in functionalized silica nanospheres and glyco-nanospheres. We also describe the application of these new luminescent nanoparticles in bioassays.

2. EXPERIMENTAL

2.1 Materials and Reagents

Cadmium oxide, lauric acid, trioctylphosphine oxide (TOPO), trioctylphosphine (TOP), selenium powder, diethyl zinc (0.1M in heptane), the siloxane surfactant of dimethyloctadecyl[3-(trimethoxysilyl)propyl] ammonium chloride, mercaptosuccinic acid, tetramethylammonium hydroxide pentahydrate, sodium silicate (27% aqueous solution), oleic acid , dextran (MW10,000), chloroacetic acid, and 3-mercaptopropyltrimethoxysilane were obtained from Aldrich. Streptavidin-maleiimide, protein A, anti protein A and Concanavalin A (Con A) were purchased from Sigma. All reagents were used as received with no further purification.

2.2 Microscopy and Spectroscopy

Emission spectra of CdSe-ZnS QDs and QDs nanoassemblies were taken in a quartz cuvette using a PTI Quanta Master luminescence spectrometer equipped with a 75 W Xenon Short-arc lamp as a light source. Luminescence images of QDs nanoassemblies were obtained using a digital luminescence imaging microscopy system. The system consists of an inverted fluorescence microscope (Olympus IX-70) equipped with a 100 W mercury lamp as a light source. The luminescence images were collected using a 40X microscope objective with NA=0.9. A filter cube containing a 470/50 nm band pass excitation filter, a 500 nm dichroic mirror and a 515 nm long-pass emission filter was used to ensure spectral imaging purity. A high performance ICCD camera (Princeton Instruments, Model BH2RFLT3) was employed for digital imaging of the QDs nanoassemblies. A PC microcomputer was used for data acquisition. The Roper Scientific software WinView/32 was used for image acquisition and analysis. Transmission electron microscope (TEM) images of QDs nanoassemblies were obtained using a JOEL 2120 electron microscope.

2.3 Preparation of QDs-doped siloxane micelles

Luminescent CdSe-ZnS QDs were prepared based the methods developed by Peng et al and Hines et al [34, 35]. To prepare QDs-doped micelles, 80 μl of 0.15μM CdSe-ZnS QDs were mixed with a chloroform solution containing 100 μl siloxane surfactant $C_{16}H_{33}$-$N^{+}(Me)_2$-$CH_2CH_2CH_2$-$Si(OMe)_3$ Cl^{-}. The mixture was then injected into 10 mL water at 75 °C under magnetic stirring. The micelles solution was incubated at 75 °C for 20 minutes. Then, the solution was cooled to room temperature and diluted with 90 ml deionized water. The pH of the micelles solution was adjusted to 9.0 to facilitate the hydrolysis of the silane head groups. The solution was heated to 70 °C for 5 minutes to initiate the condensation of the silica layer. The product was collected by slow speed centrifugation at 2000 rpm for 15 minutes and washed three times with deionized water.

2.4 Preparation of streptavidin-modified and QDs-doped silica nanospheres

To prepare QDs-doped silica nanospheres, a mixture of 50μl 5% sodium silicate and 50μl 0.2μM water soluble quantum dots was dispersed in 10ml 0.1 M AOT/heptane solution. Sodium silicate was used as silica source instead of the commonly used TEOS because of its higher water solubility. Oleic acid was used to initiate the formation of silica nanospheres in reverse micelles. The reaction mixture was sonicated for 40 minutes in a water bath sonicator. 40μl oleic acid was then added to the reverse micelles solution to neutralize sodium hydroxide. This resulted in the formation of QDs-containing silica nanospheres that were collected by centrifugation at 2000rpm for 15 minutes. The product was washed 3 times with 10ml heptane to remove unbounded AOT surfactant. The surface of the silica spheres was then functionalized with thiol groups by stirring the prepared QDs-doped silica nanospheres with 30μl 3-mercaptopropyltrimethoxysilane (MPTMS) and 30μl 29% ammonia in 20ml 95% ethanol solution for 6 hours. The thiol modified QDs-doped silica nanospheres were collected by centrifugation at 2000rpm for 15 minutes. The sample was washed with ethanol to remove free MPTMS. The thiol modified silica nanospheres were then dispersed in a pH 7.0 phosphate buffer solution for conjugating streptavidin. A sample of 0.5 mg streptavidin-maleiimide was added to the silica nanospheres solution and incubated at 4°C for 12 hrs. The streptavidin-modified, QDs-doped silica nanospheres were collected by centrifugation at 200 rpm for 15 minutes and washed 3 times with a pH 7.0 phosphate buffer solution. The functionalized silica nanospheres were stored in a pH 7.0 phosphate buffer solution until use.

2.5 Detection of anti protein A in a 'sandwich' type immunoassay

To detect anti-protein A in a 'sandwich' type immunoassay, protein A molecules were immobilized to thiol-modified glass slides. The thiol-modified slides were prepared following a procedure developed by Grabar et al [36]. The thiol-modified glass slides were immersed in a pH 7.0 phosphate buffer solution containing 0.2mg streptavidin-maleimide for 12 hours. The glass slides were rinsed with a copious pH 7.0 phosphate buffer solution, then immersed in a pH 7.0 phosphate buffer solution containing 0.2mg biotin labeled protein A at room temperature for 1hr. The slides were rinsed with a pH 7.0 phosphate buffer solution to remove loosely bound protein molecules. The protein A modified glass slides were immersed in 5μg/ml, 1 μg/ml, 0.1μg/ml, 0.01μg/ml, 1ng/ml, 0.1ng/ml anti-protein A solution for 2 hours. The slides were rinsed with a copious pH 7.0 phosphate buffer solution and then immersed in a pH 7.0 phosphate buffer solution containing QDs-doped silica nanospheres, modified with protein A at room temperature for 2 hours. The slides were washed with a pH 7.0 copious phosphate buffer solution and analyzed using a digital fluorescence imaging microscopy system.

2.6 Preparation of QDs-doped glyco-nanospheres

300 μL of 500 μM CM-dextran was mixed with 120 μL 0.19 μM mercaptosuccinic acid modified CdSe-ZnS QDs. Then 120 μL 10 μM poly-L-lysine in a pH 7.0 phosphate buffer solution were added to the CM-dextran and QDs containing solution. The reaction mixture was incubated at room temperature for 30 min. The formed glyconanospheres were washed three times using centrifugation at 4500 rpm for 15 min to remove unreacted QDs and CM-dextran.

2.7 Interaction of glyconanospheres with protein Concanavalin A (Con A)

To demonstrate the binding interactions of the surface-bound dextran to Con A, a solution of 200 μL 0.04 μM (based on the concentration of CdSe-ZnS QDs) glyconanospheres was mixed with 2 mL 0.05M HEPES pH 7.2 buffer solution containing 0.25 mg/mL Con A, 0.1 mM Mn^{2+}, and 0.1 mM Ca^{2+}. The solution was incubated at room temperature for 2 hours.

3. RESULTS AND DISCUSSION

3.1 Luminescent micelles that contain hydrophobic CdSe-ZnS QDs

Figure 1A shows the emission spectra of TOPO capped CdSe-ZnS QDs of 3 (green), 4 (yellow) and 6 (red) nm in diameter. The hydrophobic TOPO capped QDs were easily dispersed in toluene or chloroform. The samples were highly luminescent with relative emission quantum yields ranging from 0.45 to 0.65 compared to rhodamine. A photograph of the three samples is shown in figure 1b. The upper part of the image shows the three QDs solutions when illuminated with visible light. The lower part of the image shows the three bright QDs solutions when illuminated by a UV lamp.

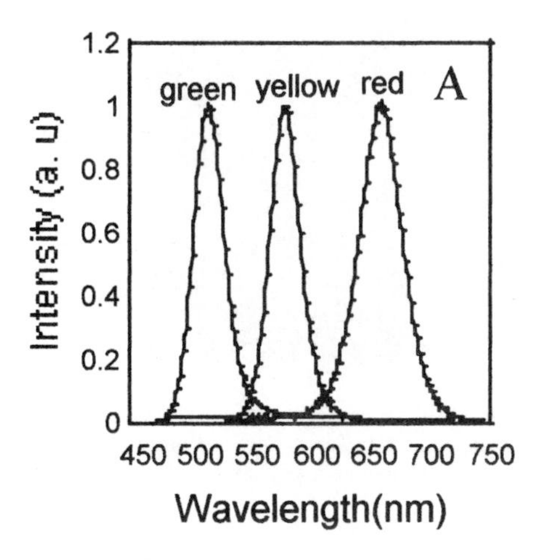

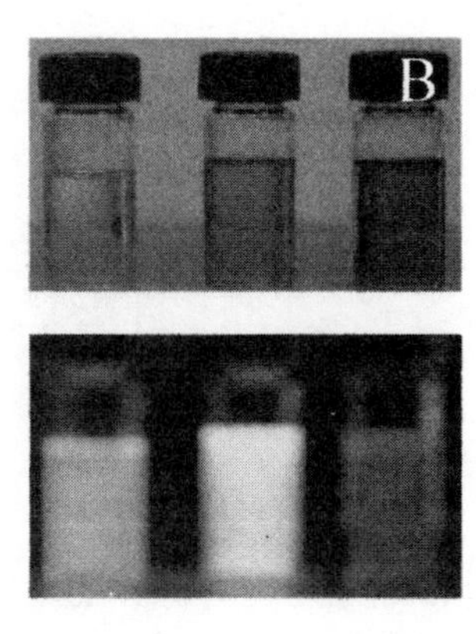

Figure 1. (**See color insert section**). (A) Emission spectra of CdSe-ZnS QDs of 3 (green), 4 (yellow) and 6 (red) nm in diameter. (B) Color photos of QDs solution illuminated with visible (up) and UV (down) light.

Surfactant molecules contain a hydrophobic tail and a hydrophilic head group. When dispersed in aqueous solution, they form different aggregation structures such as micelles, lamella and cylindrical structures depending on their concentration and their properties [37]. Micelles that contain hydrophobic cores were previously used to encapsulate organic molecules like organic dyes and hydrophobic anticancer drugs [38-39]. Based on the same principle we encapsulated TOPO capped CdSe-ZnS QDs in the hydrophobic cores of surfactant micelles. The siloxane surfactant $C_{16}H_{33}$-$N^+(Me)_2$-$CH_2CH_2CH_2$-$Si(OMe)_3$ Cl^- was used to encapsulate the QDs. It contained a hydrolysable silane head group and a long hydrophobic alkyl chain. Hydrolysis and condensation of the silane head groups formed a cross-linking silica layer on the surface of micelles. This cross-linked silica layer increased the stability of the micelles. The size and morphology of QDs doped micelles were characterized by transmission electron microscope (TEM) and fluorescence microscopy. Figure 2a shows a representative TEM image of QDs doped micelles. Since over 100 QDs were encapsulated in each micelle, the micelles were highly luminescent when irradiated with UV light. The luminescence image of quantum dots doped micelles is shown in figure 2b. The high signal to background ratio of about 50 confirms the presence of a multiple number of QDs in the stabilized micelles.

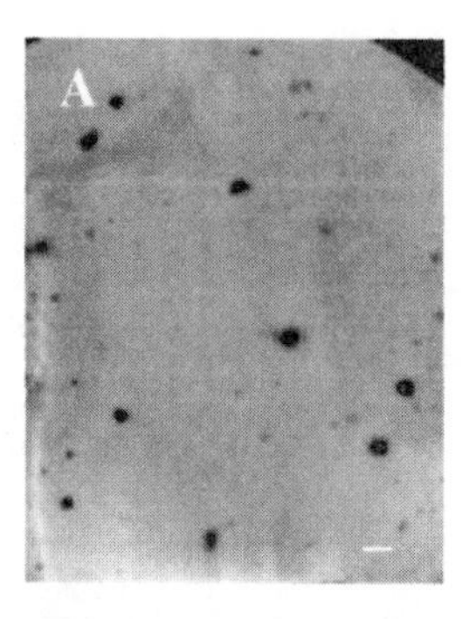

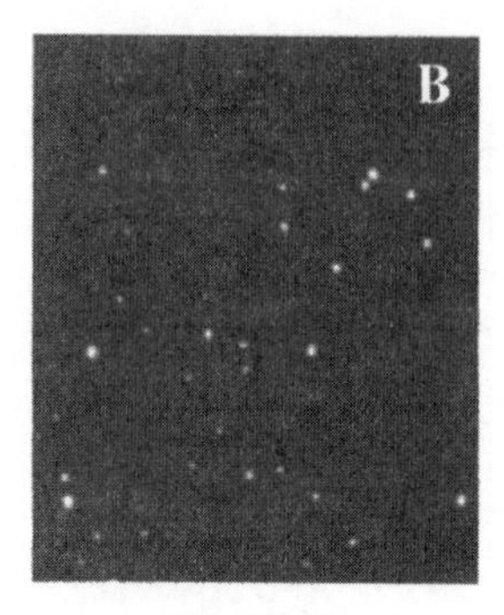

Figure 2 - (A) A TEM image of QDs-doped micelles. The scale bar is 100 nm (B) A fluorescence image of QDs-doped micelles taken using a digital fluorescence imaging microscopy system equipped with an intensified charge coupled device camera (ICCD). The excitation wavelength was 470 nm and the magnification of the microscope was 40×.

A comparison between normalized emission spectra of free QDs solutions and QDs doped-micelles solutions is shown in figure 3 for green emission (3a) and red emission QDs (3b). It can be seen that the peak width was not affected by the encapsulation and that the sharp emission spectra of both the green emission and red emission QDs were well preserved following their encapsulation in the micelles.

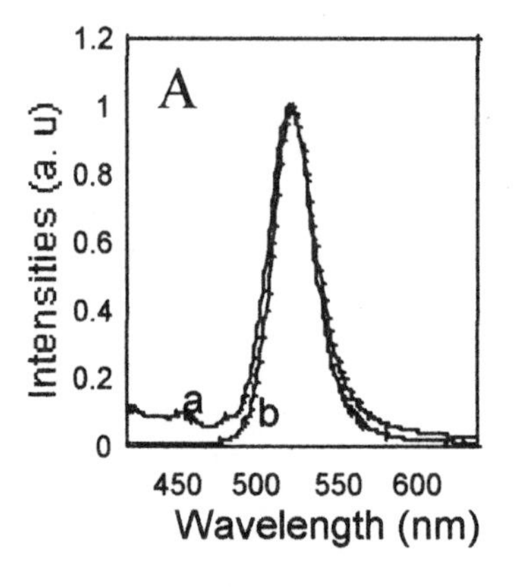

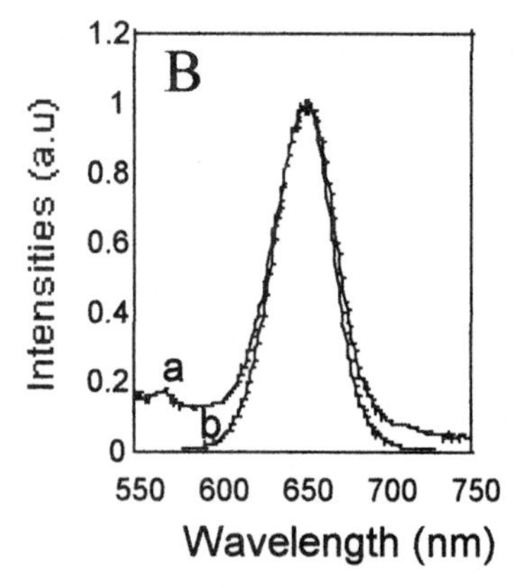

Figure 3 – A comparison between the emission spectra of solutions containing free QDs and QDs-doped micelles for -A- 3nm green emission QDs, -B- 6 nm red emission QDs. The encapsulation of QDs in the micelles did not affect their spectral properties.

While the silica micelles could be used effectively as containers in drug delivery studies there were still concerns about their applicability in bioassays that often require higher stability. Further more, there was still a need to functionalize the silica surface of the micelles to facilitate their conjugation to biomolecules. To address these issues we synthesized QDs-containing silica nanospheres and functionalized their surface with streptavidin. Unlike the silica micelles the silica spheres were solid particles and therefore exhibited higher chemical and mechanical stability.

3.2 Streptavidin functionalized silica nanospheres containing CdSe-ZnS QDs

As mentioned previously the QDs-containing silica nanospheres were synthesized in reversed micelles. Reversed micelles methods were previously used to synthesize magnetic nanoparticles and semiconductor QDs [40-42]. Chang et al recently synthesized CdS QDs-containing silica nanospheres in reverse micelles. However, their QDs-containing silica nanospheres were weakly luminescent due to low encapsulation efficiency. We used a similar synthetic method to prepare silica nanospheres that contained a large number of highly luminescent CdSe-ZnS QDs. A TEM image shown in figure 4 reveals that the average diameter of the QDs-containing silica nanospheres was about 180±20 nm. Several hundred QDs were encapsulated in a single silica sphere. A luminescence image of silica spheres containing QDs is shown in figure 4b. The synthetic method is quite versatile and allows the encapsulation of QDs of different diameters in the silica spheres. The signal to noise background of these spheres was about 150. This provided a large dynamic range for bioassays.

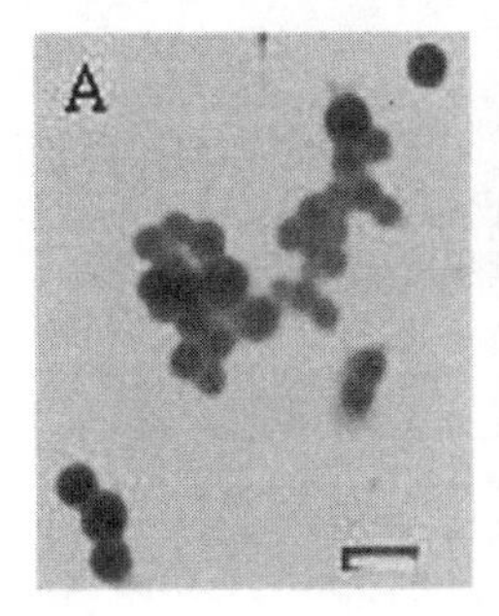

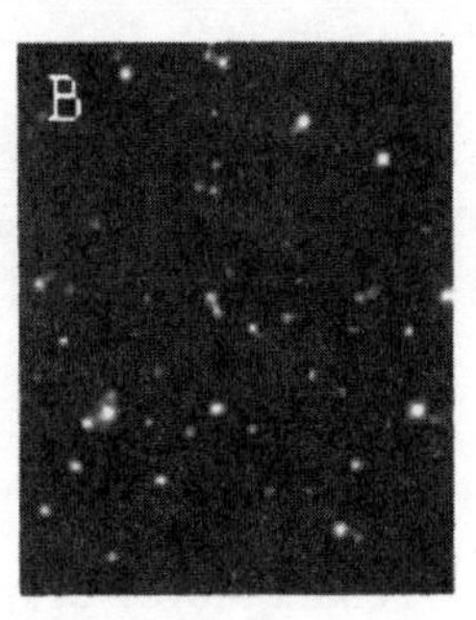

Figure 4 - (A) A TEM image of CdSe-ZnS QD doped silica nanospheres. The scale bar is 500nm. (B) A luminescence image of CdSe-ZnS QDs doped silica nanospheres. The sample was excited with 470nm and a 40X microscope objective was used for imaging. The high signal to noise ratio of 150 enabled the use of the silica spheres in bioassays that require large dynamic range.

To enable the conjugation of the QDs-containing silica spheres to biomolecules we modified their surface with thiol groups by hydrolysis and condensation of 3-mercaptopropyltrimethoxysilane in an ethanol-amonia solution of the silica spheres. The thiol groups were then used to bind maleimide labeled streptavidin to the surface of the nanospheres. The streptavidin-modified silica spheres were used to bind biotinylated protein A through avidin-biotin interactions. It should be noted that the binding constant of streptavidin to biotin is about 10^{15} M^{-1} [43]. The protein A modified QDs-containing silica nanospheres were employed as signal transducers in a sandwich immunoassay to detect anti protein A in solution [44]. As described in the experimental section protein A molecules were attached to steptavidin-modified glass slides. The protein A-modified slides were then placed in solutions of increasing anti protein A concentration. The slides were washed and placed in a solution containing protein A-modified luminescent silica nanospheres. In fact, in this assay the luminescent nanospheres replaced fluorescent labels that are commonly attached to antibodies in sandwich immunoassays. However, in the newly developed nanosphere-based assays the detection was based on digital counting of the luminescent nanospheres rather than on the measurement of integrated analog fluorescence signals. Digital luminescence images luminescent silica nanospheres bound to anti protein A on glass slides for antibody levels of 0 (negative control), 0.01 μg/ml and 1 μg/ml are shown in figure 5. The number of luminescent nanospheres clearly increases with increasing anti protein A concentration. The advantage of digital counting over analog luminescence detection is clear. The signal to background ratio remains similar regardless of the concentration of anti protein A in the analyte solution. The difference is only in the number of counted luminescent spheres. Furthermore, the technique is not sensitive to variations in the luminescent spheres size or on the number of QDs in each sphere. The detection limit of the technique was about 1 ng/ml, which was comparable to sandwich immunoassays conducted in a 96 well plate format. The new assays are simple to perform and could be easily adapted to strip-based screening assays of antibodies in complex samples like blood.

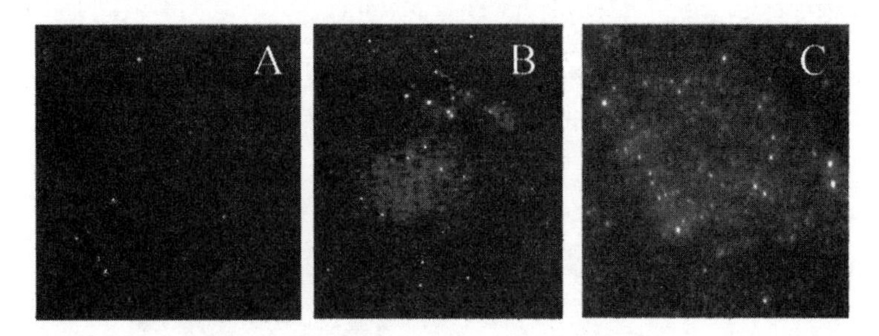

Figure 5 - Detection of anti-protein A in a 'sandwich' immunoassay. The luminescent image showed CdSe-ZnS QDs doped silica nanospheres bound to a protein A modified glass slide in the presence of anti-protein A. (A) control experiment. (B) 0.01μg/ml of antiprotein A. (C) 1μg/ml of antiprotein A

3.3 Glyconanospheres containing hydrophilic luminescent CdSe-ZnS QDs

Following the encapsulation of QDs in the core of silica micelles and in silica nanospheres we also developed a novel method to incorporate QDs in glyconanospheres. While DNA and protein molecules were already conjugated to luminescent semiconductor QDs [21, 45], this work focused for the first time on the incorporation of luminescent QDs into glyconanospheres and their application in studying lectin-carbohydrate interactions. Carbohydrate-protein interactions are involved in biological recognition processes and play a major role in the immune response to viral and bacterial infection [46]. The preparation of glyconanospheres was based on electrostatic interactions between oppositely charged molecules. Electrostatic interactions were previously used to fabricate polyelectrolyte thin films through a 'layer-by-layer' technique [13, 47-49]. In our studies CdSe-ZnS QDs were capped by mercaptosuccinic acid. The QDs were negatively charged and miscible in a pH 7.0 phosphate buffer solution. Carboxymethyldextran was also negatively charged and each dextran molecule contained about 40 carboxylic groups. Poly-L-lysine was positively charged at pH 7.0. The luminescent glyconanospheres were formed when the three charged molecules were mixed in aqueous solution. Usually, interactions between positively charged and negatively charged molecules result in the formation of a precipitate if the product of precursors' concentrations is higher than the solubility product. The precipitates are irregular in morphology since it is difficult to control their growth. Our system consisted of three charged species, the negatively charged QDs and carboxyl-dextran, and the positively charged poly-L-lysine. However, the glucose residues attached to the dextran chain remained neutral and restricted the growth of the nanospheres. As a result, the aggregates of dextran and quantum dots had finite size and morphology. Our first attempts to form stable glyconanosphres using this technique were unsuccessful. The luminescent glyconanospheres dissociated in about 10 hours when dispersed in aqueous solution at room temperature. The relative instability of the electrostatically held glyconanospheres was attributed to their large surface-to-volume ratio. The interactions between solution ions and surface charges weakened the electrostatic attraction between the positively charged poly-L-lysine and the negatively charged CM-dextran and mercaptosuccinic acid modified CdSe QDs. To stabilize the luminescent glyconanospheres, we added to the reaction mixture 1-ethyl-3-(3)-dimethylaminopropyl carbodiimide (EDC) to initiate the formation of covalent amide bonds between the carboxylic groups of carboxyl-dextran and the amino groups of poly L-Lysine. Following this treatment the glyconanospheres became stable in a pH 7.0 phosphate buffer solution for over two weeks.

A digital luminescence microscopy image of glyconanospheres that contain QDs of 3 nm in diameter (green) is shown in figure 6a respectively. A high signal to background ratio of about 100 and no aggregation are observed. The CM-dextran molecules provided the glyco-nanospheres with high surface density of glucosylic residues. Thus, the carbohydrate binding protein concanavalin A (Con A) could recognize the surface bound dextran. Con A is a lectinic protein that selectively binds to the terminal residues of α-D-glucose and α-D-mannose. Each Con A molecule contains four binding sites for carbohydrates [50]. Figure 6b shows the luminescence image of the glyconanospheres following the addition of Con A. Aggregation of the glyconanospheres due to multiple binding with Con A molecules is clearly seen. The system is now used in the development of drug screening assays to asses the potency of lectin-carbohydratye binding inhibitors.

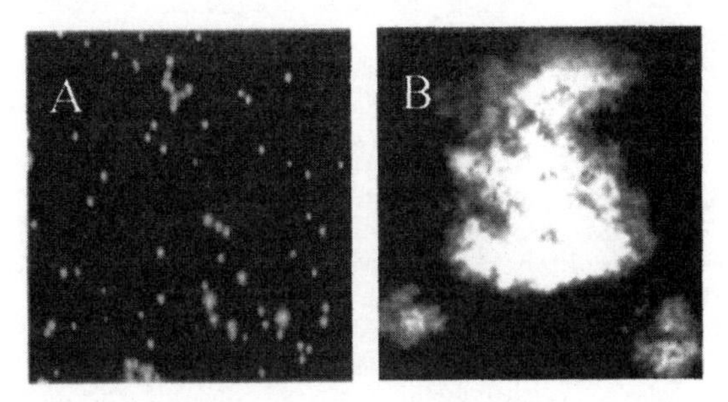

Figure 6 - (A) Luminescence image of negatively charged green emission CdSe-ZnS QDs incorporated glyconanospheres. (B) Luminescence image aggregates of glyconanospheres initiated by Con A.

4. SUMMARY AND CONCLUSIONS

We have developed three novel methods to assemble both hydrophobic and hydrophilic highly luminescent semiconductor CdSe-ZnS QDs into micelles, silica nanospheres and glyconanospheres. The emission properties of the QDs encapsulated in micelles, silica nanospheres and glyconanospheres were similar to the emission properties of individual QDs. Assembling hundreds of QDs in single nanosphere created bright and photostable particles that could be easily observed using conventional fluorescence microscopy instrumentation. By encapsulating QDs of different diameters inside the nanoparticle assemblies we prepared green, yellow and red emission nanoparticles. We further modified the surface of the nanoparticles to provide them with biomolecular recognition capabilities. These multicolor nanoparticle assemblies were found useful as biological labels in immunoassays and carbohydrate-protein interactions studies. Future studies will focus on the application of these unique luminescent nanoparticles in the simultaneous detection of antigens or antibodies in complex mixtures and in single cell assays.

5. REFERENCESS

1. Z. A. Peng, X. Peng, Formation of high-quality CdTe, CdSe, and CdS nanocrystals using CdO as precursor J. Am. Chem. Soc. **123**(1); 183-184 (2001).
2. H. Bekele, J. H. Fendler, J. W. Kelly, Self-assembling peptidomimetic monolayer nucleates oriented CdS nanocrystals J. Am. Chem. Soc. **121**(31) 7266-7267 (1999).
3. B. I. Lemon, R. M. Crooks, Preparation and characterization of dendrimer-encapsulated CdS semiconductor quantum dots. J. Am. Chem. Soc. **122**(51); 12886-12887 (2000).
4. C.-C. Chen, C.-Y. Chao, Z.-H. Lang, Simple solution-phase synthesis of soluble CdS and CdSe nanorods. Chem. Mater. **12**(6); 1516-1518 (2000).
5. S. K. Haram, B. M. Quinn, A. J. Bard, Electrochemistry of CdS nanoparticles: a correlation between optical and electrochemical band gaps. J. Am. Chem. Soc. **123**(36); 8860-8861 (2001).
6. P. Zhang, L. Gao, Synthesis and characterization of CdS nanorods via hydrothermal microemulsion. Langmuir, 19(1); 208-210 (2003).
7. W. Xu, Y. Liao, D. L. Akins, Formation of CdS nanoparticles within modified MCM-41 and SBA-15. J. Phys. Chem. B. **106**(43); 11127-11131 (2002).
8. L.-s. Li, J. Hu, W. Yang, A. P. Alivisatos, Band gap variation of size- and shape-controlled colloidal CdSe quantum rods. Nano Lett. **1**(7); 349-351 (2001).
9. M. Artemyev, B. Moller, U. Woggon, Unidirectional alignment of CdSe nanorods Nano Lett. **3**(4); 509-512 (2003).
10. L. Qu, Z. A. Peng, X. Peng, Alternative routes toward high quality CdSe nanocrystals Nano Lett. **1**(6); 333-337 (2001).
11. D. V. Talapin, A. L. Rogach, A. Kornowski, M. Haase, H. Weller, Highly luminescent monodisperse CdSe and CdSe/ZnS nanocrystals synthesized in a hexadecylamine-trioctylphosphine oxide-trioctylphospine mixture. Nano Lett. **1**(4); 207-211 (2001).
12. A. Striolo, J. Ward, J. M. Prausnitz, W. J. Parak, D. Zanchet, D. Gerion, D. Milliron, A. P. Alivisatos, Molecular weight, osmotic Second virial coefficient, and extinction coefficient of colloidal CdSe nanocrystals. J. Phys. Chem. B. **106**(21); 5500-5505 (2002).
13. D. M. Willard, L. L. Carillo, J. Jung, A. Van Orden, CdSe-ZnS quantum dots as resonance energy transfer donors in a model protein-protein binding assay. Nano Lett **1**(9); 469-474 (2001).
14. L. Manna, E. C. Scher, A. P. Alivisatos, Synthesis of soluble and processable rod-, arrow-, teardrop-, and tetrapod-shaped CdSe nanocrystals. J. Am. Chem. Soc. **122**(51); 12700-12706 (2000).
15. J. Aldana, Y. A. Wang, X. Peng, Photochemical instability of CdSe nanocrystals coated by hydrophilic thiols. J. Am. Chem. Soc. 123(36); 8844-8850 (2001).
16. W. Guo, J. J. Li, Y. A. Wang, X. Peng, Luminescent CdSe/CdS core/shell nanocrystals in dendron boxes: superior chemical, photochemical and thermal stability. J. Am. Chem. Soc. **125**(13); 3901-3909 (2003).
17. A. Schroedter, H. Weller, R. Eritja, W. E. Ford, J. M. Wessels, Biofunctionalization of silica-coated CdTe and gold nanocrystals Nano Lett. **2**(12); 1363-1367 (2002) .
18. D. V. Talapin, S. Haubold, A. L. Rogach, A. Kornowski, M. Haase, H. Weller, A novel organometallic synthesis of highly luminescent CdTe nanocrystals. J. Phys. Chem. B. **105**(12); 2260-2263 (2001).
19. S. F. Wuister, I. Swart, F. van Driel, S. G. Hickey, C. de Mello Donega, Highly luminescent water-soluble CdTe quantum dots. Nano Lett. **3**(4); 503-507 (2003).
20. S. Wang, N. Mamedova, N. A. Kotov, W. Chen, J. Studer, Antigen/Antibody immunocomplex from CdTe nanoparticle bioconjugates. Nano Lett. **2**(8); 817-822 (2002).

21. S. Pathak, S.-K. Choi, N. Arnheim, M. E. Thompson, Hydroxylated quantum dots as luminescent probes for in situ hybridization. J. Am. Chem. Soc. **123**(17); 4103-4104 (2001).
22. E. R. Goldman, E. D. Balighian, H. Mattoussi, M. K. Kuno, J. M. Mauro, P. T. Tran, G. P.Anderson, Avidin: a natural bridge for quantum dot-antibody conjugates. J. Am. Chem. Soc. **124**(22); 6378-6382 (2002).
23. S. J. Rosenthal, I. Tomlinson, E. M. Adkins, S. Schroeter, S. Adams, L. Swafford, J. McBride, Y. Wang, L. J. DeFelice, R. D. Blakely, Targeting cell surface receptors with ligand-conjugated nanocrystals. J. Am. Chem. Soc. **124**(17); 4586-4594 (2002).
24. E. R. Goldman, G. P. Anderson, P. T. Tran, H. Mattoussi, P. T. Charles, J. M. Mauro, Conjugation of luminescent quantum dots with antibodies using an engineered adaptor protein to provide new reagents for fluoroimmunoassays. Anal. Chem. **74**(4) 841-847 (2002).
25. W. C. W. Chan, S. Nie, Quantum dot bioconjugates for ultrasensitive nonisotopic detection. Science, **281**, 2016-2018 (1998).
26. M. Jr. Bruchez, M. Moronne, P. Gin, S. A. Weiss, A. P. Alivisatos, Semiconductor nanocrystals as fluorescent biological labels. Science. **281**(5385), 2013-2016 (1998).
27. Y. Chen, Z. Rosenzweig, Luminescent CdS quantum dots as selective ion probes. Anal. Chem. **74**(19); 5132-5138 (2002).
28. A. Y. Nazzal, L. Qu, X. Peng, M. Xiao, Photoactivated CdSe nanocrystals as nanosensors for gases, Nano Letters, **3**(6); 819-822 (2003).
29. M. Han, X. Gao, J. Su, S. Nie, Quantum-dot-tagged microbeads for multiplexed optical coding of biomolecules. Nature Biotech. **19**(7), 631-635 (2001).
30. N. Gapaonik, I. L. Radtchenko, G. B. Sukhorukov, H. Weller, A. L. Rogach , Toward encoding combinatorial libraries: Charge-driven microencapsulation of semiconductor nanocrystals luminescing in the visible and near IR Adv. Mater. **14**(12), 879-882 (2002).
31. S. Chang, L. Liu, S. A. Asher. Preparation and properties of tailored morphology, monodisperse colloidal silica-cadmium sulfide nanocomposites. J. Am. Chem. Soc. **116**(15), 6739-6744 (1994).
32. S. Chang, L. Liu, S. A. Asher, Creation of templated complex topological morphologies in colloidal silica. J. Am. Chem. Soc. **116**(15), 6745-6747 (1994).
33. A. L. Rogach, D. Nagesha, J. W. Ostrander, M. Giersig, N. A. Kotov, Raisin bun"-type composite spheres of silica and semiconductor nanocrystals Chem. Mater, **12**(9), 2676-2685 (2000).
34. L. Qu, X. Peng, Control of photoluminescence properties of CdSe nanocrystals in growth J. Am. Chem. Soc. **124**(9) 2049 – 2055 (2002).
35. A. M. Hines, P. Guyot-Sionnest, Synthesis and characterization of strongly luminescing ZnS-Capped CdSe nanocrystals J. Phys. Chem. **100**(2), 468–471 (1996).
36. K. C. Grabar, R. G. Freeman, M. B. Hommer, M. J. Natan, Preparation and characterization of Au colloid monolayers Anal. Chem. **67,** 735-743 (1995)
37. J. C. Fendler, E. J. Fendler, Catalysis in micellar and macromolecular systems; (Academic Press: New York, 1975).
38. S. Riegelman, N. A. Allawala, M. K. Hrenoff, L. A. Strait, The ultraviolet absorption spectrum as a criterion of the type of solubilization. J. Colloid Sci. **13**, 208-217 (1958).
39. Z. Gao, A. N. Lukyanov, A. Singhal, V. P. Torchilin, Diacyllipid-polymer micelles as nanocarriers for poorly soluble anticancer drugs Nano Lett. **2**(9), 979-982 (2002)
40. C. Liu, Z. J. Zhang, Size-dependent superparamagnetic properties of Mn spinel ferrite nanoparticles synthesized from reverse micelles. Chem. Mater.; **13**(6);2092-2096 (2001).
41. J. P. Cason, K. Khambaswadkar, C. B. Roberts, Supercritical fluid and compressed solvent effects on metallic nanoparticle synthesis in reverse micelles. Ind. Eng. Chem. Res., **39**(12) 4749-4755 (2000)
42. R. B. Khomane, A. Manna, A. B. Mandale, B. D. Kulkarni, Synthesis and characterization of dodecanethiol-capped cadmium sulfide nanoparticles in a winsor II microemulsion of diethyl ether/AOT/water Langmuir; **18**(21) 8237-8240 (2002).
43. M. Wilchek, E. A. Bayer, The avidin-biotin complex in bioanalytical applications. Anal Biochem. **5**,171(1):1-32 (1988)
44. Yongfen Chen and Zeev Rosenzweig, Synthesis and application of silica nanospheres that contain luminescent quantum dots as amplifiers in digital counting immunoassays, (to be published).
45. F. Caruso, D. Trau, H. Mohwald, R. Renneberg, Enzyme encapsulation in layer-by-layer engineered polymer multilayer capsules Langmuir **16**(4), 1485-1488 (2000).
46. M. Mammen, S. Choi, G. M. Whitesides, Polyvalent interactions in biological systems: Implications for design and use of multivalent ligands and inhibitors Angew. Chem. Int. Ed. **37**(20), 2754-2794 (1998).

47. G. Decher, J. D. Hong, Buildup of ultrathin multulayer films by a self-assembly process 1. consecutively alternating adsorption of anionic and cationic bipolar amphiphiles on charged surfaces Makromol. Chem. Macromol. Symp. **46**, 321-327 (1991).
48. G. Decher, J. D. Hong, J. Schmitt, Buildup of ultrathin multulayer films by a self-assembly process 3. consecutively alternating adsorption of anionic and cationic ployelectrolytes on charged surfaces. Thin Solid Films, **210**(1-2), 831-835 (1992).
49. J. Anzai, T. Hoshi, N. Nakamura, Construction of multilayer thin films containing avidin by a layer-by-layer deposition of avidin and poly(anion)s Langmuir, , **16**(15), 6306-6311 (2000).
50. R. Ballerstadt, J. S. Schultz, A Fluorescence affinity hollow fiber sensor for continuous transdermal glucose monitoring. Anal. Chem., **72** (17), 4185 (2000).

PHOSPHOLIPID MAIN PHASE TRANSITION ASSESSED BY FLUORESCENCE SPECTROSCOPY

Juha-Matti I. Alakoskela and Paavo K. J. Kinnunen*

*J.-M. I. A. Helsinki Biophysics and Biomembrane Group, Institute of Biomedicine /Biochemistry, P.O. Box 63, Haartmaninkatu 8, FIN-00014 University of Helsinki, Finland. P. K .J. K. is affiliated also to MEMPHYS — Center for Biomembrane Physics.

FOREWORD

Since recently phospholipid phase behavior and its biological significance have been studied almost exclusively by biophysicists. However, resurrection of the interest in the organization of lipid mixtures has attracted also cell biologists into this challenging area. It has become clear that lipid biophysics, which has been largely overlooked in cell biology, is involved in a large number of central cellular processes. Key to the understanding of lipids is their phase behavior, compiled in phase diagrams and connecting transitions. Of the latter the most thoroughly investigated is the so-called main phase transition. In the first part of this chapter we will briefly summarize its general features as well as general considerations and applications of different fluorescent probes employed in studies on this process. Almost every amphiphilic or hydrophobic fluorescent probe has at some point been used to investigate phospholipid phase behavior. We will concentrate on the properties of the probes used in our own laboratory. In the second part of the chapter we will provide an in-depth review of our recent results, which challenge some of the conventional views about the mechanisms of main phase transition.

PART I: GENERAL BACKGROUND

1. On phospholipid phases

In the modern world, liquid crystals and their phase transitions are part of everyday life. As we punch the buttons on our pocket calculators, electronic notebooks, cell phones etc., we issue orders to trigger phase transitions by electric fields, and we see the results as numbers and letters on the liquid crystal displays. Intriguingly, liquid crystals and their phase transitions are also found within us. Phospholipids, which constitute the basic fabric of cellular membranes are liquid crystals, and characteristically for this class of materials, they exhibit a great variety of different phases and connecting transitions (Kinnunen and Laggner, 1991). The phase diagrams of multicomponent lipid mixtures can be very complex as in addition to different phases possible for single component membranes there can be solid—solid (e.g. Lentz ja Litman, 1978; Untracht and Shipley, 1977), solid—fluid (Pagano et al., 1973) and fluid—fluid (Recktenwald and McConnell, 1981; Wu and McConnell, 1975) immiscibility. Such lateral heterogeneity and domain formation in biomembranes is well established, and is known to be involved in several biological processes (Kinnunen, 1991; Welti and Glaser, 1994). Quite recently a class of such membrane domains was named rafts (Simons and Ikonen, 1997). Since then these microdomains, composed of sphingolipids and cholesterol (Estep et al., 1979; Goodsaid-Zalduondo et al., 1982; Thompson and Tillack, 1985), have gained considerable attention within the field of cell biology as well as biophysics (see e.g. Epand, 2003 and references therein), and a number of proteins has been suggested to be associated to these membrane domains (Foster et al., 2003). The basis for the formation of these structures, their presence at different temperatures, as well as their interactions and phase diagrams with detergents can only be understood within the framework of membrane physics. A good example is the phase diagram of so-called raft-forming lipid mixtures and Triton X-100, showing that not only cooling but also Triton X-100 can induce the phase separation leading to the segregation of sphingolipids and cholesterol (Heerklotz, 2002).

Immiscibility of lipids and resulting lateral segregation of lipids and changes in lateral heterogeneity appear not to be the only important lipid functions in cells. The physical state of membrane lipids is known to affect the binding of peripheral membrane proteins and activity of both peripheral and integral membrane proteins (Brown, 1994; Burack et al., 1993; Epand, 1991; Holopainen et al., 2002; Kinnunen et al., 1994; McCallum and Epand, 1995; Mustonen et al., 1987; Slater et al., 1994; Zhang and Kaback, 2000). In addition to lamellar phases of the plasma membranes, also different nonlamellar lipid phases in different cell organelles appear to be ubiquitous in cells (Epand, 1996; Landh, 1995; Luzzati, 1997). In mitochondria, there is evidence for membrane cubic phases and transitions in response to environmental stimuli (Deng et al., 1999; 2002), and possible physiological role for one such transition has been suggested (Deng et al., 2002). Different cellular membranes are heterogenous in structure and appear to contain also gel-like domains, the physical state of cellular membranes depending on the metabolic state of the cell (Mamdouh et al., 1998). The physical state of plasma membranes appears to be mainly optimized to be in the liquid-crystalline or fluid state (Kinnunen, 1991), with certain propensity for nonlamellar phases (Kinnunen, 1996a; 1996b), and close to the critical temperature of the critical unilamellar state (Gershfeld and Ginsberg, 1997; Ginsberg at al., 1991; Jin et al., 1999; Tremper and Gershfeld, 1999).

It is the thermally induced phase behavior, which is most commonly studied. It should, however, be emphasized, that for the various lipids phase changes can be induced by changes in pH, ions, electric fields, water activity, as well as osmotic pressure (Cevc, 1991), i.e. parameters, which can be actively regulated by the cells. Analysis of the data measured for axons show beyond doubt that the propagation of a nerve impulse involves a transient phase change of membrane phospholipids (Kinnunen, 1991). This process has been suggested to be coupled to standing wave oscillations that would provide a mechanistic explanation for ion channel stochastic resonance as well as low frequency noise of ion channel currents (Larsson, 1997). Simultaneous optimization of several physical properties of lipids might partly explain the need for the great number of lipids present in plasma membranes (Kinnunen, 1991). Although the properties of different phases and phase diagrams of complex lipid mixtures are thus essential in order to understand biological membranes, the occurrence and mechanisms of phase transitions as such compose a fascinating problem of physics.

2. Phospholipid main phase transition

Despite the considerable effort taken over the last century phase transitions remain as a fundamental problem in physics. The very basic conceptual framework for these studies was laid down by Gibbs and Duhem, who developed the concept of phase and the criteria for their thermodynamic stability (see e.g. Duhem, 1898a; 1898b). Phase transitions of liquid crystals were initiated by the detection of two transitions for cholesterol benzoate in 1888 by Friedrich Reinitzer (see Prost and Williams, 1999). The exploration of the theoretical basis of the behavior of different properties in the vicinity of phase transitions began advancing with the introduction of phenomenological Landau theory (see Tolédano and Tolédano, 1987) and Ising model in the 1930's. Since those times a number of different theories accounting for various aspects of phase transitions have been introduced, many concentrating on explaining the critical behavior in the vicinity of the transition rather than predicting the occurrence of phase transition *a priori*. Yet there are also theories focussing on the latter issue. However, due to intervention of nucleation phenomena as well as the complexities and sensitivity of predictions to exact forms of intermolecular potentials, no single theory accounting for the occurrence and all features of phase transitions exists (Papon et al., 2002). This applies also for solid↔fluid transitions. Quoting Papon et al. (2002), "there is currently no satisfactory theory for predicting melting and crystallization phenomena in the most general manner".

Characteristics of phase transitions and particularly phospholipid main phase transition are very well known. The driving force of phospholipid main phase transition derives from the configurational entropy of the acyl chains, i.e. *trans→gauche* isomerization. There is some discrepancy for the fraction of *gauche* bonds in bilayers. For the $L_{\beta'}$ phase the fraction is most likely close to zero, while for L_α phase values ranging between 0.14—0.3 have been suggested (Marsh, 1991; Snyder et al., 2002). In MD simulations for DPPC with applied corrections the fraction of *gauche* bonds in fluid phase is 0.28, in first two bonds vicinal to the glycerol backbone approx. 0.1, then steeply increasing further along the palmitoyl chains and reaching approx. 0.3 at their methylene ends. Even at 50°C for DPPC (approx. 8 degrees above T_m and 3 to 4 degrees above the disappearance of gel phase nuclei) the fraction of all-*trans* acyl chains in MD simulations is 0.075. In reality this value could be even higher (Snyder et al., 2002). Even assuming complete lack of correlation for the order of the two acyl chains of DPPC, this would give $X \approx 0.006$ of DPPC with both acyl chains in all-*trans* configuration at 50°C. Due to the chain isomerization in the main transition the volume of the bilayer increases by 4 %, the area increases by 25 %, and the thickness decreases by 16 % (see Heimburg, 1998). Based on monolayer compression isotherms, structural comparisons, and fluorescence spectroscopic measurements of bilayers, the surface tension for lipid-water interface also increases by approx. 15 mN/m when entering the fluid phase (Konttila et al., 1988; Nagle and Tristram-Nagle, 2000), thus suggesting that associated with the phase transition there will be unfavorable surface free energy term that is overcome by the acyl chain entropy as well as other interactions. Using the suggested membrane/water effective surface tension estimates of γ_g=2.5 mN/m for the gel and γ_f=18.5 mN/m for the fluid phase (Söderlund et al., 2003), and using the mean molecular areas for DPPC in gel and fluid phases (see Table 1), we can estimate surface free energy change as $\Delta G = \gamma_f A_f - \gamma_g A_g \approx 6.4$ kJ/mol. The enthalpy change for DPPC has been reported to be 36.4 kJ/mol (Mabrey and Sturtevant., 1976), so this would suggest that 18 % of the total enthalpy is consumed for increasing both area and surface tension. Accordingly, as the volume, thickness, area, and enthalpy are coupled to smallest detail, the energetics of *gauche* conformation should contain contribution from unfavorable surface energy.

A general constant for different phospholipids couples enthalpy and volume changes, as well as heat capacity and volume (as well as area) isothermal compressibility changes, i.e. ΔH curve is identical in shape to ΔV curve and C_p curve is identical in shape to κ_T curve (Ebel et al., 2001; Halstenberg et al., 1998; Heimburg, 1998; Schrader et al., 2002). For a homologous series of phospholipids, the effects of chain length, chain unsaturation, hydration, and spontaneous curvature on the transion temperature, enthalpy, and entropy are well known (Marsh, 1991), though there is no single model applicable to all the different phospholipid classes. Interestingly, very basic thermodynamic aspects of phospholipid main transition remain a matter of discussion. In theoretical models of main phase transition all but the acyl chain degrees of freedom are assumed constant, neglecting the mobility of phospholipid headgroups as well as interfacial water (for a review see Mouritsen, 1991). Although such approaches can produce relatively good approximations, they are unlikely to fully reflect all the molecular level changes during the transition, as observations suggest that for instance the headgroup conformation during the transition differs from those in fluid and gel phase (Makino, 1991). Additionally, lipid-water interface and changes in the interface are known to affect e.g. melting temperatures, spontaneous curvature, transition enthalpies and phase diagram (Boggs, 1987; Cevc, 1986; 1991). While all the recent models assign first-order nature to the phospholipid main phase transition, the transition is quite wide, increase in enthalpy is steep but not stepwise, and no single discontinuous parameter has been described (Mouritsen, 1991). In addition phospholipid membranes display extensive fluctuations several degrees away from the transition (see Kharakoz and Shlyapnikova, 2000, and references therein), which would be typical for second-order transitions. Two different models have been put forward to explain this weak first-order nature of phospholipid main phase transition. Briefly, one of them suggests the presence of a critical point close to the transition, causing the enhancement of fluctuations and giving pseudocritical nature to the transition (Mouritsen, 1991; Lemmich et al., 1995). Alternative model suggests that low line-tension between fluid-solid interface allows for extensive heterophase fluctuations, i.e. formation and disappearance of small, unstable nuclei of the new phase, explaining the pseudocritical phenomena in terms of these fluctuations (Kharakoz and Shlyapnikova, 2000). In this respect it is of interest to consider the suggestions of Gershfeld and coworkers. They have dubbed the term critical unilamellar state and temperature, T*, for the temperature at which last gel-like nuclei disappear upon heating (Gershfeld, 1989a; 1989b). Several discontinuities in membrane properties occur at this temperature (Gershfeld and Ginsberg, 1997; Gershfeld et al., 1993; Jin et al., 1999; Koshinuma et al., 1999; Lehtonen and Kinnunen, 1994) which for at least DMPC and DPPC appear to coincide with the onset of critical swelling upon cooling, much discussed in the above theories. One of the anomalies at T* is the critical-like slowing down of heat capacity relaxation upon perturbation by a sudden increase in temperature (Gershfeld et al., 1993). With respect to the formation of gel phase nuclei at T* upon cooling it is of interest that MD simulations for the fluid phase DPPC at 50°C (≈4 to 5 degrees above T*) suggest the fraction of all-*trans* acyl chains to be 0.075 (Snyder et al., 2002).

In keeping with the above, also the view of the molecular level mechanisms of phase transition is far from complete. The basic sequence of events with increasing temperature for many phospholipids is from (quasi)crystalline phase (L_c or $L_{c'}$) to gel phase (L_β or $L_{\beta'}$) to rippled gel phase ($P_{\beta'}$) to fluid or liquid crystalline phase (L_α), the apostrophe indicating the acyl chains within the phase to be tilted. The transitions are named sub-, pre- and main transition. Unless otherwise indicated, in this review gel phase will be used to refer to $L_{\beta'}$ and $P_{\beta'}$ phases collectively, and fluid phase to L_α phase. For some phospholipids under proper circumstances $L_{\beta'}$ and $P_{\beta'}$ may exist only as metastable phases and there may be several additional metastable lamellar phases (e.g. Kodama et al., 1999). In the (quasi)crystalline, subgel $L_{c'}$ phase of DPPC tightly packed acyl chains form a sublattice that is similar to the $L_{\beta'}$ phase. In the $L_{c'}$ phase, however, there is also a lattice for the lipid headgroups, so that the phospholipids and their two acyl chains form a structure that is locally highly ordered and has a unit cell containing two DPPC molecules (Raghunathan and Katsaras, 1995). In subtransition this strict order disappears. In pretransition of multilamellar vesicles a new phase with partially melted acyl chains (or rather with some lipids with melted acyl chains) and ripples with repeat distance of approx. 14 nm form, with a sawtooth like profile (Fig. 1). The exact distribution of *gauche* bonds for the acyl chains in different part of the ripples is not known (Nagle and Tristram-Nagle, 2000). Pretransition is observed also for unilamellar vesicles, although the enthalpy associated to the pretransition of unilamellar vesicles is much smaller than with MLVs and the detection of ripple structure is not as easy (Heimburg, 1998; Lichtenberg et al., 1984; Mason et al., 1999). Yet, at low salt concentrations spontaneously forming unilamellar dimyristoylphosphatidylglycerol vesicles show large pretransition enthalpy (Riske et al., 2002). Pretransition is not only sensitive to solvent environment, but also to the stereochemistry of lipids (Eklund et al., 1984). The pretransition and associated structural changes has been suggested by Heimburg (2000) to result from line defects formed by phospholipids with melted acyl chains, and by melted and unmelted lipids segregating to surfaces of different curvature within the ripples. Heimburg's model also reproduces experimental enthalpy changes for the transition of both unilamellar and multilamellar vesicles. Only erraneous prediction of the model is that for unilamellar vesicles (with weaker coupling) T_p should shift towards T_m, while experimentally the opposite is seen (see figures in Heimburg, 2000). In regard to the line defects in large unilamellar vesicles, it is noteworthy that while the unilamellar phosphatidylcholine vesicles above the main phase transition are smooth spheres, the vesicles below T_m are only roughly spherical, composed of irregular, straight facets or plates joined by fault lines, as seen in cryo-TEM images (Andersson et al., 1995; Scheiner et al., 1999). In unilamellar vesicles these defect lines could be the place of initial melting and the first location of line defects.

Kinetic studies of the main phase transition with various detection methods (including fluorescence spectroscopy) have shown that there are five different relaxation processes following a sudden increase in temperature (Holzwarth, 1989). Fastest relaxation time of 4 ns was independent from the acyl chain length as well as lipid headgroup, and was deemed to correspond to the *trans→gauche* kink formation in individual molecules. The second relaxation time constant of 300 ns was dependent on the headgroup and was suggested to correspond to the loosening of the hexagonal lattice of the gel phase. Third relaxation time of 10—20 μs was suggested to derive from the organization of *gauche* isomers into conformations, which better allow lateral diffusion, i.e. decrease of entanglements. Fourth relaxation time of approx. 0.1—1.5 ms was

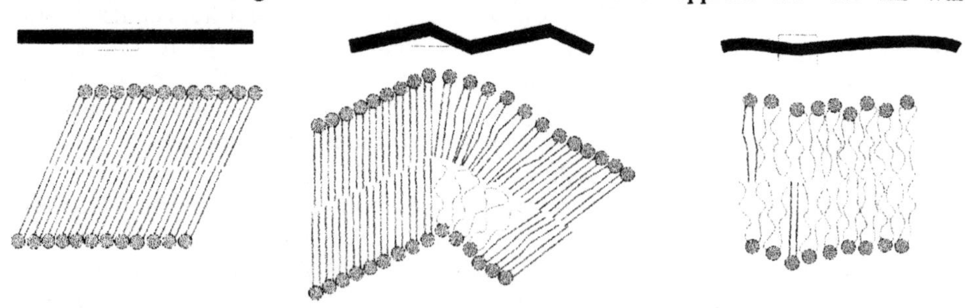

Figure 1. A schematic illustration of some phospholipid phases. The order of phases from left to right is $L_{\beta'}$, $P_{\beta'}$, and L_{α}. The upper bar shows a cross-section of membrane, and the lower image show a more detailed scheme of the ordering of acyl chains. In $L_{\beta'}$ phase the cross-section is straight, and the acyl chains highly ordered. In $P_{\beta'}$ phase there is sawtooth like periodicity with 14.2 nm periodicity for DMPC, and with the longer segments being thicker than the shorter segments (Nagle and Tristram-Nagle, 2000). It should be kept in mind that the distribution of gauche bonds within $P_{\beta'}$ phase is unknown and that therefore the scheme in lower image, showing highly disordered and ordered lipids in different areas is hypothetical. In L_{α} phase there are thermally excited undulations and protrusion of lipid molecules, the disorder of acyl chains is discussed in more detail in text.

addressed to the cluster formation with gel, intermediate, and fluid state of lipids present, and fifth relaxation time of approx. 20—30 ms to the disappearance of remaining gel state leaving fluid phase with some intermediate state lipids. All of the above relaxation times peak at approximately T_m, though some are very asymmetric, being considerably wider on the low temperature side of the peak. Most of these processes are slow compared to the fluorescence lifetimes and yet are fast compared to the timescale of steady-state experiments. One should take care not to draw too strict conclusions from the timescales of T-jump experiments, as e.g. monolayers the nucleation process has been shown to be strongly dependent on the rate of perturbation (Helm and Möhwald, 1988).

Expectedly, several properties of the phospholipid bilayers show discontinuities at main transition (see Table 1), such as membrane permeability and the lateral compressibility (Langner and Hui, 1993; Mouritsen and Kinnunen, 1996; Nagle and Scott, 1978), which have maxima at T_m. These as well as other findings have been explained in terms of dominating critical density fluctuations leading to the formation of domains of gel- and fluid-state lipids, and further to maxima in the domain boundary length and the number of defects in membranes. The above, however, is in conflict with fluorescence spectroscopic results, which suggest that this model might not be sufficient (Jutila and Kinnunen, 1997; Metso et al., 2003). Likewise, polar headgroup and interfacial water organization have been suggested to show discontinuities slightly below T_m, as found in fluorescence (Jutila and Kinnunen, 1997, Söderlund et al., 1999), IR (Mellier and Diaf, 1988; Mellier et al., 1993), and dielectric spectroscopy (Enders and Nimtz, 1984) studies on DMPC and DPPC liposomes. Indeed, although present theories reproduce most essential features to a great detail, the very pursuit of experimentation is to find results conflicting with existing theories and thus challenge the theories so as to stimulate their evolution towards full maturity. Accordingly, our own approach assessing the detailed molecular level events in the course of the phospholipid main transition has exploited fluorescence spectroscopy supplemented with other techniques such as differential scanning calorimetry, FTIR spectroscopy, and Langmuir film balance studies. Though fluorescence spectroscopy offers an inexpensive and versatile method to study membrane properties, a word of caution is that there is a great number of aspects to consider before drawing meaningful conclusions. A few of the most essential issues include an understanding of the extent and the nature of the perturbation of the membrane by the probes, knowledge of probe photophysics, the partitioning of the probe between different phases, the location of the probe as well as changes in the location of the probe, and the effect of probe itself on T_m as well as possible lateral phase separation of probe and matrix lipids within the transition zone or in one of the phases.

Table 1. Some characteristics of phosphatidylcholine phases

Property and lipid	In $L_{\beta'}$ phase	In $P_{\beta'}$ phase	At T_m	In L_α phase
rate constant for CF permeability in DPPC matrix[5], 10^{-4}/s	≈ 0.1—0.2 (20°C)	≈ 2 (37°C)	≈ 18 (42°C)	≈ 3—4 (50°C)
area / lipid, Å^2 DPPC[1]	47.9	—	—	64
hydrophobic thickess, Å DPPC[1]	34.4	—	—	28.5
volume / lipid, Å^3 DPPC[1]	(20°C) 1142—1145	—	—	(50°C) 1228—1232
thermal area expansivity, K^{-1} DMPC[2]	3	5800	—	6800
elastic area compressibility modulus, mN/m				
DMPC[2]	860	62	—	145 (at T*)
DMPC[6]	526	500	—	476 ($T-T_m$=11°C)
fraction of *gauche* bonds				
compiled data in 1991[3]	0—0.1	—	—	0.3
MD with corrections[4]	—	—	—	0.28
MD without corrections[4]	—	—	—	0.22
derived from IR spectra[4]	—	—	—	0.14±0.04
equilibrium lateral pressure, mN/m[7]	≈ 50	—	—	30—40

[1]Nagle and Tristram-Nagle, 2000. [2]Needham and Evans, 1988, direct mechanical measurement.
[3]Marsh, 1991. [4]Snyder et al., 2002.
[5]Bramhall et al., 1987. [6]Lemmich et al., 1996, based on calculations of neutron scattering data
[7]Estimates based on values in Marsh (1996), Nagle and Tristram-Nagle (2000), and Söderlund et al., 2003.

3. Fluorescent probes and main phase transition

Ever since the early days of phospholipid bilayer studies fluorescent probes have been used to study both the properties of membranes as such and to detect lipid phase transitions. Basically, most of the different fluorescence based time-resolved and steady-state strategies in studies of phospholipid phase transitions utilize some of the following processes: changes in the motion of the fluorophore detected by anisotropy, changes in the solvation energetics or kinetics probed by spectral shifts and intensity levels, changes in the excimer formation, or changes in the colocalization of different fluorescent probes or fluorescent probes and quenchers as detected by the intensity changes as the result of resonance energy transfer or collisional quenching. Of course, all of these can be also be studied by means of time-resolved fluorescence. Finally, a good strategy is to employ several fluorescent probes to highlight different aspects of the complex molecular level processes.

It is important to keep in mind that fluorescent probes report on their surroundings, and on their surroundings only. Further, they only report on changes that are significant within the fluorescence lifetime. For example, red-edge excitation shift (REES) that is related to solvent relaxation around the probe can only be detected if the timescales of solvent relaxation are close to the fluorescence lifetime of the probe (Lakowicz and Keating-Nakamoto, 1984). Accordingly, it cannot detect changes in solvent dynamics on other timescales no matter how significant they might be. Due to interactions between the surrounding matter and the fluorescent probe the properties of both the probe and surrounding matter are modified, as reflected in the slight broadening of the main transition endotherm and minor, yet clear reduction in its enthalpy (Jutila and Kinnunen, 1997; Metso et al., 2003). This might lead some to err and to believe that this prevents the use of fluorescents probes in the studies of systems delicate to perturbations, such as phospholipids undergoing phase transitions. While there *is* a valid point in this, yet, the mole fractions of fluorescent probes used in these studies are usually at the level of a few mole percent at most, commonly much less, and, therefore, the matrix lipid system as a whole is not significantly affected by the fluorescent impurity. Changes in the properties of the lipid matrix thus modify the interactions between the matrix and the immediate surroundings of the fluorophore and, further, between the fluorophore and its surroundings. In order to understand how the fluorescent probe modifies its surroundings, and how the surroundings in turn are modified by changes in the bulk matrix, it is essential to know the probe well. As a result, any study on lipid phase transitions by fluorescent probes should include thorough consideration of these interactions as well as probe photophysics. On the other hand, the very sensitivity of fluorescent probes for only certain aspects of their surroundings on a molecular scale provides a useful method to investigate just those very issues. Lastly, the probes are structurally very close to the matrix lipids, many bearing identical headgroup, glycerol backbone with the diester bonds and palmitoyl chain at *sn*-1 position. Therefore, these entities can be expected to couple quite efficiently to the organization of the matrix, in spite of the presence of the perturbing covalently linked fluorophore. This is in keeping with the fluorescence signals coinciding exactly with the progression of the transition, albeit not in a manner aligning with simple models for the transition as a strict first order process. Figure 2 shows structures for some of the probes discussed.

3.1. DPH

One of the most widely used fluorescent probes in studies of biomembranes and their models is diphenylhexatriene (DPH) and its phospholipid derivatives, e.g. 2-(3-(diphenylhexatrienyl)propanoyl)-1-hexadecanoyl-*sn*-glycero-3-phosphocholine (DPH-PC). The advantage of diphenylpolyenes compared to linear polyenes like parinaric acid is their greater chemical stability. DPH is hydrophobic and rod-like, and thus readily accommodates into the hydrocarbon part of phospholipid bilayers. Indeed, the great advantage DPH probes offer is that their rotational motion and consequently emission anisotropy correlates well with the order of the surrounding phospholipid acyl chains. The anisotropy ratios r_∞/r_0 available from time-resolved anisotropy measurements represent the square of order parameter for DPH, and values thus obtained agree very well with the order parameters from deuterium NMR for carbons 10—12 of DPPC acyl chains (Heyn, 1979). In the gel-like phase DPH molecules are mostly aligned with the acyl chains whereas in the fluid-like phase DPH becomes more disordered. Accordingly, DPH has been used to study phospholipid phase transitions ever since the early days of the field (Andrich and Vanderkooi, 1976).

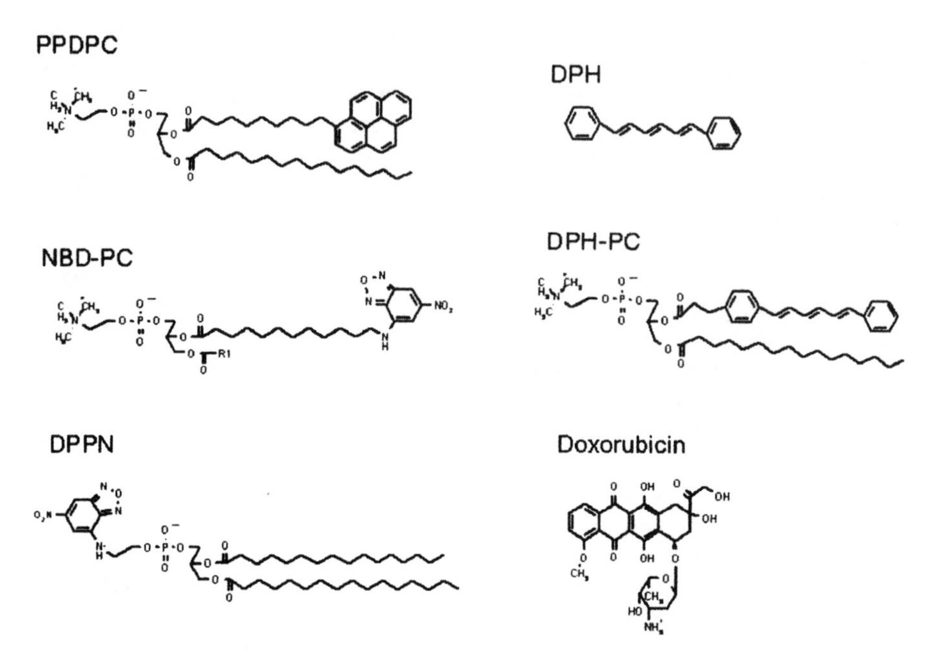

Figure 2. Structures for some the probes discussed. See text for explanations of the abbreviations.

As the emission of DPH is quenched in water, and as DPH is highly hydrophobic, the detected fluorescence originates solely from the membrane-associated probe (Lakowicz, 1999). The lowest energy $S_0 \rightarrow S_1$ transition for DPH is symmetry forbidden, the absorption corresponding to $S_0 \rightarrow S_2$ transition. However, depending on the system, none or only a small fraction of the emission comes from the relaxation of the original S_2 state to S_0 state, and most of the emission originates from S_1 state, with $S_2 \rightarrow S_0$ seen only as a slight deviation on the blue-edge of the DPH emission spectrum (Bachilo et al., 1998; Itoh and Kohler, 1987; Lakowicz, 1999). The $S_0 \rightarrow S_2$ absorption followed by $S_2 \rightarrow S_1$ internal conversion and $S_1 \rightarrow S_0$ fluorescence leads to a large Stokes' shift (the absorption maximum of DPH in membranes is approx. 354 nm and the emission maximum approx. 428 nm), long fluorescence lifetime ($\approx$10 ns) and high molar absorptivity (Lakowicz, 1999). DPH fluorescence quantum yield in various solvents increases with increasing viscosity and solvent polarizability. In addition, there is a red-shift with increasing solvent polarizability (Dupuy and Montagu, 1997, and references therein). The π electrons of excited DPH are located far from the molecule, and the excited DPH itself has high hyperpolarizability (Werncke et al., 2000).

The location, distribution, and orientation of DPH and DPH-labeled lipids in bilayers has been studied in detail. In fact, because of its rod-like shape and the favourable fluorescence properties mentioned above it has been the archaetypal test probe for different models and theories regarding the behavior of other fluorescent membrane probes. Typically, for DPH incorporated into lipid bilayers there are two decay components as opposed to one for DPH in an isotropic media (Kawato et al., 1977). If the data are interpretated in terms of restricted "wobbling-in-a-cone model" (Heyn, 1979), then upon heating the sample above phase transition, a decrease in fluorescence lifetime is seen, as well as an increase in the wobbling cone angle (20°→70°) accompanied with increased anisotropy relaxation time (≈0.5→1.2 ns) (Kawato et al., 1977). Thus decrease in steady-state anisotropy for DPH when heating through the transition is (though insignificantly) opposed by a decrease in wobbling relaxation rate, suggesting that the changes in steady-state anisotropy for DPH upon phase transition of DPPC membranes are related to the wider distribution of final angles at the moment or emission, in other words decreased r_∞, as pointed out by Kawato et al. (1977). Yet, more recent studies have suggested that while there is no stepwise or even steep change in the relaxation time constant, it decreases continuously as a function of temperature, being ≈1—2 ns at 40°C, and decreasing more steeply below T_m of DPPC (Wang et al., 1991).

Analysis of time-resolved fluorescence spectra of DPH in membranes has revealed a short-lived and red-shifted component, implicating that part of the probe resides in a relatively polar environment (Konopasek et al., 1998, and references therein). This result is also supported by more efficient energy transfer from the short-lived component to fluorescein-labeled BSA (Konopasek et al., 1998). According to depth-dependent quenching studies with DPH and its derivatives in a DOPC matrix, most of the fluorescing free DPH as well as the DPH moiety of DPH-PC are located deeply buried within the hydrophobic region, the distances from the bilayer center being 7.8 and 6.9 Å, respectively, corresponding for DOPC to the levels of carbons 9 and 8 as calculated by Kaiser and London (1998; 1999), i.e. carbons 9 and 10 according to the normal numbering (Fig. 3). The picture arising from various measurements is that some fraction of DPH is oriented in the bilayer-water interface along the bilayer plane, possibly due the favorable interactions of the aromatic ring quadrupole moments with the interfacial molecular moieties. The presence of two populations implies that the energy difference between these states is not large, and that changes in the free energies for different DPH states in membrane might lead to considerable differences in the occupation of the two states, as would be expected from Boltzman distribution. The methods based on fluorescence quenching detect only fluorescent populations, and independent data about the distribution is thus needed in order to detect also possible weakly fluorescent probes. A bimodal distribution of DPH has been verified in neutron diffraction studies (Pebay-Peroula et al., 1994). More specifically, based on Gaussian distribution these authors calculated that in the $L_{\beta'}$ phase there is one DPH population, with its long axis oriented parallel to the acyl chains, at 30° angle to the bilayer normal, buried deep within the hydrophobic region, and another, interfacial population, oriented parallel to the membrane surface. In the L_α phase the first DPH population resides between the two leaflets and is parallel to the surface, with wider distribution. The distribution of DPH-PC is likely to resemble more that of TMA-DPH, which similarly to DPH-PC has a polar group that anchors the probe to the surface. This anchorage retains the other end of the fluorophore in the interface, not allowing for the probe to reside between leaflets. Accordingly, in the case of TMA-DPH for the population aligned with the acyl chains in the $L_{\beta'}$ phase only a widening of distribution and straightening of the molecule is seen upon melting, the probe becoming aligned with the bilayer normal in the L_α phase (Pebay-Peroula et al., 1994).

The use of DPH-labeled PC derivatives (DPH-PCs) has been suggested to avoid some of the problems related to the distribution of probe in the membrane. Yet, it is of interest to note that various DPH-PCs have been shown to have bimodal distributions as well (Pap et al., 1994). This is not suprising as rotations of bonds for glycerol carbons, carbonyl bond, and propionyl carbons can easily allow for the attached probe to flip into the interface, resulting in the orientation of thr DPH-propionyl chain parallel to the bilayer plane, similarly to DPH as such.

3.2. Pyrene-labeled probes

The characteristic property of pyrene is excimer formation (Förster, 1948), which for membranes can be exploited to report on lateral organization and diffusion. When the pyrene moiety absorbs a photon (one commonly used peak is at λ=354 nm) and ends up in the excited state, it has in essence three alternative fates. First, the excited monomer may relax into the ground state by emitting a photon (at approx. λ=398 nm). Second, the relaxation to ground state can occur via nonradiative pathways. Third, the excited monomer can collide with a ground state monomer (under non-lasing conditions only a small fraction of probes at any time is in the excited state). In the last case, an excited dimer, excimer is formed. For excimer the radiative decay yields fluorescence centered at 480 nm. In excimers the lower energy results from the two pyrenes in close proximity (≈3.35 Å) being bound by transient occupation of a bonding π* orbital (for a review see Kinnunen et al., 1993).

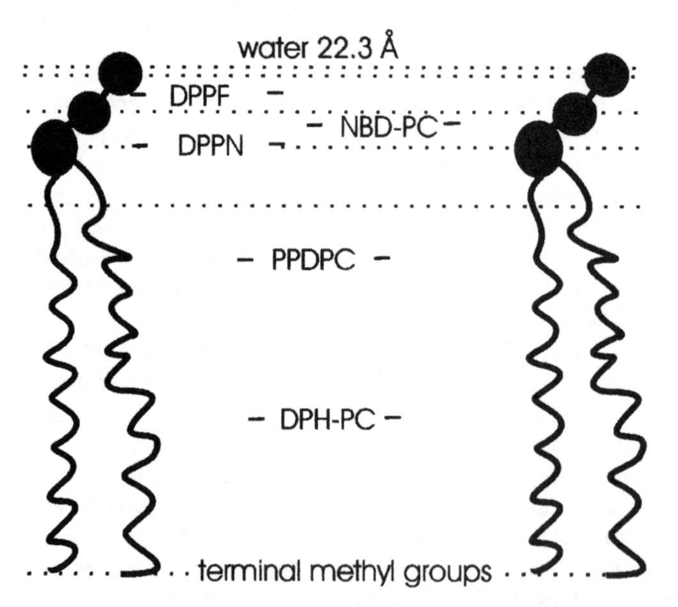

Figure 3. The depth of the fluorophore moiety of the different membrane probes as determined by depth-dependent quenching in DOPC membranes. The topmost dotted line marks the position of maximum in the distribution of interfacial water in DOPC membranes as taken from the review by White and Wimley (1998). The next lines from top to bottom are the maxima for the distribution of choline, phosphate, glycerol, carbonyl, and finally, terminal methyl groups. For the data about the probes, see text.

Early studies on membranes studies utilized pyrene as such (e.g. Galla and Sackmann, 1974; Vanderkooi and Callis, 1974). Yet, it was soon replaced by phospholipid derivatives with a pyrene moiety attached to one or both of the acyl chains (Galla et al., 1978; Sunamoto et al., 1980). The perhaps most commonly used of these phospholipid analogs is 1-palmitoyl-2-[10-(pyren-1-yl)-decanoyl]-*sn*-glycero-3-phosphocholine (PPDPC), which has been characterized in considerable detail (e.g. Somerharju et al., 1985; Kinnunen et al., 1985; 1987; Lotta et al., 1988a; 1988b; Yliperttula et al., 1988; Lemmetyinen et al., 1989; Lehtonen and Kinnunen, 1994; Lehtonen et al., 1996). The pyrene moiety of these derivatives locates within the hydrophobic core of the phospholipid bilayers. According to depth-dependent quenching analysis (Fernandes et al., 2002) the average location of the pyrene moiety of PPDPC in fluid DPPC bilayers is 14.0 Å (or 8.9 Å according to an alternative analysis judged to be less reliable by the authors) from the center of the bilayer with HWHH 5.3 Å. In most studies utilizing pyrene-containing lipids the parameter I_e/I_m derived from steady-state spectra has been measured. Importantly, I_e/I_m relates to the local concentration and rate of lipid lateral diffusion. The increase in local concentration due to factors such as segregation of pyrene-containing probes or their inequal partitioning to different coexisting phases in turn increases the collision probability. Likewise, decrease in the viscosity of probe surroundings increases I_e/I_m as the span of probe diffusion during its lifetime increases the probability it will collide with another probe. Naturally, changes in the lifetimes of monomer and excimer will also affect the intensity ratio. Recently, we have extended the use of PPDPC by measuring time-resolved changes in I_e and I_m. Interestingly, a very useful parameter appears to be the excimer fluorescence risetime (Metso et al., 2003).

Considering lateral distribution it is worthwile to notice that pyrene-labeled phospholipids have been concluded to be non-randomly distributed in the L_α phase. Accordingly, the discontinuities in I_e/I_m originally revealed the presence of regular distribution of molecules, constituting probe—lipid superlattices (Somerharju et al., 1985; Kinnunen et al., 1987; Virtanen et al., 1988). This formation of superlattices in L_α phase is explained by the repulsive interaction potential between pyrene moieties, which in turn derives most likely from the directional, radial acyl chain perturbation caused by the bulkiness of this moiety. Since then similar superlattice formation has been detected for other molecules residing in bilayers, e.g. cholesterol and phospholipids with different headgroups (for a review see Chong and Sugar, 2002). From experimental studies as well as from modelling viewpoint the time-averaged regular distribution can be seen to emerge as a general mechanism when the repulsive many-body interactions between one constituent species are sufficiently large (Sugar et al., 1994; Huang, 2002). Yet, these interactions are subtle, as for gel phase or $L_{\beta'}$ phospholipid matrix the local energy cost for accommodation of the bulky pyrene moeity into the gel phase lattice overrides the above repulsive interactions between the pyrene moieties and favors the segregation of pyrene-labeled lipids into lateral domains due to the exclusion of the probe from the crystalline matrix. Thus, comparing the two bulk phases, the pyrene-labelled lipids prefer the L_α phase. This can also be seen as pronounced reduction of I_e/I_m at T_m (e.g. Söderlund et al., 1999).

The fluorescence lifetimes for pyrene-labeled phospholipids in bilayers are typically in the range 50—100 ns (see e.g. Metso et al., 2002). The fluorescence lifetime of pyrene monomer depends not only on the intrinsic radiative relaxation and general nonradiative relaxation rates, but also by quenching by for example oxygen, proper kind of peripheral membrane proteins, such as cytochrome c (Mustonen et al., 1987) and, importantly, by ground state pyrene monomers. Accordingly, the fluorescence lifetimes are influenced by the mole fraction the of pyrene-labeled lipid in the bilayer both as a result of quenching due to diffusion and especially at higher mole fractions as a result of phase separation, i.e. local enrichment of pyrene-labeled phospholipids. It is worthwhile to notice that the quenching rates are somewhat faster for excimer than monomer and have been suggested to originate from a greater efficiency of singlet→triplet intersystem crossing for excimer (Conte, 1967).

3.3. NBD-labeled lipids

7-nitro-2,1,3-benzoxadiazol-4-yl (NBD) is another widely used fluorescent moiety in investigations on biological systems, including biomembranes (for a review see Chattopadhyay, 1990). Various NBD derivatives have been used in the visualisation of domain morphology by epifluorescence microscopy of lipid monolayers (Weis, 1991), in the detection of bilayer-to-hexagonal phase transition (Hong et al., 1988. Stubbs et al., 1989. Han and Gross, 1992), and in numerous studies of organizational changes in membranes (Chattopadhyay, 1991; Mukherjee and Chattopadhyay, 1996).

Simple NBD derivatives have three major absorbance bands in the visible and near UV region, at approx. 420, 306—360, and 225 nm (Lancet and Pecht, 1977. Fery-Forgues et al., 1993). For NBD-labeled phospholipids the two most useful absorption bands are centered at approx. 465 nm and 335 nm. The band at 465 nm originates from an intramolecular charge-transfer (ICT) type transition (Paprica et al., 1993; Fery-Forgues et al., 1993) and is associated with a large (≈ 4 Debye) change in dipole moment (Mukherjee at al., 1994). The 335 nm band corresponds to the ordinary $\pi^* \leftarrow \pi$ transition (Fery-Forgues et al., 1993). In accordance with Kasha's rule, the maximum emission wavelength λ_{max} of NBD-labeled lipids lies at 520—535 nm regardless of the absorbance band used for excitation. Absorbtivity, fluorescence quantum yield, and λ_{max} for both absorbance and emission of the ICT transition and the corresponding emission are sensitive to the polarity, hydrogen bonding, and the presence of charge transfer donors (Lancet and Pecht, 1977; Fery-Forgues et al., 1993; Lin and Struve, 1991). In fact, with increasing polarity of the surroundings, the absorptivity of the ICT transition (at 470 nm) of NBD increases strongly, whereas the absorptivity of the $\pi^* \leftarrow \pi$ transition (at 335 nm, to locally excited state) remains virtually unaffected (Fery-Forgues et al., 1993). This allows for relative selection of probes in environments of different polarity (Alakoskela and Kinnunen, 2001b). This selection applies not only to truly separate populations but also in case where there is a wide, continuous distribution of NBD, extending over regions of different polarity and with equilibration rates that are slow compared to the fluorescence lifetime of the probe. As the fluorescence with both excitation wavelengths originates from the same state, the environment induced changes in the fluorescence quantum yields can be expected to be equal. Accordingly, for probes residing in more polar surroundings, the increase in absorptivity at 470 nm partly balances the enhanced nonradiative relaxation rate, and the lack of this effect for the absorptivity at 335 nm leads to a weaker fluorescence at λ_{ex}=335 nm compared to excitation at λ_{ex}=470 nm. With excitation at 335 nm a larger proportion of the detected emission is thus derived from the probes residing in a less polar environment. Accordingly, the ratio I_{ex470}/I_{ex335} responds to changes in the environment of the NBD in its ground state, and should increase with increasing apparent polarity around the fluorophore. In some cases such effect can be useful, as the ground state environment contains contributions mainly from the solvent relaxed state, whereas changes in fluorescence intensity include the impact both from solvent dynamics and other excited state photophysics. However, changes in this ratio are very small. Additionally, the fundamental anisotropy for the different excitation bands appears to be almost the same, as reflected by the nearly identical anisotropy values regardless of excitation wavelength. This allows for an easy analysis of the anisotropy data with the two excitation wavelengths for the situations where the motional order (anisotropy) and the polarity of surroundings (basis of probe selection) are coupled (Alakoskela and Kinnunen, 2001b).

Not by any means unique but still a rare feature for fluorophores is that NBD has an —NO_2 group, which may be converted into —NH_2 by relatively mild treatment by reducing agents such as dithionite (McIntyre and Sleigth, 1991). The product of this under normal conditions irreversible reaction, 7-amino-2,1,3-benzoxadiazol-4-yl moiety, is nonfluorescent. Thus when the reducing agent is added to a suspension of unilamellar vesicles, a fast decrease in intensity corresponding to the reduction of NBD on the outer surface is seen, together with a slower decrease in intensity corresponding to diffusion of dithionite across the bilayer to reduce the NBD in the inner leaflet (McIntyre and Sleigth, 1991). For multilamellar vesicles the slower reaction has been utilized to demonstrate the increased permeability of bilayers at the temperatures of main phase transition (Langner and Hui, 1993). The rates for faster reaction report on the availability of NBD in outer leaflet. This can be affected by factors such as direct electrostatic interactions for charges as well as dipoles, hydration, orientation, and location of the NBD moiety (Alakoskela and Kinnunen, 2001a; 2001b). Based on the analysis of surface charge and hydration response of 3-hydroxyflavone probes, the (known) coupling of hydration to dipole potential has been suggested to act as the mechanistic link between the fluorescence changes of neutral probes and surface charge (Duportail et al., 2001; 2002), which should also be responsible for the response of NBD to membrane dipole potential.

A noteworthy feature of NBD is its strong selfquenching (Brown et al., 1994). In monolayers of pure NBD-labeled lipids strong hydrogen-bonding interactions between NBD moieties can be detected (Tsukanova et al., 2002). The aggregation and self-quenching of the NBD-acyl chain labelled lipid analogs in the gel phase is particularly strong, and segregation for a headgroup-labeled NBD lipid is also supported by analysis of the fluorescence lifetimes (Loura et al., 2000). The location of the NBD attached both to the headgroup and to the acyl-chain is within the interface in fluid phase lipids (Fig. 3), as suggested by studies using parallax method (Abrams and London, 1993; Chattopadhyay and London, 1987). Moreover, it appears that NBD attached via a 12 carbon spacer is located closer to the water phase than NBD attached to headgroup, their distances from the membrane center being 19.8 and 18.9 Å, respectively.

3.4. Adriamycin binding and RET studies

Adriamycin a.k.a. doxorubicin is a positively charged compound with an amino-sugar moiety and a polycyclic, aromatic moiety. Doxorubicin associates to membranes strongly by electrostatic and also by non-electrostatic interactions (Duarte-Karim et al., 1976). The electrostatic and hydrophobic membrane binding modes of doxorubicin have been suggested to correspond to less and more deeply buried positions of doxorubicin, respectively (Karczmar and Tritton, 1979). When comparing the bulk phases, doxorubicin favors the fluid phase as the tight packing of $L_{\beta'}$ phase is unfavorable for a deep penetration. Doxorubicin has a strongly absorbing band with a good overlap with the pyrene emission band. This allows for nonradiative resonance energy transfer from pyrene to doxorubicin and thus the association of doxorubicin with membranes can be studied (Mustonen and Kinnunen, 1991). We have performed extensive kinetic studies on the association of doxorubicin to membranes in the vicinity of T_m by stopped-flow technique (Söderlund et al., 1999). As discussed later, the results can be explained in terms of changes in the distribution of pyrene-lipids and by changes in electrostatics.

PART II: RECENT RESULTS

Our recent fluorescence spectroscopic studies can be grouped into two main categories. Those done with doxorubicin and NBD-labeled lipids mainly assess changes in the lipid-water interface. Those with pyrene-labeled lipids and DPH report on the order and diffusion of these impurities residing mostly in the hydrophobic region of membranes. A review of these results, as well as of selected results by other groups will follow, culminating in a novel model of phospholipid main phase transition. In contradiction to many papers published by other groups, our fluorescence spectroscopic studies of phospholipid main phase transition have been performed almost exclusively with large unilamellar vesicles (LUVs) prepared by extruding the multilamellar vesicles (MLVs) in hydrated lipid suspensions through polycarbonate filters (pore size 100 nm). Several studies suggest that this protocol yields mostly unilamellar vesicles with an average diameter close to 100 nm (e.g. Wiedmer et al., 2001). However, although most of these vesicles are unilamellar, a fraction of them has smaller vesicles inside them. The most marked differences between MLVs and LUVs is the very small molar pretransition enthalpy for LUVs compared to that of MLVs (Heimburg, 2000). Yet, DSC as well as neutron scattering do reveal the presence of pretransition also for LUVs. One of the possible reasons for the smaller enthalpy for unilamellar vesicles is that when the bilayer-to-bilayer repulsions are not present, the gel ($L_{\beta'}$) phase phosphatidylcholine vesicles the vesicle can adopt the structure of planar bilayers connected by grain boundaries / defect lines, accounting for their irregular, angular appearance in cryo-TEM images (Andersson et al., 1995; Scheiner at al., 1999). For MLVs the repulsion between bilayers forces the organization to be different, which should be reflected by more co-operative formation of line defects at the pretransition, and possibly by larger proportion of line defects after pretransition.

1. Surface properties

The changes in the membrane surface properties as detected by fluorescent probes receive contributions from several different factors. First, the mobility, orientation, and average distance of lipid headgroups change at the transition, and may be different from those both above and below the transition, as discussed in the introduction. Basically, effects on fluorescence should arise from dipole-dipole interactions and from changes in the association and penetration of ions and water into the bilayers. These effects could reflect direct electrochromism or solvatochromism (Liptay, 1969; Loew et al., 1978; 1979) or altered location of the probe induced by such interactions. A minor change in probe average orientation might greatly affect the permittivity of the environment of the probe, as it has been suggested that the permittivity of the interface is highly anisotropic, i.e. that the permittivities in the bilayer plane and in the direction of the bilayer normal are different (Raudino and Mauzerall, 1986). Based on a recent MD simulation of fluid

DPPC bilayer the in-plane relative permittivity ε_r within the hydrocarbon region is small, about 2, but within a 7 Å gradient changes to ε_r=350 for the region of PC headgroups. Normal permittivities could not be calculated because of large uncertainty in numerical values (Stern and Feller, 2003).

Second, water structure in the vicinity of bilayer is likely to be affected by phase transition. As the permittivity and refractive index of water near interfaces are different from those of bulk water (Teschke et al., 2001a; 2001b), the phase transition could also modify the refractive indices about the bilayer as well as those of the lipid phase itself, and thus have an effect on the radiative relaxation rate. This effect would derive from electrodynamics of fluorophore transition dipoles in optically thin films surrounded by medium with anisotropic local electromagnetic field, and will lead to orientation angle dependent inherent radiative relaxation rate in these thin sheets, such as lipid bilayers (Kunz and Lukosz, 1980; Lukosz, 1980; Toptygin et al., 1992). However, although data with closely-spaced measurement points are not available, a study with 5 degree temperature increments showed that the ratio of the refractive index of DPH surroundings in bilayer to that of water is approximately 1.2 and is equal for both gel and fluid phases (Toptygin et al., 1992). Significant changes arising from these effects are therefore not to be expected without probe reorientation.

Based on fluorescence spectroscopic studies of MLVs, changes in the permittivity of the interface has been suggested (Langner et al., 2000). In this study the intensity changes of fluorescein coupled to the amine group of DPPE, *N*-(fluorescein-5-thiocarbamoyl)-1,2-dipalmitoyl-*sn*-glycero-3-phosphoethanolamine (DPPF, mole fraction $X_{DPPF}=8\times10^{-4}$) were monitored as function of temperature. The authors calculated indices [I(T)-I(gel)]/I(T) and [I(T)-I(gel)]/I(gel), where I(T) represents the emission intensity at given temperature and I(gel) intensity at a temperature far below T_m. Remarkably, for DPPC in low-salt solutions ([NaCl]=14 mM) both indices start to decrease at 10 degrees below the pretransition temperature T_p, levelling at T_p, and after reaching T_m increasing steeply for a span of a few degrees, reaching a steady rate of change a few degrees above T_m. By varying probe concentration effects related to self-quenching were excluded. The amount of probe present was shown to have negligible effect on DSC scans, though higher contents of the probe were shown to widen the main phase transition and pretransition peaks and to decrease T_p. The authors ended up to suggest that the P-N dipole of phosphocholine membranes changes its orientation, with the N-end of the dipole buried more deeply into the membrane within the transition region. This in turn would result in changes of electrostatic potential as well as ionic surroundings of the fluorescein moeity, which is located approximately 21 Å from the DOPC bilayer center far above the pK_a of the probe (Kachel et al., 1998). Yet, also probe location may change as a result of membrane defects, or due to changes in the association of ions within the transition region. However, such alterations in the location of the probe necessarily reflect changes in the properties of the membrane. With respect to the ion binding, a similar conclusion based on zeta potential measurements in the transition was reached by Makino et al. (1991), i.e. the N-end of the P-N dipole being more deeply buried during the transition, however, the temperature range of suggested orientation was not quite as wide as in the previous study. Such dispecrepancies might derive from preferential location of probe into the line defects, whose formation is coupled to the formation of $P_{\beta'}$ phase in the model of pretransition by Heimburg (2000). This would also explain the large decrement of T_p at higher contents of the DPPF probe, without significant impact on T_m.

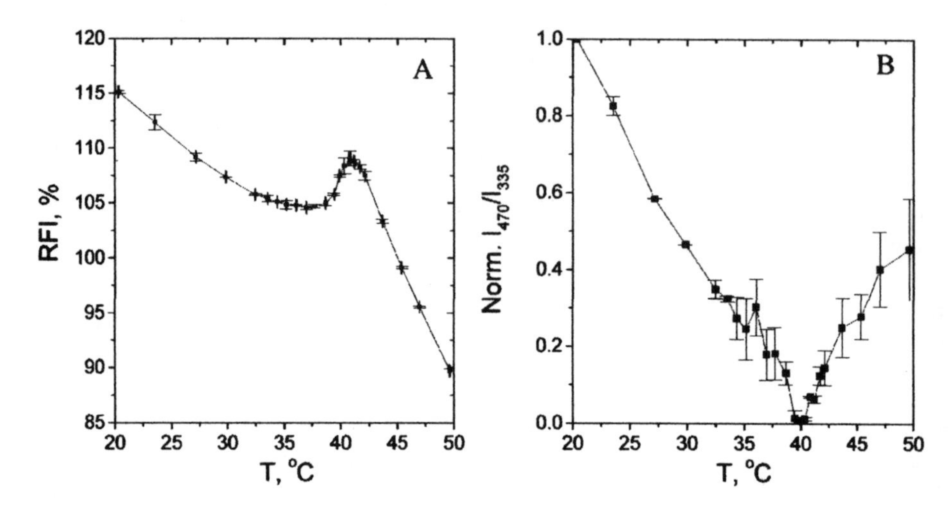

Figure 4. *Panel A*. Temperature dependence of fluorescence intensity (shown as relative to that at 45°C) of DPPN (X=0.03) in DPPC matrix. LUVs, buffer was 5 mM HEPES, pH 7.4. *Panel B*. The ratio I_{470}/I_{335} of emission intensities at excitation wavelengths 470 and 335 nm. As the changes were small and the actual ratio was dependent on LUV batch (values ranging 2.94—3.06 and 2.98—3.12 for two batches), the values have been normalized by setting the minimum and maximum to zero and one, respectively. From Alakoskela and Kinnunen, 2001b, printed with permission.

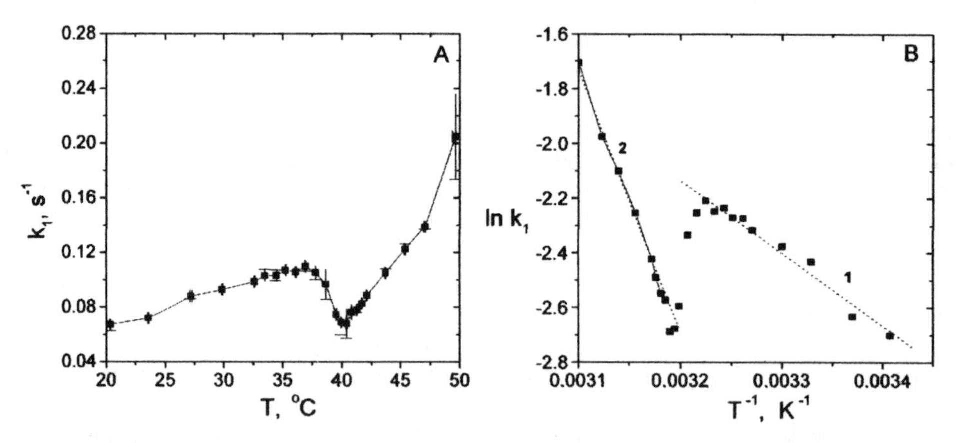

Figure 5. *Panel A*. Rate coefficient k_1 for reduction of DPPN (X=0.03) in DPPC matrix by 5.0 mM sodium dithioniote. *Panel B*. Arrhenius plots of the data in panel A. The temperature ranges below and above the discontinuity at T_m are marked by 1 and 2, respectively. Arrhenius parameters below and above T_m are $A=2.3\times10^6$ s^{-1}, E_a=22 kJ/mol, and T_m $A=1.2\times10^{28}$ s^{-1}, and E_a=80 kJ/mol, respectively. From Alakoskela and Kinnunen, 2001b, printed with permission.

In our studies of DPPC main phase transition with NBD-labeled lipids (X_{probe}=0.03) we arrived at a similar conclusion with respect to the headgroup conformations of the labeled lipid during the phase transition. For *N*-(7-nitrobenz-2-oxa-1,3-diazol-4-yl)-1,2-

dipalmitoyl-*sn*-glycero-3-phosphoethanolamine (DPPN), the derivative with NBD moiety attached to the amine of DPPE, an increase in emission intensity is seen after onset of the phase transition at approx. 37°C (Fig. 4 *A*) (Alakoskela and Kinnunen, 2001b). As DPPN is known to have a slight preference for fluid phase, this effect most likely derives for a large part from decreased self-quenching (Loura et al., 2001), suggesting that due to perturbation imposed to the gel phase by the NBD moiety, the negatively charged DPPN segregates in the bilayer, and thus these negatively charged membrane domains will dissolve into the fluid phase as the melting of the bilayer proceeds. The increased molecular spacing in the fluid phase and better shielding of DPPN charges by water may explain the slight preference of DPPN for the fluid phase, leading into a decrease in T_m (at X_{DPPN}=0.03 T_m≈40.8°C), which is the temperature at which the increase in fluorescence intensity turns into decrease. Compared to the chain-labeled NBD-PC there is very little segregation of DPPN. Yet, even slight lateral enrichment of the negatively charged DPPN below T_m makes it all the more remarkable that during the dissolution of such negatively charged segregates the rate of collisions between NBD moiety and the water soluble, negatively charged dithionite actually decreases drastically, as detected by the rate of reduction of NBD by dithionite. The rate coefficient for the fast reduction of NBD actually *decreases* steeply with increasing temperature between the temperature span of 37.0—40.3°C (Fig. 5). Yet, the error in the rate coefficients is too large and temperature dependency of rate coefficient in the fluid phase too pronounced that it could be reliably distinguished from the Arrhenius plots if there is a more rapid increase right after T_m. Support for the change above T_m is obtained from the intensity ratio with two different excitation wavelengths 470 and 335 nm corresponding to the transitions to ICT and LE states, respectively, as discussed in the section describing the photophysics of NBD. Due to the lack of absorptivity increase upon increasing polarity a larger decrease in fluorescence intensity with increasing polarity is seen with excitation set at 335 nm. Accordingly, the ratio of intensities at these different excitation wavelengths, I_{470}/I_{335} should reach a maximum when the surroundings of the ground state NBD is most polar, and a minimum when the surroundings are least polar. Within the gel phase the values for this ratio decrease continuously with temperature, and a steeper decrease beginning at approx. 37.0—37.5°C is evident, with a minimum at approx. 40°C. This is followed by a significant enhancement with increasing temperature up to approx. 41.0 to 41.5°C, followed by a further increase with temperature (Fig. 4 *B*). This should mean that whatever the cause, the apparent polarity sensed by the NBD moiety of DPPN should reach a minimum at or slightly below T_m. Changes in dipole potential and hydration associated to the orientational changes of headgroups might offer one solution, as such factors have been shown to affect the rate of the fast reduction of NBD by dithionite (Alakoskela and Kinnunen, 2001a). Yet, considering these data, simplest explanation appears to be that in the region of phase transition the reacting amino group of the NBD moiety is located most deeply within the bilayer. These alternatives are by no means exclusive, but rather support each other. Taking into account the data by Langner et al. (2000) and Makino et al. (1991), we may conclude that the fluorescence characteristics of the two probes, viz. DPPF and DPPN, and equilibrium ion association (Makino et al., 1991) as well as kinetics of ion—headgroup interactions support the notion that during the transition the P-N dipole adopts a conformation where the N-end of the dipole becomes buried into the membrane. As NBD and fluorescein are structurally very different, and as the effects of DPPN and DPPF on the matrix lipid thermal behavior are different, it is not likely that both would have a same kind of aberrant behavior implying the tilting of P-N dipole towards the bilayer core, thus suggesting that the detected behavior at least qualitatively reflects the behavior of the matrix lipid.

For the binding of doxorubicin to the liposomes and the reduction of NBD-PC by dithionite both the kinetic studies and steady-state fluorescence measurements revealed a coupling between the lateral organization of the probes and its effect on the surfaces. For NBD-PC the data are easily explained as follows. NBD-PC is strongly excluded from the gel phase of DPPC, and within gel phase matrix this probe segregates into domains. Upon heating the relative fluorescence intensity of NBD-PC thus increases continuously within the gel phase, beginning to increase more steeply at about 35—36°C ($\approx T_p$), reaching a maximum at 41.7°C ($\approx T_m$) and thereafter decreasing continuously (Alakoskela and Kinnunen 2001b). For DPPC LUVs with $X_{NBD\text{-}PC}$=0.03 the DSC scan deviates from the baseline at approx. 32°C, with a small peak at 37.7°C and a major endotherm T_m at 40.8°C, with the heat capacity returning to baseline at approx. 45°C. For the reduction of NBD-PC by dithionite the main phase transition has little effect, suggesting that the NBD moiety of NBD-PC is readily available for reduction on the outer surface in both NBD-PC-enriched domains and NBD-PC dispersed into the bulk lipid phases. For NBD-PC the most fascinating feature is that there appears to be two populations (or very wide one population) below T_m. This is evident from the steady-state anisotropy measurements with the two excitation wavelengths. By selecting a larger part of fluorescence to emit from probes in a less polar environment by employing excitation at 335 nm, one gets also higher and lower anisotropy values below and above the T_m of the parent lipid, respectively (Fig. 6), than with λ_{ex}=470 nm. For DPPN the anisotropies with λ_{ex}=335 nm are slightly lower for the whole temperature range. While the slightly lower anisotropy values at λ_{ex}=470 nm for DPPN and for NBD-PC in the fluid phase matrix most likely originate from the slightly different fundamental anisotropies at the different wavelengths, the large anisotropy difference for NBD-PC below T_m requires the presence of two populations. Yet, it is not obvious what these two populations should be. Temperature derivative of the difference $\Delta r = r_{470} - r_{335}$ reveales that the rates for the change in anisotropy difference faithfully reproduce the two peaks in the DSC scan for DPPC:NBD-PC=97:3 LUVs (Fig. 7), although the temperatures at $d(\Delta r)/dT$ peaks are approx. 0.8 degrees lower and temperatures for returning back to baseline same within experimental resolution. The bulky NBD moiety can be anticipated to be poorly accommodated into the acyl chain lattice of the gel phase, and even in fluid phase it prefers to be in the lipid-water interface. A likely scheme explaining the above fluorescence data is the following:

NBD-PC is nearly entirely enriched into domains. The effective headgroup size for NBD-PC is the size of phosphocholine headgroup plus the size of the NBD moiety, and due to the looping of the NBD-acyl chain back to surface there is effectively only one acyl chain. Accordingly, there is curvature strain in these domains, which favors the partitioning of NBD into hydrophobic part of the bilayer (see scheme in Fig. 8). Within the fluid phase the NBD moiety of NBD-PC resides solely in the interface. However, for the structurally related BODIPY-labeled PC (and to a lesser extent also DPH-PC) a bimodal distribution of the fluorophores in the fluid phase has been suggested on the basis of parallax analysis (Kaiser and London, 1998b). Thus, the free energy differences for the two positions cannot be large, and the curvature-stress-mediated effect on the partitioning appears feasible. This effect should be sensitive to the mole fraction of NBD-

PC. Indirect support for this interpretation comes from NMR studies of fluid phase PC:NBD-PS as well as PS:NBD-PS (75:25) bilayers, where NBD shows considerable partitioning into the hydrophobic region of the bilayer (Huster et al., 2003). For the melting of domains enriched in the fluorescent probe the fluorescence spectroscopic method gives larger signal than DSC. A likely explanation for this is that as the domains enriched in NBD-PC melt, the anisotropy decrement for the NBD buried in the bilayer is larger than for the NBD on the bilayer surface. This difference then vanishes completely as the buried population disappears.

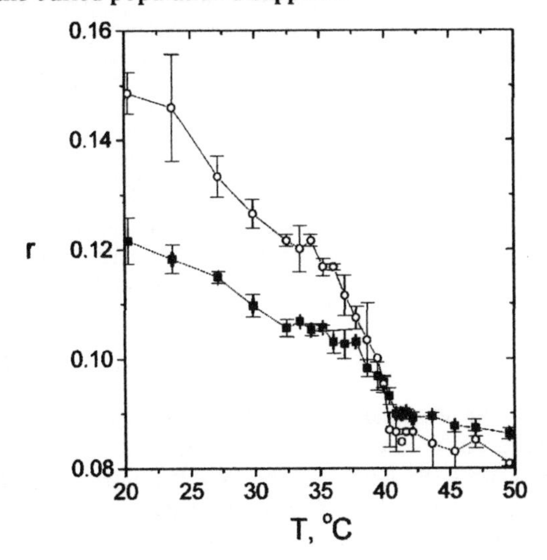

Figure 6. The fluorescence anisotropies vs. temperature acquired using excitation at 470 nm, r_{470} (■), and 335 nm, r_{335} (O). DPPC:NBD-PC=97:3 LUVs in 5 mM HEPES, pH 7.4. From Alakoskela and Kinnunen, 2001b, printed with permission.

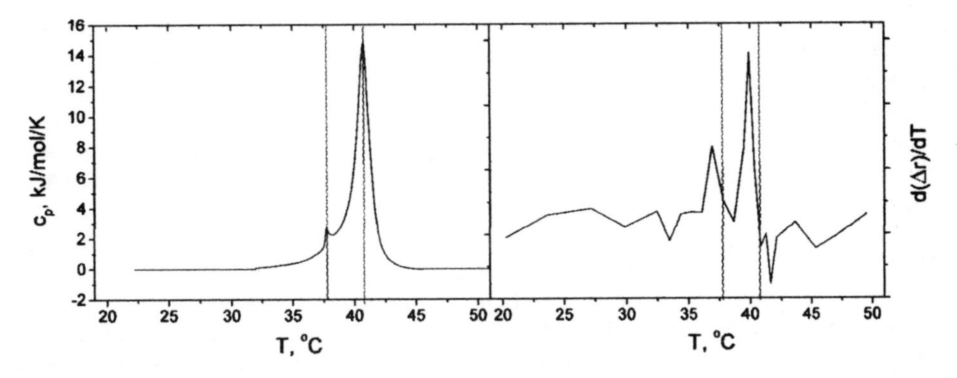

Figure 7. Panel of the left shows the DSC scan of the DPPC:NBD-PC=97:3 LUVs. Panel on the right displays numerical derivative of the difference r_{470}-r_{335} (see Fig. 6).

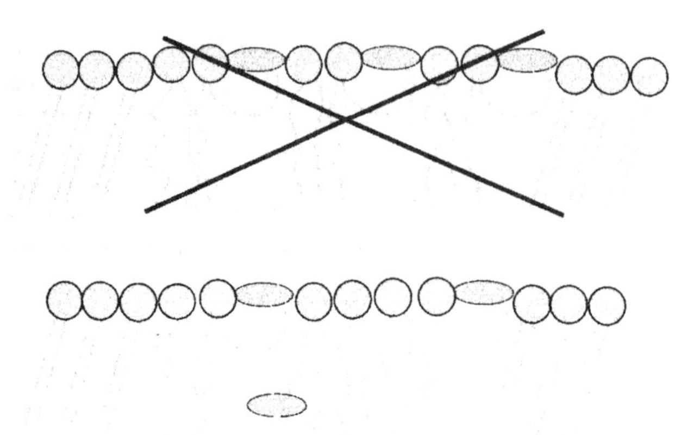

Figure 8. A schematic illustration of the distribution of the fluorophore moieties of NBD-PC in gel phase PC matrix. The NBD moieties have their well-known tendency to partition into the lipid—water interface. In NBD-PC enriched domains, however, such partioning would leave considerable free volume into the acyl chain region, thus imposing considerable curvature stress. As this stress cannot be readily relieved by a change in curvature (because of restriction impose by the opposing leaflet), it is instead compensated some of the NBD-PC partitioning deeper into the hydrocarbon core.

What does the above tell us about the main phase transition itself? The first decrement in the anisotropy difference Δr probably corresponds to a decrease in the configurational order within the NBD-PC-enriched domains. As the interfacial NBD moieties are more free to move to begin with, the effect is not that significant, and a minor shift into a more polar environment is reflected in the ratio I_{470}/I_{335} for NBD-PC. However, the more fluid-like NBD-PC-enriched domains remain trapped within the gel-phase, unable to expand. This is consistent with both the model by Heimburg (2000) as well as T-jump (Kato and Kubo, 1997) and AFM measurements (Kaasgaard et al., 2003), suggesting that melting within gel phase proceeds along life defects in order to accommodate the expansion related strain. As the more fluid-like phase finally becomes continuous, the fluorophore moiety of NBD-PC becomes nearly exclusively accommodated into the interface, evident as an increase in I_{470}/I_{335} and as the disappearance of the anomalous anisotropy difference. While the above result does not reveal novel features about the main phase transition itself, it implies that in multicomponent lipid mixtures related interactions and partitioning effects in the gel-fluid coexistence region might be expected, if the spontaneous curvatures and configurational dynamics for the lipid components are dissimilar.

For the binding of doxorubicin to liposomes there is a different story to be told. When the kinetics of doxorubicin association to membranes was assessed by stopped-flow measurements of pyrene-labeled lipid fluorescence quenching by RET to doxorubicin, it was found that for fluid bilayers with the content of anionic phosphatidylglycerol of $X_{PG} \geq 0.04$ there were two reaction components present, one slow and not responsive to the addition of salt, and the other fast and absent in the presence of [NaCl]>50 mM (Söderlund et al., 1999). Only the slow reaction could be detected for doxorubicin association to bilayers with $X_{PG} < 0.04$. Accordingly, it was concluded that the fast reaction component is mostly of electrostatic origin, the slow component being driven by hydrophobicity (Söderlund et al., 1999). However, when only $X_{PG} = 0.03$ in the form of pyrene-labeled phosphatidylglycerol (PPDPG) was included in the DPPC matrix, the fast component was found to dominate the binding at temperatures far below T_m (Fig. 9). This would be compatible with the segregation of the PPDPG probe into domains at these temperatures despite of the negative charge. More specifically, the bulky pyrene moiety is unlikely to fit into the gel phase acyl chain lattice, this exclusion from the matrix overcoming the coulombic repulsion between the PPDPG headgroups. In principle, this is compatible also with the decrease in I_e/I_m when the gel phase melts. This interpretation is further supported by the lack of the fast component when the zwitterionic PPDPC is substituted for PPDPG. For DPPC:PPDPG=97:3 LUVs at temperatures below 34°C only the fast component was found. The lack of the slow component is likely to derive from the RET method being unable to detect binding distant from donor fluorophores (far beyond Förster radius). It is possible that the onset of chain melting and formation of line defects starts at this temperature (≈34°C), which allows for the dispersion of PPDPG, these defects allowing for increased hydrophobicity-driven doxorubicin binding. This view is supported by the much slower rate for the non-electrostatic binding to the gel phase than to the fluid phase. For the enhancement of the fast, electrostatic binding between 35 to 37°C the deviation from Arrhenius like kinetics is within experimental error.

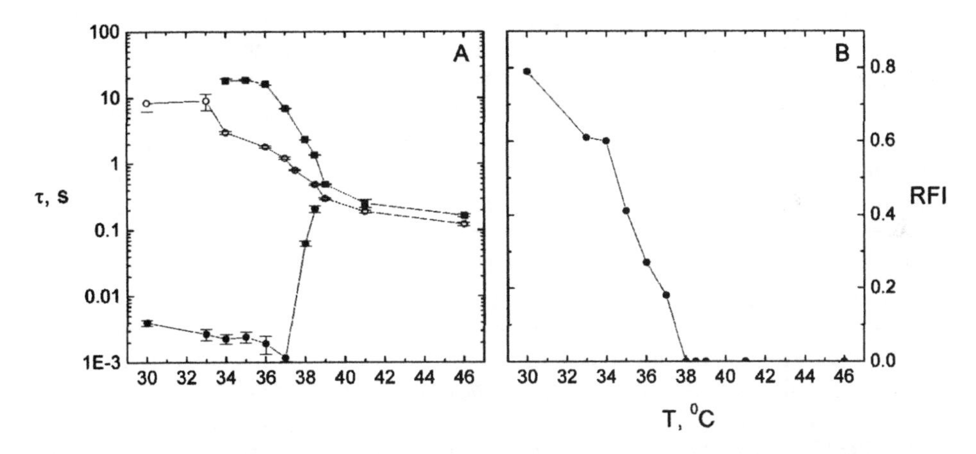

Figure 9. *Panel A* displays the time constants τ for the association of doxorubicin to DPPC:PPDPC=97:3 (▲) LUVs and for the fast electrostatic (●) and slow hydrophobic (■) reaction component for association to DPPC:PPDPG=97:3 LUVs. *Panel B* shows the relative amplitude of the faster component. From Söderlund et al., 1999, printed with permission.

However, also more curious findings about the effects of phase transition on the binding of doxorubicin to DPPC:PPDPG=97:3 LUVs were evident. The total absence of the fast component was observed at > 38.5°C, while the amplitude of this component approached values close to zero already at 38°C. T_m for the vesicles was 40.4°C. This would be compatible with the dilution of PPDPG below the mole fraction needed to promote the fast, mainly electrostatic binding. An anomalous finding is that the fast reaction slows down with increasing temperature in the range 37 to 38.5°C, the reaction time constant increasing by more than two orders of magnitude, before its disappearance at 38.5°C. Such effects are not seen if the mole fraction of the anionic lipid is varied within the fluid phase. This behavior was rationalized in terms of a change in the effective dimensionality of probe distibution, leading to pseudo-1-dimensional (p1D) arrangement of PPDPG molecules at the domains interfaces. At these p1D structures the excimer formation should still be efficient, as observed (Söderlund et al., 1999). The fitting of rate curves to pseudo-first-order kinetics might only be apparent, with the membrane binding doxorubicin itself changing significantly the local charge density in the p1D structures, and this could explain the apparent decrease in rate. Such arrangement would imply that pyrene-labelled lipids in fact favor the gel/fluid interface similarly to cholesterol, decreasing line tension and favoring formation of elongated domain structures (Soderlund et al., 1999). The slowing down of the fast reaction coincides with the decrease of I_e/I_m, and this decrement continues to approx. 41°C. Our more recent data (see below) points out that this decrease in I_e/I_m is not solely due to a decrease in the rate of excimer formation but also involves a selective loss of excimer quanta. It should be noted that experimental support for the tendency of PPDPC to partition into line defects has been obtained for DMPC (Jutila and Kinnunen, 1997), and this seems to apply also for DPPC (Metso et al., 2003). Yet, PPDPC seems to adopt p1D arrangement following pretransition, which is likely to disappear or change, when I_e/I_m begins to diminish. As I_e/I_m vs. T data are very similar for both PPDPC and PPDPG, we may assume this to apply for PPDPG as well. Close to the transition PPDPG is likely to become dispersed into the boundaries of fluctuating phases. Both the pyrene moiety as well as the negative charge of PPDPG should favor the dispersion of the probe in a fluid matrix. This dispersion process as seen from the disappearance of fast electrostatic binding appears to begin at 37°C coupled to the increase in heat capacity in DSC scans. Even by 38.5°C where the process is already completed only a small part of transition enthalpy has been consumed. As there cannot be large fluid areas present at this temperature, the decrease in electrostatic association rate could derive from breaking of the gel phase lattice without significant alterations in acyl chain order, allowing for pyrene probes to be dispersed from the p1D arrays of $P_{\beta'}$ phase into the bulk phase. At these as well as for higher anionic lipid concentrations the fast association of doxorubicin could be governed by binding to fluctuating patches, where the fraction of anionic probe is higher. In other words, when a looser lattice allows for fluctuations as well as for the dispersion of PPDPG from line defects, occasionally forming patches with sufficiently high PPDPG concentration for electrostatic attraction would appear. The decrease in reaction rate could derive from two different factors or their combination. First, the probability for the formation of patches with sufficiently high PPDPG concentration should decrease as melting proceeds. Second, in the fluid areas superlattice ordering of PPDPG could take place, and the favorable energy related to the formation of superlattice should contribute to activation energy as incoming doxorubicin molecule would necessarily perturb the superlattice upon its association to lipid bilayer. Such arrangements could explain the decrease in the fast reaction rate, while the disappearance of the fast component at temperature below the phase transition could be simply explained by dispersion of PPDPG from the line defects into the bulk bilayer. Intriguingly, with more detailed examination of phospholipid phase transition with DPH and pyrene-labeled lipids, further anomalies in the temperature-dependent behavior of the fluorescent probes emerge, as discussed in the following chapters.

Considering the data in terms of p1D arrangement already within the $P_{\beta'}$ phase, it is possible that the line defects with arrays of closely packed anionic PPDPG molecules might provide a good binding site for the cationic doxorubicin. This would account for the appearance of the fast membrane association below transition. At 38°C where the amplitude of the fast reaction reaches values close to zero and reaction rate slows down, this p1D arrangement could begin to dissolve into the bulk phases, driven by phase fluctuations. The slowing down of the remaining reaction rate for residual p1D could derive from both decreased preference of PPDPG to these line defects and decreased rate of diffusion allowing for slower formation of locally enriched PPDPG molecules. In addition, when the amount of these structures should be small, then the doxorubicin should saturate the existing PPDPG patches quickly, requiring the slow diffusion of PPDPG to form new closely spaced pairs or arrays.

2. DPH and pyrene-lipids: anomalies in the transition region

In the study on the main phase transition of DMPC vesicles with PPDPC alone and in the presence of various RET acceptors processes affecting excimer formation as well as PPDPC (donor) and acceptor colocalization are found to emerge already at temperatures far below T_m for the LUVs in question. At approximately 5 degrees below T_m where the excess heat capacity recorded by DSC deviates from the baseline for the DMPC:PPDPC=99:1 mixture a significant enhancement of excimer formation is detected in steady-state fluorescence measurements (Jutila and Kinnunen, 1997). Notably, this process was not enhanced by increasing X_{PPDPC}, but was rather attenuated. The values for I_e/I_m reach a maximum two degrees below T_m, decreasing to a minimum at T_m. Subsequently, enhancement of lateral diffusion imposes a temperature-dependent increase in I_e/I_m, though also minor post-T_m processes seem to be present during the later phase of melting (Fig. 10).

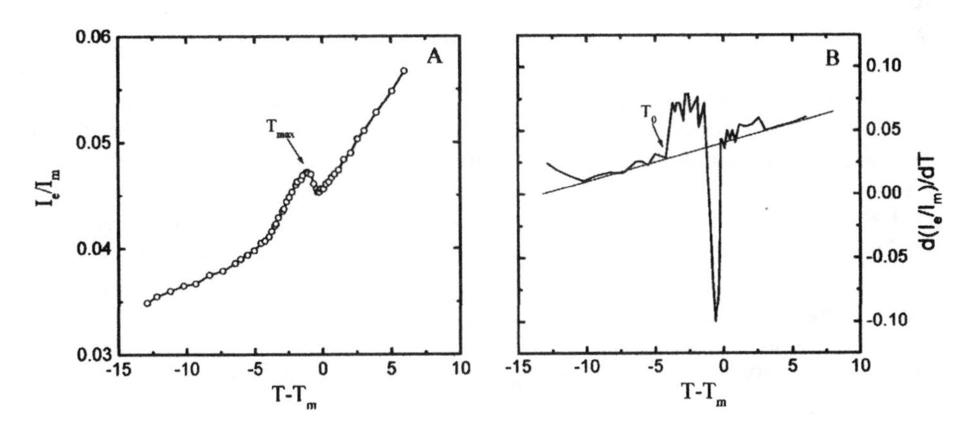

Figure 10. The ratio of excimer to monomer emission intensities I_e/I_m for PPDPC (X=0.01) in DMPC matrix (*panel A*) and the numerical temperature derivative of I_e/I_m (*panel B*). Adapted from Jutila and Kinnunen, 1997, with permission.

For colocalization studies the RET acceptors NBD-cholesterol, DPPF, and NBD-PC were chosen. The rationale for the selection of the above probes was based on hydrophobic mismatch arguments and DSC analysis. In brief, DPPF in DMPC membranes prefers gel-like phase while NBD-PC has been shown to prefer the fluid phase. Because of the preference of cholesterol for the gel/fluid interface also its NBD-labeled analog (NBD-chol) was assumed to behave in a similar manner. Yet, it is to be emphasized that at higher contents of cholesterol NBD-cholesterol is excluded from the liquid-ordered phase (Loura et al., 2001). Supporting the localization of PPDPC and NBD-chol into the interface, maximal quenching of the pyrene fluorescence was seen at approximately two degrees below T_m for DMPC:PPDPC:NBD-chol LUVs (98:1:1), while inclusion of unlabelled cholesterol at X_{chol}=0.01 had no significant effect on the behavior of system. Curiously, minima in the colocalization of PPDPC with DPPF and NBD-PC were detected at T_m. From the formation of superlattices in fluid phase we expect a repulsive potential between PPDPC in the fluid phase and the bulky pyrene moiety is likely to fit even more poorly into the gel phase lattice. Accordingly, the interface where the fluid phase chain movements are disturbed by the vicinity of hard gel phase the boundary can be thought of as preferred site for pyrene-labelled lipid. Another unfavorable interaction within the fluid phase would be the hydrophobic mismatch of PPDPC with the constituent lipids, as in matrices where the number of acyl chain carbons $N \neq 20$ PPDPC shows fluid-fluid immiscibility at sufficiently high mole fractions (Lehtonen et al., 1996).

In our more detailed time-resolved studies on the characteristics of excimer formation and fluorescence in the vicinity of the phase transition of DPPC similar behavior was found (Metso et al., 2003). More specifically, for PPDPC four different temperature ranges of interest could be detected. For the first region the excimer fluorescence rise time τ_R (corresponding to the rate of excimer formation), as well as the excimer fluorescence decay time decreased slightly with increasing temperature (Fig. 11). This is likely to reflect the behavior of the probe in the bulk $L_{\beta'}$ phase, with pronounced increase in I_e/I_m. For the second phase, beginning at approx. $(T-T_m)$=-10 degrees, there is a continuous increase in excimer risetime (Fig. 11) together with an increase in I_e/I_m up to a maximum a few degrees below T_m (Fig. 12), similarly to the maximum observed for PPDPC in DMPC vesicles (Jutila and Kinnunen, 1997). As the increase in excimer risetime coincides with the pretransition of unilamellar vesicles (Heimburg, 1998; 2000), it seems likely that the suggested p1D organization of pyrene-labeled lipids (Söderlund et al., 1999; Metso et al., 2003) would explain also the results of Jutila and Kinnunen (1997), corresponding to a trapping of the probe into line defects at the pretransition and in the $P_{\beta'}$ phase, coupled to a continuous increase in *trans*→*gauche* isomerization between pretransition and main transition. The above would match perfectly the model of Heimburg (2000), AFM observations by Kaasgaard et al. (2003), and T-jump measurements with detection by transmitted light and by freeze-fracture EM (Kato and Kubo, 1997).

After this growth of ripples another process emerges, beginning at roughly $(T-T_m)$=-4, and leading to a considerable decrease in I_e/I_m (Jutila and Kinnunen, 1997; Metso et al., 2003) due to a decrease in excimer intensity (Metso et al., 2003). This process is centered at approximate $(T-T_m)$=-2 and is completed at $(T-T_m)$=1 in DPPC (Metso et al., 2003) and at T_m in DMPC membranes (Jutila and Kinnunen, 1997). As observed previously, the acyl chain order and DPH anisotropy correlate well, and accordingly, also Metso et al.

found that for DPH-labeled phospholipid DPH-PC the anisotropy values reflect accurately the heat capacities, the most significant decrement in anisotropy occurring at approx. T_m and the faster decrease continuing until $(T-T_m)=3$, corresponding to the temperature where the heat capacities recorded by DSC approach the baseline (Metso et al., 2003). In this respect, it must be emphasized, that the lifetime of pyrene is significantly longer (almost by an order of magnitude) than that of DPH, which could cause differences in probe responses. It is obvious, however, that this cannot be the only reason for the observed behavior, when the width of the lifetime distribution of DPH in DPPC collapses from approx. 0.6 ns to approx. 0.05 ns within the range from 4 to 2 degrees below T_m, respectively, remaining constant after that (Gratton and Parasassi, 1995). This was interpretated as water penetration into bilayers, changing the permittivity distribution around the DPH, and having fast dynamics compared to the fluorescence lifetime of this probe.

In the light of above data, it is of interest that for phosphatidylcholines with 17 to 20 carbon acyl chains an additional heat capacity peak between T_p and T_m was found by DSC (Jørgensen, 1995). For distearoyl-PC (DSPC) the temperature of this sub-main transition, T_{sm}, is at $(T-T_m)=-1$. For disaturated PCs with longer acyl chains T_{sm} approaches T_m, merging with the T_m peak at acyl chain lengths longer than 20 carbons. The enthalpy of the sub-main transition becomes vanishingly small for lipids with acyl chains shorter than 17 carbons. Interestingly, the enthalpy for this peak is strongly enhanced by the presence of KCl, and in combined small- and wide-angle X-ray studies it was found that the sub-main transition corresponds to a transition between two ripple phases (Pressl et al., 1997). The sub-main transition in DSPC MLVs is accompanied by partial melting of acyl chains, changes in the acyl chain lattice, and possibly by changes in bilayer corrugations or modification in the size and shape of domains (Pressl et al., 1997). The sub-main transition like heat capacity peaks can be simulated, if bilayers are allowed to have regions with different equilibrium lateral pressures. These effects have suggested to correspond to curvature stress and hydration interactions in the coexistence region (Trandum et al., 1999). Yet, the molecular level details of the sub-main transition remain elusive. One proposal for the origin of sub-main transition is decoupling of lattice order and acyl chain order (Nielsen et al., 1996) in a manner similar to that formation of liquid-ordered phase by inclusion of sufficiently high mole fractions of cholesterol (Ipsen et al., 1987). In this scheme the sub-main transition was suggested to correspond to a loss of lateral lattice order already at $(T-T_m)=-1$, followed by the more complete loss of acyl chain configurational order at T_m.

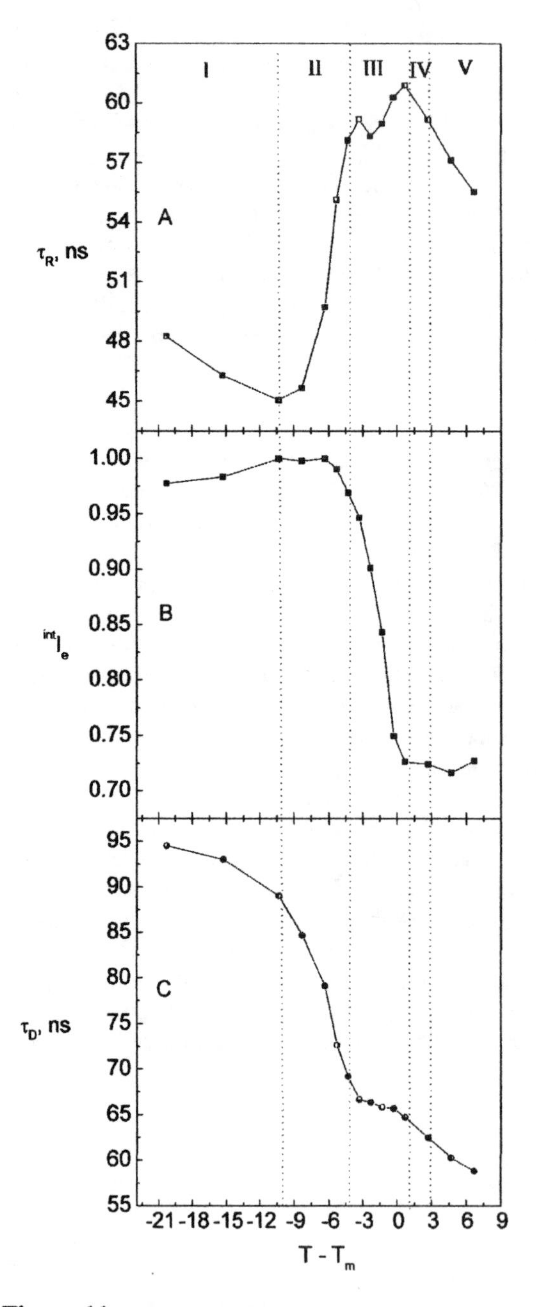

Figure 11. Time-resolved fluorescence data for pyrene excimer formation in DPPC:PPDPC=99:1 LUVs. *Panel A* shows the excimer risetime reflecting rate of excimer formation, *panel C* shows the decay time for excimer. *Panel B* shows the integrated excimer intensity curve that is essentially identical to excimer intensity curve measured by steady-state fluorescence measurements. Adapted from Metso et al., 2003, with permission.

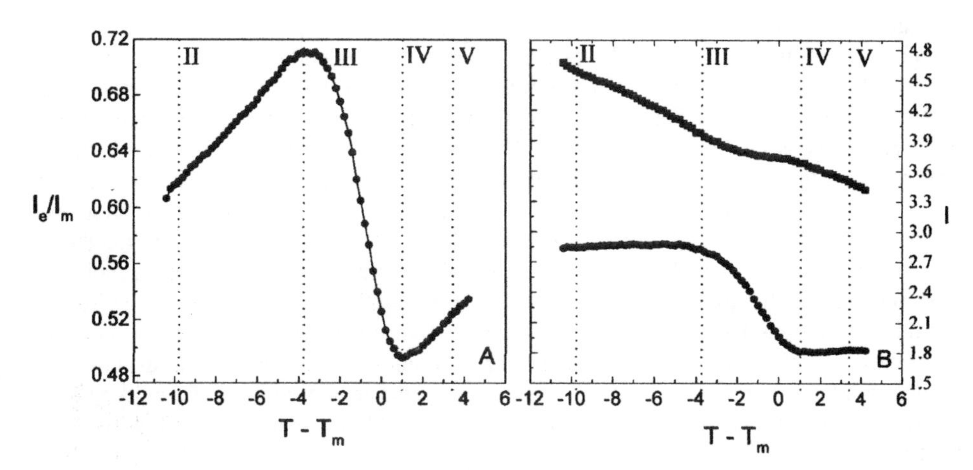

Figure 12. In *panel A* the steady-state excimer to monomer ratio for PPDPC (X=0.01) in DPPC matrix is shown. *Panel B* displays the changes in monomer (■, upper trace) and excimer (●) intensity. From Metso et al., 2003, with permission.

It is not inconcievable that processes similar to the DSPC sub-main transition, for instance, could be present in phospholipids with shorter acyl chains, such as DPPC and DMPC. Although not necessarily evident as narrow transitions For the latter it would have to be a low enthalpy process, as expected from the gradual (nonlinear) attenuation of the sub-main transition enthalpy with decreasing acyl chain length (Jørgensen, 1995). The difference (T_m-T_{sm}) should thus exceed one degree. From the available four data points measured by Jørgensen (1995), one would expect with exponential or linear (in parenthesis) extrapolation T_{sm} roughly 1.8 (1.6) degrees below T_m for DPPC and 3.5 (2.1) degrees below T_m for DMPC. However, it must be emphasized that these extrapolations should not be given much weight as they are based on very limited data.

Enhancement of fluctuations in the vicinity of the line defects or grain boundaries together with the remaining DPPC gel phase lattice becoming more loose with diminishing acyl chain order may allow pyrene-lipids to become dispersed from the line defects of $P_{\beta'}$ phase into the L_α phase. In the $P_{\beta'}$ phase with both gel- and fluid-like lipids, the gel-like lipids are likely to have less lateral order and also to have diminished acyl chain order compared to the gel-phase lipids further below T_m. In fact, for fluctuations to bring about small L_α phase domains into the $P_{\beta'}$ bulk and not only as fluid-like lipids packed at certain positions at corrugations, the surrounding phase indeed should be able to expand, as the average area for lipid in the L_α phase is larger by approx. 25 %. The heterophase fluctuation model (Kharakoz et al., 1993; Kharakoz and Shlyapnikova, 2000) concentrates on the formation of gel phase upon cooling the fluid phase assembly and shows that the apparently critical-like processes could be explained in terms of heterophase fluctuations without the proximity of a close-lying critical temperature. These authors estimated the dense, bulk gel phase like domains in these fluctuations to consist on average of nine molecules (Kharakoz and Shlyapnikova, 2000). Further, the time scale for these fluctuations was suggested to be of the order of 10—60 ns, which is comparable to the timescale of phase fluctuations assessed by ultrasound measurements (Kharakoz et al., 1993) as well as spectroscopically by the solvation-water sensitive fluorophore Laurdan (Parasassi et al, 1990). In the heterophase fluctuation model a gel/fluid line tension of $0.7k_BT$ per boundary lipid was suggested. This value is large enough to allow the transition to be of first order, yet sufficiently small for the transition to be weakly first-order. However, a question about the sharpness of the boundary of such small domains formed by heterophase fluctuations arises when taking into account that for larger domains the gel/fluid boundary is wide instead of a sharp zone. NSOM studies have revealed that in phospholipid monolayers the domain boundary can have width up to 100—200 nm at low surface pressures (Hwang et al., 1995). This is an immense distance considering that it corresponds to 125—250 phospholipids. Accordingly, if outside the nucleus of 9 molecules an outer boundary extends not further than a radius of three molecules, the ratio of boundary/core lipids would be roughly 8. Likewise, an outside diffuse boundary extending at a radius of five lipids from the core the apparent boundary/core lipid ratio would be roughly 17. The above mechanism could explain why the whole bilayer appears to become boundary- or intermediate-like at T_m as suggested by Metso et al. (2003). With diffuse boundary one would also expect rather low gel/fluid line tensions.

The decrease in excimer intensity in the interval (T_m-4) to (T_m+1) is accompanied by only a modest decrease in the rate of excimer formation (Fig. 11 and 12), and is not accompanied by a significant increase in the rate of excimer decay or by an increase in monomer intensity (in fact showing small decrease also in monomer intensity), thus revealing decreased quantum yield. The previously suggested trapping of PPDPC as a substitutional impurities into the lattice at nodes corresponding to frequencies of lattice phonons (Jutila and Kinnunen, 1997; Metso et al., 2003) would be compatible with the slight decrement in the rate of excimer formation. The origin of the selective decrease in excimer quantum yield is elusive, and Occam's razor appears a double-edged sword when considering the process. The decrease in quantum yield would be compatible with superlattice formation for pyrene-lipids in the new phase that allows for long-range quantum mechanical coupling of pyrenes, leading to resonance between several pyrenes and decay at longer wavelengths (Kinnunen et al., 1987; Metso et al., 2003). Yet, the mechanistic basis for the diminishing quantum yield is more complicated. In order to elucidate the gedanken experiments which lead us to the model we are about to present, we will list a few requirements. Any process which time-dependently quenches excimers, should be accompanied by a decrease in excimer lifetime. This was not observed. Almost intantaneous quenching for some population would require, first, the presence of two non-interconvertible populations, as otherwise intantaneous quenching would lead to zero intensity, and second, the instantaneous quenching to restrict itself exclusively on excimers. While some quenchers do indeed have larger quenching rates for excimers, no exclusive excimer quenching has ever been reported. Considerable dilution of PPDPC in turn should mainly impose its effect via the increased excimer formation time constant, whereas only a minor increment was observed, not necessarily large enough to produce the decrease in excimer intensity. While several models could be constructed by combining various counteracting effects, we present the following model incorporating all the different detected features to a framework of single distribution change. Yet, even this model requires several assumptions. Essentially, the key event of the model is the formation of regions having PPDPC and matrix PC lipids in a superlattice arrangement. Consider a membrane with an area A_0 and PPDPC mole fraction x_0. The formation of a membrane patch with mole fraction $x_0+\Delta x$ of PPDPC and area A_1 would leave (assuming constant area) remaining area $A=A_0-A_1$ and mole fraction $x=(A_0x_0-A_1x_0-A_1\Delta x)/(A_0-$

A_1)=x_0–$A_1\Delta x/(A_0-A_1)$. Accordingly, if we have superlattice for area A_1 which is small compared to A_0, we would expect $x \approx x_0$, even when Δx is large compared to x_0=0.01. As $x \approx x_0$, the properties in pyrene fluorescence within area A should remain almost constant. Yet, a slight increase in τ_R would be expected with corresponding increase in monomer lifetime. For the population within the superlattice, lateral diffusion of pyrene would be required to be slow compared to the lifetime. The origin of this slow diffusion should reflect the fact that fast diffusion of PPDPC would destroy the superlattice, the superlattice stabilizing free energy acting as a free energy barrier for excimer formation. In the absence of other processes this would decrease excimer formation within A_1 and lead to enhancement of monomer intensity. However, according to previous observations (Kinnunen et al., 1987) quantum mechanical coupling in regular lattices resonance structure should lead to loss of excitation quanta to wavelenghts longer than those of excimer. This would explain that also monomer intensity is slightly diminished.

As the superlattice formation as such forms at specific mole fractions, x_1 in the above example, these pyrene-derivative responses should be more sensitive to the average mole fraction x_0 than the other features of the matrix lipid transition. This allows one testpoint. Another would be to determine the probe diffusion—obviously by means other than by fluorescence spectroscopy—within the fluid phase as a function of X_{PPDPC}, expecting the superlattice mole fractions to deviate from smooth behavior. As the quantum yield sustains approximately steady level after complete melting, it is evident that these quantum mechanical coupling processes should initially be present also in the fluid phase PPDPC superlattice and to dissolve only gradually. Such an arrangement would require coexistence of PPDPC in superlattices and PPDPC outside superlattices in the fluid phase. Such coexistence is implied by the time-averaged nature of PPDPC superlattices (Somerharju et al., 1985). Remarkably, were this model true, it would suggest that when the superlattice structure forms in the fluid phase its lifetime far exceeds the lifetime of pyrene.

An intriguing question concerning the transition is the lack of symmetry. Although the T_m-4 for the onset of the anomalous behavior is symmetric to T_m+4 where the ultrasound relaxations start to deviate strongly from the baseline, the behavior for PPDPC is by no means symmetric. The causes for this could be an overlap of the outer boundary regions of fluctuations, with a decrease in the effective boundary/core ratio, and ultimately the asymmetric behavior of PPDPC itself with respect to gel and fluid phase, allowing for PPDPC to better evade from fluctuations when the fluid phase becomes the continuous phase and gel phase nuclei appear as heterophase fluctuations within the fluid phase. To this end, the longer lifetime and stronger exclusion from gel phase nuclei for pyrene would allow for such evasion, whereas the lifetime of DPH is too short compared to the timescales of fluctuations. DPH and DPH-PC would thus be sensitive to the fluctuations, thus explaining why the anisotropy of DPH changes until the fluctuations become extinct, and why DPH anisotropy more faithfully reproduces the endotherms recorded by DSC. The possible weak post-T_m processes favorable to excimer formation could originate from increased diffusion as the diffusion obstacles, i.e. gel-like domains, gradually disappear. Additionally, it seems likely that T* for the critical unilamellar state, suggested to correspond to the disappearance of last gel phase nuclei (Gershfeld, 1989), is essential also for the segregation of PPDPC induced by osmotic stress (Lehtonen and Kinnunen, 1995). Increasing dehydration by the inclusion of polyethyleneglycol can induce increased segregation of PPDPC, as detected by I_e/I_m, up to a temperature at which the I_e/I_m curves corresponding to different osmotic pressures converge back to the zero osmotic pressure curve. For DPPC this temperature T_0 corresponds exactly with T*. (44°C) and for DMPC they are close (T_0=32°C, T*=29°C). For POPC T_0=10°C (T_m=4°C) while there is no reference for value of T* for this lipid. For DOPC with two unsaturated chains T*=7°C, 27—28 degrees above T_m (Gershfeld, 1989). It seems possible that pyrenes as substitutional impurities could be a preferred site of nucleation.

3. Towards a new model of phospholipid main phase transition

The mechanistic findings presented above can be compiled into a novel model for the actual course of the thermally driven phase transitions of saturated phosphatidylcholines, as follows. In the initial low-energy phase the inherent coupling between molecular area and acyl chain order requires the first melting lipids to be packed into line defects, following the formation of $P_{\beta'}$ gel phase of the original $L_{\beta'}$ gel phase, as suggested previously (see Heimburg, 2000). With increasing temperature there is a continuous decrease in the acyl chain order, leading to modification of the relative number of ordered and disordered lipids in the corrugations of the $P_{\beta'}$ phase, and explaining the elevated basal heat capacities between pretransition and main transition (Heimburg, 2000). In this phase pyrene-labeled probes partition into the line defects of corrugations, accounting for the decreased rate of excimer formation.

Subsequently, the excitations in gel-like lipids of the straight surfaces of the corrugations (with saw-tooth like cross-sections) decrease the order within these regions, and lead to a looser lattice (loss of translational order) without a significant change in the chain ordering, in effect following the lines of thought by Nielsen et al. (1996). Alternatively, the looser lattice could form because of a change in acyl chain tilt angle, or simply due to a cross-point of gradual processes leading to a loosening of gel phase lattice caused by defects and increased formation of fluid phase domains. Whatever the cause, this looser lattice would allow for an expansion and formation of fluid-like nuclei within the gel phase. Accordingly, the breakdown of strict lattice order would account for the observed collapse of DPH lifetime distribution, resulting from the penetration of water into the bilayer (Gratton and Parasassi, 1995) at $T-T_m$ of -4 to –2 degrees, as well as for the onset of decrease in I_e/I_m for PPDPC (Jutila and Kinnunen, 1997; Söderlund et al., 1999; Metso et al., 2003).

As fluctuations become more intense an increasing number of nuclei of the new phase emerge surrounded with diffuse, large outer boundary, the fraction of the apparent "boundary-like" or intermediate lipids reaching high values already before transition, and beginning to decrease when the boundary areas begin to overlap. In addition to the previous process this might be the origin of apparearance for changes in water ordering preceding T_m by 0.5—1.5 degrees (Enders and Nimtz, 1984; Mellier and Diaf, 1988; Mellier et al., 1993), as well as for the changes in lipid headgroup orientation at the transition (Makino et al., 1991; Langner et al., 2000; Alakoskela and Kinnunen, 2001b). Specifically, the cause for the bending of the P-N dipoles towards the hydrophobic part could originate from combined effects of interfacial defects, allowing for choline to penetrate deeper into the membrane, and from electrostatic energy minimization at the gel phase boundary where the gel phase P-N dipoles point to the opposite direction, i.e. towards water. The rate of temperature change and local membrane curvature should affect the nucleation process and could explain some of the temperature-related discrepancies between different studies. It is also possible, that domains formed due to fluctuations occasionally coalesce, and remain trapped due to kinetic barriers, forming larger domains while the active process still retains the nature of heterophase fluctuations. The preference of PPDPC for the domain boundary together with the lifetime of excimer being equal or longer than the timescale of heterophase fluctuations could also contribute to PPDPC fluorescence behavior.

As the fluid phase (possibly with some almost fluid boundary lipids) becomes continuous the response of pyrene-lipids for the presence of gel phase nuclei gets weak, as the lifetime of pyrene far exceeds the timescales of fluctuations and as the pyrene-lipid is strongly excluded from the gel phase, the pyrene-labeled lipid can thus evade the remaining gel phase nuclei. Instead, DPH with lifetime shorter than timescale of fluctuations may still become entrapped into the nuclei or their immediate boundary and thus report the presence of diminishing number of nuclei by decrease in anisotropy (see Fig. 13 for a very schematic illustration of the changes).

It seems possible that the remaining gel phase areas and finally gel phase nuclei in the bilayers have a regular distribution because of dipole—dipole repulsion. Such regular distribution of the condenced domains is seen in PC monolayers, where the distances are vast compared to molecular or even vesicle scale, i.e. on the scale of tens of micrometers (Helm and Möhwald, 1988). The larger effective range for dipole—dipole repulsion in monolayers would be expected due to the low dielectricity of air. Such superlattices should affect the fluctuation profile, as the energetics for formation of a nucleus close to a pre-existing nucleus should be energetically unfavorable compared to the situation where the nuclei maximally separated, which would be the case for a hexagonal lattice.

The above model is tentative only. Yet, it is compatible with the fluorescence data obtained for different probes, and closely courts two models successful in explaining several features of the pretransition and pretransition—main transition coupling (Heimburg, 2000), and the apparently critical behavior in the vicinity of weakly first-order main transition (Kharakoz et al., 1993; Kharakoz and Shlyapnikova, 2000). Should our model indeed resemble the real course of events, this would demonstrate that while the use of fluorescence spectroscopy requires great care in the interpretation of the data, and is complicated by the fact that the fluorescent phospholipid analogs have to be considered as substitutional impurities in the bilayer matrix and have different timescales, by thorough examination and consideration given to probe photophysics and impurity—matrix interactions the correspondence of fluorescence signals to processes involved in phase transitions can be deciphered, in spite of the fact that no exact structural information can be obtained. In some cases the apparent lack of evidence by other methods could simply be explained by the greater sensitivity of fluorescence spectroscopy.

The well-known effect of cholesterol is to decrease gel—fluid line tension, which leads to decrease in cooperativity of transition and therefore increased width of transition zone, with cholesterol-containing membranes displaying larger compressibility below and above main transition (Halstenberg et al., 1998). Due to the decrease in gel-fluid line tension one would expect smaller fraction of boundary lipids per domain for cholesterol-containing membranes. Accordingly, as future prospects it would be interesting to test the effect of cholesterol on the fluorescence of pyrene-labeled lipids in the vicinity of main transition. Another interesting aspect is to repeat similar fluorescence measurements with multilamellar vesicles that have higher cooperativity and therefore display narrower transition range. Such measurements would help to resolve whether the suggested loosening of lattice is a separate entity or simply results when some critical level of acyl chain melting has been reached.

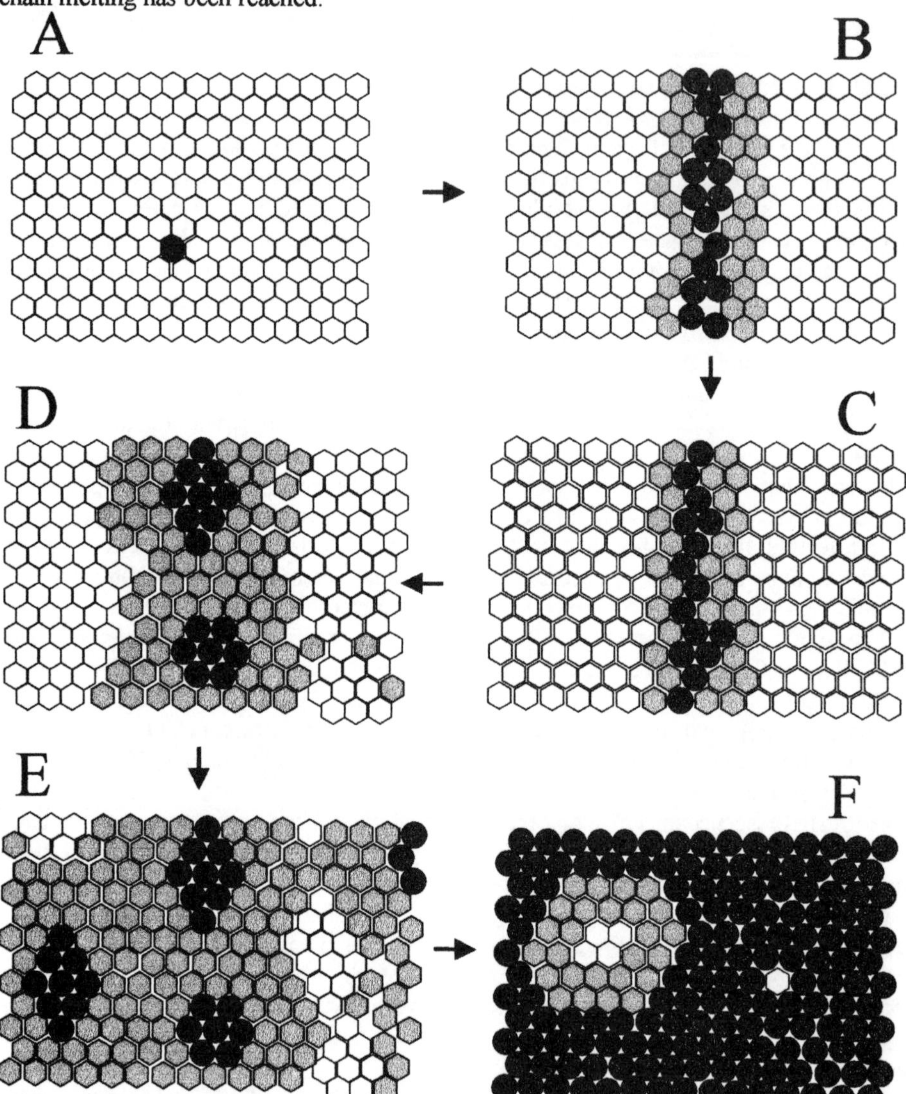

Figure 13. A scheme of changes in the vicinity of main phase transition. Panel A shows $L_{\beta'}$ phase with single excited lipid with disordered acyl chains. The larger area required by excited lipid imposes strain on the $L_{\beta'}$ phase lattice. As shown in panel B, this strain is relieved by packing the excited lipids into line defects, which leads to the formation $P_{\beta'}$ phase (Heimburg, 2000; see also Fig. 1). Along the line defects there should exist lipids with an intermediate order or boundary-lipid like nature. With further increase in temperature lattice becomes more loose, yet without a sudden increase in area (such a sudden increase not taking place in reality is shown in panel C). Rather this allows for the fluid-like domains to become dispersed into the bulk from the line defects, and new fluid phase domains to form (and disappear) due to phase fluctuations (panel D). The domains are likely to be surrounded by layers of boundary lipids, giving the whole bilayer a boundary-like nature due to the large fraction of boundary lipids (panel E). Gradually the fraction of fluid phase increases, leaving only gel-phase domain heterofluctuations and single all-*trans* phospholipids in a fluid matrix (panel F). With further increase in temperature the fluctuations will cease and the fraction of all-*trans* lipids approaches zero.

ACKNOWLEDGEMENTS

We thank Antti Metso for several discussions. JMIA thanks the MD/PhD programme of Helsinki Biomedical Graduate School and Orion Research Foundation for support. HBBG is funded by the Finnish Academy and MEMPHYS by the Danish National Research Council.

REFERENCES

Abrams, F.S., and London, E., 1993, Extension of the parallax analysis of membrane penetration depth to the polar region of model membranes: use of fluorescence quenching by a spin-label attached to the phospholipid polar headgroup, *Biochemistry* 32: 10826—10831.

Andersson, M., Hammarstroem, L., and Edwards, K., 1995, Effect of bilayer phase transitions on vesicle structure, and its influence on the kinetics of viologen reduction. *J. Phys. Chem.* 99: 14531—14538.

Andrich, M. P., and Vanderkooi, J. M., 1976, Temperature dependence of 1,6-diphenylhexatriene fluorescence in phospholipid artificial membranes, *Biochemistry* 15: 1257—1261.

Bachilo, S. M., Spangler C. W., and Gillbro, T., 1998, Excited state energies and internal conversion in diphenylpolyenes: from diphenylbutadiene to diphenyltetradecaheptaene, *Chem. Phys. Lett.* 283: 235—242.

Boggs, J. M., 1987, Lipid intermolecular hydrogen bonding: influence on structural organization and membrane function, *Biochim. Biophys. Acta* 906: 353—404.

Brown, M. F., 1994, Modulation of rhodopsin function by properties of the membrane bilayer, *Chem. Phys. Lipids* 73: 159—180.

Brown, R. S., Brennan, J. D., and Krull, U. J., 1994, Self-quenching of nitrobenzoxadiazole labeled phospholipids in lipid membranes, *J. Chem. Phys.* 100: 6019—6027.

Burack, W. R., Yuan, Q., and Biltonen, R. L., 1993, Role of lateral phase separation in the modulation of phospholipase A_2 activitity, *Biochemistry* 32: 583—589.

Cevc, G., 1986, How membrane chain melting properties are regulated by the polar surface of the lipid bilayer, *Biochemistry* 26: 6305—6310.

Cevc, G., 1991, Isothermal lipid phase transitions, *Chem. Phys. Lipids* 57: 293—307.

Chattopadhyay, A., 1990, Chemistry and biology of *N*-(7-nitrobenz-2-oxa-1,3-diazol-4-yl)-labeled lipids: fluorescent probes of biological and model membranes, *Chem. Phys. Lipids* 53: 1—15.

Chattopadhyay, A., and London, E., 1987, Parallax method for direct measument of membrane penetration depth utilizing fluorescence quenching by spin-labeled phospholipids, *Biochemistry* 26: 39—45.

Chattopadhyay, A., and London, E., 1988, Spectroscopic and ionization properties of *N*-(7-nitrobenz-2-oxa-1,3-diazol-4-yl)-labeled lipids in model membranes, *Biochim. Biophys. Acta* 938: 24—34.

Chattopadhyay, A., and Mukherjee, S., 1993, Fluorophore environments in membrane-bound probes: a red edge excitation shift study, *Biochemistry* 32: 3804—3811.

Chong, P. L.-G., and Sugar, I. P., 2002, Fluorescence studies of lipid regular distribution in membranes, *Chem. Phys. Lipids* 116: 153—175.

Conte, J. C., 1967, Oxygen quenching and energy transfer in pyrene solutions, *Rev. Port. Quim.* 9: 13—21.

Davenport, L., Knutson, J. R., and Brand, L., 1986, Anisotropy decay associated fluorescence spectra and analysis of rotational heterogeneity. 2. 1,6-Diphenyl-1,3,5-hexatriene in lipid bilayers, *Biochemistry*, 25: 1811—1816.

Deng, Y., Kohlwein, S. D., and Mannella, C. A., 2002, Fasting induces cyanide-resistant respiration and oxidative stress in the amoeba Chaos carolinensis: implications for the cubic structural transition in mitochondrial membranes, *Protoplasma* 219: 160—167.

Deng, Y., Marko, M., Buttle, K. F., Leith, A., Mieczkowski, M. and Mannella, C. A., 1999, Cubic membrane structure in amoeba (Chaos carolinensis) mitochondria determined by electron microscopic tomography, *J. Struct. Biol.* 127: 231—239.

Duarte-Karim, M., Ruysschaert, J. M., and Hildebrand, J., 1976, Affinity of adriamycin to phospholipids. A possible explanation for cardiac mitochondrial lesions, *Biochem. Biophys. Res. Commun.* 71: 658—663.

Duhem, P., 1898a, On the general problem of chemical statics, *J. Phys. Chem.* 2: 1—42.

Duhem, P., 1898b, On the general problem of chemical statics, *J. Phys. Chem.* 2: 91—115.

Duportail, G., Klymchenko, A., Mély, Y., and Demchenko, A., 2001, Neutral fluorescence probe with strong ratiometric response to surface charge of phospholipid membranes, *FEBS Lett.* 508: 196—200.

Duportail, G., Klymchenko, A., Mély, Y., and Demchenko, A. P., 2002, On the coupling between surface charge and hydration in biomembranes: experiments with 3-hydroxyflavone probes, *J. Fluoresc.* 12: 181—185.

Dupuy, B., and Montagu, M., 1997, Spectral properties of a fluorescent probe, all-trans-1,6-diphenyl-1,3,5-hexatriene. Solvent and temperature effects, *Analyst* 122: 783—786.

Ebel, H., Grabitz, P., and Heimburg, T., 2001, Enthalpy and volume changes in lipid membranes. I. The proportionality of heat and volume changes in the lipid melting transition and its implication for the elastic constants, *J. Phys. Chem. B* 105: 7353—7360.

Eklund, K. K., Virtanen, J. A., and Kinnunen P. K. J., 1984, Involvement of phospholipid stereoconfiguration in the pretransition of dimyristoylphosphatidylcholine, *Biochim. Biophys. Acta* 793: 310—312.

Enders, A., and Nimtz, G., 1984, Dielectric relaxation study of dynamic properties of hydrated phospholipid bilayers, *Ber. Bunsenges. Phys. Chem.* 88: 512—517.

Epand, R. M., 1996, Functional roles of non-lamellar forming lipids, *Chem. Phys. Lipids* 81: 101—104.

Epand, R., ed., 2003, Cholesterol-rich domains, special issue of *Biochim. Biophys. Acta* 1610: 155—290.

Estep, T. N., Mountcastle, D. B., Barenholz, Y., Biltonen, R. L., and Thompson, T. E., 1979, Thermal behavior of synthetic sphingomyelin-cholesterol dispersions, *Biochemistry* 18: 2112—2117.

Fernandes, M. X., García de la Torre, J., and Castanho, M. A. R. B., 2002, Joint determination by Brownian dynamics and fluorescence quenching of the in-depth location profile of biomolecules in membranes, *Anal. Biochemistry* 307: 1—12.

Fery-Forgues S., Fayet, J.-P., and Lopez, A., 1993, Drastic changes in the fluorescence properties of NBD probes with the polarity of the medium: involvement of a TICT state? *J. Photochem. Photobiol. A: Chem.* 70: 229—243.

Förster, T, 1948, Zwischenmolekulare Energiewanderung und Fluoreszenz, Ann. Physik 2: 55—75.

Foster, L. J., de Hoog, C. L., and Mann, M., 2003, Unbiased quantitative proteomics of lipid rafts reveals high specificity for signaling factors, *Proc. Natl. Acad. Sci.* 100: 5813—5818.

Fujiwara, Y., and Amao, Y., 2003, Optical oxygen sensor based on controlling the excimer formation of pyrene-1-butylic acid chemisorption layer onto nano-porous anodic oxidized aluminium plate by myristic acid, *Sens. Actuator B—Chem.* 89: 58—61.

Galla, H. J., Hartmann, W., and Sackmann, E., 1978, Lipid-protein-interaction in model membranes: binding of melittin to lecithin bilayer vesicles, *Ber. Bunsenges.* 82: 917—921.

Galla, H. J., and Sackmann, E., 1974, Lateral diffusion in the hydrophobic region of membranes. Use of pyrene excimers as optical probes, *Biochim. Biophys. Acta* 339: 103—115.

Gershfeld, N. L., 1989a, The critical unilamellar lipid state: a perspective for membrane bilayer assembly, *Biochim. Biophys. Acta* 988: 335—350.

Gershfeld, N. L., 1989b, Spontaneous assembly of a phospholipid bilayer as a critical phenomenon: influence of temperature, composition, and physical state, *J. Phys. Chem.* 93: 5256—5261.

Gershfeld, N. L., and Ginsberg, L., 1997, Probing the critical unilamellar state of membranes, *J. Membrane Biol.* 156: 279—286.

Gershfeld, N. L., Mudd, C. P., Tajima, K., and Berger, R. L., 1993, Critical temperature for unilamellar vesicle formation in dimyristoylphosphatidylcholine dispersions from specific heat measurements. *Biophys. J.* 65: 1174—1179.

Ginsberg, L., Gilbert, D. L., and Gershfeld, N. L., 1991, Membrane bilayer assembly in neural tissue of rat and squid as a critical phenomenon: influence of temperature and membrane proteins, *J. Membrane Biol.* 119: 65—73.

Goodsaid-Zalduondo, F., Rintoul, D. A., Carlson, J. C., and Hansel, W., 1982, Luteolysis-induced changes in phase composition and fluidity of bovine luteal cell membranes, *Proc. Natl. Acad. Sci. USA* 79: 4332—4336.

Gratton, E., and Parasassi, T., 1995, Fluorescence lifetime distributions in membrane systems, *J. Fluoresc.* 5: 51—57.

Halstenberg, S., Heimburg, T., Hianik, T., Kaatze, U., and Krivanek, R., 1998, Cholesterol-induced variations in the volume and enthalpy fluctuations of lipid bilayers, *Biophys. J.* 75: 264—271.

Han, X., and Gross, R. W., 1992, Nonmonotonic alterations in the fluorescence anisotropy of polar head group labeled fluorophores during the lamellar to hexagonal phase transition of phospholipids, *Biophys. J.* 63: 309—316.

Heerklotz, H., 2002, Triton promotes domain formation in lipid raft mixtures, *Biophys. J.* 83: 2693—2701.

Heimburg, T., 1998, Mechanical aspects of membrane thermodynamics. Estimation of the mechanical properties of lipid membranes close to the chain melting transition from calorimetry, *Biochim. Biophys. Acta* 1415: 147—162.

Heimburg, T., 2000, A model for the lipid pretransition: coupling of ripple formation with the chain-melting, *Biophys. J.* 78: 1154—1165.

Helm, C. A., and Möhwald, H., 1988, Equilibrium and nonequilibrium features determining superlattices in phospholipid monolayers, *J. Phys. Chem.* 92: 1262—1266.

Heyn, M. P., 1979, Determination of lipid order parameters and rotational correlation times from fluorescence depolarization experiments, *FEBS Lett.* 108: 359—364.

Holopainen, J. M., Angelova, M. I., Söderlund, T., and Kinnunen, P. K. J., 2002, Macroscopic consequences of the action of phospholipase C on giant unilamellar liposomes, *Biophys. J.* 83: 932—943.

Holzwarth, J. F., 1989, Structure and dynamics of phospholipid membranes from nanoseconds to seconds, *NATO ASI Ser. A* 178: 383—412.

Hong, K., Baldwin, P. A., Allen, T. M., and Papahadjopoulos, D., 1988, Fluorimetric detection of the bilayer-to-hexagonal phase transition in liposomes, *Biochemistry* 27: 3947—3955.

Hoover, R. L., Dawidowicz, E. A., Robinson, J. M., and Karnovsky, M. J., 1983, Role of cholesterol in the capping of surface immunoglobulin receptors on murine lymphocytes, *J. Cell Biol.* 97: 73—80.

Huang, J., 2002, Exploration of molecular interactions in cholesterol superlattices: effect of multibody interactions, *Biophys. J.* 83: 1014—1025.

Huster, D., Müller, P., Arnold, K., and Herrmann, A., 2003, Dynamics of lipid chain attached fluorophore 7-nitrobenz-2-oxa-1,3-diazol-4-yl (NBD) in negatively charged membranes determined by NMR spectroscopy, *Eur. Biophys. J.* 32: 47—54.

Hwang, J., Tamm, L. K., Böhm, C., Ramalingam, T. S., Betzig, E., and Edidin, M., 1995, Nanoscale complexity of phospholipid monolayers investigated by near-field scanning optical microscopy, *Science* 270: 610—614.

Ipsen, J. H., Karlstroem, G., Mouritsen, O. G., Wennerstroem, H., and Zuckermann, M. J., 1987, Phase equilibria in the phosphatidylcholine-cholesterol system, *Biochim. Biophys. Acta* 905: 162—172.

Itoh, T. and Kohler, B. E., 1987, Dual fluorescence of diphenylpolyenes, *J. Phys. Chem.* 91: 1760—1764.

Jin, A. J., Edidin, M., Nossal, R., and Gershfeld, N. L., 1999, A singular state of membrane lipids at cell growth temperatures, *Biochemistry* 38: 13275—13278.

Jutila, A., and Kinnunen, P. K. J., 1997, Novel features of the main transition of dimyristoylcholine bilayers revealed by fluorescence spectroscopy, *J. Phys. Chem. B.* 101: 7635—7640.

Kaasgaard, T., Leidy, C., Crowe, J. H., Mouritsen, O. G., and Jørgensen, K., 2003, Temperature-controlled structure and kinetics of ripple phases in one- and two-component supported lipid bilayers, *Biophys. J.* 85: 350—360.

Kachel, K., Asuncion-Punzalan, E., and London, E., 1998, The location of fluorescent probes with charged groups in model membranes, *Biochim. Biophys.Acta* 1374: 63—76.

Kaiser, R. D., and London, E., 1998, Location of diphenylhexatriene (DPH) and its derivatives within membranes: comparison of different fluorescence queching analyses of membrane depth, *Biochemistry* 37: 8180—8190.

Kaiser, R. D., and London, E., 1998b, Determination of the depth of BODIPY probes in model membranes by parallax analysis of fluorescence quenching, *Biochim. Biophys. Acta* 1375: 13—22.

Kaiser, R. D., and London, E., 1999, Correction to: Location of diphenylhexatriene (DPH) and its derivatives within membranes: comparison of different fluorescence queching analyses of membrane depth, *Biochemistry* 38: 2610.

Karczmar, G. S., and Tritton, T. R., 1979, The interaction of adriamycin with small unilamellar vesicle liposomes. A fluorescence study, *Biochim. Biophys. Acta* 557: 306—319.

Kato, S., and Kubo, T., 1997, Relaxation process after the cooling jump across the pretransition of dipalmitoylphosphatidylcholine bilayers, *Chem. Phys. Lipids* 90: 31—44.

Kawato, S., Kinosita, K. Jr., and Ikegami, A., 1977, Dynamic structure of lipid bilayers studied by nanosecond fluorescence techniques, *Biochemistry* 16: 2319—2324.

Kharakoz, D. P., Colotto, A., Lohner, K., and Laggner, P., 1993, Fluid-gel interphase line tension and density fluctuations in dipalmitoylphosphatidylcholine multilamellar vesicles. An ultrasonic study, *J. Phys. Chem.* 97: 9844—9851.

Kharakoz, D. P., and Shlyapnikova, E. A., 2000, Thermodynamics and kinetics of the early steps of solid-state nucleation the fluid bilayer, *J. Phys. Chem. B.* 104: 10368—10378.

Kinnunen, P. K. J., 1991, On the principles of functional ordering in biological membranes, *Chem. Phys. Lipids* 57: 375—399.

Kinnunen P. K. J., 1996a, On the molecular-level mechanisms of peripheral protein-membrane interactions induced by lipids forming inverted non-lamellar phases, *Chem. Phys. Lipids* 81: 151—166.

Kinnunen, P. K. J., 1996b, On the mechanisms of the lamellar → hexagonal HII phase transition and the biological significance of the HII propensity, in: *Handbook of Nonmedical Applications of Liposomes* 1, D. D. Lasic and Y. Barenholz, eds., CRC, Boca Raton, pp. 153—171.

Kinnunen, P. K. J., Kõiv, A., Lehtonen, J. Y. A., and Mustonen, P., 1994, Lipid dynamics and peripheral interactions of proteins with membrane surfaces, *Chem. Phys. Lipids* 73: 181—207.

Kinnunen, P. K. J., Kõiv, A., and Mustonen, P., 1993, Pyrene-labelled lipids as fluorescent probes in studies on biomembranes and membrane models, in: *Methods and applications of fluorescence spectroscopy*, O. Wolfbeis, ed., Springer Verlag, New York, pp. 159—171.

Kinnunen, P. K. J., and Laggner, P., Editors, 1991, *Phospholipid Phase Transitions*. Special issue of *Chem. Phys. Lipids,* 57: 109—399.

Kinnunen, P. K. J., Tulkki, A. P., Lemmetyinen, H., Paakkola, J., and Virtanen, J. A., 1987, Characteristics of excimer formation in Langmuir-Blodgett assemblies of 1-palmitoyl-2-pyrenedecanoylphosphatidylcholine and dipalmitoylphosphatidylcholine, *Chem. Phys. Lett.* 136: 539—545.

Kinnunen, P. K. J., Virtanen, J. A., Tulkki, A. P., Ahuja, R. C., and Moebius, D., 1985, Pyrene-fatty acid-containing phospholipid analogues: characterization of monolayers an Langmuir-Blodgett assemblies, *Thin Solid Films* 132: 193—203.

Kodama M., Aoki, H., and Miyata, T., 1999, Effect of Na^+ concentration on the subgel phases of negatively charged phosphatidylglycerol, *Biophys. Chem.* 79: 205—217.

Konopasek, I., Kvasnicka, P., Herman, P., Linnertz, H., Obsil, T., Vecer, J., Svobodova, J., Strzalka, K., Mazzanti, L., and Amler, E., 1998, The origin of the diphenylhexatriene short lifetime component in membranes and solvents, *Chem. Phys. Lett.* 293: 429—435.

Konttila, R., Salonen, I., Virtanen, J. A., and Kinnunen P. K. J., 1988, Estimation of the equilibrium lateral pressure in liposomes of 1-palmitoyl-2-[10-(pyren-1-yl)-10-ketodecanoyl]-*sn*-glycero-3-phosphocholine and the effect of phospholipid phase transition, *Biochemistry* 27: 7443—7446.

Kunz, R. E., and Lukosz, W., 1980, Changes in fluorescence lifetimes induced by variable optical environments, *Phys. Rev. B* 21: 4814—4828.

Lakowicz, J. R., 1999, Principles of Fluorescence Spectroscopy, 2nd ed., Kluwer Academic, New York.

Lakowicz, J. R., and Keating-Nakamoto, S., 1984, Red-edge excitation of fluorescence and dynamic properties of proteins and membranes, *Biochemistry* 23: 3013—3021.

Lancet D., and Pecht, I., 1977, Spectroscopic and immunochemical studies with nitrobenzoxadiazolealanine, a fluorescent dinitrophenyl analogue, *Biochemistry* 16: 5150—5157.

Langner, M., and Hui, S. W., 1993, Dithionite penetration through phospholipid bilayers as a measure of defects in lipid molecular packing, *Chem. Phys. Lipids* 65: 23—30.

Langner, M., Pruchnik, H., and Kubica, K., 2000, The effect of the lipid bilayer state on fluorescence intensity of fluorescein-PE in a saturated lipid bilayer, *Z. Naturforsch.(C)* 55: 418—424.

Larsson, K., 1997, Standing wave oscillations in axon membranes and the action potential, *Colloid Surf. A* 129—130: 267—272.

Lehtonen, J. Y. A., Holopainen, J. M., and Kinnunen, P. K. J., 1996, Evidence for the formation of microdomains in liquid crystalline large unilamellar vesicles caused by hydrophobic mismatch of the constituent phospholipids, *Biophys. J.* 70: 1753—1760.

Lehtonen, J. Y. A., and Kinnunen, P. K. J., 1994, Changes in the lipid dynamics of liposomal membranes induced by poly(ethylene glycol): free volume alterations revealed by inter- and intramolecular excimer-forming phospholipid analogs, Biophys. J. 66: 1981—1990.

Lemmetyinen, H., Yliperttula, M., Mikkola, J., Virtanen, J. A., and Kinnunen, P. K. J., 1989, Kinetic study of monomer and excimer fluorescence of pyrene-substituted phosphatidylcholine in phosphatidylcholine bilayers, *J. Phys. Chem.* 93: 7170—7175.

Lemmich, J., Mortensen, K., Ipsen, J. H., Hønger, T., Bauer, R., and Mouritsen, O. G., 1995, Pseudocritical behavior and unbinding of phospholipid bilayers, *Phys. Rev. Lett.* 75: 3958—3961.

Lentz, B. R., and Litman, B. J., 1978, Effect of head group on phospholipid mixing in small, unilamellar vesicles: mixtures of dimyristoylphosphatidylcholine and dimyristoylphosphatidylethanolamine, *Biochemistry* 17: 5537—5543.

Lichtenberg, D., Menashe, M., Donaldson, S., and Biltonen, R. L., 1984, Thermodynamic characterization of the pretransition of unilamellar dipalmitoyl-phosphatidylcholine vesicles, *Lipids* 19: 395—400.

Lin, S., and Struve, W. S., 1991, Time-resolved fluorescence of nitrobenzoxadiazole-aminohexanoic acid: effect of intermolecular hydrogen-bonding on non-radiative decay, *Photochem. Photobiol.* 54: 361—365.

Liptay, W., 1969, Electrochromism and solvatochromism, *Angew. Chem.—Int. Edit.* 8: 177—188.

Loew, L. M., Bonneville, G. W., and Surow, J., 1978, Charge shift optical probes of membrane potential. Theory, *Biochemistry* 17: 4065—4071.

Loew, L. M., Scully, S., Simpson, L., and Waggoner, A. S., 1979, Evidence for a charge-shift electrochromic mechanism in a probe of membrane potential, *Nature* 281: 497—499.

Lotta, T. I., Laakkonen, L. J., Virtanen, J. A., and Kinnunen, P. K. J., 1988a, Characterization of Langmuir-Blodgett films of 1,2-dipalmitoyl-sn-glycero-3-phosphatidylcholine and 1-palmitoyl-2-[10-(pyren-1-yl)decanoyl]-*sn*-glycero-3-phosphatidylcholine by FTIR-ATR, *Chem. Phys. Lipids*, 46: 1—12.

Lotta, T. I., Virtanen, J. A., and Kinnunen, P. K. J., 1988b, Fourier transform infrared study on the thermotropic behavior of fully hydrated 1-palmitoyl-2-[10-(pyren-1-yl)decanoyl]-*sn*-glycero-3-phosphatidylcholine, *Chem. Phys. Lipids* 46: 13—23.

Loura, L. M. S., Fedorov, A., and Prieto, M., 2000, Membrane probe distribution heterogeneity: a resonance energy transfer study, *J. Phys. Chem. B* 104: 6920—6931.

Loura, L. M. S., Fedorov, A., and Prieto, M., 2001, Exclusion of a cholesterol analog from the cholesterol-rich phase in model membranes, *Biochim. Biophys. Acta* 1511: 236—243.

Lukosz, W., 1980, Theory of optical-environment-dependent spontaneous-emission rates for emitters in thin layers, *Phys. Rev. B* 22: 3030—3038.

Luzzati, V., 1997, Biological significance of lipid polymorphism: the cubic phases, *Curr. Opin. Struct. Biol.* 7: 661—668.

Mabrey, S., and Sturtevant, J. M., 1976, Investigation of phase transitions of lipids and lipid mixtures by high sensitivity differential scanning calorimetry, *Proc. Natl. Acad. Sci. USA* 73: 3862—3866.

Makino, K., Yamada, T., Kimura, M., Oka, T., Ohshima, H., and Kondo, T., 1991, Temperature- and ionic strength-induced conformational changes in the lipid head group region of liposomes as suggested by zeta potential data, *Biophys. Chem.* 41: 175—183.

Mamdouh, Z., Giocondi, M. C., and Le Grimellec, C., 1998, In situ determination of intracellular membrane physical state heterogeneity in renal epithelial cells using fluorescence ratio microscopy, *Eur. Biophys. J.* 27: 341—351.

Marčelja, S., and Wolfe, J., 1979, Properties of bilayer membranes in the phase transition or phase separation region, *Biochim. Biophys. Acta* 557: 24—31.

Marsh, D., 1991, General features of phospholipid phase transitions, *Chem. Phys. Lipids* 57: 109—120.

Marsh, D., 1996, Lateral pressures in membranes, *Biochim. Biophys. Acta* 1286:183—223.

Mason, P. C., Gaulin, B. D., Epand, R. M., Wignall, G. D., and Lin J. S., 1999, Small angle neutron scattering and calorimetric studies of large unilamellar vesicles of the phospholipid dipalmitoylphosphatidylcholine, *Phys. Rev. E* 59: 3361—3367.

McIntyre, J. C., and Sleight, R. G., 1991, Fluorescence assay for phospholipid membrane asymmetry, *Biochemistry* 30: 11819—11827.

Mellier, A., and Diaf, A., 1988, Infrared study of phospholipid hydration. Main phase transition of saturated phosphatidylcholine/water multilamellar samples, *Chem. Phys. Lipids* 46: 51—56.

Mellier, A., Ech-Chahoubi, A., and Le Roy, A., 1993, Infrared kinetic study of the main phase transition of saturated phosphatidylcholines/water multilamellar systems: determination of the cooperative unit and the activation energies, *J. Chim. Phys.* 90: 51—62.

Metso, A., Jutila, A., Mattila, J.-P., Holopainen, J. M., and Kinnunen, P. K. J., 2003, Nature of the main transition of dipalmitoylphosphocholine bilayers inferred from fluorescence spectroscopy, *J. Phys. Chem. B* 107: 1251—1257.

Mouritsen, O. G., 1991, Theoretical models of phospholipid phase transitions, *Chem. Phys. Lipids* 57: 179—194.

Mouritsen, O. G., and Kinnunen P. K. J., 1996, Role of lipid organization and dynamics for membrane functionality, in: *Biological Membranes — A Molecular Pespective from Computation and Experiment*, K. Merz Jr. and B. Roux, eds., Birkhäuser, Boston, pp. 463—502.

Mukherjee, S., and Chattopadhyay, A., 1996, Membrane organization at low cholesterol concentrations: A study using 7-nitrobenz-2-oxa-1,3-diazol-4-yl-labeled cholesterol, *Biochemistry* 35: 1311—1322.

Mukherjee, S., Chattopadhyay, A., Samanta, A., and Soujanya, T., 1994, Dipole moment change of NBD group upon excitation studied using solvatochromic and quantum chemical approaches: Implications in membrane research, *J. Phys. Chem.* 98: 2809—2812.

Mustonen, P., Virtanen, J. A., Somerharju, P. J., and Kinnunen, P. K. J., 1987, Binding of cytochrome c to liposomes as revealed by the quenching of fluorescence from pyrene-labeled phospholipids, *Biochemistry* 26: 2991—2997.

Mustonen, P., and Kinnunen, P. K. J., 1991, Activation of phospholipase A2 by adriamycin in vitro. Role of drug-lipid interactions, *J. Biol. Chem.* 266: 6302—6307.

Nagle, J. F., and Scott, H. L., Jr., 1978, Lateral compressibility of lipid mono- and bilayers. Theory of membrane permeability, *Biochim. Biophys. Acta* 513: 236—243.

Nagle, J., and Tristram-Nagle, S., 2000, Structure of lipid bilayers, *Biochim. Biophys. Acta* 1469: 159—195.

Needham, D., and Evans, E., 1988, Structure and mechanical properties of giant lipid (DMPC) vesicle bilayers from 20 °C below to 10 °C above the liquid crystal—crystalline phase transition at 24 °C, *Biochemistry* 27: 8261—8269.

Nielsen, M., Miao, L., Ipsen, J. H., Jørgensen, K., Zuckermann, M. J., and Mouritsen O. G., 1996, Model of a sub-main transition in phospholipid bilayers, *Biochim. Biohys. Acta* 1283: 170—176.

Pagano, R. E., Cherry, R. J., and Chapman, D., 1973, Phase transitions and heterogeneity in lipid bilayers, *Science* 181: 557—559.

Pap, E. H. W., ter Horst, J. J., van Hoek, A., and Visser, A. J. W. G., 1994, Fluorescence dynamics of diphenyl-1,3,5-hexatriene-labeled phospholipids in bilayer membranes, *Biophys. Chem.* 48: 337—351.

Papon, P., Leblond, J., and Meijer, P. H. E., 2002, The Physics of Phase Transitions. Springer-Verlag, Berlin Heidelberg, Germany.

Paprica, P. A., Baird, N. C., and Petersen, N. O., 1993, Theoretical and experimental analyses of optical transitions of nitrobenzoxadiazole (NBD) derivatives, *J. Photochem. Photobiol. A: Chem.* 70: 51—57.

Pebay-Peyroula, E., Dufourc, E. J., and Szabo, A. G., 1994, Location of diphenylhexatriene and trimethylammonium-diphenyl-hexatriene in dipalmitoylphosphatidylcholine bilayers by neutron diffraction. *Biophys. Chem.* 53: 45—56.

Pressl, K., Jørgensen, K., and Laggner, P., 1997, Characterization of the sub-main- transition in distearoylphosphatidylcholine studied by simultaneous small- and wide-angle X-ray diffraction, *Biochim. Biophys. Acta* 1325: 1—7.

Prost, J., and Williams, C. E., 1999, Liquid crystals: between order and disorder, in: *Soft Matter Physics*, M. Daoud and C. E. Williams, eds., Springer-Verlag, New York, pp. 289—315.

Raghunathan, V. A., and Katsaras, J., 1995, Structure of the $L_{c'}$ phase in a hydrated lipid multilamellar system, *Phys. Rev. Lett.* 74: 4456—4459.

Raudino, A., and Mauzerall, D., 1986, Dielectric properties of the polar head group region of zwitterionic lipid bilayers, *Biophys. J.* 50: 441—449.

Recktenwald, D. J., and McConnell, H. M., 1981, Phase equilibriums in binary mixtures of phosphatidylcholine and cholesterol, *Biochemistry* 20: 4505—4510.

Riske, K. A., Döbereiner, H.-G., and Lamy-Freund, M. T., 2002, Gel-fluid transition in dilute vs. concentrated DMPG aqueous solutions, *J. Phys. Chem. B* 106: 239—246.

Saha, S., and Samanta, A., 1998, Photophysical and dynamic NMR Studies on 4-amino-7-nitrobenz-2-oxa-1,3-diazole derivatives: elucidation of the nonradiative deactivation pathway, *J. Phys. Chem. A* 102: 7903—7912.

Scheiner, M. F., Marsh, D., Jahn, W., Kloesgen, B., and Heimburg, T., 1999, Network formation of lipid membranes: triggering structural transitions by chain melting. *Proc. Natl. Acad. Sci.* 96: 14312—14317.

Schrader, W., Ebel, H., Grabitz, P., Hanke, E., Heimburg, T., Hoeckel, M., Kahle, M., Wente, F., and Kaatze, U., 2002, Compressibility of lipid mixtures studied by calorimetry and ultrasonic velocity measurements, *J. Phys. Chem. B* 106: 6581—6586.

Simons, K., and Ikonen, E., 1997, Functional rafts in cell membranes, *Nature* 387: 569—572.

Slater, S. J., Kelly, M. B., Taddeo, F. J., Ho, C., Rubin, E., and Stubbs C. D., 1994, The modulation of protein kinase C activity by membrane lipid bilayer structure, *J. Biol. Chem.* 269: 4866—4871.

Snyder, R. G., Tu, K., Klein, M. L., Mendelssohn, R., Strauss, H. L., and Sun, W., 2002, Acyl chain conformation and packing in dipalmitoylphosphatidylcholine bilayers from MD simulation and IR spectroscopy, *J. Phys. Chem. B* 106: 6273—6288.

Söderlund, T., Jutila, A., and Kinnunen, P. K. J., 1999, Binding of adriamycin to liposomes as a probe for membrane lateral organization, *Biophys. J.* 76: 896—907.

Söderlund, T., Alakoskela J.-M. I., Pakkanen, A., and Kinnunen P. K. J., 2003, Comparison of the effects of surface tension and osmotic pressure on the interfacial hydration of a fluid phospholipid bilayer, *Biophys. J.* 85: 2333—2341.

Somerharju, P. J., Virtanen, J. A., Eklund, K. K., Vainio, P., and Kinnunen, P. K. J., 1985, 1-Palmitoyl-2-pyrenedecanoyl glycerophospholipids as membrane probes: evidence for regular distribution in liquid-crystallinen phosphatidylcholine bilayers, *Biochemistry* 24: 2773—2781.

Stern, H. A., and Feller, S. E., 2003, Calculation of the dielectric permittivity profile for a nonuniform system: application to a lipid bilayer simulation, *J. Chem. Phys.* 118: 3401—3412.

Stubbs, C.D., Williams, B. W., Boni, L. T., Hoek, J. B., Taraschi, T. F., and Rubin, E., 1989, On the use of *N*-(7-nitrobenz-2-oxa-1,3-diazol-4-yl)phosphatidylethanolamine in the study of lipid polymorphism, *Biochim. Biophys. Acta* 986: 89—96.

Sugar, I. P., Tang, D., and Chong, P. L. G., 1994, Monte Carlo simulation of lateral distribution of molecules in a two-component lipid membrane: effect of long-range repulsive interactions, *J. Phys. Chem.* 98: 7201—7210.

Sunamoto, J., Kondo, H., Nomura, T., and Okamoto, H., 1980, Liposomal membranes 2. Synthesis of a novel pyrene-labeled lecithin and structural studies on liposomal bilayers, *J. Am. Chem. Soc.* 102: 1146—1152.

Teschke, O., Ceotto, G., and de Souza, E. F., 2001a, Dielectric exchange force. A convenient technique for measuring the interfacial water relative permittivity profile, *Phys. Chem. Chem. Phys.* 3: 3761—3768.

Teschke, O., Ceotto, G., and de Souza, E. F., 2001b, Interfacial water dielectric-permittivity-profile measurements using atomic force microscopy, *Phys. Rev. E* 64: 011605/1—011605/10.

Thompson, T. E., and Tillack, T. W., 1985, Organization of glycosphingolipids in bilayers and plasma membranes of mammalian cells, *Annu. Rev. Biophys. Biophys. Chem.* 14: 361—386.
Tolédano, J.-C., and Tolédano, P., 1987, The Landau theory of phase transitions. World Scientific Publishing Co Pte Ltd., Singapore.
Toptygin, D., Svobodova, I., Konopasek, I., and Brand, L., 1992, Fluorescence decay and depolarization in membranes, *J. Chem. Phys.* 96: 7919—7930.
Trandum, C., Westh, P., and Jørgensen, K., 1999, Slow relaxation of the sub-main transition in multilamellar phosphatidylcholine vesicles, *Biochim. Biophys. Acta* 1421: 207—212.
Tremper, K. E., and Gershfeld, N. L., 1999, Temperature dependence of membrane lipid composition in early blastula embryos of *Lytechinus pictus*: selective sorting of phospholipids into nascent plasma membranes, *J. Membrane Biol.* 171: 47—53.
Tsukanova, V., Grainger, D. W., and Salesse, C., 2002, Monolayer behavior of NBD-labeled phospholipids at the air/water interface, *Langmuir* 18: 5539—5550.
Untracht, S. H., and Shipley, G. G., 1977, Molecular interactions between lecithin and sphingomyelin. Temperature- and composition-dependent phase separation, *J. Biol. Chem.* 252: 4449—4457.
Vanderkooi, J. M., and Callis, J. B., 1974, Pyrene. A probe of lateral diffusion in the hydrophobic region of membranes, *Biochemistry* 13: 4000—4006.
Virtanen, J. A., Somerharju, P., and Kinnunen, P. K. J., 1988, Prediction of patterns for the regular distribution of soluted guest molecules in liquid crystalline phospholipid membranes, *J. Mol. Electronics* 4: 233—236.
Wang, S., Beechem, J. M., Gratton, E., and Glaser, M., 1991, Orientational distribution of 1,6-diphenyl-1,3,5-hexatriene in phospholipid vesicles as determined by global analysis of frequency domain fluorimetry data, *Biochemistry* 30: 5565—5572.
Weis, R. M., 1991, Fluorescence microscopy of phospholipid monolayer phase transitions, *Chem. Phys. Lipids* 57: 227—239.
Welti, R., and Glaser, M., 1994, Lipid domains in model and biological membranes, *Chem. Phys. Lipids* 73: 121—137.
Werncke, W., Hogiu, S., Pfeiffer, M., Lau, A., and Kummrow, A., 2000, Strong S_1—S_2 vibronic coupling and enhanced third order hyperpolarizability in the first excited singlet state of diphenylhexatriene studied by time-resolved CARS, *J. Phys. Chem. A* 104: 4211—4217.
White, S. H., and Wimley, W. C., 1998, Hydrophobic interactions of peptides with membrane interfaces, *Biochim. Biophys. Acta* 1376: 339—352.
Wiedmer, S. K., Hautala, J., Holopainen, J. M., and Kinnunen, P. K. J., Riekkola, M.-L., 2001, Study on liposomes by capillary electrophoresis, *Electrophoresis* 22: 1305—1313.
Wu, S. H. W., and McConnell, H. M, 1975, Phase separations in phospholipid membranes, *Biochemistry* 14: 847—854.
Yliperttula, M., Lemmetyinen, H., Mikkola, J., Virtanen, J., and Kinnunen, P. K. J., 1988, Stationary and time-resolved fluorescence anisotropy of pyrene lecithin in LB films, *Chem. Phys. Lett.* 152: 61—66.
Zhang, W., and Kaback, H. R., 2000, Effect of lipid phase transition on the lactose permease from *Escherichia coli*, *Proc. Natl. Acad. Sci. USA* 85: 7202—7205.

NEW ANALYSIS OF SINGLE MOLECULE FLUORESCENCE USING SERIES OF PHOTON ARRIVAL TIMES

Eugene Novikov, Johan Hofkens, Mircea Cotlet, Frans C. De Schryver, Noël Boens*

ABSTRACT

Until recently, single molecule fluorescence experiments have been made by dividing the time into a set of intervals and to observe the number of fluorescence photons arriving in each interval. It is obvious that the detected photons per time interval carry less information than the arrival times of the photons themselves. Indeed, from the arrival times, one can still calculate the number of photons in any user-defined interval, whereas when only the number of photons in an interval is recorded, information about their positions in time is lost. In this chapter we present a new analysis of single molecule fluorescence data based on the positions in time of the detected fluorescence photons. We derive mathematically different statistical characteristics describing the single molecule fluorescence experiment assuming an immobilized molecule. The theory of random point processes using the generating functionals formalism is ideally suited for a consistent description, linking the statistical characteristics of the excitation and detected photons to the statistical characteristics of the single motionless molecule. The following statistical characteristics are described: the probability density distributions of the single and first photocount time positions in a user-defined detection interval, the probability distribution of the number of photocounts per user-defined detection interval, the time correlation function, and the inter-arrival time probability density distribution. The new analysis is

* To whom correspondence should be addressed. E-mail: Noel.Boens@chem.kuleuven.ac.be. Eugene Novikov, Institut Curie, Service Bioinformatique, 26 Rue d'Ulm, Paris Cedex 05, 75248 France. Johan Hofkens, Mircea Cotlet, Frans C. De Schryver, Noël Boens, Katholieke Universiteit Leuven, Department of Chemistry, Celestijnenlaan 200F, 3001 Heverlee – Leuven, Belgium.

illustrated using the traces of photon arrival times of individual rhodamine 6G (R6G) molecules to obtain information on their photophysics.

1. INTRODUCTION

Fluorescence spectroscopy of single immobilized molecules[1,2,3,4,5,6] is a very sensitive method for the investigation of the properties of single molecules. A typical single molecule fluorescence experiment[1-5] consists of three events as represented schematically in **Figure 1**: (i) excitation of and (ii) emission from a single immobilized molecule, and (iii) detection of the emitted photons.

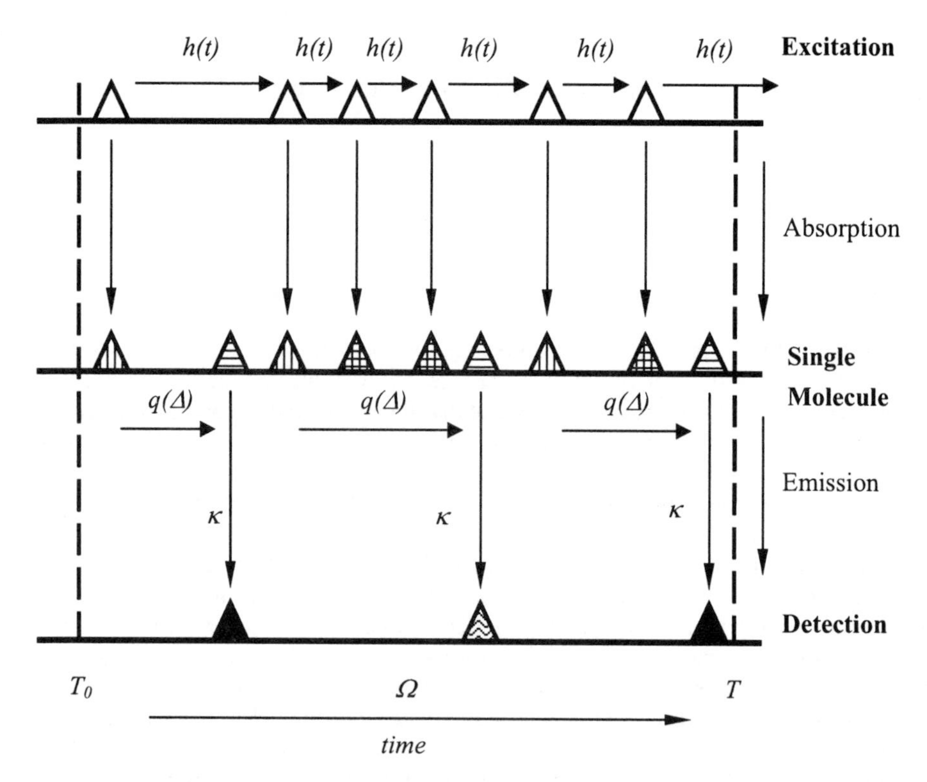

Figure 1: Scheme of a single molecule fluorescence experiment in a thin film or on a surface. Open symbols: excitation photons. Symbols with vertical line pattern: absorbed photons. Symbols with horizontal line pattern: emitted photons. Symbols with cross-line pattern: photons arriving at the single molecule in the excited state. Black symbols: detected photons (photocounts). Symbols with wave-line pattern: photons which are lost due to limited detection efficiency.

The time occurrences of excitation photons (open symbols in **Figure 1**) are completely defined by the probability density distribution (PDD) of inter-photon distance $h(t)$. When a well-stabilized laser is used as continuous excitation source, the duration between consecutive photons is approximated by an exponential distribution[7]. In the case of pulsed excitation (each pulse is composed of several photons, but only one of them can be absorbed by the single molecule), the duration between consecutive photons can be assumed as equal, and $h(t)$ is approximated by a delta-function.

A single molecule in the ground state can absorb an excitation photon (symbols with vertical lines in **Figure 1**) bringing the molecule into the excited state. If other excitation photons arrive at this excited molecule, they are lost (symbols with cross-lines in **Figure 1**). After some time the molecule relaxes to the ground state either by emitting a photon (symbols with horizontal lines in **Figure 1**) or non-radiatively. Emitted photons arriving at the detector within the detection interval $\Omega = [T_0;T]$ are detected (black symbols in **Figure 1**) with probability κ, the efficiency of the detection set-up, or go undetected (wavy symbols in **Figure 1**).

In general, the molecule remains in an excited state for a random time interval Δ, defined by the PDD $q(\Delta)$, which we call the single molecule impulse response function (SMIRF). It should be emphasized that the SMIRF is not equivalent to the fluorescence delta response function usually obtained from the analysis of single-photon timing data. For example, in the simplest case of a single molecule with an excited singlet and an excited triplet state, the SMIRF is bi-exponential and incorporates both decay times and the intersystem crossing yield, while the fluorescence delta response function is mono-exponential and only yields information on the fluorescence lifetime.

The statistical characteristics of the detected photons (photocounts) contain information about the excitation photons and the single molecule. The detection procedure is dependent on the relationship between the time interval between excitation photons and the characteristic time scales of intra-molecular relaxation. If the intensity λ of excitation (average number of photons per time unit impinging on the single molecule) is much lower than the rate of relaxation, the time between the excitation and emitted photon can be measured and collected into a histogram, thus building up the SMIRF. When the single molecule fluorescence experiment is performed at higher excitation powers, such that the time interval between excitation photons is comparable to or smaller than the characteristic time scale of intra-molecular relaxation, any of the statistical characteristics of the detected photons will also contain the characteristics of the single molecule. Parametric fitting of an appropriate model to the measured data can extract these characteristics without direct measurement of the time positions of the excitation photons. The statistical characteristics of the photocounts that are most widely used for probing intra-molecular relaxation are the time correlation function[8,9,10,11], the PDD of the time interval between two consecutive photocounts[11,12,13] and the probability distribution (PD) of the number of photocounts per detection interval[14,15].

Performing experiments at high excitation powers may result in an increased probability of photobleaching[16,17]. Therefore, it is advantageous to decrease the excitation power to prevent chromophores from photobleaching. Low light level excitation, however, will lead to a decrease of the detection count rates. In this case problems connected to the lower statistical accuracy of the measured time correlation function[18] and the PD of the number of photocounts per detection interval may arise. This normally results in poor statistical stability of the estimated characteristics of the SMIRF. That is the reason why for processes with low photon count rates other detection procedures, which can utilize the measured information in a more efficient way, should be developed. From this point of view such statistical characteristics as the PDD of the single and first photocount time occurrence in the detection interval can be of practical interest[19]. Note that a low average number of photocounts per detection interval (yielding an appreciable number of intervals with zero photocounts) can be caused not only by low excitation intensities λ, but also by a low efficiency of the detection set-up κ, a short detection interval Ω and/or a low fluorescence quantum efficiency ϕ_f of the molecule. As the photon detection level depends on λ, κ, Ω, and ϕ_f, it is impossible to define quantitatively the term "low" (or "high") light level detection. Indeed, one can switch from a "high" photon detection level (practically all detection intervals contain multiple photocounts) to a "low" one (detection intervals with mostly zero and one photocounts) by simply shortening the duration of Ω while leaving the other parameters (λ, κ, and ϕ_f) of the experiment unchanged.

In this chapter, we present a unified statistical description of the single molecule fluorescence experiment, provided that the molecule is motionless during the time scale of the experiment. Since it can be assumed that each photon has a well-defined time position, single molecule experiments can be described adequately in terms of the theory of point processes[20,21,22,23]. In this case, each excitation photon can be considered as a "point" on the time axis, and the single molecule represents a system for point process transformations. It is intuitively clear that *the time positions of the photocounts in the detection interval* are more informative than *the number of observed photocounts in the interval*. From the time positions of photocounts one can always calculate the number of photocounts in any user-defined detection interval. When only the number of photocounts in a detection interval is recorded, information about their position in time is lost. The usefulness of the random point process theory for the description of single molecule fluorescence experiments has already been shown[12,13]. In the present chapter, we extend this approach to cover a broader range of the statistical characteristics of photocounts, models and experimental configurations. For the description of point process transformations, we apply the mathematical formalism of generating functionals (GFL)[20,22]. GFL formalism leads to a set of functional equations, linking the statistical characteristics of excitation and detected photons via the characteristics of the single molecule. The following statistical characteristics are described: (i) the PDD of the first photocount time occurrence in the (user-defined) detection interval, (ii) the PDD of the single photocount time occurrence in the (user-defined) detection interval, (iii) the PD of the number of photocounts n per (user-defined) detection interval, (iv) the PDD of the time interval between two consecutive photocounts, and (v) the measurement of the time correlation function. Using the new analysis, detailed insight in the photophysical properties of individual molecules can be obtained[24].

2. RANDOM POINT PROCESSES

2.1. Basic Definitions

2.1.1. *Probabilistic Description of Random Point Processes*

We assume that a random point process (RPP) can be completely defined by the time positions ($\tau_1, \tau_2, \ldots, \tau_n$) of the points in a time interval $\Omega = [T_0;T]$ of interest (**Figure 2**). For the statistical description of such RPP we can use a set of joint PDDs:

$$\{\pi_i(t_1,\ldots,t_i;\Omega)\}_{i=0,1,\ldots} \tag{1}$$

where $\pi_i(t_1,\ldots,t_i;\Omega)$ is the joint PDD of obtaining *exactly* i points in the detection interval Ω and each of these points belongs to the subinterval $[t_j,t_j+dt_j)\in\Omega$, $j = 1,\ldots,i$. For example, $\pi_0(\Omega)$ is the probability to obtain zero events within Ω, the PDD to obtain exactly one event within Ω is $\pi_1(t_1, \Omega)$, etc. For the densities $\{\pi_i(t_1,\ldots,t_i;\Omega)\}_{i=0,1,\ldots}$ the normalization condition is fulfilled:

$$\sum_{i=0}^{\infty}\frac{1}{i!}\int_{\Omega^i}\pi_i(t_1,\ldots,t_i;\Omega)dt_1\ldots dt_i = 1 \tag{2}$$

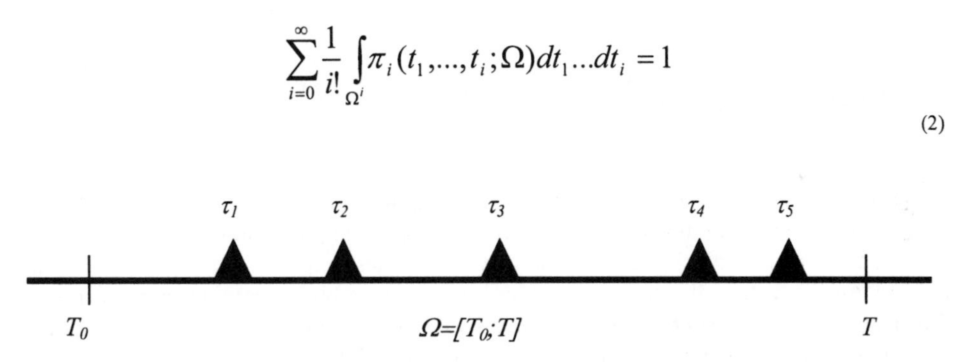

Figure 2: Definition of the random point processes and generating functional.

The π-set is not the only set that can be used for the description of a RPP. A useful set of PDDs is the a-set. $a_i(t_1,\ldots,t_i)$ denotes the joint PDD to detect the *first i points* in the detection interval Ω and each of these points belongs to the subinterval $[t_j,t_j+dt_j)\in\Omega$, $j = 1,\ldots,i$, irrespective of the time occurrences of the *subsequent* points in Ω. These densities are derived via π-densities (1) as

$$a_i(t_1,\ldots,t_i) = \pi_i(t_1,\ldots,t_i;[0;t_i]) \tag{3}$$

The a-set is important from a practical point of view: the necessity to detect a few earlier arrived points (remarkably, only the first one) in the detection interval leads to simplified hardware design requirements. Another set of PDDs frequently used in applications is the f-set:

$$\{f_i(t_1,\dots,t_i)\}_{i=1,2,\dots} \tag{4}$$

where $f_i(t_1,\dots,t_i)$ defines the joint PDD of obtaining i points in the detection interval Ω and each of these points belongs to the subinterval $[t_j,t_j+dt_j)\in\Omega$, $j = 1,\dots,i$, irrespective of the time occurrences of the other points in Ω. The functions of special interest from this set are the intensity $f_1(t)$ (average number of points per time unit) and the time correlation function $f_2(t_1,t_2)$.

We note that for the complete statistical description of a RPP, it is sufficient to have all functions from only one density set (π-, f- or a-sets, other sets can still be defined). The other density sets can be derived from a known one. As an example, we define the f- and a- sets via the π- set[22]:

$$f_i(t_1,\dots,t_i)=\sum_{k=0}^{\infty}\frac{1}{k!}\int_{\Omega^k}\pi_{i+k}(t_1,\dots,t_i,t_{i+1},\dots,t_{i+k};\Omega)dt_{i+1}\dots t_{i+k},\ i=1,2,\dots \tag{5}$$

$$a_i(t_1,\dots,t_i)=\sum_{k=0}^{\infty}\int_{t_i}^{T}\int_{t_{i+1}}^{T}\int_{t_{k+i-1}}^{T}\pi_{i+k}(t_1,\dots,t_i,t_{i+1},\dots,t_{i+k};\Omega)dt_{i+1}\dots t_{i+k},\ i=0,1,\dots \tag{6}$$

The possibility to use different PDD sets is important from a practical point of view: mathematical derivation for one set of functions may be simpler than for the others, or the detection of one statistical characteristic may require less expensive (hardware/software) technologies than the others. For example, it is always advantageous to work with one-dimensional functions such as π_1, f_1, a_1, compared to the multi-dimensional counterparts.

To derive all functions, one by one, for any PDD set is a very tedious task. Fortunately there is a very powerful means allowing treatment of all RPP statistical characteristics in a coherent way, namely the generating functional (GFL). In the next section we proceed with the definition, properties and application of generating functionals.

2.1.2. Generating Functional

The GFL for the RPP is defined as[20,22]

$$L[u;\Omega]=\left\langle \prod_{j=1}^{n}\left[1+u(\tau_j)\right]\right\rangle_{n;\tau_1,\ldots,\tau_n\in\Omega} \tag{7}$$

where u is a trial function [in general, an arbitrary function used as a formal parameter for the functional (7)] and $\langle\rangle_{n,\tau_1,\ldots,\tau_n}$ denotes averaging over the number n of points and times $\tau_1,\ldots,\tau_n$ of their occurrences in the interval Ω. Performing averaging in Eq. (7) yields[20,22]

$$L[u;\Omega]=\sum_{i=0}^{\infty}\frac{1}{i!}\int_{\Omega^i}\pi_i(t_1,\ldots,t_i;\Omega)\prod_{j=1}^{i}[1+u(t_j)]dt_1\ldots dt_i \tag{8}$$

or, using the f-set [Eq. (4)]:

$$L[u;\Omega]=1+\sum_{i=1}^{\infty}\frac{1}{i!}\int_{\Omega^i}f_i(t_1,\ldots,t_i)\prod_{j=1}^{i}u(t_j)dt_1\ldots dt_i \tag{9}$$

One can immediately see that with $u\equiv 0$, the GFL represents a normalization condition [Eq. (2)]. We also define the conditional GFL, which will be further used in modeling:

$$L[u;\Omega|\tau_1,\ldots,\tau_n]=\sum_{i=0}^{\infty}\frac{1}{i!}\int_{\Omega^i}\pi_i(t_1,\ldots,t_i;\Omega|\tau_1,\ldots,\tau_n)\prod_{j=1}^{i}[1+u(t_j)]dt_1\ldots dt_i \tag{10}$$

where $\pi_i(t_1,\ldots,t_i;\Omega)$ is the joint PDD of obtaining *exactly i points* in the detection interval Ω and each of these points belongs to the subinterval $[t_j,t_j+dt_j)\in\Omega$, $j=1,\ldots,i$ provided n other points occurred at the time instances $\tau_1,\ldots,\tau_n$.

It should be stressed that a trial function $u(t)$ does not have any physical meaning. The GFL takes this function as an argument and produces a number as a result. Since this function does not have any physical meaning, neither does the GFL. A GFL can be considered as a formal structure used to combine the RPP statistical characteristics (PDD sets). As any other mathematical abstraction (e.g. a number), it is useless until a set of operations applicable for that abstraction is defined.

2.1.3. *Functional Derivatives*

If one has a method to combine the RPP statistical characteristics into one mathematical object, then there should be a formal way to extract those characteristics back. For this aim we use the operation of functional differentiation[20]. For example, the π PDD [Eq. (1)] can be extracted from GFL [Eq. (8)] by means of functional differentiation when $u(t)=-1$:

$$\pi_i(t_1,...,t_i;\Omega) = \left.\frac{\delta^i L[u;\Omega]}{\delta u(t_1)...\delta u(t_i)}\right|_{u(t)=-1} \quad (11)$$

Performing i-fold functional differentiation of (8) or (9) with $u(t)=0$, yields:

$$f_i(t_1,...,t_i) = \left.\frac{\delta^i L[u;\Omega]}{\delta u(t_1)...\delta u(t_i)}\right|_{u(t)=0} \quad (12)$$

Further we outline the basic rules of the functional differentiation[22]:

If $L[u;\Omega] = a(t)$, where $a(t)$ is independent of $u(t)$, then $\delta L[u;\Omega]/\delta u(t) = 0$.

If $L[u;\Omega] = \int_\Omega a(\tau)u(\tau)d\tau$, then $\delta L[u;\Omega]/\delta u(t) = a(t)$.

If $L[u;\Omega] = a(\tau)u(\tau)$, then $\delta L[u;\Omega]/\delta u(t) = a(t)\delta(t-\tau)$, where $\delta(t)$ is the Dirac delta function.

If $L[u;\Omega] = \int_{\Omega^2} f(t_1,t_2)u(t_1)u(t_2)dt_1dt_2$, then $\frac{\delta L[u;\Omega]}{\delta u(t)} =$
$= \int_\Omega f(t,t_2)u(t_2)dt_2 + \int_\Omega f(t_1,t)u(t_1)dt_1$.

If a trial function for a GFL is itself a GFL: $L[u;\Omega] = L_0[I[u;\Omega|\cdot];\Omega]$, then $\frac{\delta L[u;\Omega]}{\delta u(t)} = \int_\Omega \frac{\delta L_0[I[u;\Omega|\cdot];\Omega]}{\delta I[u;\Omega|\tau]} \frac{\delta I[u;\Omega|\tau]}{\delta u(t)} d\tau$.

2.1.4. Generating Function

A generating function (GF) is yet another formal construction, which is frequently used in statistics[25]. A GF can be obtained from GFL [Eq. (8)] by replacing the trial function u by a formal variable z-1.

$$\Theta[z;\Omega] = \langle z^n \rangle_n = \sum_{i=0}^{\infty} P_i(\Omega)z^i = L[z-1;\Omega] \quad (13)$$

The PD of the number of points in Ω is obtained via multi-fold differentiation of GF $\Theta[z;\Omega]$ with respect to z, when $z = 0$:

$$P_i(\Omega) = \left.\frac{d^i \Theta[z;\Omega]}{i!\,dz^i}\right|_{z=0} = \frac{1}{i!}\int_{\Omega^i} \pi_i(t_1,...,t_i;\Omega)dt_1...t_i \tag{14}$$

A somewhat easier way to obtain the PD is to perform inverse Fourier transform of the characteristic function (CF), which can be obtained from GF $\Theta[z;\Omega]$, Eq. (13) by substituting the formal variable z by the complex exponent $\exp(-j\varpi)$:

$$\Theta[z;\Omega] = \sum_{i=0}^{\infty} P_i(\Omega)z^i = \sum_{i=0}^{\infty} P_i(\Omega)\exp(ij\varpi) = \Theta[\varpi;\Omega] \tag{15}$$

where j is the complex unit and $\varpi \in [0;2\pi]$ is the frequency of the discrete Fourier transform. Thus, CF $\Theta[\varpi,\Omega]$ (15) is the discrete Fourier transform of the PD of the number of points in Ω.

We note that the transformation of GFL into GF eliminates information about the time positions of the points. Afterwards, what one can recover are only the characteristics of the number of points per interval (probability to obtain certain number of points in Ω, or average amount of points in Ω, etc). It is however, impossible to derive any of PDD sets (π, f or a) using just only GF.

2.2. Examples of Random Point Processes

To exemplify the GFL technique, in this section we consider several examples of RPP.

2.2.1. One-Point RPP and RPP With Multiple Points

One-point RPP consists of only one point, whose position is distributed according to the PDD $f(t)$. The GFL for such a process takes the form:

$$L[u;\Omega] = 1 + \int_{\Omega} f(t)u(t)dt \tag{16}$$

If several (multiple) points appear simultaneously at the same time instance, GFL [Eq. (16)] is transformed to:

$$L[u;\Omega] = 1 + \int_{\Omega} f(t)\langle u(t)^n \rangle_n dt \tag{17}$$

where $<>_n$ denotes averaging over the number of points occurring simultaneously.

2.2.2. Poisson Random Point Process

Poisson random point process (PRPP) is frequently used in different applications. For example, when a well-stabilized laser is used as a continuous excitation source, the duration between consecutive photocounts is approximated by an exponential distribution and the train of the generated photocounts possesses Poisson statistics[7]. Another important implication is that a superposition of a reasonably large amount of any independent RPP leads to a PRPP[21]. The GFL of the PRPP process takes the form:

$$L[u;\Omega] = \exp\left\{ \int_{\Omega} \lambda(t) u(t) dt \right\} \tag{18}$$

where $\lambda(t)$ is the intensity of the process.

Mathematical notation (18) is very compact, but contains all information about the process. For example, one can show independence of PRPPs occurring in different (non-intercrossing) time intervals Ω_i, $i=1,\dots,n$:

$$L\left[u;\bigcup_{i=1}^{n}\Omega_i\right] = exp\left\{ \int_{\bigcup_{i=1}^{n}\Omega_i} \lambda(t)u(t)dt \right\}$$

$$= \prod_{i=1}^{n} exp\left\{ \int_{\Omega_i} \lambda(t)u(t)dt \right\} = \prod_{i=1}^{n} L[u;\Omega_i] \tag{19}$$

The resulting GFL [Eq. (19)] is a product of GFLs for each interval, implying independence of the processes occurring at these intervals. Functional differentiation with $u(t) = -1$ generates a π-set:

$$\pi_i(t_1,\dots,t_i;\Omega) = \prod_{j=1}^{i} \lambda(t_j) \exp\left\{ -\int_{\Omega} \lambda(t)dt \right\}, i = 0,1,\dots \tag{20}$$

The GF of the PRPP takes the form

$$\Theta[z;\Omega] = \exp\{\Lambda(\Omega)(z-1)\} \tag{21}$$

where $\Lambda(\Omega) = \int_{\Omega} \lambda(t)dt = \langle n \rangle = \sigma^2(\Omega)$; $\langle n \rangle$ and $\sigma^2(\Omega)$ are the average and the variance of the number of points n in Ω. The PD of the number n is given by:

$$P_n(\Omega) = <n>^n \exp(-<n>)/n!, n = 0,1,...$$

(22)

2.2.3. *Doubly Stochastic Poisson Point Process*

It is well known that the random process of photocounts is only Poisson when the intensity of light is non-fluctuating (coherent radiation). The photocount statistical characteristics for the incident light with fluctuating intensity can be described by the doubly stochastic random point process (DSRPP)[7]. A DSRPP is a PRPP with the intensity being a random process[23] (**Figure 3**). The GFL of a DSRPP can be found by averaging the Poisson GFL (18) over the random process $\lambda(t)$:

$$L[u;\Omega] = \left\langle \exp\left\{ \int_{\Omega} \lambda(t) u(t) dt \right\} \right\rangle_{\lambda(t)}$$

(23)

where we assume that for any time points $t \in \Omega$, $\lambda(t) \geq 0$.

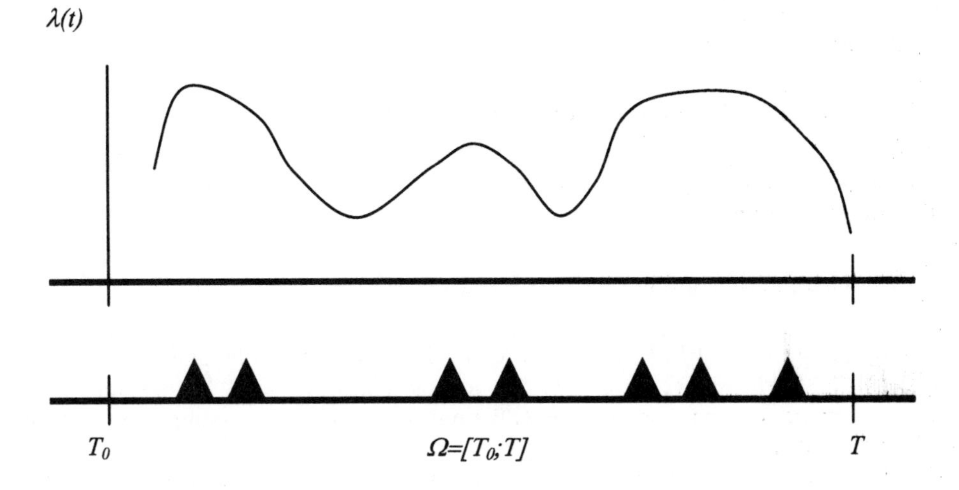

Figure 3: Doubly stochastic RPP

Functional differentiation of Eq. (23) with respect to $u(t)$ provided $u(t)=0$ yields:

$$f_i(t_1,...,t_i) = \langle \lambda(t_1)...\lambda(t_i) \rangle, \quad i = 1,2,...$$

(24)

As follows from Eq. (24), the moment correlation functions of the DSRPP correspond to moment correlation functions of the random intensity of PRPP. The PD of the number of points per counting interval Ω can be obtained from Eqs. (13) and (23):

$$P_k(\Omega) = \langle E^k exp(-E)/k! \rangle_E, \; k=0,1,...$$

(25)

where $E = \int\limits_{\Omega} \lambda(t)dt$.

GFL (23) can be written explicitly if the statistical characteristics of the random process $\lambda(t)$ are known. For example, if the intensity of the PRPP is a gaussian random process, the DSRPP will be a pair-wise random point process (PWRPP) with the following GFL:

$$L[u;\Omega] = \exp\left\{ \int\limits_{\Omega} g_1(t)u(t)dt + \int\limits_{\Omega}\int\limits_{\Omega} g_2(t_1,t_2)u(t_1)u(t_2)dt_1dt_2 \right\}$$

(26)

where $g_1(t)=\langle\lambda(t)\rangle$, $g_2(t_1,t_2)=\langle[\lambda(t_1)-g_1(t_1)][\lambda(t_2)-g_1(t_2)]\rangle$.

2.3. Models for Random Point Processes Transformations

In this section we consider application of the GFL technique to model basic types of RPP transformations.

2.3.1. Superposition of Random Point Processes

A superposition of RPPs can be represented as a mixture of the points, belonging to different RPPs (in **Figure 4**: mixture of open and black symbols). We assume that the time occurrences of the points of one RPP are independent of the point time occurrences of the other RPPs.

In terms of the GFL technique, superposition of n independent RPPs with GFL $L_i[u_i;\Omega]$, $i = 1,...,n$ is defined as a product[20,22]:

$$L[u_1,...,u_n;\Omega] = \prod_{i=1}^{n} L_i[u_i;\Omega]$$

(27)

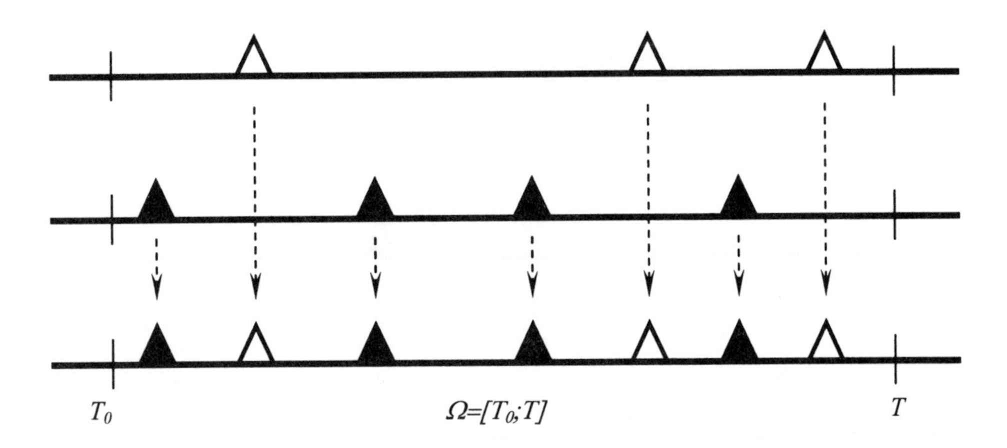

Figure 4: Superposition of random point processes

Note that in Eq. (27), we assign different trial functions u_i to the different RPPs. This difference let us *distinguish* points coming from one RPP from the points of the other RPPs in the resulting GFL. For example, to extract GFL of the k-th RPP, we should put all u_i equal to zero, except for u_k:

$$L_k[u_k;\Omega]=L[0,\ldots,0,u_k,0,\ldots,0;\Omega] \tag{28}$$

In general, if we want to combine in one GFL points with different physical characteristics or even of different physical nature (like photons and electrons), we should use different trial functions in order to preserve that difference.

To make a superposition into a mixture of *indistinguishable* points, we identify all trial functions: $u = u_1 = \ldots = u_n$:

$$L[u;\Omega]=\prod_{i=1}^{n} L_i[u;\Omega] \tag{29}$$

Once the resulting functional $L[u;\Omega]$ is known, statistical characteristics of the resulting RPP can be obtained by means of functional differentiation.

2.3.2. *Shift of Points*

If each point of the RPP with GFL $L_1[u;\Omega]$ can change its position independently of the other points from the same RPP (**Figure 5**), so that the new time position of a point i becomes $t_i = \tau_i + \xi_i$, where τ_i is the old position and ξ_i is the time shift, the GFL of the resulting RPP takes the form[22]:

$$L[u;\Omega] = L_1\left[\int_{\Omega} f(t-\cdot)u(t)dt;\Omega\right] \tag{30}$$

where $f(t)$ is the PDD of the shift ξ.

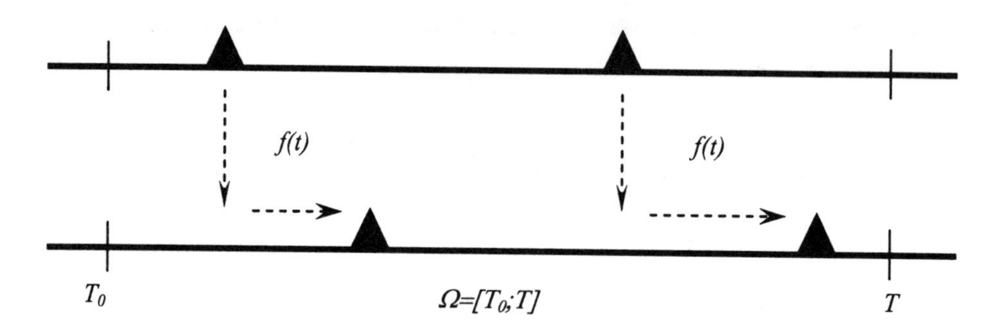

Figure 5: Shift of time points

2.3.3. Rejection of Points

Rejection of points results in the loss (or disappearance) of points from a RPP (**Figure 6**). If each point from the RPP with GFL $L_1[u;\Omega]$ can be lost (*independently* of the other points) with the probability $q(t)$, the resulting GFL takes the form:

$$L[u;\Omega] = L_1[q(\cdot)u(\cdot);\Omega] \tag{31}$$

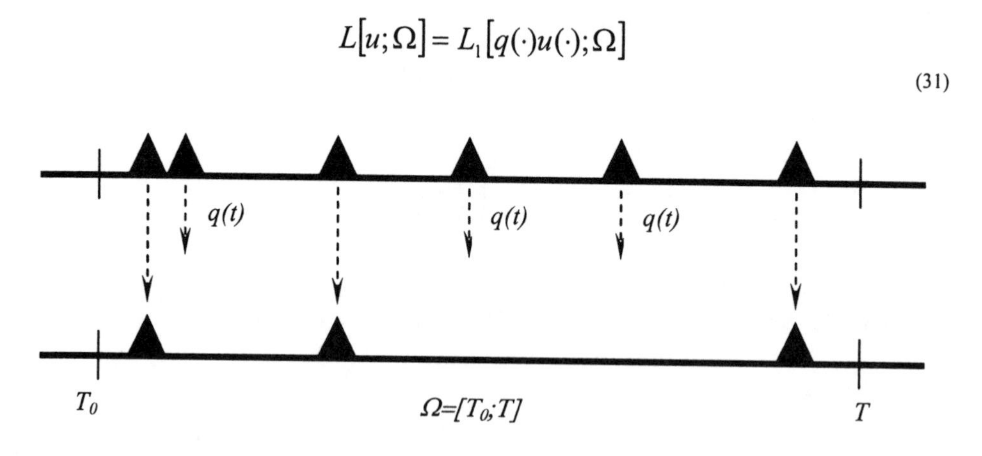

Figure 6: Rejected RPP

2.3.4. Cluster Random Point Processes

Cluster random point process (CRPP) is an example of the so-called genetically linked RPP (**Figure 7**). Each of the points from the primary RPP (open symbols in **Figure 7**) generates one or several secondary points (black symbols in **Figure 7**), i.e., points of the second generation. Both primary and secondary points contribute to the resulting process. We assume that: (i) each point of the primary RPP produces points of the second generation independently of the other points; (ii) the laws of generation for every primary point may only depend on the coordinate of the primary point.

We assign GFL $I[v;\Omega]$ to the primary RPP and GFL $L_0[u;\Omega|\tau]$ to each secondary RPP. Note that the latter GFL is conditioned on τ – the coordinate of the initiating point occurrence. Superposition of a primary RPP and all secondary RPPs forms the cluster RPP with the following GFL[22]

$$L[v,u;\Omega] = I[(1+v(\cdot))L_0[u;\Omega|\cdot]-1;\Omega] \tag{32}$$

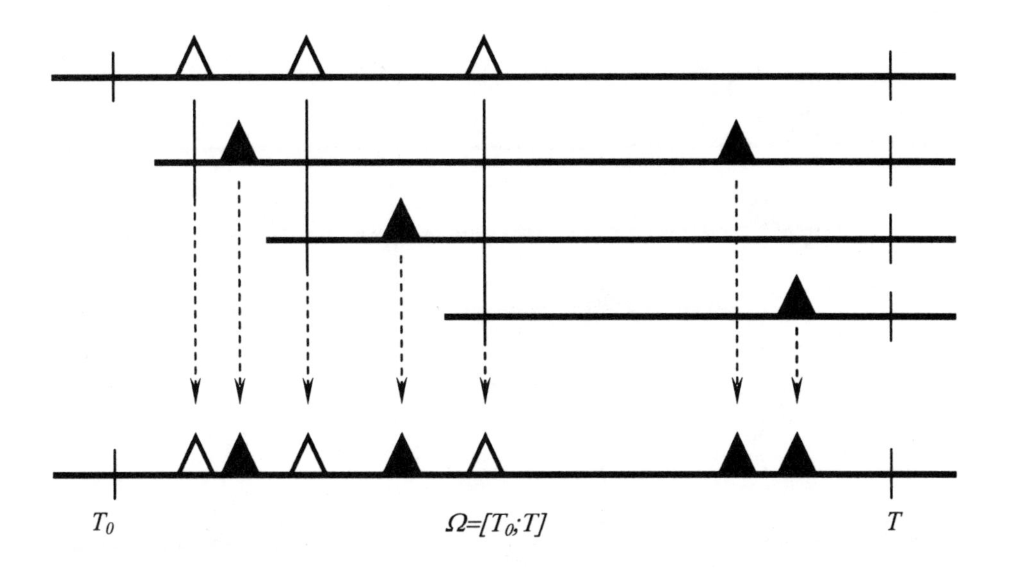

Figure 7: Cluster RPP

As in the case of RPP superposition (section 2.3.1), we preserve the difference between the points of the primary RPP and secondary RPPs. Thus, to eliminate the points of the primary RPP from the resulting RPP (**Figure 8**) it is sufficient to set the trial function v equal to zero:

$$L[u;\Omega] = I\left[L_0\left[u;\Omega|\cdot\right] - 1;\Omega\right] \tag{33}$$

If the trial functions v and u are equivalent ($w = v = u$), the mixture of the points from the primary and secondary generations becomes indistinguishable:

$$L[w;\Omega] = I\left[(1 + w(\cdot))L_0\left[w;\Omega|\cdot\right] - 1;\Omega\right] \tag{34}$$

Functional differentiation of the functional Eqs. (32), (33), and (34) establishes the link between the statistical characteristics of the primary process I, the characteristics of the secondary points L_0 and the characteristics of the resulting processes L.

We consider an example, where the primary RPP is the PRPP (section 2.2.2) and the secondary RPP is a one-point RPP (section 2.2.1). The GFL for the one-point RPP (16) is conditioned on the time occurrence τ of the point from the primary PRPP:

$$L_0\left[u;\Omega|\tau\right] = 1 + \int_{\Omega} f(t|\tau)u(t)dt \tag{35}$$

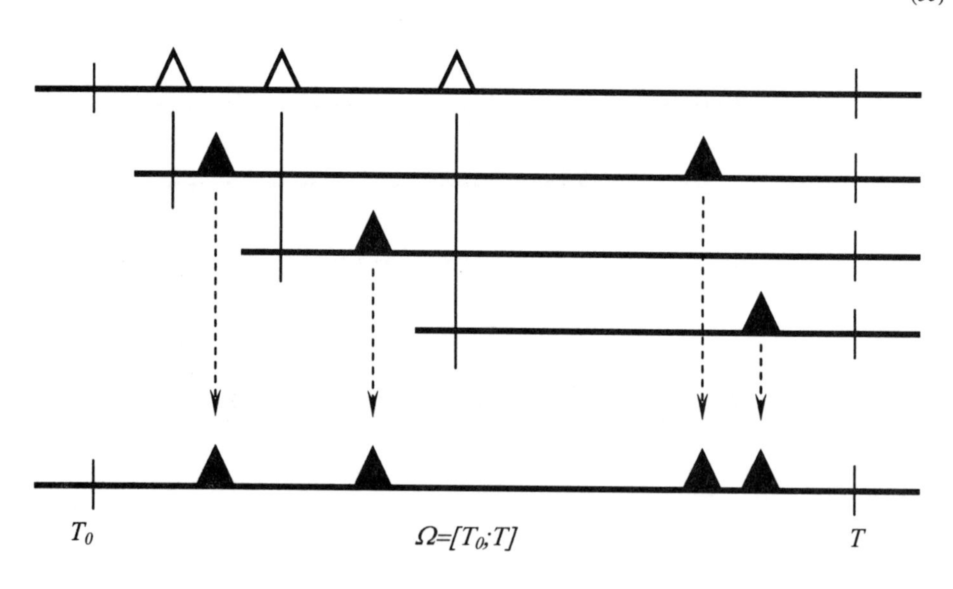

Figure 8: Superposition of secondary random point processes

Substitution of GFL (18) and (35) into the model of CRPP (34) yields the resulting GFL:

$$L[w;\Omega] = exp\left\{ \int_{\Omega} \lambda(t)w(t)dt + \int_{\Omega}\int_{\Omega} \lambda(\tau)f(t-\tau)[1+w(\tau)]w(t)dtd\tau \right\} \tag{36}$$

Note that GFL (36) is equivalent to the GFL of the PWRPP (26), meaning that completely different mechanisms may lead to RPPs with very similar statistical properties.

2.3.5. Branching Random Point Processes

The models for the branching random point process (BRPP) are based on the CRPP. To ensure branching, we allow secondary points to produce points of the following generations (**Figure 9**).

We will consider points of two types: generating (wavy symbols in **Figure 9**) and detected (black symbols in **Figure 9**). A generating point is a point that can generate further generating points (zero, one, or several); otherwise it is transformed into a single detected point. The characteristics of the newly generated points are dependent on the characteristics of the generating point. Detected points are considered as points of the resulting process. We suppose that after a random (in general, infinite) number of generations only detected points will exist in the system. We assume that the algorithms of generation and transformation for every generating point may only be dependent on the coordinate of the generating point.

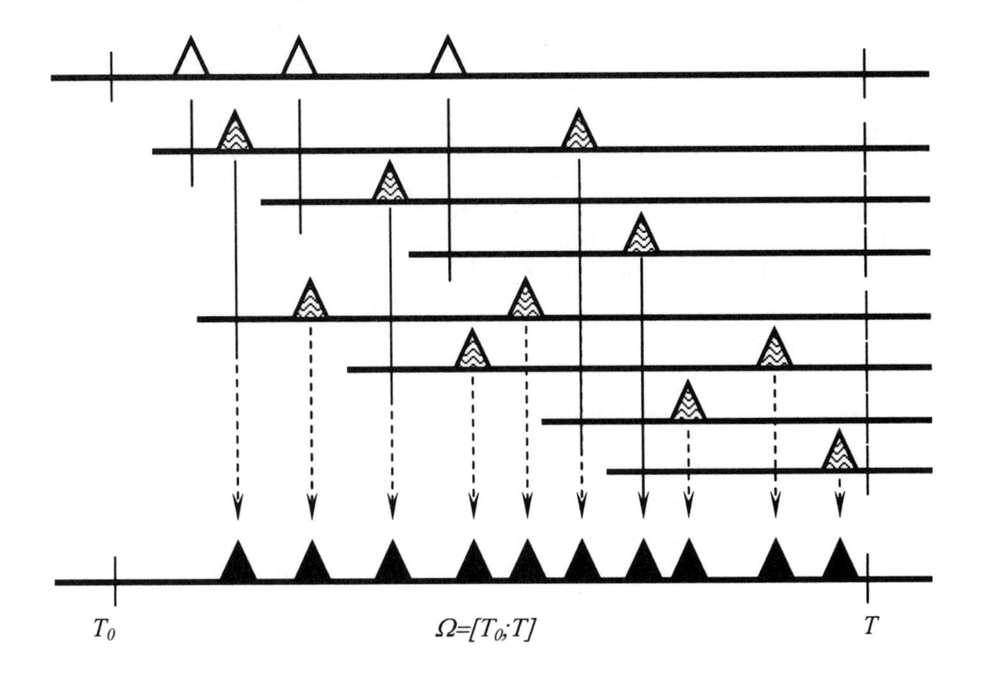

Figure 9: Branching RPP

As indicated earlier, to model the process with different types of points in terms of the GFL formalism, one should assign different trial functions to the different point types. In this section, we will use the trial function w for the detected points and the trial function u for the generating points. Let $L_0[w,u;\Omega|t]$ be the GFL of the generating and detected points, belonging to a single generation, produced by one generating point with coordinate t *independently* of the other points existing in the system. It can be shown[26] that the GFL $L[w;\Omega|t]$ of the resulting RPP of the detected points after an infinite number of generations, initiated by a generating point with coordinate t, can be written as

$$L[w;\Omega \mid t] = L_0[w, L[w;\Omega \mid \cdot] - 1;\Omega \mid t] \tag{37}$$

The statistical characteristics of the points belonging to the first (initiating) generation (open symbols in **Figure 9**) are usually different from the ones at the intermediate stages. Therefore, we use the special GFL $I[w,u;\Omega]$ to take into account the specific behavior of the initiating points. Combining the statistical characteristics of the first-stage points and points produced during the multi-stage generation yields the expression for the resulting GFL[26]:

$$L[w;\Omega] = I[w, L[w;\Omega \mid \cdot] - 1;\Omega] \tag{38}$$

Equations (37) and (38) are the set of functional equations for obtaining the statistical characteristics of the detected RPP when single generation parameters are known and *vice versa*: to obtain the statistical parameters of a single generation based on the characteristics of the detected process.

In this case we do not have a straightforward expression for the GFL of the resulting BRPP. The GFL of the resulting BRPP is a solution of the set of functional equations, which can very rarely be obtained analytically. Normally, these equations are functionally differentiated yielding a set of integral equations connecting the statistical characteristics of the output RPP, first-stage RPP and the statistical characteristics of the secondary points, belonging to a single generation.

2.3.6. Multi-Stage Transformations

If several types of points participate in branching, so that points of one type at one stage can be transformed into points of another type at the following stages, functional equations must be written for each point type.

We consider an example (**Figure 10**) with two types of points (wavy and cross symbols), besides the initial one (open symbols). In our case each point can produce points of its own type and points of a different type. For two types of points we get a set of two functional equations:

$$L_B\left[w,u_B,u_G;\Omega \mid \tau_B\right]$$
$$= L_{0B}\left[w, L_B\left[w,u_B,u_G;\Omega \mid \cdot_B\right]-1, L_G\left[w,u_B,u_G;\Omega \mid \cdot_G\right]-1;\Omega \mid \tau_B\right] \tag{39}$$

$$L_G\left[w,u_B,u_G;\Omega \mid \tau_G\right]$$
$$= L_{0G}\left[w, L_B\left[w,u_B,u_G;\Omega \mid \cdot_B\right]-1, L_G\left[w,u_B,u_G;\Omega \mid \cdot_G\right]-1;\Omega \mid \tau_G\right] \tag{40}$$

where $L_B\left[w,u_B,u_G;\Omega \mid \tau_B\right]$ and $L_G\left[w,u_B,u_G;\Omega \mid \tau_G\right]$ are GFL of the resulting RPP of a mixture of two different types B (cross symbols) and G (wavy symbols) produced by a point of type B, occurring at time τ_B, or of type G, occurring at time τ_G, correspondingly; $L_{0B}\left[w,u_B,u_G;\Omega \mid \tau_B\right]$ and $L_{0G}\left[w,u_B,u_G \mid \tau_G\right]$ are the correspondent single-stage GFLs. As in the previous case the resulting GFLs $L_B\left[w,u_B,u_G;\Omega \mid \tau_B\right]$ and $L_G\left[w,u_B,u_G;\Omega \mid \tau_G\right]$ must be averaged on the positions of the points belonging to the first (initiating) generation.

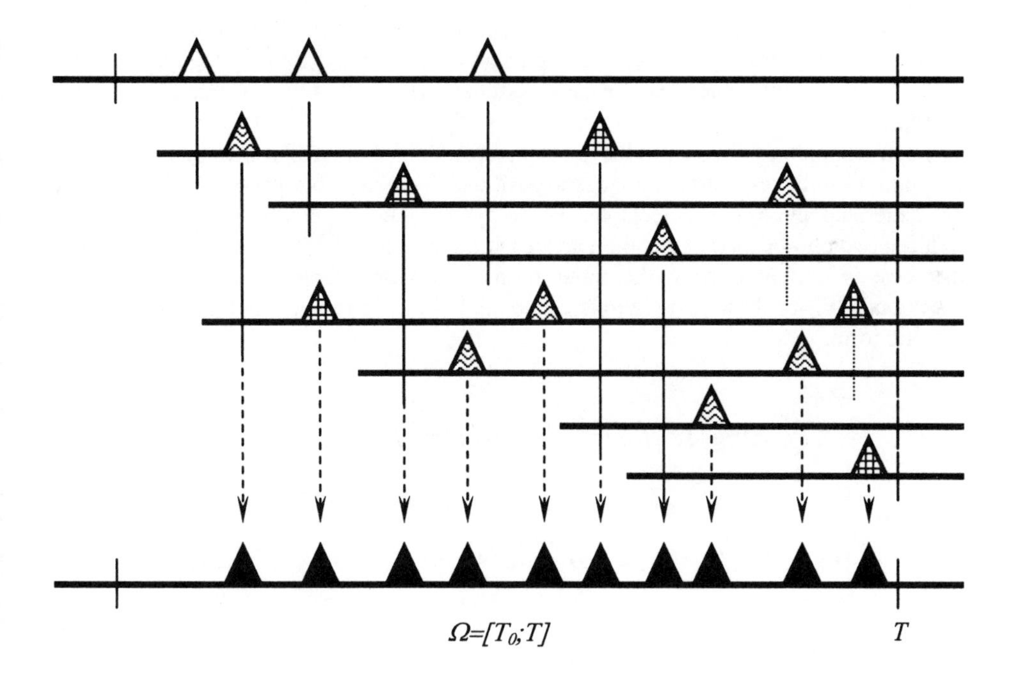

Figure 10: Branching RPP with two types of points

3. APPLICATION TO SINGLE MOLECULE FLUORESCENCE

3.1. Model

For the purposes of an adequate description of the single molecule fluorescence experiment, an appropriate adaptation and further elaboration of the model presented in Sections 2.3.5 and 2.3.6 is necessary. Hence, we start by using the term "photon" instead of the term "point". We will further assume that all photons from the initial train of excitations can be subdivided into two groups: detectable (corresponding to symbols with vertical lines in **Figure 1**) and undetectable photons (the symbols with cross lines in **Figure 1**). The detectable photons yield photons reaching the detector (i.e. detected photons or photocounts, the black symbols in **Figure 1**). The undetectable photons correspond to photons that are lost during the time when the molecule remains in an excited state. As in Section 2.3.5, we will assign different trial functions to the different types of photons: the trial functions v and u are associated with detectable and undetectable photons, respectively. Note that in the context of single molecule fluorescence, detectable and undetectable photons act as generating points (as shown in Section 2.3.5).

Since the input train of excitation photons is defined by the PDD $h(t)$ of inter-photon distance, we can consider the excitation process as a process of multi-stage generations, where each photon generates the next photon with a time occurrence distributed according to $h(t)$.

Each detectable photon produces an undetectable photon, if it occurs within the Δ time-interval with respect to the current position of a detectable photon. Otherwise, the new generated photon is detectable. To summarize, each detectable photon formally produces two photons: (i) one photon arrives at the detector (i.e. it is detected) and (ii) the other one is a photon from the input train (either detectable or undetectable). The conditional GFL of these two photons generated by one detectable photon, occurring at t, takes the form:

$$L_{0v}\left[w,v,u;\Omega\,|\,t,z\right]=[1+\kappa w(z)]\left\{1-\int_t^T h(x-t)dx+\right.$$
$$\left.+\int_z^T h(x-t)\langle 1+v(x,x+\Delta)\rangle_\Delta dx+\int_t^z h(x-t)[1+u(x,z)]dx\right\} \tag{41}$$

where the trial function w is assigned to detected photons, z represents the time occurrence of the detected photon and $\langle\rangle_\Delta$ denotes averaging over Δ.

As for detectable photons, each undetectable photon produces the next excitation photon with the PDD of time occurrence $h(t)$. If this new photon occurs within the Δ time-interval started at the time occurrence of the previous detectable photon, it becomes undetectable; otherwise, it is detectable. In contrast to detectable photons, an undetectable photon does not produce a detected photon. The conditional GFL of one photon (either detectable or undetectable) generated by one undetectable photon, occurring at t, takes the form:

$$L_{0u}\left[w,v,u;\Omega\mid t,z\right]=\frac{L_{0v}\left[w,v,u;\Omega\mid t,z\right]}{[1+\kappa w(z)]}=1-\int_t^T h(x-t)dx+$$

$$+\int_z^T h(x-t)\langle 1+v(x,x+\Delta)\rangle_\Delta dx+\int_t^z h(x-t)[1+u(x,z)]dx$$

(42)

where z denotes the time instance where the Δ time interval – started at the time occurrence of the previous detectable photon – ends, i.e., z corresponds to the occurrence time of a detected photon.

The point generation develops in time so that finally all photons in Ω will be either detected or lost. The detected characteristics are enclosed in the conditional GFLs $L_v[w;\Omega|t,z]$ and $L_u[w;\Omega|t,z]$ of the detected photons, initiated by one detectable or undetectable photon, respectively, occurring at t, provided that the next detected photon will arrive at time z. When one extends the formalism, outlined in Sections 2.3.5 and 2.3.6, to processes with three types of points (i.e., detectable, undetectable, and detected photons), one can show that the GFLs $L_v[w;\Omega|t,z]$ and $L_u[w;\Omega|t,z]$ are the solutions of the set of functional equations, represented by

$$L_v\left[w;\Omega\mid t,z\right]=L_{0v}\left[w,L_v\left[w;\Omega\mid\cdot,\cdot\right]-1,L_u\left[w;\Omega\mid\cdot,\cdot\right]-1;\Omega\mid t,z\right]$$

(43)

$$L_u\left[w;\Omega\mid t,z\right]=L_{0u}\left[w,L_v\left[v_d;\Omega\mid\cdot,\cdot\right]-1,L_u\left[v_d;\Omega\mid\cdot,\cdot\right]-1;\Omega\mid t,z\right]$$

(44)

Substitution of Eqs. (41) and (42) in the set given by Eqs. (39) and (40) yields:

$$L_v\left[w;\Omega\mid t,z\right]=[1+\kappa w(z)]\times$$

$$\times\left\{1-\int_t^T h(x-t)dx+\int_z^T h(x-t)\langle L_v\left[w;\Omega\mid x,x+\Delta\right]\rangle_\Delta dx+\right.$$

$$\left.+\int_t^z h(x-t)L_u\left[w;\Omega\mid x,z\right]dx\right\}$$

(45)

$$L_u\left[w;\Omega \mid t,z\right]$$

$$= 1 - \int_t^T h(x-t)dx + \int_z^T h(x-t)\left\langle L_v\left[w;\Omega \mid x, x+\Delta\right]\right\rangle_\Delta dx +$$

$$+ \int_t^z h(x-t)L_u\left[w;\Omega \mid x,z\right]dx$$

(46)

By averaging Eq. (45) over the time occurrences of the first detectable photon, occurring in Ω, one obtains the expression for the resulting GFL:

$$L\left[w;\Omega\right] = 1 - \int_{T_0}^T a_1(x)dx + \int_{T_0}^T a_1(x)\left\langle L_v\left[w;\Omega \mid x,x+\Delta\right]\right\rangle_\Delta dx$$

(47)

where $a_1(t)$ is the PDD of the arrival time of the **first *detectable* photon** in Ω.

The temporal characteristics of the detected photons are obtained by functional differentiation (see Section 2.1.3) of functional Eqs. (45), (46) and (47). The PD of the number of photons in the detection interval Ω can be calculated via GF $\Theta[z;\Omega]$ (see Section 2.1.4). Integral equations for $\Theta[z;\Omega]$ are derived from Eqs. (45), (46), and (47) using the formal variable $(z - 1)$ instead of the trial function w.

If the detected signal (i.e., the number of counts per detection interval along with their time occurrences) is an indistinguishable mixture of counts, either produced by photons emitted by the investigated molecule or due to background, the GFL of the detected process can be written as the product

$$\hat{L}\left[w;\Omega\right] = L\left[w;\Omega\right]L_B\left[w;\Omega\right]$$

(48)

where $L_B[w;\Omega]$ is the GFL of background counts. In Eq. (48) it is assumed that the time occurrences of background counts are independent of those of photocounts coming from the molecule. Addition of the "background" GFL $L_B[w;\Omega]$ makes the calculations of the output distribution more complicated, but does not change the general principle of the procedure: one needs to perform a functional differentiation of relationship (48), yielding equations related to the characteristics of the single molecule, the excitation and detected photons as well as the characteristics of the background radiation (see Appendix 1 for more details). In most cases, background radiation is represented by a stationary PRPP, which is completely defined by its intensity (see Section 2.2.2). Its value, along with the other system parameters, can be estimated. However, the estimation procedure becomes more stable, if the background level is known from an independent experiment.

The model represented by Eqs. (45), (46), and (47) is appropriate for different types of molecular behavior and types of excitation. Next, we shall consider continuous excitation as the one most frequently used in single molecule fluorescence.

3.2. Continuous Excitation

In the case of continuous excitation, when a laser with a well-stabilized intensity is used, the time interval between consecutive photons is approximated by an exponential distribution[7]:

$$h(t) = \lambda \exp(-\lambda t) \tag{49}$$

where λ is the intensity of the input process.

Substituting Eq. (49) into (46) and subsequent differentiation of Eq. (46) with respect to t yields

$$\frac{\partial}{\partial t} L_u\left[w;\Omega \mid t,z\right] = 0 \tag{50}$$

with the initial condition

$$L_u\left[w;\Omega \mid t,z\right]_{t=z}$$

$$= \exp\{-\lambda(T-z)\} + \lambda \int_z^T \exp\{-\lambda(x-z)\} \left\langle L_v\left[w;\Omega \mid x, x+\Delta\right]\right\rangle_\Delta dx \tag{51}$$

Since the solution of differential equation (50) is independent of t, the initial condition (51) represents the solution for Eq. (50). Substitution of Eq. (51) into (45) yields

$$L_v\left[w;\Omega \mid t,z\right] = [1+\kappa w(z)] \times$$

$$\times \left\{ \exp\{-\lambda(T-z)\} + \lambda \int_z^T \exp\{-\lambda(x-z)\} \left\langle L_v\left[w;\Omega \mid x, x+\Delta\right]\right\rangle_\Delta dx \right\} \tag{52}$$

Since the right-hand side of Eq. (52) is independent of t, we can omit t from the conditional part of GFL $L_v[w;\Omega|t,z]$. Then Eq. (52) can be rewritten as

$$L_v\left[w;\Omega \mid z\right] = [1+\kappa w(z)] \times$$

$$\times \left\{ \exp\{-\lambda(T-z)\} + \lambda \int_z^T \exp\{-\lambda(x-z)\} \left\langle L_v\left[w;\Omega \mid x+\Delta\right]\right\rangle_\Delta dx \right\} \tag{53}$$

or

$$L_v^*[w;\Omega \mid z] = \left\langle exp\{-\lambda(T - z - \Delta)\} + \right.$$

$$\left. + \lambda \int_{z+\Delta}^{T} exp\{-\lambda(x - z - \Delta)\}[1 + \kappa w(x)] L_v^*[w;\Omega \mid x] dx \right\rangle_{\Delta} \tag{54}$$

where

$$L_v^*[w;\Omega \mid z] = \frac{L_v[w;\Omega \mid z]}{[1 + \kappa w(z)]} \tag{55}$$

Substituting Eq. (55) into (47) yields

$$L[w;\Omega] = 1 - \int_{T_0}^{T} a_1(x)dx + \int_{T_0}^{T} a_1(x) \left\langle [1 + \kappa w(x + \Delta)] L_v^*[w;\Omega \mid x + \Delta] \right\rangle_{\Delta} dx \tag{56}$$

After explicitly performing the averaging over Δ, Eq. (56) can be rewritten as

$$L[w;\Omega] = 1 - \int_{T_0}^{T} a_1^*(x)dx + \int_{T_0}^{T} a_1^*(x)[1 + w(x)] L_v^*[w;\Omega \mid x] dx \tag{57}$$

where $a_1^*(t)$ is the PDD of the arrival time of the **first *detected* photon** in Ω.

The functional Eqs. (54) and (57) represent the basic set of equations, that links the statistical characteristics of the detected photons to the distribution of the Δ time interval for continuous excitation.

3.3. Detection

The temporal statistical characteristics of the detected photons can be obtained by means of functional differentiation (Section 2.1.3) applied to Eqs. (54) and (57).

3.3.1. First and Single Photocount Time Occurrence in the Detection Interval

Single-fold functional differentiation of Eqs. (54) and (57) with respect to w yields, respectively,

$$\frac{\delta L_{v}^{*}[w;\Omega\,|\,z]}{\delta w(t)}=\Big\langle \kappa\lambda\, exp\{-\lambda(t-z-\Delta)\}L_{v}^{*}[w;\Omega\,|\,t]+$$

$$+\lambda\int_{z+\Delta}^{T} exp\{-\lambda(x-z-\Delta)\}[1+\kappa w(x)]\frac{\delta L_{v}^{*}[w;\Omega\,|\,x]}{\delta w(t)}dx\Big\rangle_{\Delta} \tag{58}$$

$$\frac{\delta L[w;\Omega]}{\delta w(t)}=a_{1}^{*}(t)L_{v}^{*}[w;\Omega\,|\,t]+\int_{T_0}^{T}a_{1}^{*}(x)[1+w(x)]\frac{\delta L_{v}^{*}[w;\Omega\,|\,x]}{\delta w(t)}dx \tag{59}$$

Substituting $w = 0$ in Eq. (59) leads to the intensity $f_1(t)$ of the output process:

$$f_1(t)=a_1^{*}(t)+\int_{T_0}^{t}a_1^{*}(x)f_{v1}(t-x)dx \tag{60}$$

where $f_{v1}(t)$ is the solution of the integral equation:

$$f_{v1}(t)=\Big\langle \kappa\lambda\, exp\{-\lambda(t-\Delta)\}+\lambda\int_{0}^{t-\Delta}exp\{-\lambda(t-\Delta-x)\}f_{v1}(x)dx\Big\rangle_{\Delta} \tag{61}$$

obtained from Eq. (58) with $w = 0$. For stationary processes, which is the case for well-stabilized laser radiation, the rate of the detected process is easily accessible:

$$f_1(t)=f_{v1}(\infty)=const \tag{62}$$

This suggests an easy way for the determination of the PDD $a_1^{*}(t)$ of the first detected photon in Ω by solving the integral equation (60) provided that $f_{v1}(t)$ is obtained from the solution of Eq. (61).

Substituting $w = -1$ in Eq. (59) yields the PDD of the single photon occurrence in Ω:

$$\pi_1(t;\Omega)=a_1^{*}(t)\pi_{v0}([t;T]) \tag{63}$$

where $\pi_{v0}([t;T])$ is the probability of zero photocounts in the interval $[t;T]$, obtained from the solution of integral equation (54) when $w = -1$:

$$\pi_{v0}([t;T]) = \left\langle exp\{-\lambda(T-t-\Delta)\} + (1-\kappa)\lambda \int_{t+\Delta}^{T} exp\{-\lambda(x-t-\Delta)\}\pi_{v0}([x;T])dx \right\rangle_{\Delta} \tag{64}$$

The schemes of the detection of the PDD of the arrival time of the first (A) or single photon (B) during the detection interval Ω are shown in **Figure 11**. In the first case (A), the time interval t between the first arriving photon and T_0 (T_0 should be provided by an independent timer) is measured and stored in a histogram. Other photons occurring in $[T_0;T]$ are ignored. When the PDD of the single photon arrival time is measured (B), the arrival time t of the first photon occurring in Ω is measured *provided there are no other photons in the detection interval.* If additional photons arrive during the detection interval, this detection cycle should not be taken into account.

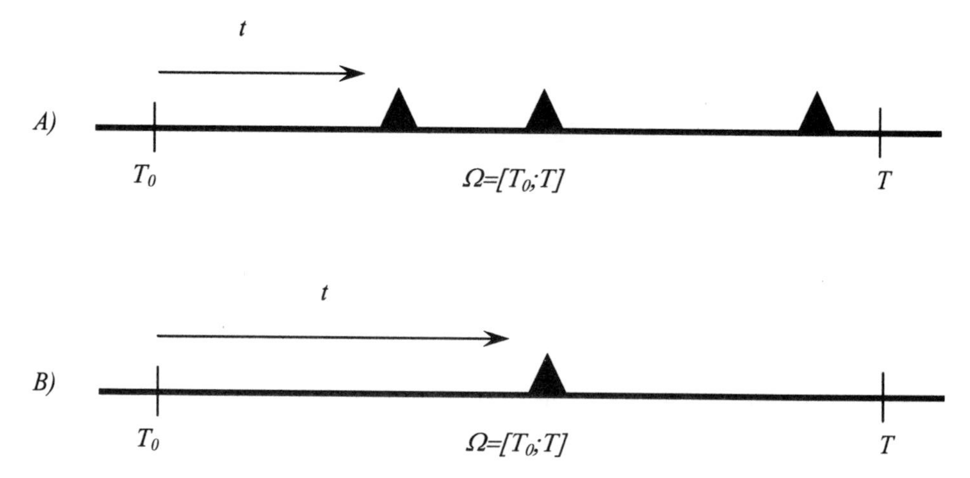

Figure 11: First photon (A) and single photon (B) detection.

3.3.2. Correlation Function and Inter-Arrival Time PDD

Second-order differentiation of Eqs. (54) and (57) with respect to w, yields, respectively,

$$\frac{\delta L_v^*[w;\Omega \mid z]}{\delta w(t_1)\delta w(t_2)}$$

$$= \lambda \left\langle \int_{z+\Delta}^{T} exp\{-\lambda(x-z-\Delta)\}[1+\kappa w(x)] \frac{\delta L_v^*[w;\Omega \mid x]}{\delta w(t_1)\delta w(t_2)} dx + \right.$$

$$+ \kappa\lambda \, exp\{-\lambda(t_1-z-\Delta)\} \frac{\delta L_v^*[w;\Omega \mid t_1]}{\delta w(t_2)} +$$

$$\left. + exp\{-\lambda(t_2-z-\Delta)\} \frac{\delta L_v^*[w;\Omega \mid t_2]}{\delta w(t_1)} \right\rangle_{\Delta} \tag{65}$$

$$\frac{\delta^2 L[w;\Omega]}{\delta w(t_1)\delta w(t_2)} = \int_{T_0}^{T} a_1^*(x)[1+w(x)] \frac{\delta^2 L_v^*[w;\Omega \mid x]}{\delta w(t_1)\delta w(t_2)} dx +$$

$$+ a_1^*(t_1) \frac{\delta L_v^*[w;\Omega \mid t_1]}{\delta w(t_2)} + a_1^*(t_2) \frac{\delta L_v^*[w;\Omega \mid t_2]}{\delta w(t_1)} \tag{66}$$

The integral equations for the time correlation function of the output process are obtained by setting $w = 0$ in Eqs. (65) and (66):

$$f_{v2}(t_1,t_2 \mid z)$$

$$= \left\langle \lambda \int_{z+\Delta}^{t_1} exp\{-\lambda(x-z-\Delta)\} f_{v2}(t_1,t_2 \mid x) dx + \right.$$

$$\left. + \kappa\lambda \, exp\{-\lambda(t_1-z-\Delta)\} f_{v1}(t_2-t_1) \right\rangle_{\Delta} \tag{67}$$

$$f_2(t_1,t_2) = \int_{T_0}^{t_1} a_1^*(x) f_{v2}(t_1,t_2 \mid x) dx + a_1^*(t_1) f_{v1}(t_2-t_1) \tag{68}$$

where it is assumed that $f_2(t_1,t_2) = 0$ and $f_{v2}(t_1,t_2|z) = 0$ when $t_2 < t_1$. Taking into account that the detected process is stationary, and combining Eqs. (60), (61), (62), (67), and (68) one obtains the solution for the time correlation function

$$f_2(t_1,t_2)=f_2(t_2-t_1)=f_2(t)=f_{v1}(t)f_{v1}(\infty) \tag{69}$$

It should be emphasized that $f_2(t)$ is not the intensity correlation function that is usually calculated for fluorescence intensity traces of single molecules[5,14], but a correlation function calculated with the arrival times within a user-defined detection interval. $f_2(t)$ can be built up by histogramming the time intervals between the running and all subsequent photocounts for which the time positions do not exceed the length of the user-defined time interval Ω. Note that the origin of the interval Ω does coincide with the time position of the running photocount.

Substituting w = -1 in Eq. (58) and assuming that $\Omega = [0;t]$ yields the integral equation for the PDD of the time interval between two consecutive photocounts:

$$g(t)=\left\langle \kappa\lambda \exp\{-\lambda(t-\Delta)\}+(1-\kappa)\lambda\int_0^{t-\Delta}\exp\{-\lambda(t-\Delta-x)\}g(x)dx\right\rangle_\Delta \tag{70}$$

3.3.3. PD of the Number of Photocounts per Detection Interval

The functional equation for GF $\Theta[z;\Omega]$ of the number of photocounts per detection interval is obtained by substituting the formal variable $(z-1)$ instead of the trial function w in Eqs. (54) and (57):

$$\Theta_v^*[z;\Omega\,|\,t]=\Big\langle exp\{-\lambda(T-t-\Delta)\}+1+ \\ +\lambda[\kappa(z-1)]\int_{t+\Delta}^{T}exp\{-\lambda(x-t-\Delta)\}\Theta_v^*[z;\Omega\,|\,x]dx\Big\rangle_\Delta \tag{71}$$

$$\Theta[z;\Omega]=1-\int_{T_0}^{T}a_1^*(x)dx+z\int_{T_0}^{T}a_1^*(x)\Theta_v^*[z;\Omega\,|\,x]dx \tag{72}$$

Performing inverse Fourier transform of the characteristic function, obtained from GF $\Theta[z;\Omega]$ (72) by substituting the formal parameter z by the complex exponent $exp(-j\varpi)$ (see Section 2.1.4), yields the PD of the number of photocounts in Ω.

A note should be made concerning the implementation of the procedure for solving integral equations [for example, Eqs. (60), (61), (64), (70), (71)]. In general, it is always possible to find a numerical solution[27]. An analytical solution can often be found using the Laplace transforms method[28,29], or (e.g. in the case of multi-exponential SMIRF) the integral equations can be simplified directly. As an illustration, in Appendix 2 we demonstrate application of the Laplace transforms method to Eqs. (60), (61), (64), (70), and in Appendix 3, we show how the integral equation (71) for GF $\Theta[z;\Omega]$ can be solved for a bi-exponential SMIRF.

4. Experimental

4.1. Set-Up and Sample Preparation

To demonstrate the practical usefulness of the analysis method, traces with photocount time positions from a single R6G molecule on a glass surface were recorded using an experimental set-up as described in reference[24]. For each photocount the time position from the start of the experiment was stored with 50 ns time resolution. The information of the time positions of the photocounts (the experimental time position trace) can then be used to software-construct arrays of the fluorescence intensity *vs.* time with different user-defined detection intervals.

4.2. Analysis

In **Figure 12**, we show the fluorescence intensity of R6G *vs.* time for different user-defined detection intervals, respectively. If the chosen detection interval is short, only zero and one photocounts per detection interval are observed (**Figure 12A**). The values of Ω where only zero and one photocounts per detection interval occur are dependent on λ, κ, and the molecule (absorption cross section, emission quantum yield). The lower λ, κ, and the excitation-emission efficiency of the molecule are, the wider the detection interval can be before detection intervals with two or more photocounts will be observed. For intermediate values Ω, the number of photocounts per detection interval increases (**Figure 12B**). The typical single molecule trajectory is found for long detection intervals (**Figure 12C**).

Next, we applied the five before-defined statistical characteristics of the photocounts [the time-correlation function $f_2(t)$, the PDDs of the single $\pi_1(t,\Omega)$ and first $a_1^*(t)$ photocount arrival times in the detection interval, the PD of the number of photocounts per detection interval $P_n(\Omega)$, and the PDD of the time interval between two consecutive photocounts $g(t)$] to probe the intra-molecular photophysical behavior of R6G.

The largest value of Ω used in the analysis was chosen to ensure a sufficient number of detection intervals with single photocounts for building up $\pi_1(t,\Omega)$. If the chosen detection interval is too long, one can hardly obtain detection intervals with a single photocount [i.e., $\pi_1(t,\Omega)$ will show only a few spikes]. On the other hand, if the detection interval is too short, the dynamic range of $a_1^*(t)$ and $f_2(t)$ is very small and the change of $\pi_1(t,\Omega)$ can hardly be seen while $P_n(\Omega)$ will contain only a few non-zero channels. Based on these criteria, three Ω values were chosen for the analysis of experimental time position trace: $\Omega = 0.7$, 0.35, and 0.175 ms.

The total time of the analyzed trajectory was 12 s. In **Figure 12C**, the measured trajectory in the selected region does not show any systematic changes, i.e., the average number of photocounts per detection interval is practically constant.

The background level was estimated from the fluorescence intensity trajectory after photobleaching and was found to be 0.65 counts ms^{-1} (see **Figure 12C**). These background values were held constant in the parameter estimation.

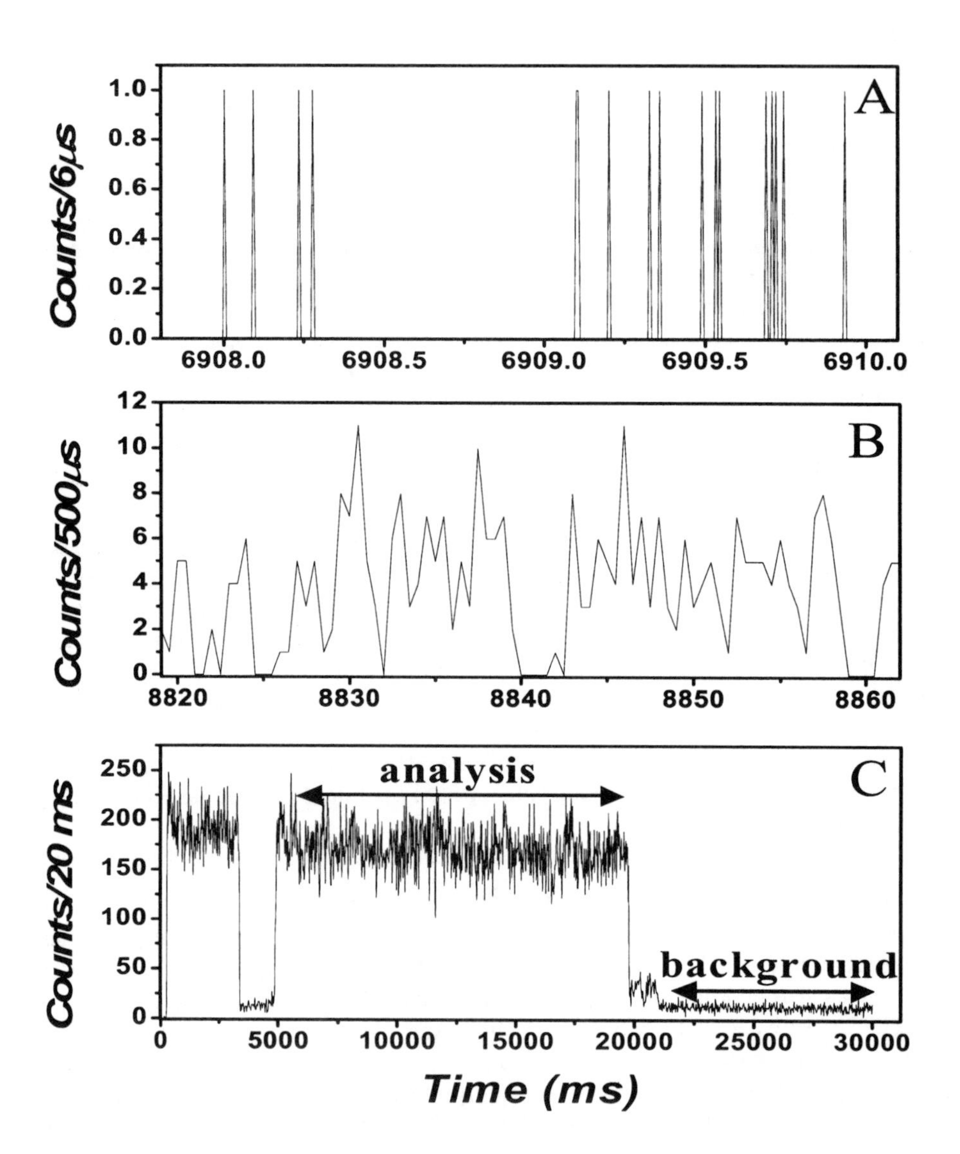

Figure 12: Fluorescence intensity trajectories of R6G for different user-defined detection intervals Ω. (A) Ω = 6 μs, (B) Ω = 0.5 ms, (C) Ω = 20 ms. The time regions used for the analysis and the calculation of the background are indicated in **Figure 12C**. Note that in **Figure 12A** and **Figure 12B** only a small fraction of the total trajectory is shown. The time-axis values in these figures correspond to those of **Figure 12C**.

A bi-exponential probability density function $q(\Delta)$ (system with singlet and triplet excited states) was used to model the single molecule fluorescence of R6G:

$$q(\Delta) = [(1-p)/\tau_s]exp(-\Delta/\tau_s) + [p/\tau_t]exp(-\Delta/\tau_t), \tag{73}$$

where τ_s and τ_t are the average lifetimes of the singlet and triplet excited states, respectively, and p is the probability for a single molecule to undergo intersystem crossing from the singlet excited state to the triplet excited state. As long as the molecule remains in the triplet excited state, no photocounts are observed during that time (the excitation-emission cycle is interrupted). Since the average lifetime τ_s for R6G[30] is much smaller than the time resolution of the experimental set-up (i.e., 50 ns), Eq. (73) can be rewritten as

$$q(\Delta) = (1-p)\delta(\Delta) + [p/\tau_t]exp(-\Delta/\tau_t), \tag{74}$$

where $\delta(\Delta)$ is the delta-function and the probability p will approximate the intersystem crossing yield[31] in this case.

The histograms of the derived characteristics were put into the Marquardt nonlinear least-squares optimization procedure[32], which searches for the parameter values minimizing the sum of squared differences between the experimentally derived statistical characteristic and the estimated one. Only visual inspection of the measured and estimated statistical characteristics was used to judge the quality of the fits.

The parameters that should be estimated are the excitation intensity λ (impinging on the single molecule), the efficiency κ of the detection set-up, and the parameters of the single molecule itself (p and τ_t). Since the estimated parameters are independent of the statistical characteristics and detection intervals used in the analysis, a global analysis procedure with linked parameters is the most appropriate. All statistical characteristics [$f_2(t)$, $\pi_1(t,\Omega)$, $a_1^*(t)$, $P_n(\Omega)$, and $g(t)$] derived from a single experimental time position trace at various detection intervals were analyzed simultaneously with the same set of unknown parameters (λ, κ, p, and τ_t). The values obtained for R6G were λ = 99.5 photons ms^{-1}, κ = 10 %, p = 0.61 % and τ_t = 0.41 ms[30]. The measured and recovered statistical characteristics of R6G are shown in **Figure 13**. The PDD $\pi_1(t,\Omega)$ and the PD $P_n(\Omega)$ demonstrate clearly the dependence on the width of Ω. For the time-correlation function $f_2(t)$, the PDD $a_1^*(t)$ and the PDD $g(t)$, the shape of the experimentally derived characteristic is independent of the width of Ω. Using shorter detection intervals only decreases the range of the statistical characteristics $f_2(t)$, $a_1^*(t)$, and $g(t)$. That is the reason why for $f_2(t)$, $a_1^*(t)$, and $g(t)$ only the curves at the longest detection interval are reproduced in the figures.

The excellent quality of the fits – as judged by visual inspection of **Figure 13** – shows that the developed analysis approach is suitable for an adequate description of single molecule fluorescence. Each individual statistical characteristic [$f_2(t)$, $\pi_1(t,\Omega)$, $a_1^*(t)$, $P_n(\Omega)$, and $g(t)$] extracts its own piece of information from the measured data.

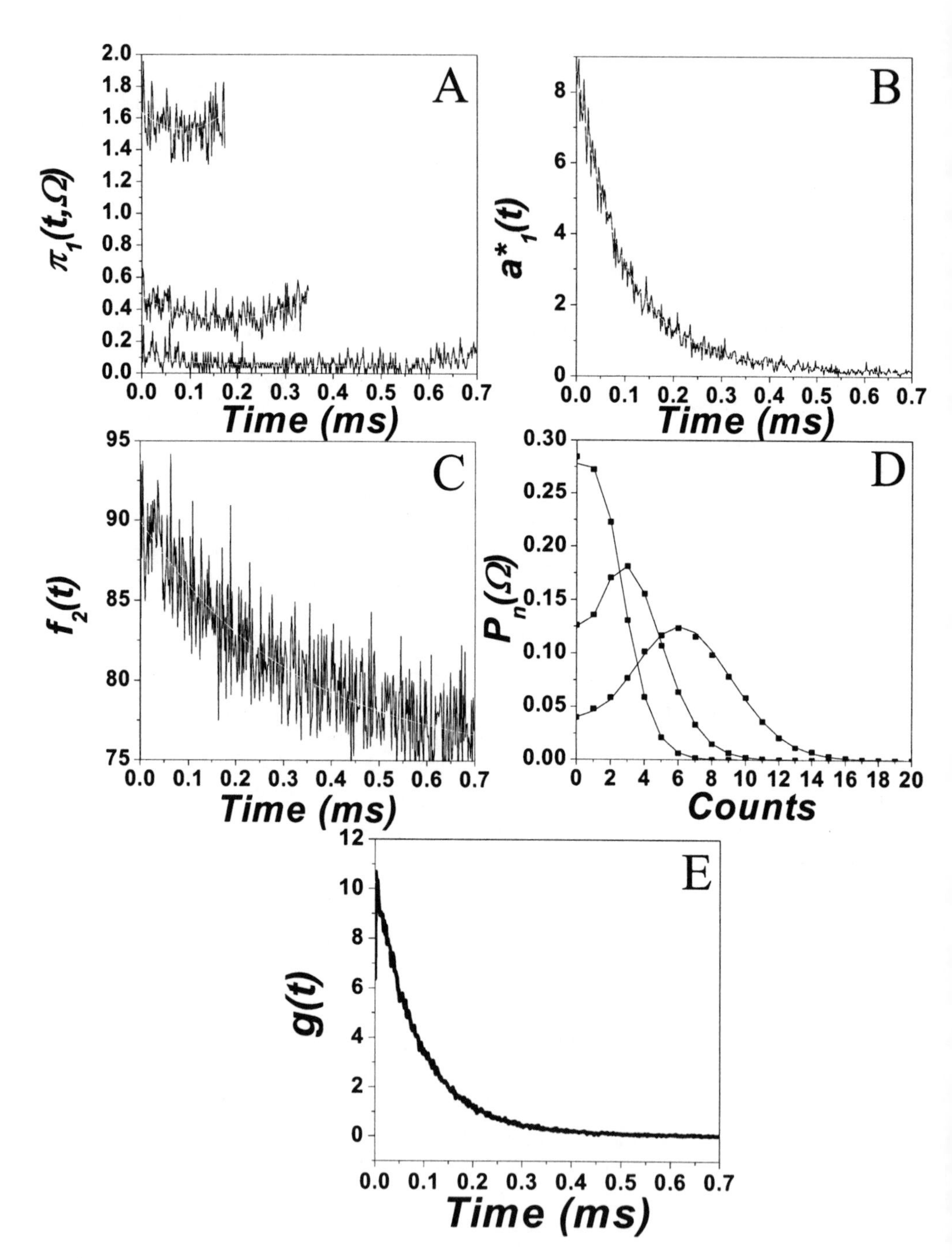

Figure 13: Experimentally derived and recovered statistical characteristics of R6G: (A) $\pi_1(t,\Omega)$, plotted for three Ω values (0.7, 0.35, and 0.175 ms); (B) $a_1^*(t)$; (C) $f_2(t)$; (D) $P_n(\Omega)$, plotted for three Ω values (0.7, 0.35, and 0.175 ms); (E) $g(t)$.

Indeed, the use of longer detection intervals increases the dynamic range of $f_2(t)$, $a_1^*(t)$, and $g(t)$ and this leads to a more reliable parameter recovery. Although the PD $P_n(\Omega)$ is wide in this case, the standard deviation on the number of photocounts per detection interval is large. $\pi_1(t,\Omega)$ becomes gradually less useful under this condition because the number of detection intervals with a single photocount decreases. Conversely, the use of short detection intervals favors $\pi_1(t,\Omega)$ because there are more detection intervals with a single photocount. In this case $P_n(\Omega)$ contains just a few channels with non-zero photocounts per detection interval but with small standard deviation. When these statistical characteristics at different detection intervals are combined in a global analysis, a more consistent picture of the molecule emerges with higher reliability of the estimated single molecule parameters. It is important to stress the appropriateness of the global analysis approach: since the molecular parameters are independent of the width of the detection interval and the used statistical characteristics of the photocounts, it is recommended to analyze the whole set of statistical characteristics at different Ω simultaneously with the parameters λ, κ, p, and τ_t linked.

It must be emphasized that the shot-noise of the excitation light does not influence the parameter estimation because the Poisson nature of the excitation is naturally incorporated into the model.

5. CONCLUSIONS

Up to now, single molecule fluorescence experiments were performed by dividing the time into a set of intervals and to observe the number of fluorescence photons arriving in each interval. It is obvious that the observed photons carry less information than the arrival times of the photons themselves. From the arrival times, one can still calculate the number of photons in any user-defined interval, whereas when only the number of photons in an interval is recorded, information about their positions in time is lost. Therefore, we have presented a new analysis method of single molecule fluorescence data based on the positions in time of the detected fluorescence photons. The theory of point processes using the generating functionals formalism is ideally suited for a consistent description, linking the statistical characteristics of the excitation and detected photons to the statistical characteristics of the single motionless molecule. We have shown in detail how the developed theory can be implemented for continuous and pulsed excitation. Furthermore, the method allows for straightforward incorporation of various photophysical models by defining the appropriate functional form of $q(\Delta)$.

The new method for the analysis of single molecule fluorescence time position traces has been demonstrated on experimental data of R6G. The time positions of the detected photons were registered with an experimental time resolution of 50 ns which allows one to construct fluorescence intensity trajectories with user-defined Ω of 50 ns and higher. Finally, the paper shows the first use of global analysis of single molecule fluorescence intensity trajectories, enhancing the reliability of the estimated parameters.

We have concentrated on $f_2(t)$, $\pi_1(t,\Omega)$, $a_1^*(t)$, $P_n(\Omega)$, and $g(t)$ because these characteristics are represented by single-dimensional arrays of numbers, which can be built easily and processed quickly. As was stated in section 2.1.1, a random point process is completely defined by the set of joint PDD $\{\pi_i(t_1,...,t_i;\Omega)\}_{i=0,1,...}$, or equivalently by the sets $\{a_i(t_1,...,t_i)\}_{i=1,2,...}$ or $\{f_i(t_1,...,t_i)\}_{i=1,2,...}$. In principle, if one can measure the time positions of all photocounts occurring within Ω, it is possible to create a set of multi-dimensional arrays $\{\pi_i(t_1,...,t_i;\Omega)\}_{i=0,1,...}$, which afterwards can be analyzed globally. This alternative set of statistical characteristics allows one to extract all information, which is enclosed in the measured time position trace. The other statistical characteristics, such as $f_2(t)$, $a_1^*(t)$, and $P_n(\Omega)$ can be derived if one knows all probability density functions of the set $\{\pi_i(t_1,...,t_i;\Omega)\}_{i=0,1,...}$ (see Section 2.1.1).

6. ACKNOWLEDGEMENTS

The authors are indebted to the *Fonds voor Wetenschappelijk Onderzoek – Vlaanderen* and the DWTC (Belgium) through IAP-V-03 for continuing financial support.

7. REFERENCES

1. S. Nie and R. Zare, Optical detection of single molecules, *Annu. Rev. Biophys. Biomol. Struct.* **26**, 567-596 (1997).
2. X. S. Xie and J. K. Trautman, Optical studies of single molecules at room temperature, *Annu. Rev. Phys. Chem.* **49**, 441-480 (1998).
3. W. E. Moerner and M. Orrit, Illuminating single molecules in condensed matter, *Science* **283**, 1670-1676 (1999).
4. S. Weiss, Fluorescence spectroscopy of single biomolecules, *Science* **283**, 1676-1683 (1999).
5. K. D. Weston, P. J. Carson, J. A. DeAro and S. K. Buratto, Single-molecule detection fluorescence of surface-bound species in vacuum, *Chem. Phys. Lett.* **308**, 58-64 (1999).
6. A. M. Berezhkovskii, A. Szabo and G. H. Weiss, Theory of single-molecule fluorescence spectroscopy of two-state systems, *J. Chem. Phys.* **110**, 9145-9150 (1999).
7. B. Saleh, *Photoelectron Statistics* (Springer, Berlin, 1978).
8. K. D. Weston and S. K. Buratto, Millisecond intensity fluctuations of single molecules at room temperature, *J. Phys. Chem.* A **102**, 3635-3638 (1998).
9. H. Yang and X. S. Xie, Probing single-molecule dynamics photon by photon, *J. Chem. Phys.* **117**, 10965-10979 (2002).
10. H. Yang and X. S. Xie, Statistical approaches for probing single-molecule dynamics photon-by-photon, *Chem. Phys.* **284**, 423-437 (2002).
11. R. Verberk and M. Orrit, Photon statistics in the fluorescence of single molecules and nanocrystals: Correlation functions versus distributions of on- and off-times, *J. Chem. Phys.* **119**, 2214-2222 (2003).
12. L. Fleury, J. M. Segura, G. Zumofen, B. Hecht and U. P. Wild, Nonclassical photon statistics in single-molecule fluorescence at room temperature, *Phys. Rev. Lett.* **84**, 1148-1151 (2000).
13. A. Molski, J. Hofkens, T. Gensch, N. Boens and F. De Schryver, Theory of time-resolved single-molecule fluorescence spectroscopy, *Chem. Phys. Lett.* **318**, 325-332 (2000).
14. W.-T. Yip, D. Hu, J. Yu, D. A. Vanden Bout and P. F. Barbara, Classifying the photophysical dynamics of single- and multiple-chromophoric molecules by single molecule spectroscopy, *J. Phys. Chem.* **102**, 7564-7575 (1998).
15. S. Jang and R. J. Silbey, Theory of single molecule line shapes of multichromophoric macromolecules, *J. Chem. Phys.* **118**, 9312-9323 (2003).

16. C. Eggeling, J. Widengren, R. Rigler and C. A. M. Seidel, Photobleaching of fluorescent dyes under conditions used for single-molecule detection: evidence of two-step photolysis, *Anal. Chem.* **70**, 2651-2659 (1998).
17. S. Wennmalm and R. Rigler, On death numbers and survival times of single dye molecules, *J. Phys. Chem.* **103**, 2516-2519 (1999).
18. D. E. Koppel, Statistical accuracy in fluorescence correlation spectroscopy, *Phys. Rev. A* **10**, 1938-1945 (1974).
19. E. Novikov, N. Boens and J. Hofkens, New strategies for low light level detection in single molecule spectroscopy, *Chem. Phys. Lett.* **338**, 151-158 (2001).
20. D. J. Daley and D. Vere-Jones, *An Introduction to the Theory of Point Processes* (Springer, New York, 1988).
21. D. R. Cox and V. Isham, *Point Processes* (Chapman and Hall, London, 1980).
22. V. V. Apanasovich, A. A. Koljada and A. F. Chernjavski, *The Statistical Analysis of Series of Random Events in Physical Experiment* (University Press, Minsk, 1988) (in Russian).
23. J. Grandell, *Double stochastic point processes* (Berlin, Springer, 1978).
24. E. Novikov, J. Hofkens, M. Cotlet, M. Maus, F.C. De Schryver and N. Boens, A new analysis method of single molecule fluorescence using series of photon arrival times: theory and experiment, *Spectrochim. Acta A* **57**, 2109-2133 (2001).
25. W. Feller, *An Introduction to Probability Theory and Its Applications*, Volume 1 (John Wiley & Sons, New York, 1968)
26. V. V. Apanasovich and E. G. Novikov, Branching point processes with independent transformations, *J. Phys. A: Math. Gen.* **28**, 433-443 (1995).
27. K. E. Atkinson, *The Numerical Solution of Integral Equations of the Second Kind* (Cambridge University Press, Cambridge, 1997).
28. A. J. Jerri, *Introduction to Integral Equations with Applications* (John Wiley and Sons, New York, 1999).
29. H. Stehfest, Numerical inversion of laplace transforms, *Communications of the ACM* **13**, 47-49 (1970).
30. J. Enderlein, P. M. Goodwin, A. Van Orden, W. P. Ambrose, R. Erdmann and R. A. Keller, A maximum likelihood estimator to distinguish single molecules by their fluorescence decays, *Chem. Phys. Lett.* **270**, 464-470 (1997).
31. J. A. Veerman, M. F. Garcia-Parajo, L. Kuipers,and N. F. van Hulst, Time-varing triplet state lifetimes of single molecules, *Phys. Rev. Lett.* **83**, 2155-2158 (1999).
32. P. R. Bevington, *Data Reduction and Error Analysis for the Physical Sciences* (McGraw-Hill, New York, 1969).

8. LIST WITH ABBREVIATIONS

CF: *c*haracteristic *f*unction
GF: *g*enerating *f*unction
GFL: generating *f*unctiona*l*
PD: *p*robability *d*istribution
PDD: *p*robability *d*ensity *d*istribution
R6G: *r*hodamine *6G*
RPP: *r*andom *p*oint *p*rocess
BRPP: *b*ranching *r*andom *p*oint *p*rocess
CRPP: *c*luster *r*andom *p*oint *p*rocess
DSRPP: *d*oubly *s*tochastic *r*andom *p*oint *p*rocess
PRPP: *P*oisson *r*andom *p*oint *p*rocess
PWRPP: *p*air-*w*ise *r*andom *p*oint *p*rocess
SMIRF: *s*ingle *m*olecule *i*nstrument *r*esponse *f*unction

9. APPENDIX 1. BACKGROUND RADIATION

First-order functional differentiation of Eq. (48) with respect to the trial function w yields

$$\frac{\delta\hat{L}[w;\Omega]}{\delta w(t)} = \frac{\delta L[w;\Omega]}{\delta w(t)} L_B[w;\Omega] + L[w;\Omega]\frac{\delta L_B[w;\Omega]}{\delta w(t)} \tag{75}$$

The PDD of the arrival times of the single and first count in a mixture of photocounts arriving from the molecule and background counts takes, respectively, the form

$$\hat{\pi}_1(t;\Omega) = \pi_1(t;\Omega)\pi_0^B(\Omega) + \pi_0(\Omega)\pi_1^B(t;\Omega) \tag{76}$$

$$\hat{a}_1(t) = a_1^*(t)\left\{1 - \int_{T_0}^{t} a_1^B(x)dx\right\} + \left\{1 - \int_{T_0}^{t} a_1^*(x)dx\right\}a_1^B(t) \tag{77}$$

where $\pi_1(t;\Omega)$ and $a_1^*(t)$ are respectively the PDD of the arrival time of the single and first photocounts originating from the molecule; $\pi_1^B(t;\Omega)$ and $a_1^B(t)$ are respectively the PDD of the arrival time of the single and first background counts; $\pi_0(\Omega)$ and $\pi_0^B(\Omega)$ are the probabilities to obtain zero counts coming from the molecule and the background, respectively.

Substituting the formal variable $(z - 1)$ instead of the trial function w in Eq. (48) yields the expression for the GF of the number of counts of a mixture of molecular photocounts and background counts:

$$\hat{\Theta}[z;\Omega] = \Theta[z;\Omega]\Theta_B[z;\Omega] \tag{78}$$

Further elaboration of Eq. (78) is straightforward, provided the functional representation of GF $\Theta_B[z;\Omega]$ is known.

The second functional derivative of Eq. (48) with respect to the trial function w yields

$$\frac{\delta^2\hat{L}[w;\Omega]}{\delta w(t_1)\delta w(t_2)} = \frac{\delta L[w;\Omega]}{\delta w(t_1)}\frac{\delta L_B[w;\Omega]}{\delta w(t_2)} + \frac{\delta L[w;\Omega]}{\delta w(t_2)}\frac{\delta L_B[w;\Omega]}{\delta w(t_1)} +$$

$$+ \frac{\delta^2 L[w;\Omega]}{\delta w(t_1)\delta w(t_2)} L_B[w;\Omega] + L[w;\Omega]\frac{\delta^2 L_B[w;\Omega]}{\delta w(t_1)\delta w(t_2)} \tag{79}$$

Substituting $w = 0$ into Eq. (79) yields the time correlation function of the detected counts:

$$\hat{f}_2(t_1,t_2) = f_2(t_1,t_2) + f_1(t_1)f_1^B(t_2) + f_1(t_2)f_1^B(t_1) + f_2^B(t_1,t_2)$$
(80)

where $f_2(t_1,t_2)$ is the time correlation function of molecular photocounts; $f_2^B(t_1,t_2)$ is the time correlation function of the background counts; $f_1(t)$ and $f_1^B(t)$ are the intensities of molecular photocounts and background counts, respectively.

It is commonly assumed that background counts can be represented by a stationary Poisson random point process. In that case the GFL takes the form[7]

$$L_B[v;\Omega] = exp\left\{\lambda_B \int_\Omega v(t)dt\right\}$$
(81)

where λ_B is background time-independent intensity. Substituting the statistical characteristics of the Poisson random point process, obtained from Eq. (81), into Eqs. (76), (77), (78), and (80) yields, respectively,

$$\hat{\pi}_1(t) = [\pi_1(t,\Omega) + \lambda_B \pi_0(\Omega)]exp\{-\lambda_B T\}$$
(82)

$$\hat{a}_1(t) = \left[a_1^*(t) + \lambda_B - \lambda_B \int_0^t a_1^*(x)dx\right]exp\{-\lambda_B t\}$$
(83)

$$\hat{\Theta}[z;\Omega] = \Theta[z;\Omega]exp\{\lambda_B T(z-1)\}$$
(84)

$$\hat{f}_2(t_1,t_2) = f_2(t_1,t_2) + f_1(t_1)\lambda_B + f_1(t_2)\lambda_B + \lambda_B^2$$
(85)

The influence of the background radiation on the PDD of the time interval between two consecutive photocounts can be modeled using the conditional GFL of photocounts, provided a count has been detected at time z:

$$\hat{L}[w;\Omega|z] = p_p L_{|p}[w;\Omega|z] L_{B|p}[w;\Omega|z] + p_b L_b[w;\Omega|z] L_{B|b}[w;\Omega|z]$$
(86)

where $L_{|p}[w;\Omega|z]$ and $L_{|b}[w;\Omega|z]$ are the GFLs of molecular photocounts provided a molecular photocount or background count, respectively, has been detected at time z; $L_{B|p}[w;\Omega|z]$ and $L_{B|b}[w;\Omega|z]$ are the GFLs of background counts provided a molecular photocount or background count, respectively, has been detected at time z, p_p and p_b $(= 1 - p_p)$ are the probabilities to detect a molecular photocount or background count, respectively, at time z. p_p represent the ratio of the intensity of molecular photocounts to the total intensity of the detected counts (including background counts). Since the processes are assumed to be stationary, the time instance z can be arbitrary (for simplicity, 0) and can be omitted from expression (86):

$$\hat{L}[w;\Omega] = p_p L_{|p}[w;\Omega] L_{B|p}[w;\Omega] + p_b L_b[w;\Omega] L_{B|b}[w;\Omega] \tag{87}$$

One can show that the PDD of the time interval between two consecutive counts in a mixture of molecular photocounts and background counts is given by:

$$\hat{g}(t) = -\frac{d\hat{L}[-1;[0;t]]}{dt}$$

$$= -\frac{d}{dt}\left\{p_p \pi_{v0|p}([0;t]) \pi_{B0|p}([0;t]) + p_b \pi_{0|b}([0;t]) \pi_{Bv0|b}([0;t])\right\} \tag{88}$$

where $\pi_{v0|p}([0;t])$ and $\pi_{0|b}([0;t])$ are the probabilities of zero molecular photocounts in the interval $[0;t]$ provided a molecular photocount or background count, respectively, has been detected at time 0, and $\pi_{Bv0|p}([0;t])$ and $\pi_{B0|b}([0;t])$ are the probabilities of zero background counts in the interval $[0;t]$, provided a molecular photocount or background count, respectively, has been detected at time 0. $\pi_{0|b}([0;t])$ can be calculated from Eqs. (56) or (57) by substituting $w = -1$ and $\Omega = [0;t]$. $\pi_{v0|p}([0;t])$ can be obtained from Eq. (54) (with $w = -1$ and $\Omega = [0;t]$) or using PDD $g(t)$ of the time interval between two consecutive photocounts [Eq. (70)] via the relation

$$\pi_{v0|p}([0;t]) = 1 - \int_0^t g(x)dx \tag{89}$$

If background counts are represented by a stationary Poisson random point process (81), Eq. (88) is simplified to

$$\hat{g}(t) = -\frac{d}{dt}\left\{exp\{-\lambda_B t\}\left(p_p \pi_{v0|p}([0;t]) + p_b \lambda_B \pi_{0|b}([0;t])\right)\right\} \tag{90}$$

which can be further elaborated using Laplace transforms[28].

10. APPENDIX 2. SOLUTIONS FOR $a_1^*(t)$, $\pi_1(t;\Omega)$, $f_2(t)$ AND $g(t)$

Explicit analytical solutions for the statistical characteristics $a_1^*(t)$, $\pi_1(t;\Omega)$, $f_2(t)$ and $g(t)$ can be elaborated in terms of Laplace transforms[28]. The Laplace transform of the Eq. (61) yields:

$$F_{v1}(s)=\frac{\kappa+F_{v1}(s)}{s+\lambda}\lambda Q(s)$$

(91)

where $F_{v1}(s)$ and $Q_{v1}(s)$ are the Laplace transforms of $f_{v1}(s)$ and $q(\Delta)$ respectively, and s is the Laplace variable. The solution of the algebraic Eq. (91) is straightforward:

$$F_{v1}(s)=\frac{\kappa\lambda Q(s)}{s+\lambda-\lambda Q(s)}$$

(92)

Substituting Eq. (92) into Eq. (69) yields the Laplace transform of the time correlation function:

$$F_2(s)=f_{v1}(\infty)F_{v1}(s)$$

(93)

The Laplace transform of the PDD $a_1^*(t)$ is a solution of the integral equation (60):

$$A_1(s)=\frac{F_1(s)}{1+F_{v1}(s)}$$

(94)

where $F_1(s)$ is the Laplace transform of the intensity $f_1(t)$. For stationary processes ($f_1(t)\equiv const$), we obtain

$$A_1(s)=\frac{const}{[1+F_{v1}(s)]s}$$

(95)

The Laplace transform of the PDD of the time interval between two consecutive photocounts can be obtained as a solution of the integral equation (70):

$$G(s)=\frac{\kappa\lambda Q(s)}{s+\lambda+(\kappa-1)\lambda Q(s)}(s)$$

(96)

The PDD of the single photon occurrence in Ω $\pi_1(t;\Omega)$ is defined by Eq. (63), where $a_1^*(t)$ can be found by the inverse Laplace transform of Eq. (95) and $\pi_{v0}([t;T])$ represent a solution of the integral equation (64), which can be written using the PDD of the time interval between two consecutive photocounts as

$$\pi_{v0}([t;T])=1-\int_0^{T-t}g(x)dx$$

where $g(x)$ is obtained by the inverse Laplace transform of Eq. (96).

The Laplace transforms (92), (93), (95) and (96) can be further expanded provided the form of the SMIRF $q(\Delta)$ is known. For example, for the bi-exponential SMIRF:

$$q(\Delta) = p_1 \, exp(-\Delta / \tau_1) + p_2 \, exp(-\Delta / \tau_2) \tag{97}$$

where τ_1 and τ_2 are the decay times and p_1 and p_2 are the associated contributions of the first and second exponential components, respectively, the Laplace transform takes the form:

$$Q(s) = \frac{p_1}{s + 1/\tau_1} + \frac{p_2}{s + 1/\tau_2} \tag{98}$$

The inverse Laplace transforms have been performed numerically using the algorithm described in reference[29].

11. APPENDIX 3. GF FOR CONTINUOUS EXCITATION

Performing averaging in Eq. (71) yields

$$\Theta_v^*[z;\Omega \,|\, t] = \int_T^\infty q(\Delta) d\Delta + p(T - t) + \alpha(z)\lambda \int_t^T p(x - t) \Theta_v^*[z;\Omega \,|\, x] dx \tag{99}$$

where $\alpha(z) = [1 + \kappa(z - 1)]$ and

$$p(x) = \int_0^x q(\Delta) \, exp\{-\lambda(x - \Delta)\} d\Delta \tag{100}$$

Further analytical evaluation of the obtained integral equation (99) requires the explicit form for the SMIRF $q(\Delta)$. As an example, we will consider a bi-exponential SMIRF:

$$q(\Delta) = p_1 \, exp(-\Delta / \tau_1) + p_2 \, exp(-\Delta / \tau_2) \tag{101}$$

where τ_1 and τ_2 are the decay times and p_1 and p_2 are the associated contributions of the first and second exponential components, respectively. Substituting Eq. (101) into Eq. (100) one obtains the expression for $p(x)$:

$$p(x) = \sum_{i=1}^{3} A_i \, exp\{-x / \xi_i\} \tag{102}$$

where $\xi_1 = 1/\lambda$, $\xi_2 = \tau_1$, $\xi_3 = \tau_2$, and

$$A_1 = \frac{p_1\tau_1}{1-\lambda\tau_1} + \frac{p_2\tau_2}{1-\lambda\tau_2}, \quad A_2 = -\frac{p_1\tau_1}{1-\lambda\tau_1}, \quad A_3 = -\frac{p_2\tau_2}{1-\lambda\tau_2} \tag{103}$$

One can show that the solution of the integral equation (99) with a triple-exponential function $p(x)$ (102) will also be triple-exponential:

$$\Theta_v^*[z;\Omega\,|\,t] = \sum_{i=1}^{3} B_i(z)\exp\{\mu_i(z)t\} \tag{104}$$

where the coefficients $B_i(z)$ and $\mu_i(z)$ are determined as follows. The coefficients $\mu_i(z)$ are the roots of the polynomial equation

$$\mu^3(z) + \mu^2(z)(C_0 - \rho_0) + \mu(z)(C_1 - \rho_1) + (C_2 - \rho_2) = 0 \tag{105}$$

where the coefficients C_i, $i = 0,1,2$ are represented as

$$\begin{aligned} C_0 &= -(\xi_1\xi_2\xi_3)^{-1} \\ C_1 &= (\xi_1\xi_2)^{-1} + (\xi_1\xi_3)^{-1} + (\xi_2\xi_3)^{-1} \\ C_2 &= -\xi_1^{-1} - \xi_2^{-1} - \xi_3^{-1} \end{aligned} \tag{106}$$

and the coefficients ρ_i, $i=0,1,2$ take the form

$$\rho_k = -\lambda\eta(z)\sum_{j=0}^{k} C_{j-1}\sum_{i=1}^{3}\frac{A_i}{\xi_i^{k-j}}, \quad (C_{-1} = 1) \tag{107}$$

The coefficients $B_i(z)$ are obtained as a solution of the following set of algebraic equations:

$$\Theta_v^*[z;\Omega\,|\,T] = \sum_{i=1}^{3} A_i^* = \sum_{i=1}^{3} B_i(z)\exp\{\mu_i(z)T\} \tag{108}$$

$$\left.\frac{d\Theta_v^*[z;\Omega\,|\,t]}{dt}\right|_{t=T} = \sum_{i=1}^{3}\frac{A_i^*}{\xi_i} - \lambda\eta(z)\Theta_v^*[z;\Omega\,|\,T]\sum_{i=1}^{3} A_i$$

$$= \sum_{i=1}^{3}\mu_i(z)B_i(z)\exp\{\mu_i(z)T\} \tag{109}$$

$$\left.\frac{d^2\Theta_v^*[z;\Omega\,|\,t]}{dt^2}\right|_{t=T}$$

$$=\sum_{i=1}^{3}\frac{A_i^*}{\xi_i^2}-\lambda\eta(z)\left\{\Theta_v^*[z;\Omega\,|\,T]\sum_{i=1}^{3}\frac{A_i}{\xi_i}+\left.\frac{d\Theta_v^*[z;\Omega\,|\,t]}{dt}\right|_{t-T}\sum_{i=1}^{3}A_i\right\}$$

$$=\sum_{i=1}^{3}\mu_i^2(z)B_i(z)exp\{\mu_i(z)T\} \tag{110}$$

where $A_1^* = A_1$, $A_2^* = \lambda A_2\xi_2$, $A_3^* = \lambda A_3\xi_3$. The set of Eqs. (108) - (110) is obtained from the original integral equation (99) by equating the corresponding derivatives of the left-hand and right-hand side of this equation at time *T*.

By averaging Eq. (72) the obtained solution (104) over the time occurrences of the first detectable photons in Ω, one finds the final expression for the GF of the number of photons per detection interval.

SEMICONDUCTOR LIGHTS SOURCES IN MODULATION FLUOROMETRY USING DIGITAL STORAGE OSCILLOSCOPES

Stephan Landgraf*

1. INTRODUCTION

The idea of building up a new method for the determination of fluorescence lifetime using semiconductor light sources, such as laser diodes (LDs) and ultrabright light emitting diodes (LEDs), was born in 1995. It had been shown earlier that LDs and LEDs can be used as pulsed light sources in a single photon timing (SPT) setup (LD: [1], LED: [2]). The integration of semiconductor light sources in commercial modulation fluorometers had also been demonstrated successfully [3]. But the main point was that all of these applications are rather expensive and difficult to handle. So a simpler setup had to be developed.

The following items should be included:

a) The method should be based on standard electronics, mainly on devices present in a typical physical chemistry laboratory. This excludes SPT techniques and commercial phase fluorometers based on two frequency generators with cross-correlation.

b) The number of components should be as small as possible. This means to use modulation fluorometry with a single frequency. The stepwise improvements are discussed in more detail in section 4.

c) The light emitting devices (mainly LDs and LEDs) should be as exchangeable as possible. Where ever possible changing of the light source itself should be enough to create a new wavelength.

* Stephan Landgraf, Graz University of Technology, Institute for Physical and Theoretical Chemistry, Technikerstr. 4/I, Graz, Austria A-8010

d) This leads directly to a modular setup that is very flexible when adapting to a given experiment. On the other hand the setup should be very robust when mounted together.
e) To increase applicability the connection to light guides should be integrated from the very beginning.
f) Finally the procedure of adjusting the apparatus should be as simple as possible. This lead to a setup where only the intensity of the light source is controlled (by an iris diaphragm). All other components are in the proper position when mounted together by only two screws per module.

For the first setup a frequency generator, a digital storage oscilloscope, a simple dc source or LD/LED and a photomultiplier (PM) with power supply is enough. Depending on the shortest lifetime wanted the price for this type of setup ranges from 2000 $ to 20000 $. For sensitivity reasons the PM should not be replaced by photodiodes. For more information on the technique and the history of LD/LED modulation fluorometry in our institute see section 4.

2. SPECTRAL PROPERTIES OF SEMICONDUCTOR LIGHT SOURCES

During the last few years major improvements in light emitting semiconductors have been made. Especially new materials, like III-V semiconductor materials, allow more and more short wavelength emission down to the UV region. All successful developments concerning laser diodes are connected to one name and one company, S. Nakamura [4] und Nichia Corporation (Japan). Now four different blue-UV LDs are available, but still very expensive (2000-3000 $). The four LDs cover a range of nearly 100 nm but they also have significant disadvantages: They are still much too expensive, the lifetime is unsatisfactory for continuos operation, and the 375 nm type has a low visible emission like the 370 nm LED (UG 1 filter necessary for low emission intensities). All types of LDs are summarized in Fig. 1. LDs are almost monochromatic light sources with narrow linewidth of typical 1 nm depending on the emission power. The curve is nearly Gaussian type and the long wavelength cut-off is sharp which is ideal for fluorescence applications (except LD 375). Alternatively a huge number of different LEDs exist covering a spectral region from 350

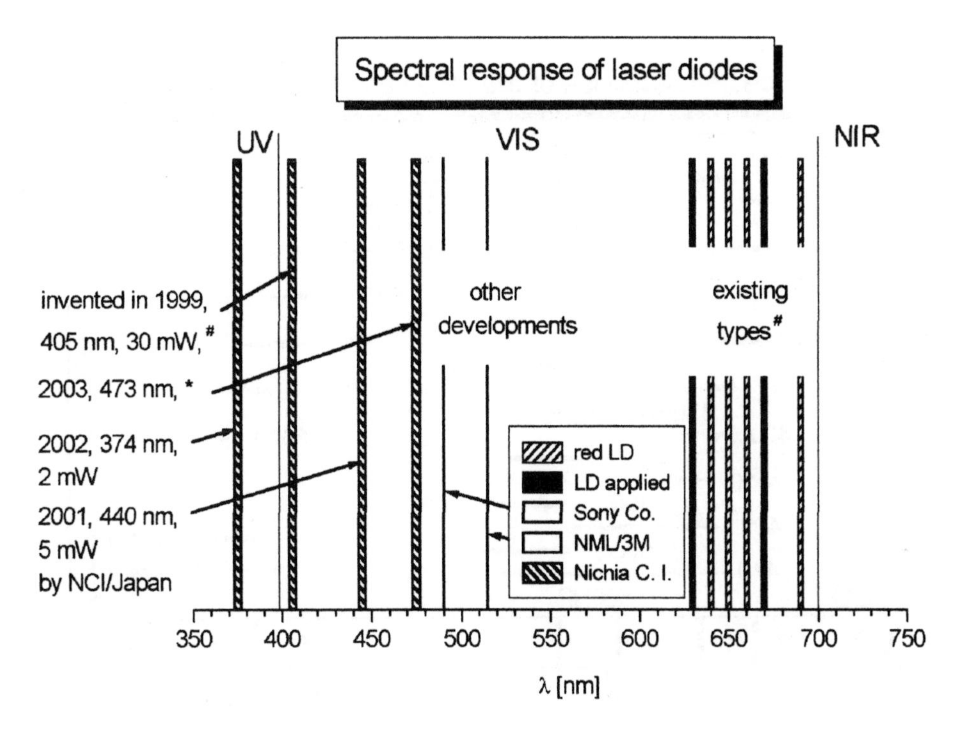

Fig. 1: Scheme of the optical spectra of LDs. Developments during the past 10 years. Red LDs are available for a longer time but new intensive ones have been built. Only LDs marked with # are commercially available, other UV-VIS types are engineering samples, $ no data present.

nm to the NIR. The spectral width of the LED emission is much wider compared to LD. The full width at half maximum (FWHM) is between 16 and 64 nm depending on the type of LED. Additionally the curve is not a Gaussian like one but has a tail to long wavelengths. This can be a problem in fluorescence applications due to overlap of scattered light and fluorescence emission. Therefore one quality criteria is the long wavelength cut-off, e.g. 1 % emission intensity as shown in Fig. 2 together with other optical information. This value depends very much on the type of LED. Additional filters are also very helpful here, especially band pass filters such as U330, violet additive, blue additive, green additive, cyan substrative, and magenta substrative (Edmund Industrial Optics, Barrington, NJ, USA), UG1, UG5, and BG25 (Schott, Mainz, Germany), and 400, 450, 500, and 550 nm short pass filter (Melles Griot, Carlsbad, CA, USA). The combination of LEDs with the filters mentioned above allows almost any fluorescence application conceivable without necessity of a monochromator.

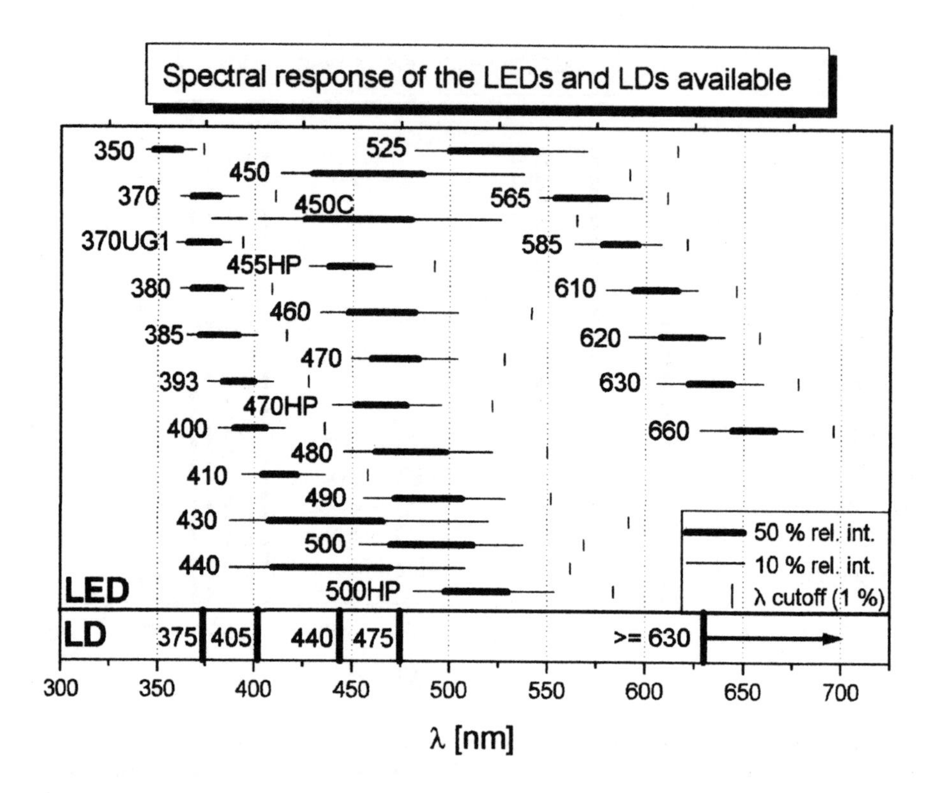

Fig. 2: Optical output spectra of almost all commercially available LEDs, including some HPLEDs. The thick lines indicate the spectral width (FWHM) and the thin lines indicate the range with 10 % or more output intensity. For fluorescence applications the long wavelength cutoff (1 % relative intensity) is also marked with |. For comparison all LDs are shown by a short vertical line at the bottom of the figure.

Since 2002 LEDs are also used for illumination. These types are called high power LEDs (HPLEDs). The most interesting HPLEDs are in the blue-green region having up to 100 mW optical output power. There spectral response is shown in Fig. 3.

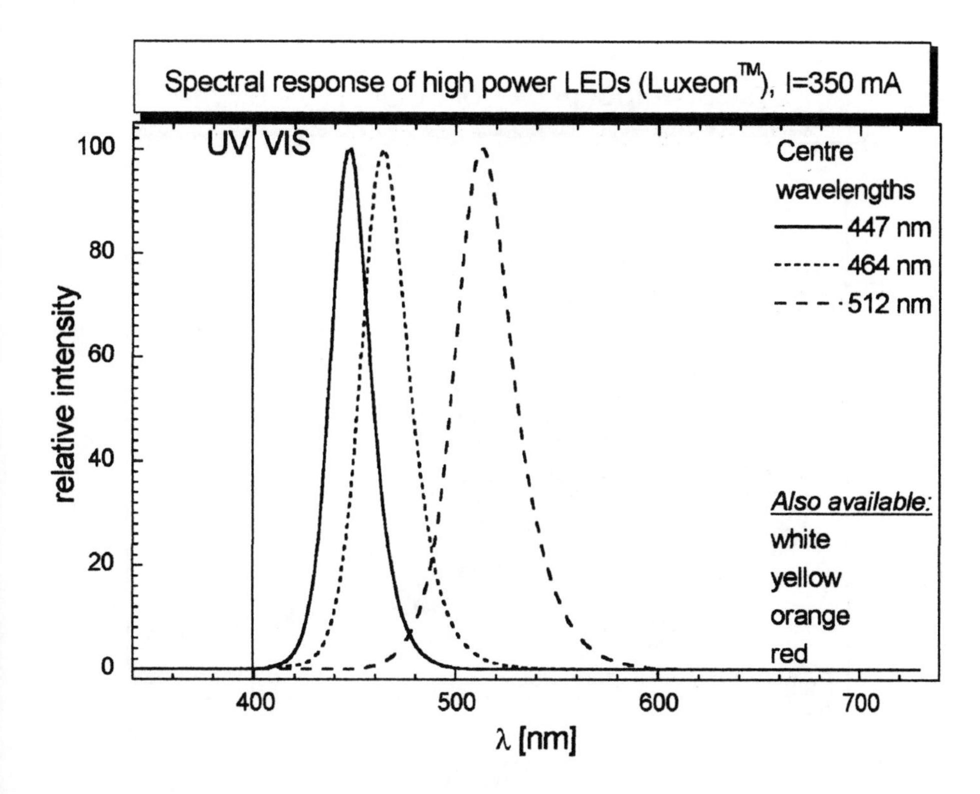

Fig. 3: Blue-green HPLEDs (solid = royal-blue type 455, dotted = blue type 470, and dashed = cyan type 500). Spectra measured with UV-VIS diode array spectrometer (DAS, Zeiss, Jena, Germany).

Optical output power and wavelength of almost all LEDs, including HPLEDs are summarized visually in Fig. 4. Most ultrabright LEDs are located in the 1 to 10 mW region, much brighter than normal LEDs used as indicators in electronic devices (0.01 to 0.1 mW). HPLEDs have one order of magnitude higher emission intensity (at higher supply current). This gives some improvements in low intensity measurements when using additional short or band pass filters in the excitation path to cut the long wavelength emission. HPLEDs are also very useful in lecture hall demonstrations. Very recently new LEDs, named violet and pink, became commercially available with mixed emission to create the final colour. They are not useful for fluorescence experiments.

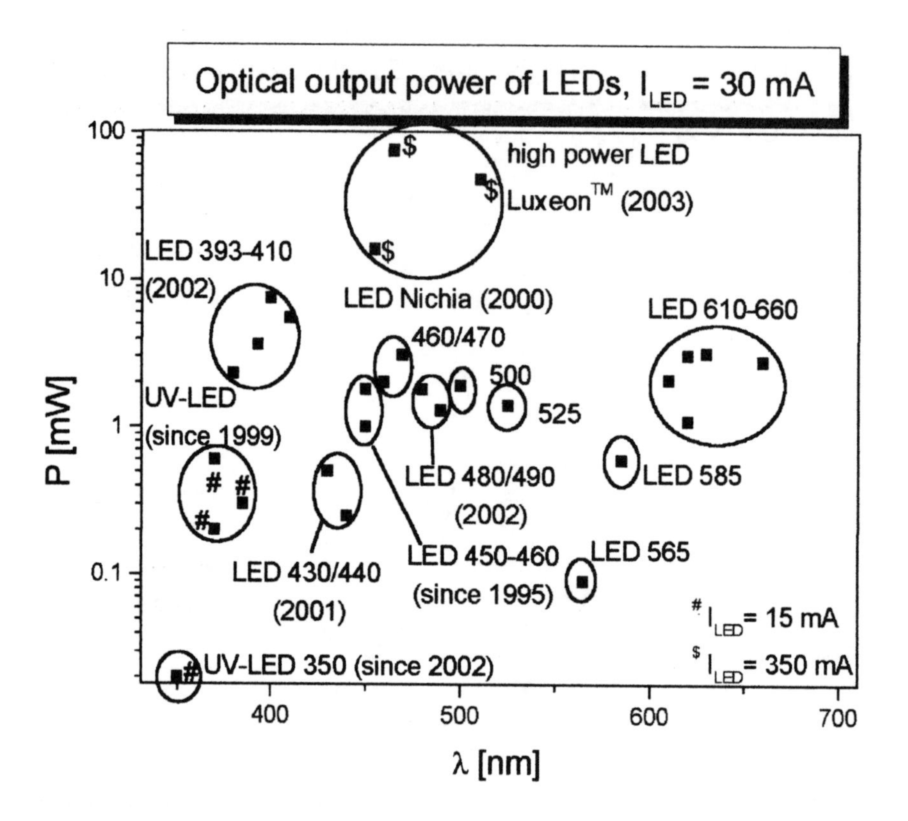

Fig. 4: Schematic representation of the optical output power vs. the center wavelength of emission of almost all commercially available LEDs, including some HPLEDs. LEDs not further specified are from local distributors. Output power determined with Newport 1825 power meter with 818-UV detector (Newport, Irvine, CA, USA).

A lot of static applications are already known, such as absorption and fluorescence methods for chemical, biological, medical and environmental investigations. For more details, see [5].

Lecture hall demonstration experiment: In a small box (15x8x5 cm) a row of LEDs is mounted with 7.54 mm (0.3 inch) spacing. To cover the whole visible region from violet to red the following types are included: LED 380, 393, 400, 430, 450, 480, 500, 525, 585, 610, 620, 630, 660 nm. Each LED has one resistor in series to limit the current to approximately 30 mA (formula : $R[\Omega] = (6V-U_{LED})/0.03A$, e.g. 68 Ω for LED 380 nm and 150 Ω for LED 660 nm). As power supply 8 NiCd or NiMH accumulators (~1000 mAh for 2 h continuous operation) are used together with a low drop voltage regulator (6 V=) with a small heat sink. All components are available from local suppliers and can be mounted easily on a prototype board. The only modifications of the box are 13 holes (5 mm) on top for LEDs and two holes (on switch and plug for charging) on one side. Cuvettes with different fluorophores with blue, green yellow and red fluorescence are moved from one side to the other close to the LEDs. As soon as the colour of the LED is the same as the colour of the fluorescence the emission disappears (correlation: energy ↔ wavelength ↔ colour). This is a very simple and reliable experiment for undergraduated chemistry courses and a good introduction into fluorescence (absorption, emission, lifetime, quantum yield, ...).

3. TIME-RESOLVED PROPERTIES OF LEDS

More interesting for physical chemistry is the study of kinetics in photochemical reactions schemes. One major investigation is the determination of the fluorescence lifetime. This can be done in three different ways: First, using a strong short laser pulse and transient recording of the fluorescence intensity. This is a complicated and expensive method with strong decomposition of the sample. Deconvolution at short lifetimes is difficult and inaccurate. Secondly, excitation with a weak but high repetition light source and counting of the first (and single) fluorescence photon, mostly named SPT or TCSPC. The apparatus is much less difficult to handle and much less expensive. Deconvolution is much simpler and more accurate here, especially when pulse semiconductor light sources are used (highly reproducible pulse profile). Thirdly, the modulation technique, where the sample is radiated continuously while the intensity of the excitation source is changed in a sinusoidal way. The fluorescence is phase shifted and decreased in intensity (ac part only). More details are given in [6].

Everybody who already has an SPT apparatus in his laboratory is strongly encouraged to use pulsed LDs and LEDs. There are different ways to built a ns pulsed system as well as a large number of suppliers for pulsed semiconductor light sources. A brief overview is given in Tab. 1. Main advantages are: High repetition rate (short measuring time), monochrome or narrow band emission, perfectly reproducible pulse profile for deconvolution, low voltage supply, and long lifetime without changes in pulse intensity or shape. Frequency-doubled LDs using second harmonic generation of NIR-LEDs are no longer of interest for fluorescence measurements of organic substances. The devices have a very narrow emission with excellent beam quality, but are very expensive and have a very low level of intensity (typical some μW). They have been replaced by UV and blue LDs except 335 nm emission from an LD 670 nm (16 ps, 70 μW, 0.1 % efficiency, $LaIO_3$ crystal [7]) and 308 nm emission from a cooled LD 635 nm (616 nm at 165 K, cw, 50 pW, 0.005 % efficiency, $LaIO_3$ crystal [8]).

Table 1: Overview on pulsed LDs and LEDs including recent commercial systems. The values given are the pulse width in ns (upper value) and the repetition rate in kHz (lower value). More references are given in [5] and [6]. For comparison a pulsed Xe lamp is also given.

System **LD** [f]	Direct [a]	High current [a]	Pico-Quant [b]	Hama-matsu [c]	APHS [d]	Becker& Hickl [e]
375			0.06 $0 - 4 \cdot 10^4$			0.06 $2/5/8 \cdot 10^4$
405			0.06 $0 - 4 \cdot 10^4$	0.06 0 - 2000	0.04 0 - 1000 0.05 $0 - 10^5$	0.06 $2/5/8 \cdot 10^4$
445			0.06 $0 - 4 \cdot 10^4$		0.06 0 - 1000	
473			0.06 $0\text{-}4 \cdot 10^4$			
635 to NIR	1.7 0 - 2000		0.06 $0 - 4 \cdot 10^4$	0.05 0 - 2000	0.035 0 - 1000 0.045 $0 - 10^5$	0.05 $2/5/8 \cdot 10^4$ 0.09 $5 \cdot 10^4$
LED [f]						
380			0.75 $0 - 4 \cdot 10^4$			
450C/U330	3.1 0 - 2000	1.75				
450C		1.75				
460			0.75 $0 - 4 \cdot 10^4$			
510			0.75 $0 - 4 \cdot 10^4$			
600			0.75 $0 - 4 \cdot 10^4$			
Xe-lamp, only useful in the UV up to 360 nm						

+UG1	4.5
360 nm max. [a]	4.35
without filter	6.5
470 nm max. [a]	4.35

[a] Direct connection to a frequency generator, high current pulser similar to Araki et al., for technical details see [6]
[b] see http://www.picoquant.com/, LD475 will be available at the end of 2003.
[c] see http://www.hpk.co.jp/eng/products/Syse/PLPE.htm
[d] see http://www.aphs.de
[e] see http://www.becker-hickl.de/lasers.htm
[f] values of center wavelength can vary due to manufacturing tolerances.

Single frequency modulation fluorometry is a very precise and simple to handle alternative to SPT. Tab. 2 shows advantages and disadvantages of modulation fluorometry and SPT. As shown both methods comply each other in a perfect way.

Table 2: Comparison of modulation fluorometry and SPT. The winner (if any) is underlined.

Experimental detail	Modulation fluorometry	SPT/TCSPC
Single exponential precision	<u>very high</u>	high
Detection of weak second exponential	<u>very sensitive</u>	difficult
Two exponential analysis	difficult	<u>simple</u>
Complex kinetic analysis	very difficult	<u>possible</u>
Minimum measuring time	down to 1-10 s	<u>less than 1 s</u>
Typical measuring time	some min	some min
Decomposition of sample	slow	<u>very slow</u>
Optimum conditions for 1 ns lifetime	<u>30 MHz frequency</u>	1 ns pulse
Experimental precautions	<u>detector saturation</u>	s/s ratio s/s delay

Time-resolved properties of LD/LED pulses are mainly dominated by the electronics. As shown in Tab. 2 typical pulse width are 40-100 ps for LDs and ~1 ns for LEDs. In modulation fluorometry the transition frequency is the important value. So for any new LED the high frequency behaviour has to be determined. Tables of these values are given in [5] and [6]. Time-resolved data for most recent LEDs are given in Tab. 3. Note the remarkable output power and frequency response of the three violet LEDs (393, 400, and 410 nm).

Table 3: Spectral properties and frequency response of the most recent LEDs. λ is the central wavelength of emission, FWHM is the full width at half maximum of emission, P_{MAX} is the maximum usable continuous output power at I_{LED} =30 mA measured with a 1825 power meter (Newport) with 818-UV detector, f_{TR} is the transition frequency, f_{MAX} is the maximum usable frequency at highest possible modulation amplitude from the frequency generator, U_{MOD} is the modulation amplitude needed for 50 % optical modulation, measured with PM 5600U (Hamamatsu) at 900 V (50 Ω, 1 MHz).

LED [mcd] [a]	λ (nm)	FWHM [d] (nm)	P_{MAX} (mW)	f_{TR} (MHz)	f_{MAX} (MHz)	U_{MOD} (dBm)
350/0.03 [b]	354	16	0.02	120	200	+6.0
380/3.0 [b]	376	17	2.3	45	120	+6.3
393/200	392	17	3.6	96	180	+5.3
400/200	397	18	7.5	57	120	+5.3
410/200	412	20	5.5	36	100	+7.5
HP470	464	28	75 [c]	36	80	+14 [e]
480/2000	480	38	1.8	90	160	+8.5
490/1400	488	37	1.3	52	120	+9.5

[a] Values (in mcd) given by the manufacturer.
[b] Same as [a] but in mW.
[c] I_{LED} = 350 mA.
[d] Measured with Zeiss MMS diode array spectrometer.
[e] With a special circuit.

In Fig. 5 the typical values of transition frequency and maximum usable frequency are shown for almost all LEDs, including most recent types and one HPLED. As indicated on the right side most LEDs lie near or over the limit of a typical PM (R928, Hamamatsu, Herrsching, Germany). To avoid limitations of the detector a faster type, like the R5600 (Hamamatsu) has to be used, even for LEDs. Also remarkable is the high frequency response of the new HPLED 470 nm. This makes HPLEDs also useful for time-resolved fluorescence and not only for illuminations.

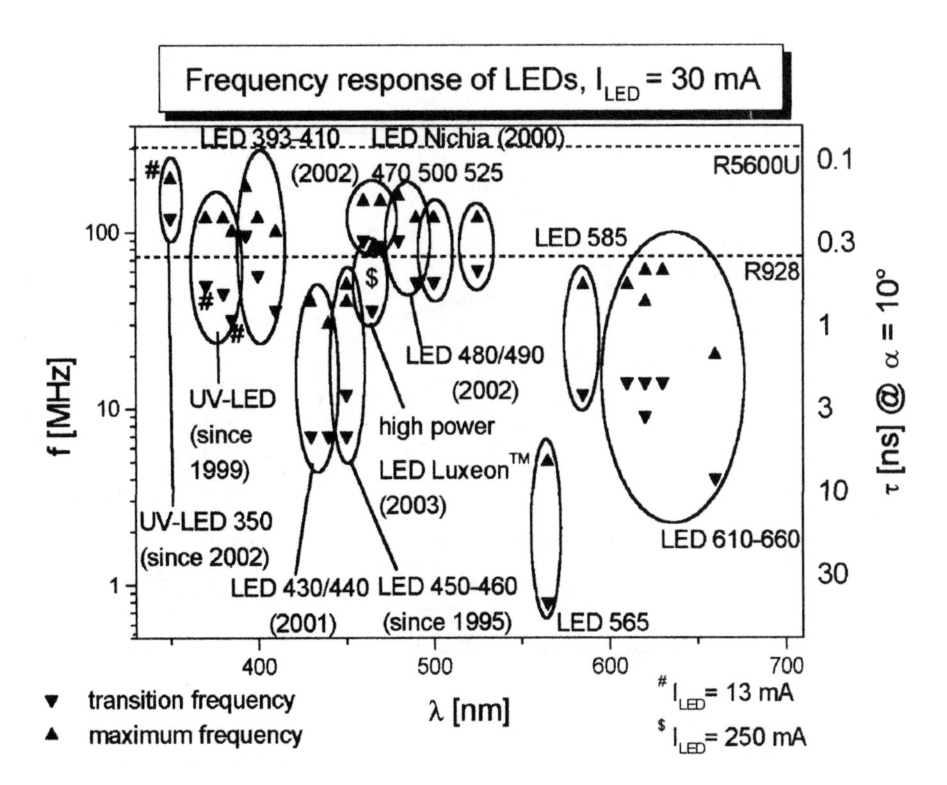

Fig. 5: Schematic representation of the frequency response vs. the center wavelength of emission of almost all commercially available LEDs, including also one HPLEDs. The right scale indicates the fluorescence lifetime measured at 10° phase angle. The frequency limit of two often used PMs are also included as dashed lines. LEDs not further specified are from local distributors. Frequencies determined with 9370 digital storage oscilloscope (LeCroy) and 6071A frequency generator (Fluke) using peak-to-peak analysis after 1000 sweeps.

To understand the two frequency values of Fig. 5 some more information has to be added. First, the transition frequency is defined similar to electronics: The ac part of a source is reduced to one half of the low frequency response at the transition frequency (-3 dB). Secondly, the maximum usable frequency is affected by two effects. The decrease of the ac signal can be solved by increasing modulation depth at the frequency generator (see section 4). But overtones and non harmonics are also increased. So from a certain frequency upward the LED can no longer be used in a proper way. The behaviour of each LED is different. Two cases (slow and fast LED) are shown in Fig. 6.

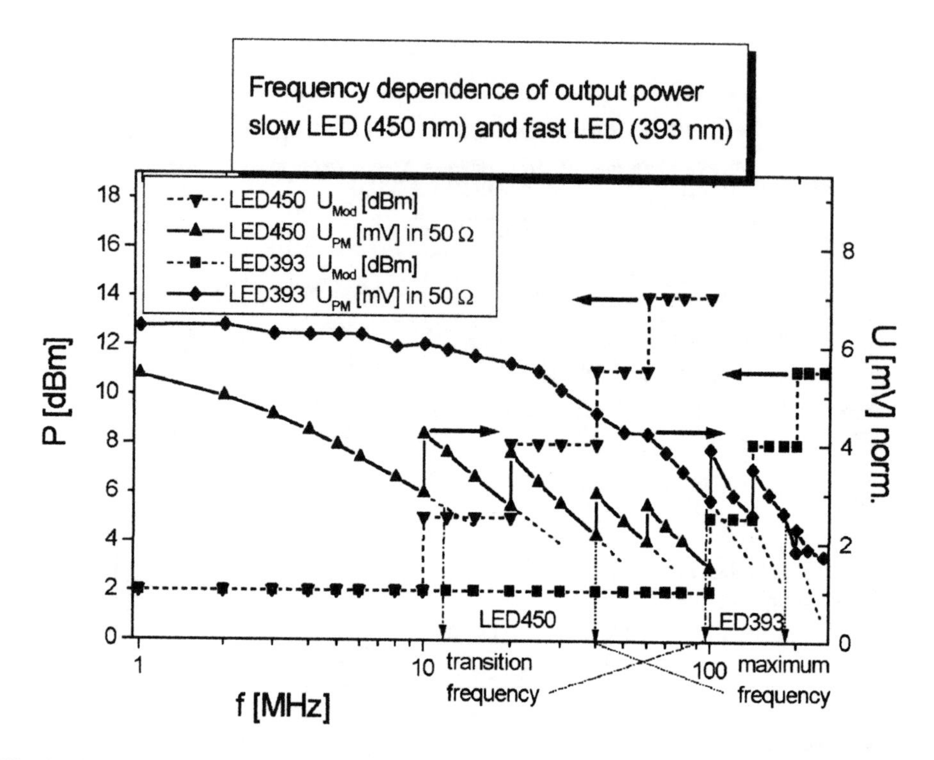

Fig. 6: Measurement of the ac part of the light emission in relative voltage units vs. the frequency applied for two LEDs (450 nm, older type and 393 nm, newer type). The left axis shows the modulation amplitude (dashed curves) whereas the right scale indicates the ac part of the emission signal (solid curves). Modulation amplitudes higher than shown in this figure leads to strong deviation from sine waveform. Transition frequency arrows are dashed-dotted and maximum usable frequency arrows are dotted. Amplitudes determined with 9370 digital storage oscilloscope (LeCroy) and 6071A frequency generator (Fluke) using peak-to-peak analysis after 1000 sweeps.

Slow LEDs with a transition frequency of about 10 MHz or less behave like an RC low pass filter in electronics. This means that the ac part of the signal drops with 3 dB/octave. Over a large frequency range the amplitude of the frequency generator can be increased to enhance the range of application. Fast LEDs instead behave more like an LC low pass filter with a drop of 6 dB/octave (ac part). This makes the difference of the two frequencies smaller. Nevertheless LEDs can be used up to 200 MHz depending on the type of LED. This is still 100 MHz apart from the limit of the PM5600 (300 MHz, -3 dB). Taking into account both the wavelength and the frequency range of LEDs really a lot of applications are possible with an increasing number each year due to new technical developments of semiconductor light sources.

4. MODULATION FLUOROMETRY WITH DSOS - BASICS AND APPLICATIONS

The first practical application of an LED was published in 1973 by Flaschka et al. [9]. He demonstrated the applicability of an LED for photometric measurements in a flow cell. From that time on an increasing number of publications appeared describing applications of LDs and LEDs. An extensive review with 244 references is given in [5].

Changing the light intensity of a lamp by controlling the supply current is often used. But only LDs and LEDs have a mA power consumption, a ns response, and dynamic range from zero to the maximum frequency at the same time. This makes semiconductor light sources an ideal tool for modulation experiments. A sinusoidal change of the light output power is achieved by a sinusoidal change of the supply current, as shown in Fig. 7 for an LED and an LD.

The P vs. I curve for an LD is almost linear whereas for LEDs slight deviations are observed. Nevertheless the distortions of the optical signal as measured by FFT analysis of the digital storage oscilloscope are very small over a wide range of experimental settings. Tests have been made to show that distortions (from 2nd and 3rd harmonic) don't influence the result as long as they are below -20 db from the fundamental modulation frequency. These values are available for all LEDs.

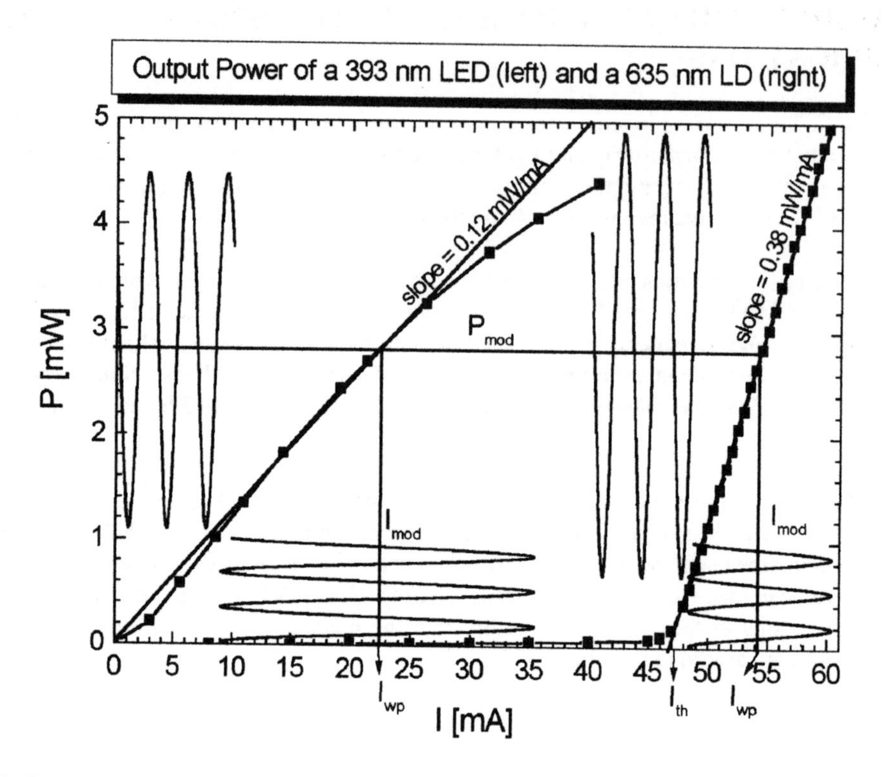

Fig. 7: Output power of a 393 nm LED (left) and a 635 nm LD (right) vs. the supply current. Note that the slope is different comparing LD and LED. Therefore modulation amplitude is always larger for LEDs than for LDs and optical modulation is smaller. On the other hand LD optical output power depends much more on the temperature. Sine functions in the graph are schematic but very close to reality. Typical value for LDs is the threshold current I_{th} whereas LEDs have only the working current I_{wp}. Output power determined with Newport 1825 power meter with 818-UV detector (Newport).

In 1995 the idea came up to try building a modulation fluorometer from parts available in a typical physical chemistry laboratory. External parts were: Digital storage oscilloscope 9410 (LeCroy), PM 928 (Hamamatsu) with power supply, and a frequency generator model 166 (Wavetek, now Willtek, Ismaning, Germany). Additionally we bought an inexpensive KN120S (GFO, Drakenburg, Germany) power supply and modified it for better stability (thermal insulation between transformer and electronics) and better air flow through the device). All optical components like lenses, filters, sample holder, etc. were already present and mountable on an optical rail. The first module to be built was the excitation source containing a 635 nm LD and a coupling unit to mix ac and dc supply. This module has been modified and optimized during the last years. Finally an LED unit with internal current stabilization (from a single 9-12 V= supply) and ac/dc coupling has been built. All variations are summarized in Fig. 8.

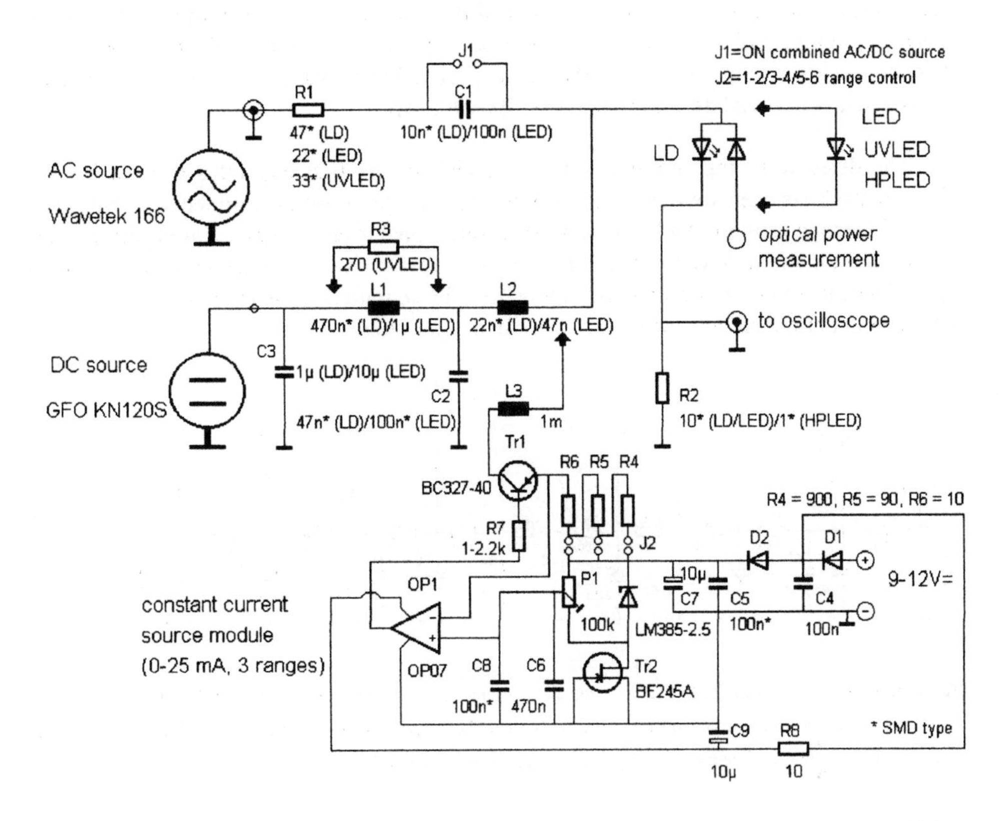

Fig. 8: All models of LD/LED excitation modules. The frequency generator has been replaced by a 6071A (Fluke). The external dc source has been integrated into the module as shown in the lower part of the figure. Different configurations of LD, LED, and HPLED are also indicated.

The minimum necessary stability of the generator is 0.01 % during the measurements for standard experiments. Best precision is achieved with 0.001 %. Part of the recent work to improve the method is to built up an inexpensive alternative to the frequency generator. First tests with a 20 MHz DDS generator (ELV, Leer, Germany) are very promising. DDS technology is available up to 62.5 MHz (AD9850, Analog Device, Norwood, MA, USA) which is enough for ns lifetime measurements using LDs and LEDs. The only thing to be done is the computer control of the DDS generator for automatic frequency scans (see later).

With the basic setup it could be demonstrated that this simple combination works down to a lifetime of 2 ns with good accuracy and reproducibility (±0.2 ns). The results were first presented on the Photochemistry Meeting of the German Chemical Society in Dresden in 1995 and published later in the conference proceedings [10]. There was a lot of interest, that encouraged us to continue the development of the new method. First improvement was to replace the optical rail by a modular setup containing all optics, sample holder, excitation and detection. This was published in the same proceedings two years later [11] and already contained an improved detection system still used today.

The detector in the modular setup is a PM. In the first apparatus a normal R928 (Hamamatsu) has been applied. Its transition frequency has been determined at 83 MHz and the usable range is up to 150 MHz with decreasing signal level. To have a good performance of the detector and to have the possibility to use it for SPT without change of the detector a special bleeder circuit has been built, similar to [12]. The circuit is shown in Fig. 9. The detector has been in permanent use for 8 years without any problems.

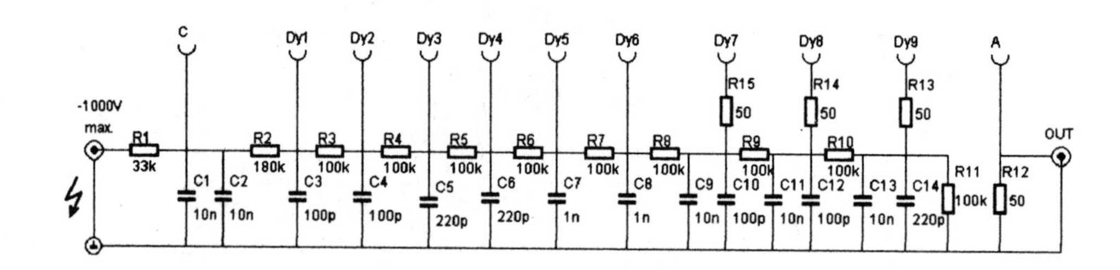

Fig. 9: Bleeder circuit for an R928 PM (or equivalent) usable for modulation fluorometry and SPT. All capacitors are ceramic type with 1 kV max. voltage. All resistors are SMD type in 1206 outline (1/4 W). The corresponding circuit board is 70 mm in diameter (5 mm ground contact when mounted).

As already shown in section 3 the newer LEDs are quite fast with a frequency range up to 200 MHz. The LDs are also much faster than the R928 PM. To increase the lifetime range and to decrease the size of the modules a faster R5600 PM (Hamamatsu) has been integrated (again for modulation fluorometry and SPT). The optimized bleeder circuit is shown in Fig. 10. Due to one dynode less in the R5600 PM the sensitivity is slightly lower compared to R928 (about one half).

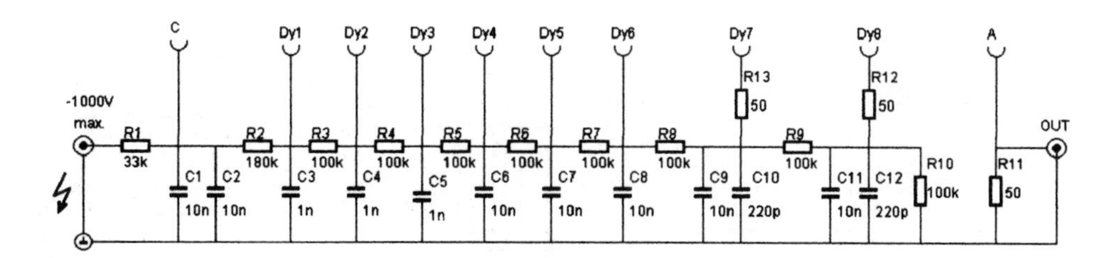

Fig. 10: Bleeder circuit for an R5600 PM (or equivalent) usable for modulation fluorometry and SPT. All capacitors are ceramic type with 2 kV max. voltage (except 10 nF/1 kV). All resistors are SMD type in 1206 outline (1/4 W), except R6, R7, and R9 that are standard metal film resistors (1/4 W) and work as the necessary bridges. The corresponding circuit board is 60x60 mm (5 mm ground contact when mounted).

With this design a module of only 80x80x30 mm size with two external connections (BNC: Out, HV-BNC: -1000 V_{max}) has been built. The signal response (fall time) has been measured (650 ps [13]). Transition frequency is close to 300 MHz (experiments up to 350 MHz). Three samples of the module exist and all of them work reliably without any failure up to now.

The PM power supply is a critical point. We made good experience with two types of HV power supplies. First, simple linearly regulated power supply from a 1.5 kV source (550 V~ transformer with voltage doubling rectifier diode network). This works fine when capacitors resist at least 1.8 kV and a BUY71 (TO-3, NPN Darlington, 2.2 kV, 2 A, 40 W) is used. Secondly, HV modules from Matsusada (HPMR-1.1N, Kusatsu-City, Japan) have very low ripple and high stability at a moderate price. This module is the better choice for people not so familiar with high voltage devices due to an integrated output plug and low voltage supply (24 V=).

The 1998 setup also used complete computerisation for the measurement and data recording. This has been achieved by replacing the frequency generator and the oscilloscope by a 6071A (Fluke) and 9370 (LeCroy), respectively. This also increased the response of the apparatus to 1 GHz. The first data evaluation was to find the zero crossing points, optimizing them by a linear fit and averaging over all points. This routine had some problems with noisy signals and gave no amplitude information. Later a second routine has been added that makes a global sine function fit over all data points. Now the results are much more reproducible and amplitude data are also calculated. Finally a table with frequencies, time delays and relative amplitudes is created. From this all kinetic evaluation can be done. More information about the data evaluation is given in [6]. The whole setup as used now (2003) is shown in Fig. 11.

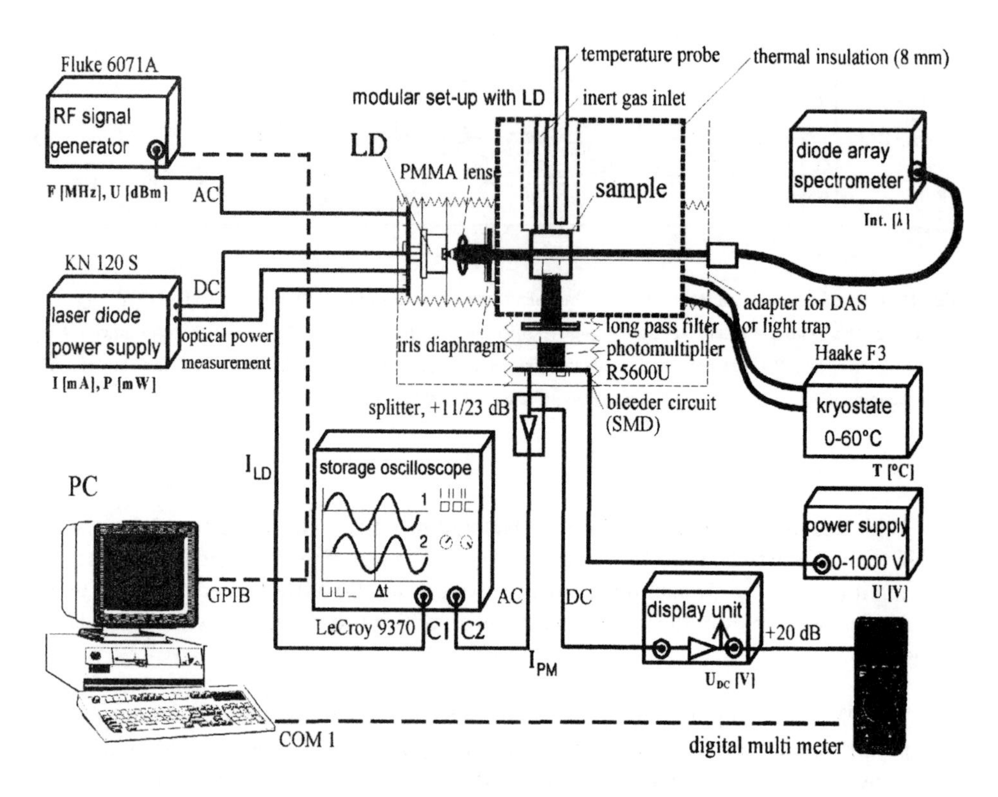

Fig. 11: Schematic diagram of the present experimental setup, shown for LD excitation. Other configurations are also available, see text. Central unit for surface analysis has also been built.

Note: to switch from modulation fluorometry to SPT only the excitation module has to be changed. All other components of the modular setup remain the same. Main other configurations are: Change to LED excitation sources, change DAS to power measurement (1825 with 818-UV head, Newport) to determine the actual power in the sample. The output splitter is also available in a passive design, that can be combined with external amplifiers, such as HP8447A (Hewlett Packard, now Agilent, Palo Alto, CA, USA), PA-25 (Conrad Electronic, Hirschau, Germany), and RFA-401/402 (ELV). The results do not alter with the amplifier applied.

One important test is the reproducibility of the measurement reference vs. itself. The apparent lifetime should be exactly zero. This tests have been done, as shown in Fig. 12.

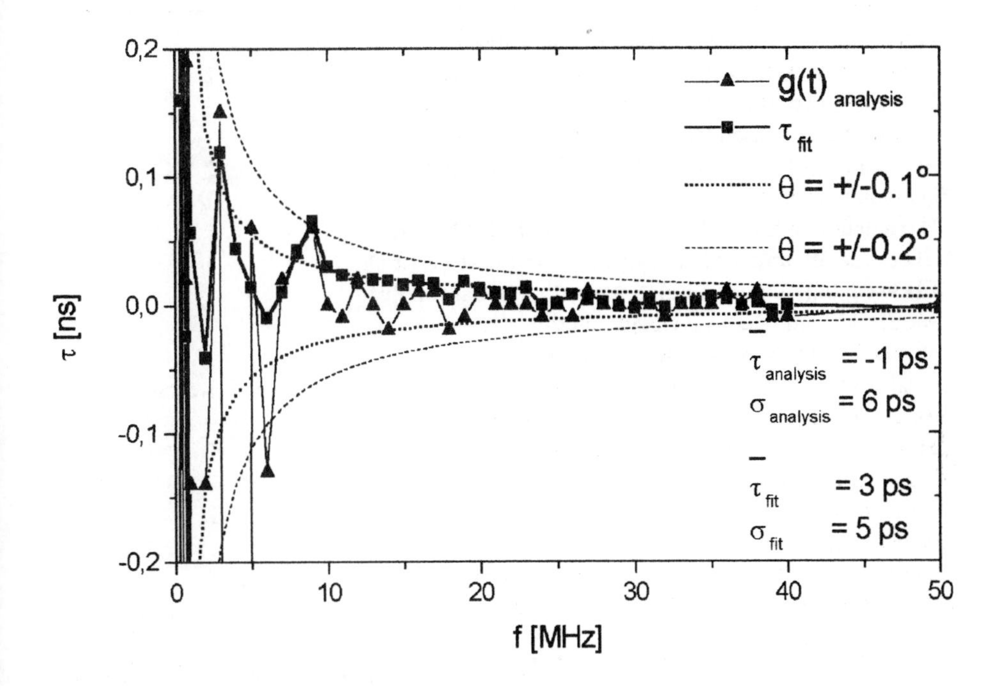

Fig. 12: Result of the analytic (thin line from zero crossings) and the numerical (thick line from sine fit) data evaluation of a measurement reference vs. itself. To higher frequencies the result converged to zero with a small deviation of -1 and 6 ps for the fit and the analysis, respectively. The standard deviation is 3 and 5 ps, respectively. The error in the phase angle of ±0.1° (dotted) and ±0.2° (dashed) is also included.

It can be seen clearly that the fit routine gives better results compared to the analysis of the zero crossings. The deviations can be used directly to calculate the systematic error for fluorescence lifetime experiments (ideal sample). As in many other experiments the precision can be improved by increasing the measuring time (sampling) but sample decomposition is also increased. Experimental errors of less than 1% (absolute and relative) can be achieved with approximately 1 min per frequency. The limit of precision has been estimated from a large number of results to be about 0.2 % in the present setup.

To have the possibility to perform a huge number of different experiments a large number of modules has been built, as summarized in Tab. 4.

Table 4: List of components present in modular setup.

Module	Quantity
Central unit for 10x10 and 10x20 mm cuvettes	6
Central unit surface analysis up to 60 mm diameter	2
PM module with R5600 with integrated bleeder circuit one with Peltier-based temperature stabilization	3
PM module with R928 with integrated bleeder circuit	2
PD module for intensity measurement, student course	2
LD excitation unit with 635 nm LD, with PMMA lense	1
LED excitation unit for 5 mm types for 350-660 nm LEDs	3
LED excitation unit for 5 mm types with internal supply	1
UVLED 370 nm fixed wavelength module	1
HPLED 470 nm fixed wavelength module	1
Signal splitter, 0 dB, +11 dB, and +23 dB	1 each
Insulation setup with heat exchanger for liquid media	4
Combined temperature probe holder and inert gas inlet	1
Lense holder	2
Polarizer modules	2
Iris diaphragm with 0.6 to 5 mm ⌀	6
Filter holder for 40x40, 50x50, and ⌀50 mm filter	8
Adapter for 818-UV power head, light guides 2.2/5/7 mm [a]	several
Cover plates, light traps	several

[a] 2.2 mm for PMMA fibre (1 mm core diameter, Conrad) and 5/7 mm for liquid light guides (3/5 mm core diameter, Lumatec, Deisenhofen, Germany).

There have already been performed numerous fluorescence lifetime measurements with the method described above. In the first paper [10] only two red emitting dyes (nile blue and oxazine 4 have been used to show the applicability of the setup. The results agree very well with literature values with a relative error of about 10 %. Later more measurements with the LD 635 nm (including oxazine 1 [13]) and with the LED 450 nm (Ru(II)-tris-bipyridylhexafluorophosphate in various solvents have been made [11]. Introducing the UVLED 370 nm fluorescence lifetimes of phenylene-1,4,diethylene-bis(N-methylpyridium) diiodide, phenylene-1,4,diethylene-bis(N-methylquinolinium) diiodide, and phenylene-1,4,diethylene-bisquinoline have been measured in different solvents in a range of 0.11 to 1.2 ns (with up to 120 MHz modulation frequency). Quenching experiments have been performed using a Stern-Volmer evaluation and the results have been discussed in the frame of Burshtein's finite lifetime theory [14]. Similar experiments using perylene as the fluorophore and N,N'-dimetylaniline and N,N',2,4,6-pentamethylaniline as a quencher have been made in various solvents and solvent mixtures of different viscosity, namely acetonitrile-PEG 10000, acetonitrile-ethyleneglycole, acetonitrile-tatramethylurea [15], methanol-glycerol, methanol-cyclohexanol, and pure n-alcohols [16]. Later practical applications of LED modulation fluorometry have been tested with the analysis of the fluorescence of crude oil samples. It has been shown that down to a concentration of 1 ppm in cyclohexane the typical frequency dependent decay times can be measured [6]. Recent results show that also solid samples, like contaminated sand, can be measured with a sensitivity of 100 to 1000 ppm depending on the size of the sand particles [17]. In a recent work on the fluorescence and electron transfer properties of 2,5-dicyano-N,N,N',N'-tetrametyl-p-phenylenediamine the connection to an external apparatus (400 °C high temperature cuvette holder) has been applied successfully measuring the fluorescence lifetime of the fluorophore in the gas phase [18].

One major advantage of the modular setup is the simple integration of new light sources. The excitation modules for LEDs are compatible with all 5 mm round type plastic housing. Simply removing the LED itself results in a new wavelength of the apparatus. No other changes have to be made. Only devices with metal housings, like UVLED 370 nm or LDs, have their own excitation module. This procedure has been shown in an earlier photochemistry work [14]. In Tab. 5 main properties of modulation fluorometry in the recent form is summarized.

Table 5: Summary of the properties of modulation fluorometry.

Property	Value
Wavelength range (LD + LED)	350 to 670 nm
Modulation frequency LED	100 kHz to 200 MHz
Modulation frequency LD	100 kHz to 700 MHz
Detector response, PM R5600	up to 350 MHz
Sensitivity in 10x10 cuvettes, full scale intensity	$3\cdot10^{-8}$ M fluorescein
Sensitivity in 10x10 cuvettes, limit lifetime measurements	$1\cdot10^{-9}$ M fluorescein
Lifetime range (LD + LED)	1 ppm crude oil
Minimum pulse intensity in TCSPC mode *	100 ps to 1 μs
Lifetime range (LD + LED) in TCSPC mode	2500 photons/pulse
	100 ps to 1 μs

* $1\cdot10^{-5}$ M perylene (quantum yield close to unity).

Very recent improvements: Temperature control of the center block (0 to 60 °C, ±0.1 °C), temperature measurement very close to the cuvette (Mo sensor or Ni/CrNi thermocouple with 3 mm diameter, ±0.1 °C), and inert gas atmosphere inside the center block to avoid oxygen entering into the sample and any type of condensed water or ice on the cuvette or on the optics. Active coupling units: a series of active and passive coupling units for the PM signal has been built. Active units are based on one or two MSA-0311 (Hewlett Packard) cascadable high frequency amplifiers with 2.3 GHz transition frequency and 11 or 23 dB gain from a single 12 V= power supply, respectively. In all units the dc part of the signal is connected to a display unit with internal 10 dB amplification. HPLED holder for Batwing type housing. The HPLED is cooled by the whole module and the heat (1.2 W max.) is dissipated into the excitation units. The units also support combined and splitted ac and dc sources.

Next improvements will be to simplify the PC controlled frequency generator (see above), to integrate UV and blue LDs into the modulation fluorometry, to improve detection response in TCSPC mode with an MCP-PM (R3809, Hamamatsu) and a new PC card, and to built a simple automatic iris diaphragm control.

5. CONCLUSION AND OUTLOOK

Static and time-resolved applications of semiconductor light sources in physical, analytical, bioanalytical, clinical and technical chemistry are already a laboratory standard. LDs and LEDs have several advantages due to their small size, low power consumption, high and stable light intensity, and long lifetime. Recent developments to shorter wavelengths, faster response, and higher intensity can be integrated immediately into the modular experimental setup and increase the number of possible applications rapidly. This principle of miniaturization also leads to easy to handle apparatuses. It is very important to continue expansion and optimization of the experimental setups to increase the accuracy and the possible applications and to decrease costs and size. During the last seven years we have developed a new method of fluorescence lifetime determination that offers an alternative to pulse techniques, like SPT. Describing the methods and experimental results we have already published nine articles and one contribution to a book. It is still an amazing task to follow all new applications with LDs and LEDs over the next years especially in the UV region.

REFERENCES

1. a) D.L. Farrens, P.-S. Song, Subnanosecond single photon timing measurements using a pulsed diode-laser, *Photochem. Photobiol.* **54**, 313-317 (1991); b) J.A. Tatum, J.W. Jennings, III, D. L. MacFarlane, Compact, inexpensive, visible diode laser source of high repetition rate picosecond pulses, *Rev. Sci. Instrum.* **63**, 2950-2953 (1992).

2. a) T. Araki, H. Misawa, Light eimitting diode-based nanosecond ultraviolet light source for fluorescence lifetime measurements, *Rev. Sci. Instrum.* **66**, 5469-5472 (1995); b) T. Araki, Y. Fujisawa, M. Hashimoto, An ultraviolet nanosecond light pulse generator using a light emitting diode for test of photodetectors, *Rev. Sci. Instrum.* **68**, 1365-1368 (1997).

3. a) K.W. Berndt, I. Gryczynski, and J.R. Lakowicz, Phase-modulation fluorometry using a frequency-doubled pulsed laser diode light source, *Rev. Sci. Instrum.* **61**, 1816-1820 (1990); b) K.W. Berndt, I. Gryczynski, and J.R. Lakowicz, Phase-modulation fluorometry using a frequency-doubled pulsed laser diode light source, *Proc. SPIE-Int. Soc. Opt. Eng.* **1204**, 253-261 (1990); c) R.B. Thompson, J.K. Frisoli, J.R. Lakowicz, Phase fluorometry using a continuously modulated laser diode, *Rev. Sci. Instrum.* **64**, 2075-2078 (1992).

4. S. Nakamura, S. Pearton, and G. Fasol, *The Blue Laser Diode, The Complete Story*, (Springer, Berlin, 2000).

5. S. Landgraf, Application of laser diodes and ultrabright light emitting diodes for static and time-resolved optical methods in physical chemistry, in : *Handbook of Luminescence, Display Materials and Devices,* edited by H.S. Nalwa and L.S. Rohwer (American Scientific Publishers, CA, USA, 2003), Vol. 3, Chapter 9, 371-398.

6. S. Landgraf, Application of semiconductor light sources for investigations of photochemical reactions, *Spectrochimica Acta A* **57**, 2029-2048 (2001) .

7. Y. Uchiyama, M. Tsuchiya, Generation of ultraviolet (335-nm) light by intracavity frequency doubling from active mode-locking action of an external-cavity AlGaInP diode laser, *Opt. Lett.* **24**, 1148-1150 (1999).

8. H.R. Barry, B. Bakowski, L. Corner, T. Freegrade, O.T.W. Hawkins, G. Hancock, R.M.J. Jacobs, R. Peverall, G.A.D. Ritchie, OH detection by absorption of frequency-doubled diode laser radiation at 308 nm, *Chem. Phys. Lett.* **319**, 125-130 (2000).

9. H. Flaschka, C. McKeithan, and R. Barnes, Light emitting diodes and phototransistors in photometric modules, *Anal. Lett.* **6**, 585-594 (1973).

10. S. Landgraf, G. Grampp, Application of cw-laserdiodes for the determination of fluorescence lifetimes, *J. Inf. Rec. Mats.* **23**, 203-207 (1996).

11. S. Landgraf, G. Grampp, Application of laser diodes and ultrabright light emitting diodes for the determination of fluorescence lifetimes in the nano- and subnanosecond region, *J. Inf. Rec. Mats.* **24**, 141-148 (1998).

12. R.M.S. Bindra, R.E. Imhof, D.J.S. Birch, Developement of a fibre optic luminescence lifetime spectrometer, *Proc. Indian Acad. Sci. (Chem. Sci.)* **104**, 339-350 (1992).

13. S. Landgraf, G. Grampp, A subnanosecond time-resolved fluorescence lifetime spectrometer applying laser diodes *Chemical Monthly* **131**, 839-848 (2000).

14. T.A. Fayed, G. Grampp, S. Landgraf, Photoinduced electron transfer fluorescence quenching of different diolefinic laser dyes, *J. Inf. Rec. Mats* **25**, 367-380 (2000).

15. J. Sobek, S. Landgraf, G. Grampp, Investigations on the viscosity dependence of photoinduced electron transfer processes in liquid solutions, *J. Inf. Rec. Mats.* **24**, 149-156 (1998).

16. G. Angulo, G. Grampp, S. Landgraf, J. Sobek, Experimental investigations on the viscosity effects on photoinduced electron transfer reactions in solution, *J. Inf. Rec. Mats.* **25**, 381-389 (2000) .

17. S. Landgraf, Use of ultrabright LEDs for the Determination of Static and Time-resolved Fluorescence Information of Crude Oil Samples, *J. Biochem. Biophys. Meth.* (2004), accepted.

18. to be published.

NOBLE-METAL SURFACES FOR METAL-ENHANCED FLUORESCENCE

Chris D. Geddes*[1, 2], Kadir Aslan[1], Ignacy Gryczynski[2], Joanna Malicka[2], and Joseph R. Lakowicz*[2]

1. INTRODUCTION

Noble metal nanoparticles exhibit strong absorption bands, which known as the surface plasmon resonances, result in strong absorption and scattering, and create an enhanced local electromagnetic field near-to the surface of the particles. The surface plasmon resonances are highly dependent on the size and the shape of the metal and the dielectric properties of the surrounding medium. These near field enhancements have given rise to surface-enhanced resonant Raman scattering (SERRS) and metal-enhanced fluorescence (MEF) (Figure 1). Unlike SEERS, the optimal MEF signal occurs at a certain distance from the surface of the metal nanoparticles. The fluorophores in direct contact with the metal surface are typically quenched. Theoretical and experimental work using rough surfaces and particles has suggested that the distance-dependent enhancement fluorescence intensity is more pronounced for low quantum yield fluorophores.[1-8] This enhancement is accompanied by a significantly reduced lifetime. The increased fluorescence intensities accompanied by reduced lifetimes suggest an increased radiative decay rate for the fluorophores interacting with the metals.

The presence of a nearby metal surface (m) can increase the radiative decay rate by addition of a new rate Γ_m (Figure 2, Right). The metallic surface can cause Förster-like quenching with a rate (k_m), can concentrate the incident field (E_m), and can importantly for sensing applications, increase the radiative decay rate (Γ_m). These new phenomena typically occur at different distances from the metal surface as depicted in Figure 3. As the value of Γ_m increases, that is, the spontaneous rate at which a fluorophore emits photons, the quantum yield increases while the lifetime decreases. This is unusual to most fluorescence spectroscopists, as the quantum yield and lifetime usually change in unison.

[1]-Institute of Fluorescence, University of Maryland Biotechnology Institute, [2] - Center for Fluorescence Spectroscopy, 725 W. Lombard St., Baltimore, MD 21201 USA,
* Corresponding authors, chris@cfs.umbi.umd.edu, and lakowicz@cfs.umbi.umd.edu

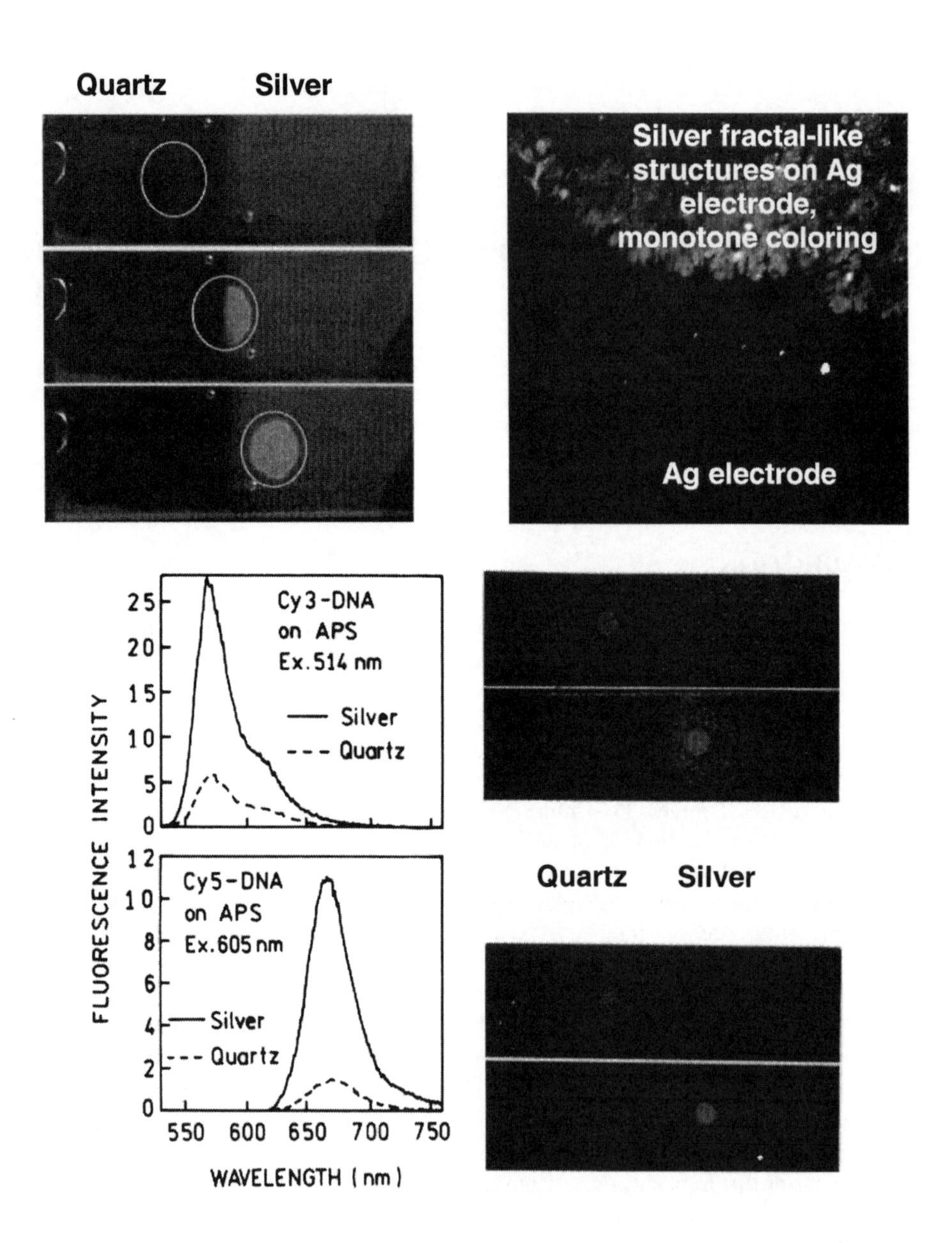

Figure 1. (See color insert section) Photograph of fluorescein-labeled HSA (molar ratio of fluorescein/HSA =7) on quartz (Top-Left) and on SiFs (Top-Left) as observed with 430-nm excitation and a 480-nm long-pass filter. The excitation was progressively moved from the quartz side to the silver, Top to Bottom, respectively (Reference 10). Silver fractal-like structures grown on silver electrodes (Top-Right) (Reference 38). Emission spectra of Cy3-DNA (Bottom-Left) and Cy5-DNA (Bottom-Left) between quartz plates with and without SiFs. Photographs of corresponding fluorophores (Bottom-Right) (Reference 11).

To illustrate this point we calculated the lifetime and quantum yield for fluorophores with an assumed natural lifetime τ_N = 10 ns, $\Gamma=10^8$ s^{-1} and various values for the non-radiative decay rates and quantum yields. The values of k_{nr} varied from 0 to 9.9 x 10^7 s^{-1}, resulting in quantum yields from 1.0 to 0.01. Suppose now the metal results in increasing values of Γ_m. Since Γ_m is a rate process returning the fluorophore to the ground state, the lifetime decreases as Γ_m becomes comparable and larger than Γ (Figure 4, Left). In contrast, as Γ_m / Γ increases, Q_m increases, but no change is observed for fluorophores where Q_0 = 1 (Figure 4, Right).

As a result of these calculations, we predicted that the metallic surfaces can create new unique fluorophores with increased quantum yields and shorter lifetimes. Figure 5 illustrates that the presence of a metal surface within close proximity of a fluorophore with low quantum yield (Q_0 = 0.01) increases its quantum yield ~10-fold resulting in brighter emission, while reducing its lifetime 10-fold, resulting in an enhanced photostability of the fluorophore due to spending less time in an excited state. i.e. less time for oxidation and other processes, etc.

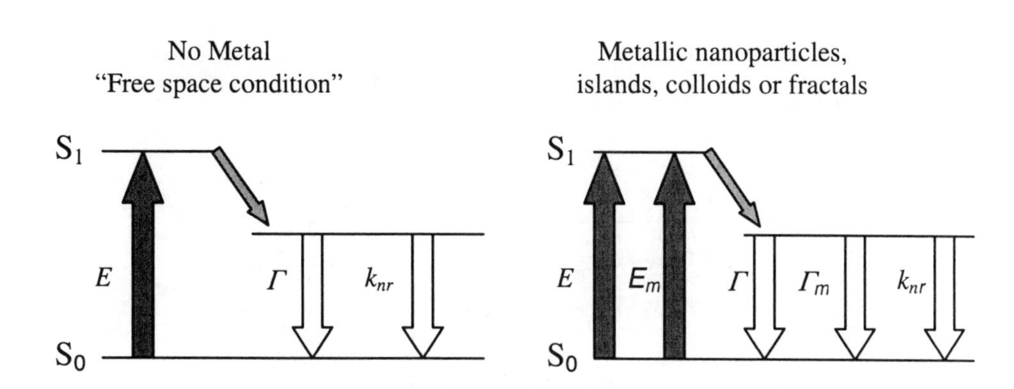

Figure 2. Classical Jablonski diagram for the *free space condition* and the modified form in the presence of metallic particles, islands, colloids or silver nanostructures. E-excitation, E_m, Metal-enhanced excitation rate; Γ_m, radiative rate in the presence of metal. For our studies, we do not consider the effects of metals on k_{nr}.

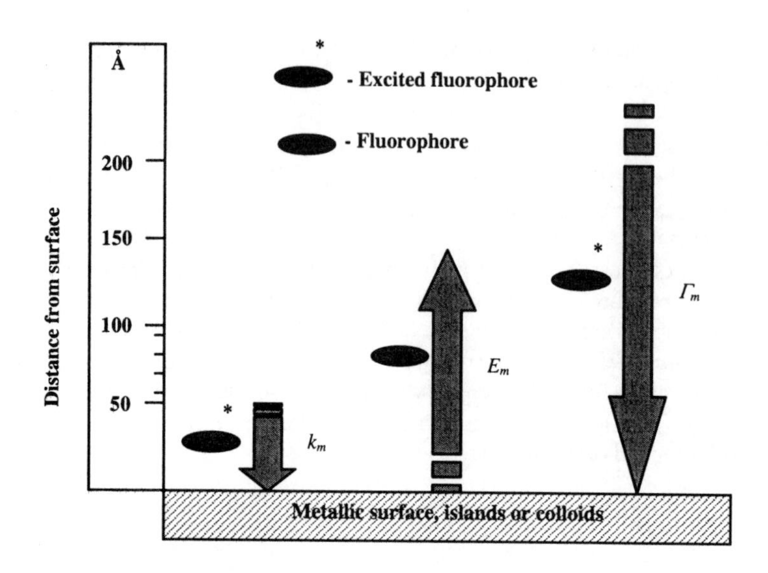

Figure 3. Predicted distance dependencies for a metallic surface on the transitions of a fluorophore. The metallic surface can cause Förster-like quenching with a rate (k_m) can concentrate the incident field (E_m) and can increase the radiative decay rate (Γ_m). Adapted from reference 6.

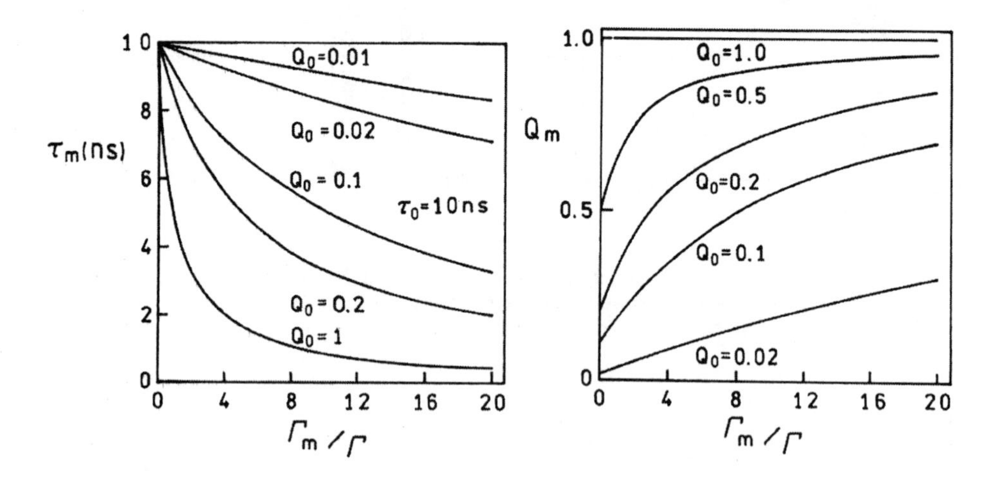

Figure 4. The effect of an increase in radiative decay rate on the lifetime and quantum yield. Adapted from reference 7.

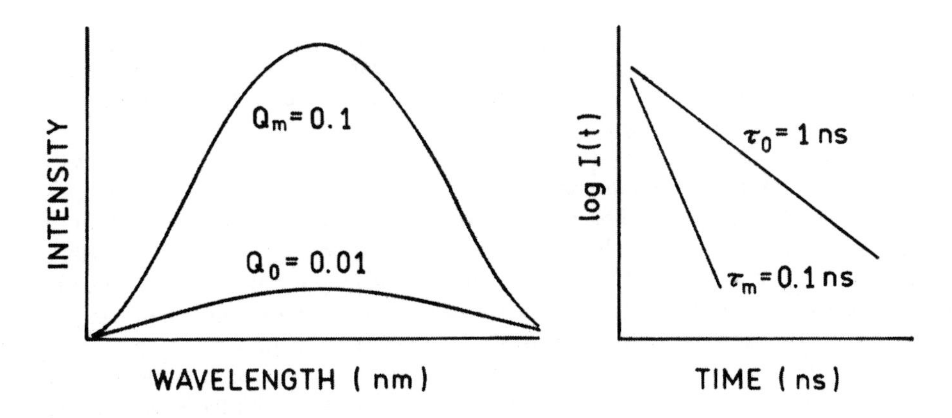

Figure 5. Metallic surfaces can create unique fluorophores with high quantum yields and short lifetimes. Adapted from reference 5.

In this chapter, we summarize our work on the preparation of solid substrates coated with silver and platinum particles to be used in metal-enhanced fluorescence (MEF). We have started our studies on MEF using relatively facile deposition of silver particles onto glass slides; Silver Island films (SiFs). We have explored the possibility of using silver colloids as well as anisotropic particles such as nanorods and triangles. We have also investigated the light-deposition and electroplating of silver onto electrodes/glass slides in order to obtain localized enhancement of fluorescence. Finally, we discuss the possibility of using MEF in drug discovery, one possible application of MEF.

2. SILVER ISLAND FILMS

In a typical SiFs preparation, firstly, substrates (glass or quartz slides) are soaked in a 10:1 (v/v) mixture of H_2SO_4 (98%) and H_2O_2 (30%) overnight, and then washed with deionized water and dried with air. A clean 30-ml beaker equipped with a Teflon-coated stir bar is used for the deposition of silver particles. To a fast stirring silver nitrate solution (0.22 g in 26 ml of deionized water), 8 drops of freshly prepared 5% NaOH solution are added. This results in formation of dark-brownish precipitates of silver particles. Then, approximately 1 ml of ammonium hydroxide is added drop by drop to re-dissolve the precipitates. The clear solution is cooled to 5°C by placing the beaker in an ice bath, followed by soaking the cleaned and dried quartz slides in the solution. While keeping the slides at 5°C, a fresh solution of D-glucose (0.35 g in 4 ml of water) is added. Subsequently, the temperature of the mixture is allowed to warm up to 30°C. As the color of the mixture turns from yellow green to yellowish brown and the color of the slides become greenish; the slides are removed from the mixture, washed with water, and sonicated for 1 min at room temperature. The SiFs deposited slides are rinsed with deionized water several times, and are stored in deionized water for several hours prior to the florescence experiments. Typical absorption spectrum of SiFs is given in Figure 6.[8]

In our studies, we have used two fluorophores that are conjugated to proteins, namely, indocyanine green (ICG) and fluorescein (Figure 7). These fluorophores were ideally chosen to demonstrate metal-enhanced fluorescence due to the fact that they are FDA approved for use in humans.

As indicated earlier, the presence of a nearby metallic surface can modify the radiative rate of an excited fluorophore. This results in an increase in the fluorescence intensity, a reduction in the lifetime and an in increase in quantum yield of the fluorophores. It is expected that the properties of the fluorophores with various quantum yields will be affected differently by the presence of a nearby metal due to the difference in radiative rate modifications. This hypothesis was tested by investigating the changes in emission properties of two fluorophores with similar absorption/emission spectra but with different quantum yield (Rhodamine B, quantum yield, $Q_0 = 0.48$ and Rose Bengal, $Q_0 = 0.02$). Figure 8 (Top and Middle) summarizes our observations with these fluorophores. The emission from Rhodamine B on SiFs was 20 % higher as compared to the emissions from the glass side. On the other hand, emissions from Rose Bengal on SiFs were ≈ 5-fold higher than those on the glass side.[8] We also tested the effect of SiFs on ICG which is bound to Human Serum Albumin (HSA) (Figure 8). The intensity of ICG is increased approximately 10-fold on the SiFs as compared with the quartz. The emission spectrum is similar both on quartz and SiFs. We found the same amount of increase in the emission of ICG whether the surfaces were coated with HSA, which already contained bound ICG, or if the surfaces were first coated with HSA followed by exposure to a dilute solution of ICG. From on-going studies of albumin-coated surfaces, we estimated that the same amount of HSA binds to each surface, with the difference in binding being less than a factor of two. Hence the observed increase in the intensity on SiFs is not due to the increased ICG-HSA binding but rather to a change in the quantum yield and rate of excitation of ICG near particles.[9]

We also investigated the emission spectral properties of the fluorophore labeled-DNA. Emission spectra of Cy3-DNA and Cy5-DNA are shown in Figure 1 (Bottom). The emission intensity is increased 2- to 3- fold between SiFs as compared to between the quartz slides for Cy3-DNA and Cy5-DNA, respectively. The slightly larger increase in emission intensity for Cy5-DNA compared to Cy3-DNA is consistent with the results where larger enhancements are observed with low quantum yield fluorophores. Figure 1 also shows the photographs of the labeled oligomers on quartz and on SiFs. The emission from the labeled-DNA on quartz is almost invisible and is brightly visible on the SiFs. This difference intensity is due to an increase in the photonic mode density near the fluorophore, which in turn results in an increase in the radiative decay rate and quantum yield of the fluorophores. We note that the photographs are taken through emission filters and the increase in emission intensity is not due to an increased excitation scatter from the silvered plates. [10]

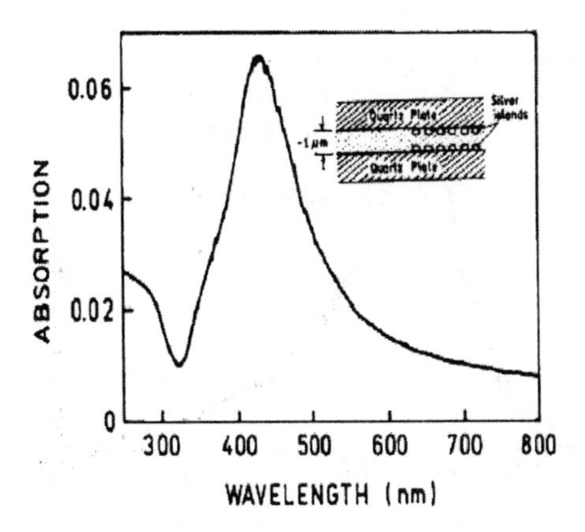

Figure 6. Typical absorption spectrum of silver island films (SiFs). Schematic for a fluorophore solution between two SiFs (Inset). Adapted from reference 8.

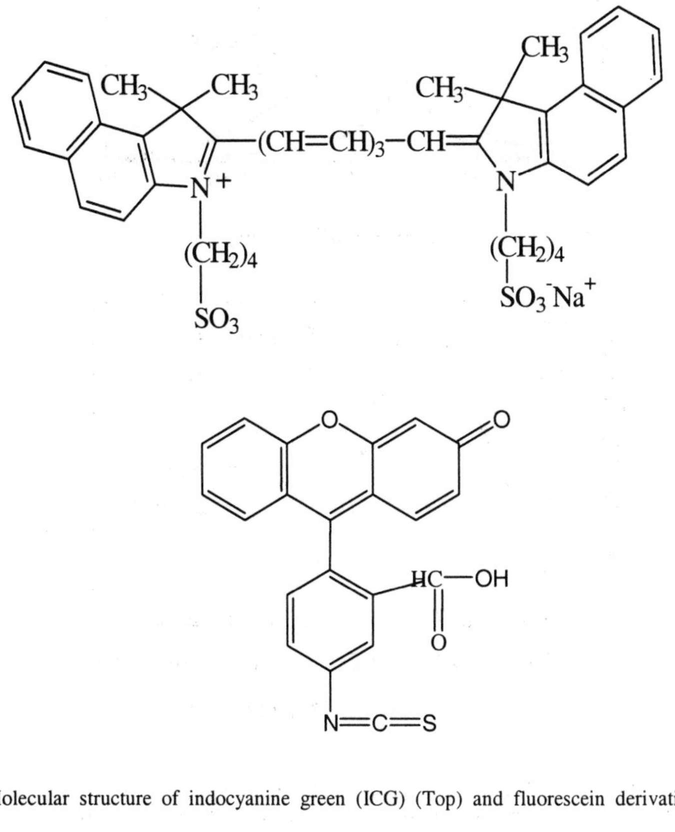

Figure 7. Molecular structure of indocyanine green (ICG) (Top) and fluorescein derivative that are used (Bottom).

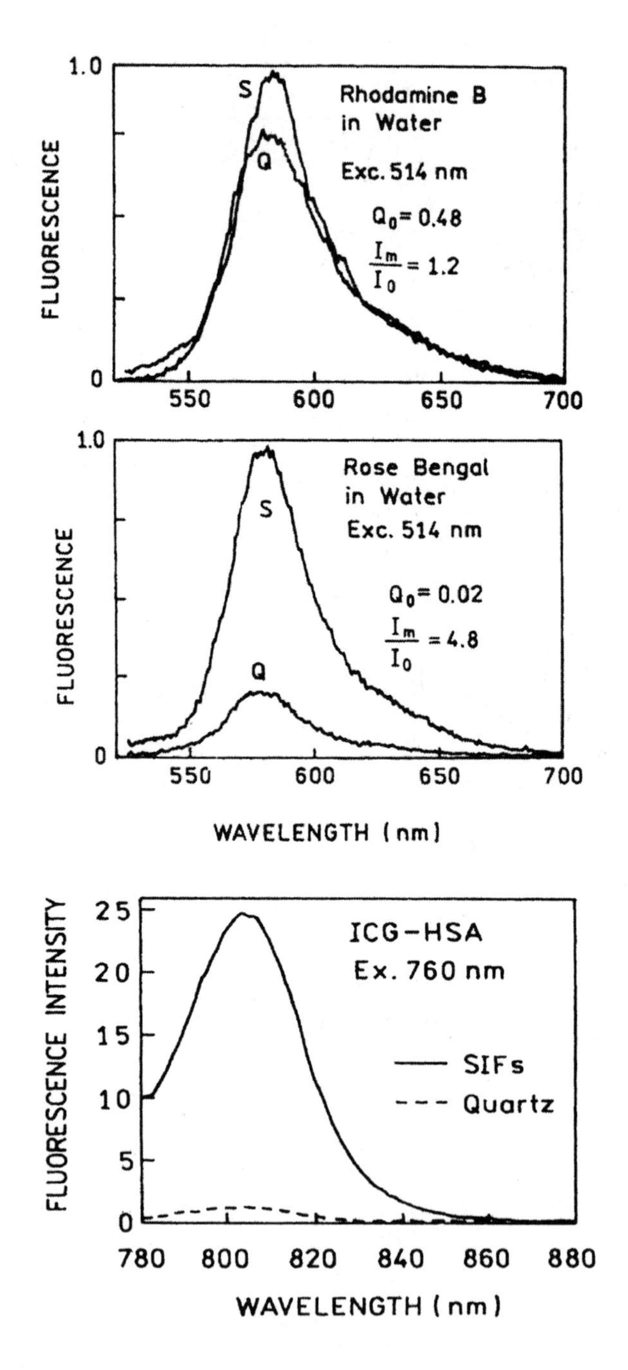

Figure 8. The effect of silver island films on the emission spectra of Rhodamine B (Top) and Rose Bengal (Middle) (Reference 8). Fluorescence spectra of ICG-HSA on SiFs and on quartz (Bottom) (Adapted from reference 9).

Photostability of the labeled-DNA was studied by measuring the emission intensity during continuous illumination at a laser power of 20 mW (Figure 9). The intensity initially dropped rapidly, but became more constant at longer illumination times. Although not a quantitative result, examination of these plots visually suggests slower photobleaching at longer times in the presence of silver particles compared with quartz slides. We also questioned the nature of emission from the fluorophores remaining after 300 s illumination. The emission spectra of Cy3-DNA and Cy5-DNA were identical before and after illumination, both in the absence and presence of silver islands (data not shown). This result indicates that the detected emission, even after intense illumination, is still due to Cy3 and Cy5, and not a photolysis byproduct. [11]

The results from Rhodamine B and Rose Bengal, Figure 8, were consistent with our expectations that the presence of metal increases the emission intensity and quantum yields, and decreases the lifetime of the fluorophores. Nonetheless, one could be concerned with possible artifacts due to dye binding to the surfaces or other unknown effects. For this reason, we examined a number of additional fluorophores between uncoated quartz plates and between silver island films. In all cases, the emission was more intense for the solution between the silver islands. For example, [Ru(bpy)$_3$] and [Ru(phen)$_2$dppz] have quantum yields near 0.02 and 0.001, respectively. A larger enhancement was found for [Ru(phen)$_2$dppz] than for [Ru(bpy)$_3$]. The enhancements for 10 different fluorophore solutions are shown in Figure 10. In all cases, lower bulk-phase quantum yields result in larger enhancements for samples between silver island films.

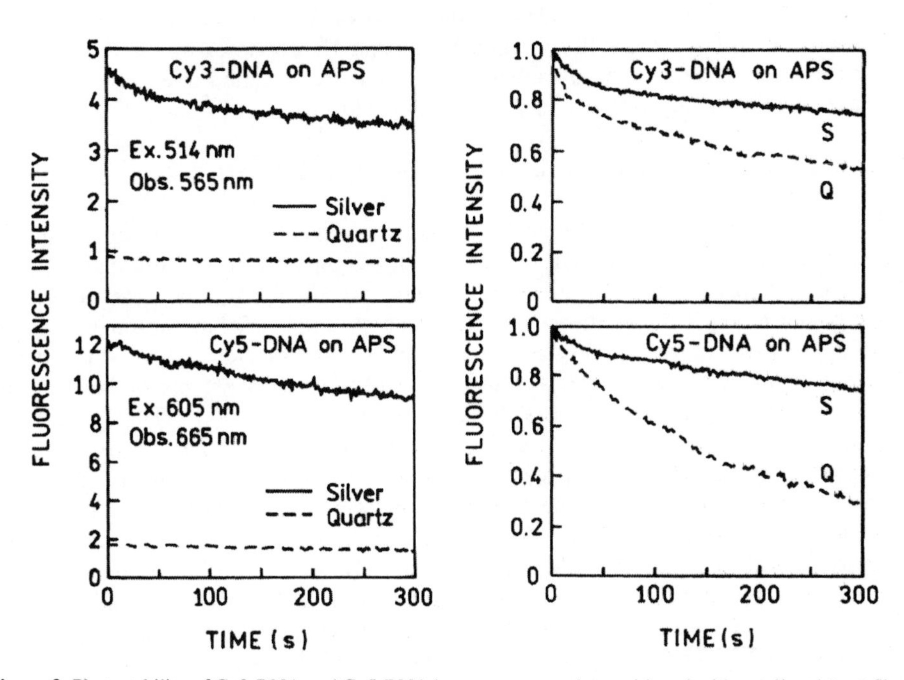

Figure 9. Photostability of Cy3-DNA and Cy5-DNA between quartz plates with and without silver island films. The laser power was 20 mW. Adapted from reference 11.

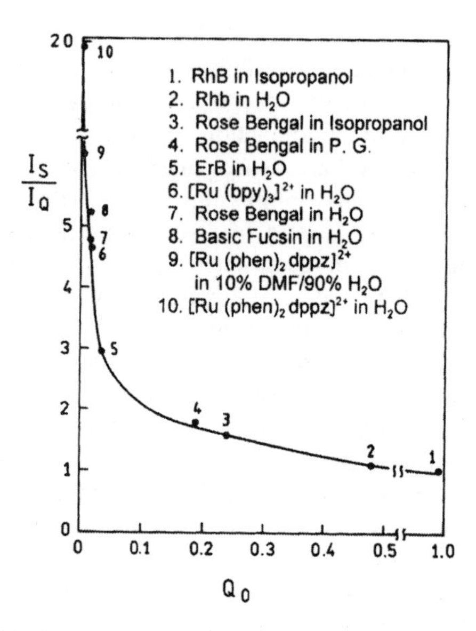

Figure 10. The effect of silver island films on the quantum yields of fluorophores. Adapted from reference 8.

The results in Figures 8–11 *provide strong support* for our assertion that the proximity of a fluorophore to the metal islands resulted in increased quantum yields (i.e., a modification in Γ_m). It is unlikely that these diverse fluorophores would all bind to the silver islands or display other unknown effect results that resulted in enhancements that increased monotonically with decreased quantum yields.

2.1. Ultra-bright Over-labeled Proteins: A New Class of Probes

Proteins covalently labeled with fluorophores are widely used as reagents, such as immunoassays or immuno-staining of biological specimens with specific antibodies. In these applications, fluorescein is one of the most widely used probes. An unfortunate property of fluorescein is self-quenching, which is due to Forster resonance energy transfer between nearby fluorescein molecules (homo-transfer).[12] As a result, the intensity of labeled protein does not increase with increased extents of labeling, but actually decreases. Figure 11 shows the spectral properties of FITC–HSA with molar labeling ratios (L) ranging from 1-to-1 (L=1) to 1-to-9 (L=9). The relative intensity decreased progressively with increased labeling. The insert in Figure 11 shows the intensities normalized to the same amount of protein, so that the relative fluorescein concentration increases nine-fold along the x- axis. It is important to note that the intensity per labeled protein molecule does not increase and in fact decreases, as the labeling ratio is increased from 1 to 9. We found that the self-quenching could be largely eliminated by the close proximity to SiFs, [9] as can be seen from the emission spectra for labeling ratios of 1 and 7 (Figure 12) and from the dependence of the intensity on the extent of labeling (Figure 13. We speculate that the *decrease* in self-quenching is due to an increase in the rate of radiative decay, Γ_m

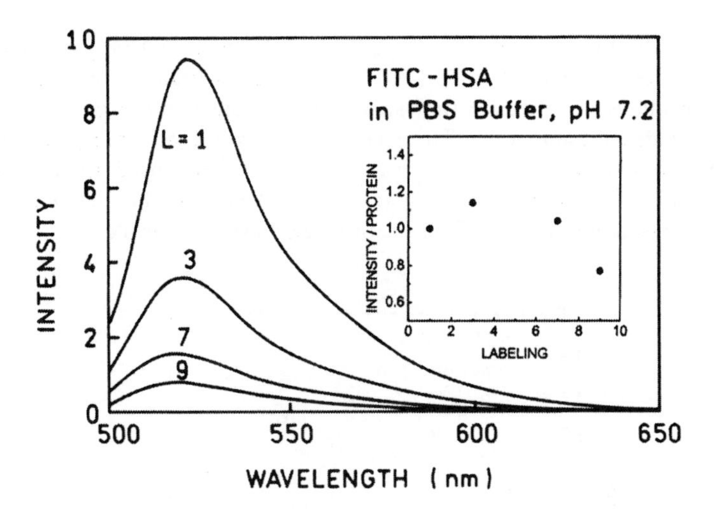

Figure 11. Dependence of emission intensity on the degree of labeling. Adapted from reference 10.

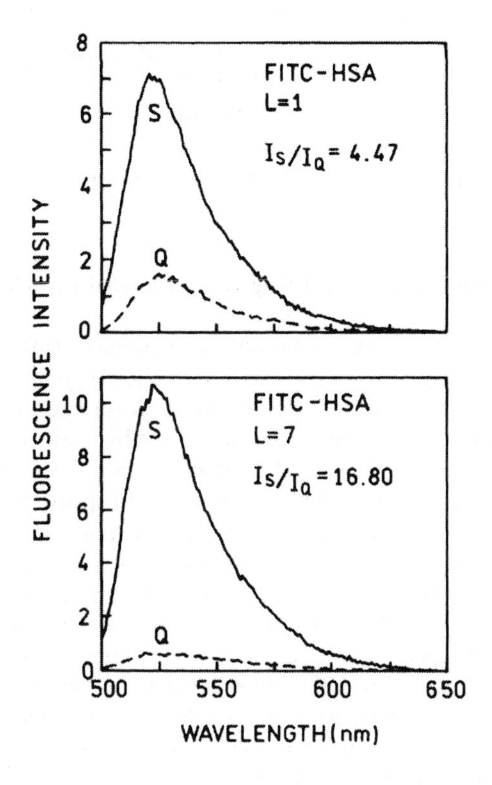

Figure 12. Emission spectra of FITC-HSA with different degrees of labeling on quartz (Q) and on SiFs (S). Adapted from reference 10.

The dramatic difference in the intensity of heavily labeled HSA on glass and on SiFs is shown pictorially in Figure 1 (Top-Left). The effect is dramatic as seen from the nearby invisible intensity on quartz (left side) and the bright image on the SiFs (right side) in this unmodified photograph. These results suggest the possibility of ultra bright labeled proteins based on high labeling ratios and metal-enhanced fluorescence. We conclude that SiFs, and most probably colloidal silver, can be utilized to obtain dramatically increased intensities of fluorescein-labeled macromolecules.

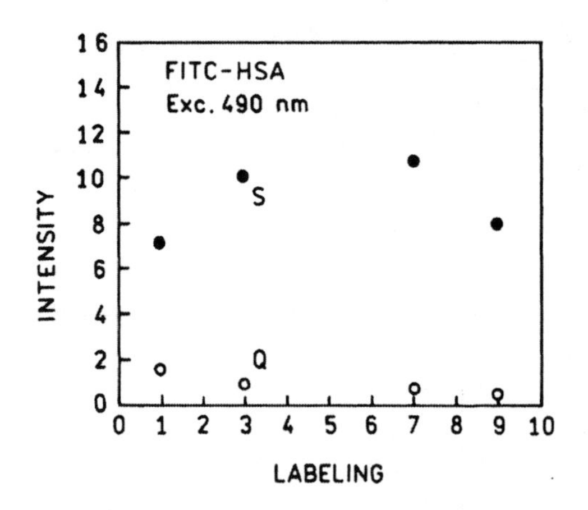

Figure 13. Emission intensity of FITC-HSA at 520 nm vs. different degrees of labeling on quartz and on SiFs (S). Adapted from reference 10.

2.2. Reduced Signal-to-Background Using Over-Labeled Proteins

We hypothesized that metal-enhanced fluorescence (MEF) can improve the sensitivity of fluorescence-based assays, particularly the assays in which the over-labeled proteins are employed, by reducing the signal to noise ratio due to release of self quenching of fluorophores. This hypothesis was tested by a simple experiment involving a mixture of an over-labeled HSA and another fluorophore that is used to provide background signal. Figure 14 shows emission spectra of a quartz plate coated with FITC–HSA to which we adjusted the concentration of Rhodamine B (0.25 μM) to result in an approximate 1.5-fold larger Rhodamine B intensity. One can consider the Rhodamine B to be simple auto-fluorescence or any other interference signal. When the same conditions are used for FITC–HSA on silver with L=1 the fluorescein emission is now two- to three-fold higher than that of Rhodamine B (Figure 14, Right-top). When using the heavily labeled sample (L=7) the fluorescein emission becomes more dominant (Figure 14, Right-bottom). Thus, we conclude that silver particles, when bound to a protein heavily labeled with fluorophores, can provide significantly higher intensities due to a decrease in the extent of self-quenching.

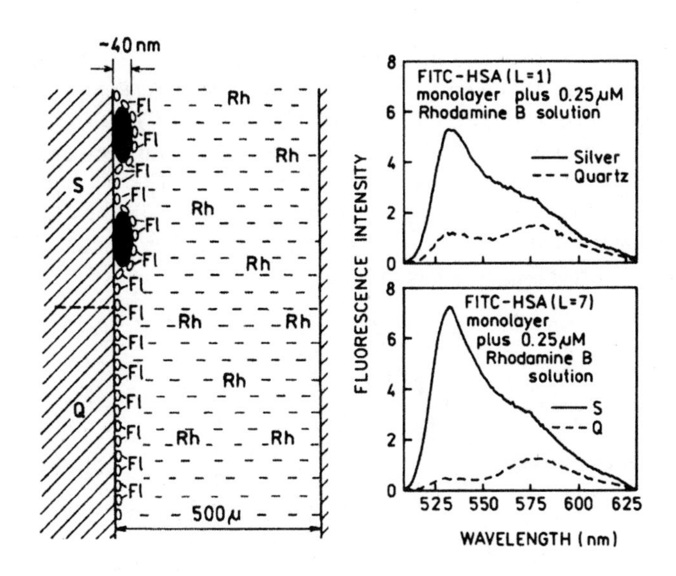

Figure 14. Emission spectra of a monolayer of FITC–HSA L=1 (Right-Top) and L=7 (Right-Bottom) containing 0.25 μM Rhodamine B between the quartz plates (Q) or one SIF (S). Schematic of the sample with bound fluorescein and free Rhodamine B (Left). Adapted from reference 10.

3. SURFACE IMMOBILIZED SILVER COLLOIDS

We have shown that silver island films (SiFs) increase the emission intensity of ICG or FITC conjugated to HSA. Silver island films are formed by chemical reduction of silver with direct deposition onto a glass substrate with random shapes, which is difficult to control. In contrast, preparation of colloidal suspensions of silver is rather standard and easily controlled to yield homogeneously sized spherical silver particles. An advantage of a colloidal suspension is the possibility of injection for medical imaging. Thus, we also investigated the effects of spherical silver particles on the emissions of ICG. In this regard, we deposited already prepared silver colloids onto glass slides. The general procedure is give below:

Glass microscope slides are cleaned as described in Section 2. The glass surfaces are coated with amino groups by soaking the slides in a solution of (3-aminopropyl)-trimethoxysilane (APS). Silver colloids are prepared by the reduction of a warmed solution of silver nitrate and sodium citrate. This procedure is reported to yield homogeneously sized colloids near 20-30 nm in diameter.[13-16] The APS treated slides are soaked in the colloid suspension overnight, followed by rinsing with deionized water. Binding the ICG-HSA to the surfaces, whether quartz or silver, is accomplished by soaking both the quartz and colloid coated slides in a 30 μM ICG, 60 μM HSA solution overnight, followed by rinsing with water to remove the unbound material.

Figure 15 (Top) shows the schematic representation of silver colloids deposited on glass slides. Figure 15 (Middle) also shows an absorption spectrum, typical of our-colloid coated APS glass slides. The absorption peak centered near 430 nm is typical of colloidal silver particles with sub-wavelength dimensions but not completely at the small particle limit. An AFM image of silver colloid coated glass slides shows that the size of the silver colloids was smaller than 50 nm with partly aggregated sections (Figure 15-Bottom). The surfaces were incubated with ICG-HSA to obtain a monolayer surface coating. The emission spectra showed a ≈ 30-fold larger intensity on the surfaces coated with silver colloids (Figure 16). We also measured the lifetimes of ICG on both surfaces and observed a significant reduction on the lifetimes on the silver colloids (data not shown) providing additional evidence that the increases in intensity is in fact due to modification of the radiative decay rate, Γ_m, by silver colloids.

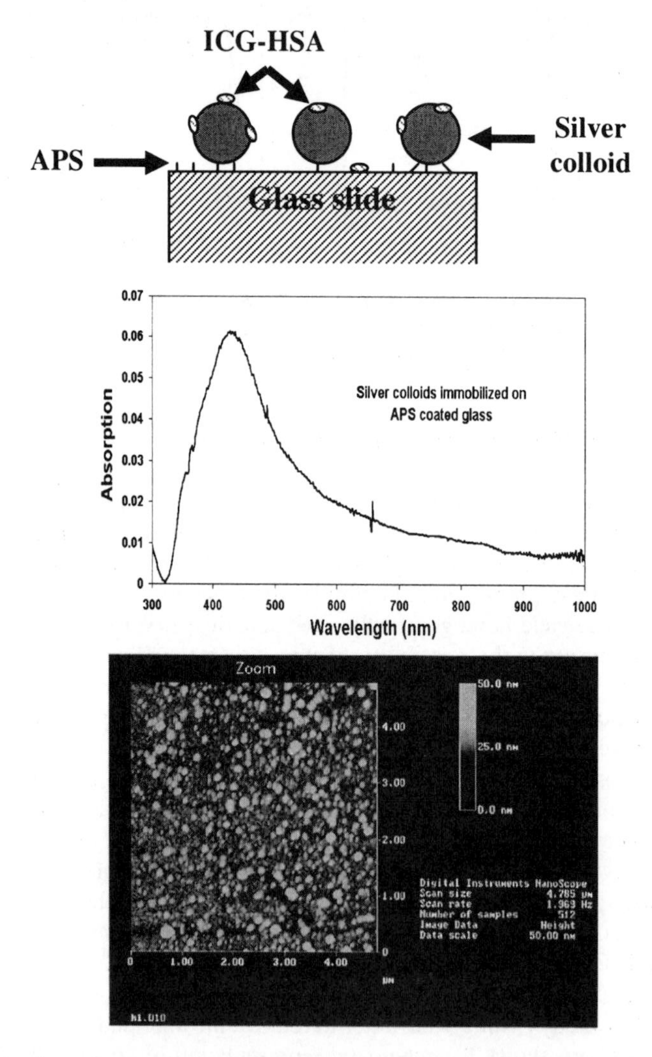

Figure 15. Glass surface geometry (Top); APS is used to functionalize the surface of the glass with amine groups which readily bind silver colloids, absorption spectrum of silver colloids on APS-coated glass (Middle), and AFM image of a silver colloid coated glass (Bottom). APS; 3-aminopropylethoxy silane. Adapted from reference 17.

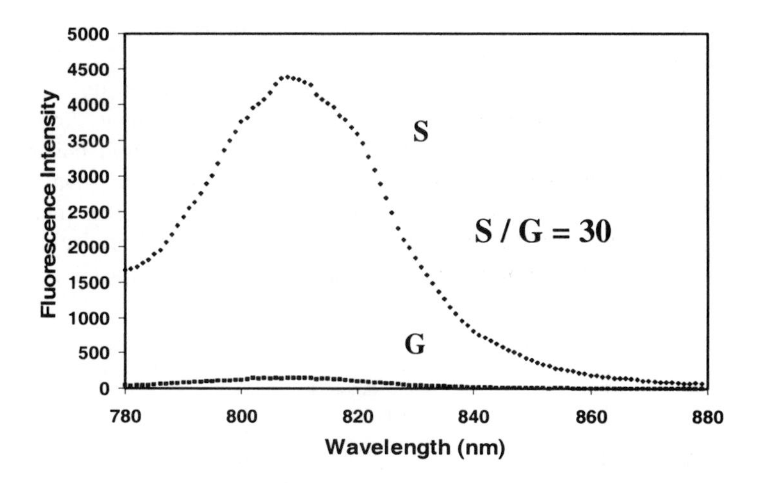

Figure 16. Fluorescence spectra of ICG-HSA on silver colloids and on quartz. S = Silver, G = Glass/Quartz. Adapted from reference 17.

4.0. SURFACE IMMOBILIZED SILVER NANORODS

Elongated silver particles are predicted to enhance the emission of fluorescence due to the increased local excitation fields around the edges of the particles. Recently, we have developed a methodology for depositing silver nanorods with controlled sizes and loadings onto glass substrates. In this method, firstly, APS-coated substrates are coated with silver colloids with sizes less than 4 nm. Then, the silver seed-coated glass slides were immersed in a cationic surfactant (CTAB) solution for 5 minutes. One ml of 10 mM $AgNO_3$ and 2 ml of ascorbic acid were added. 0.4 ml of 1 M NaOH was immediately added and the solution was mixed gently to accelerate the growth process. The silver nanorods were formed on the glass slides within 10 minutes. In order to increase the loading of silver nanorods on the surface, the silver nanorods-coated glass substrates were immersed in CTAB again, and same amounts of $AgNO_3$, ascorbic acid and NaOH were added as in the first step. This process is repeated until the desired loading of silver nanorods on the glass slides was obtained. Binding of the ICG-HSA to the surfaces were accomplished by soaking the slides in a 30 μM ICG, 60 μM HSA solution overnight, followed by rinsing with deionized water to remove the unbound material.[18]

The absorption spectra of silver nanorods deposited on glass substrates are shown in Figure 17 (Top). Silver nanorods display two distinct surface plasmon peaks; transverse and longitudinal, which appear at ≈ 420 and ≈ 650 nm, respectively. In our experiments, the longitudinal surface plasmon peak shifted and increased in absorbance as more nanorods are deposited on the surface of the substrates. In parallel to these measurements, we have observed an increase in the size of the nanorods (by Atomic Force Microscopy, data not shown). In order to compare the extent of enhancement of fluorescence with respect to the extent of loading of silver nanorods deposited on the surface, we have arbitrarily chosen the value of absorbance at 650 nm as a means of loading of the nanorods on the surface. This is because the 650 nm is solely attributed to the longitudinal absorbance of the nanorods.

Figure 17 (Bottom) shows the fluorescence emission intensity of ICG-HSA measured from both glass and on silver nanorods, and the enhancement factor versus the loading density of silver nanorods. We have obtained up to 50-fold enhancement in emission of ICG on silver nanorods when compared to the emission on glass (Figure 17-Bottom). We also measured the lifetime of ICG on both glass and silver nanorods, and observed a significant reduction in the lifetime of ICG on silver nanorods, providing the evidence that an increased emission is due to radiative rate modifications, Γ_m.

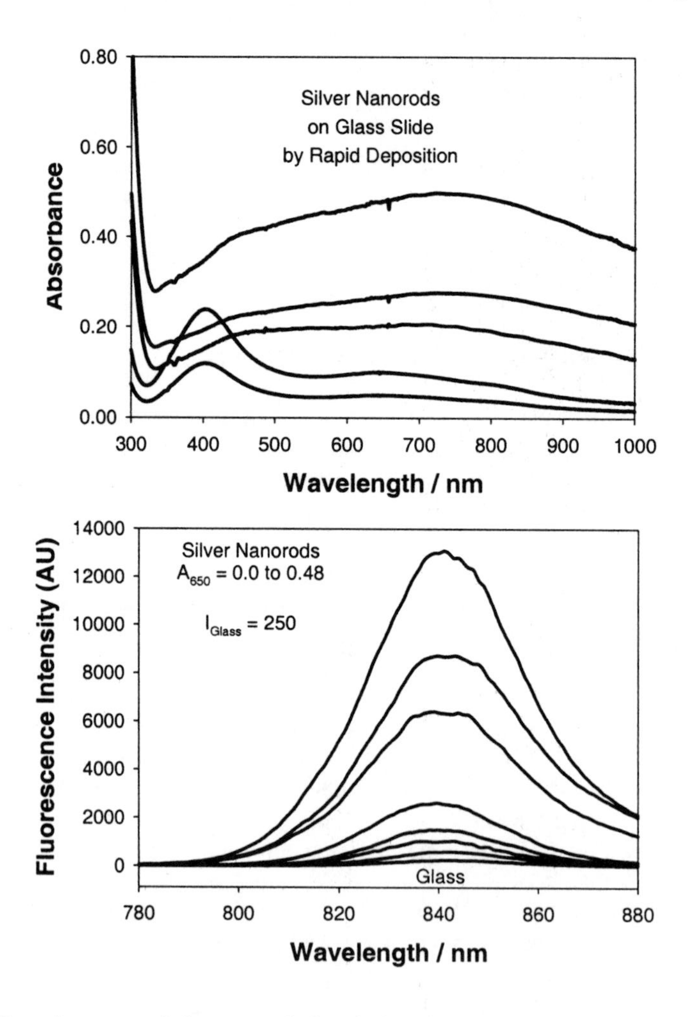

Figure 17. Absorption spectra of silver nanorods deposited on glass substrates (Top). Emission spectra of ICG-HSA on silver nanorods (Bottom).

5.0. SURFACE IMMOBILIZED SILVER TRIANGLES

Similar to the elongated silver nanoparticles, triangular silver particles have also increased local excitation fields around the edges of the particles, and thus they can be used for metal-enhanced fluorescence. Triangular silver particles display four distinct plasmon resonance peaks: 352, 431, 465 and 550 nm. According to calculation of Schatz and coworkers [19] for the triangular prisms with discrete dipole approximation method, the 552 and 465 nm peaks are in-plane dipole and quadrupole plasmon resonance, respectively. While the 431 and 352 nm peaks are out-of-plane dipole and quadrupole plasmon resonance, respectively. [20] We have deposited triangular silver nanoparticles on to glass slides using the same procedure that we have used for the deposition of silver nanorods on glass slides; the difference is the time of the growth process. We have observed that the triangular particles are formed within 5 minutes after the growth process is started with the addition of NaOH. The absorption spectra for the triangular silver particles deposited on glass slides are shown in Figure 18 (Top). The plasmon resonance peaks characteristic of triangular silver particles is observed. The triangular silver particles appear red in color, in contrast to green color observed for silver nanorods.

Figure 18 (Bottom) shows the fluorescence emission intensity of ICG-HSA measured from both glass and on triangular silver particles. We have obtained up to 16-fold enhancement in emission of ICG on silver nanorods when compared to the emission on glass (Figure 18-Bottom). We also measured the lifetime of ICG on both glass and silver nanorods, and observed a significant reduction in the lifetime of ICG on silver nanorods, providing the evidence that an increased emission is due to radiative rate modifications, Γ_m.

6.0. LIGHT-DEPOSITION OF SILVER FOR METAL–ENHANCED FLUORESCENCE

It would be useful to obtain metal-enhanced fluorescence (MEF) at desired locations in the measurement device for use in medical and biotechnology applications, such as diagnostic or micro-fluidic devices. While a variety of methods could be used, we reasoned that the light-directed deposition of silver would be widely applicable. In recent years, a number of laboratories have reported light-induced reduction of silver salts to metallic silver. [21-24] Typically, a solution of silver nitrate is used that contains a mild potential reducing agent such as a surfactant[23] or dimethylformamide. [24] Exposure of such solutions to ambient or laser light typically results in the formation of silver colloids in suspension or on the glass surfaces. These results suggested the use of light-induced silver deposition for locally enhanced fluorescence.

In a typical preparation of light-induced deposition of silver on glass slides; the silver-colloid-forming solution was prepared by adding 4 mL of 1% trisodium citrate solution to a warmed 200 mL 10^{-3} M $AgNO_3$ solution. This warmed solution already contains some silver colloids as seen from a surface plasmon absorption optical density near 0.1. A 180 mL aliquot of this solution was syringed between the glass microscope slide and the plastic cover slip, which created a micro-sample chamber 0.5 mm thick (Figure 19). For all experiments a constant volume of 180 mL was used.

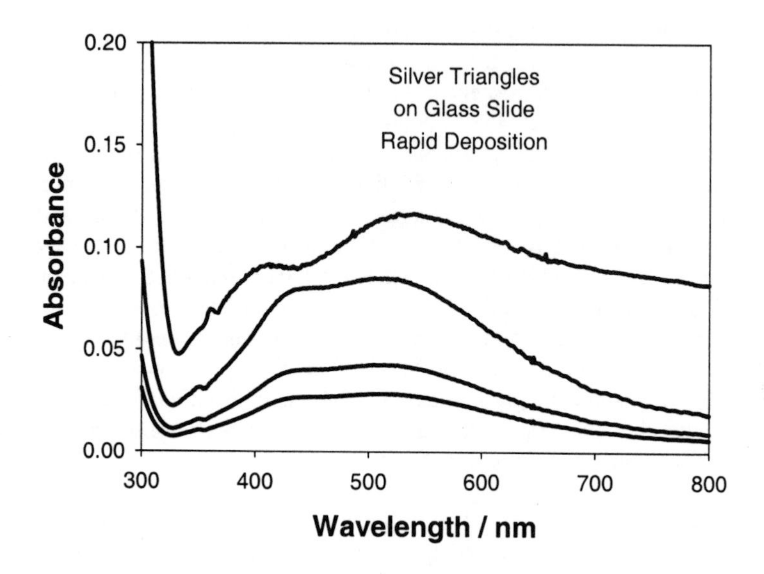

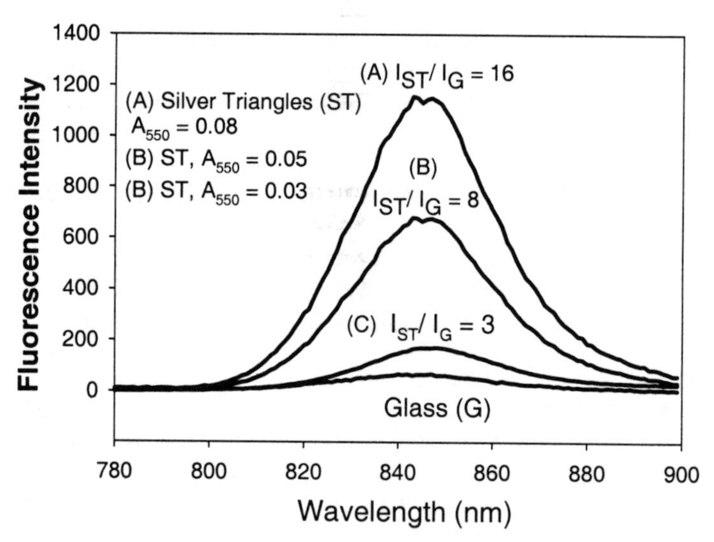

Figure 18. Absorption spectra of silver triangles deposited on glass substrates (Top). Emission spectra of ICG-HSA on silver triangles (Bottom).

Irradiation of the sample chamber was undertaken using a HeCd laser, with a power of ~8 mW, which was collimated and defocused using a 10x microscope objective, numerical aperture (NA) 0.40, to provide illumination over a 0.5 mm diameter spot. [25]

We examined the emission spectrum of ICG–HSA when bound to illuminated or non-illuminated regions of the APS treated slides. For APS treated slides (Figure 20) the intensity of ICG was increased about 7-fold in the regions with laser-deposited silver.

We examined the photostability of ICG–HSA when bound to glass or laser-deposited silver. We reasoned that ICG molecules with shortened lifetimes should be more photostable because there is less time for photochemical processes to occur. The intensity of ICG–HSA was recorded with continuous illumination at 760 nm. When excited with the same incident power, the fluorescence intensities, when considered on the same intensity scale, decreased somewhat more rapidly on the silver (Figure 21, top). However, the difference is minor. Since the observable intensity of the ICG molecules prior to photobleaching is given by the area under these curves, it is evident that at least 10-fold more signal can be observed from ICG near silver as compared to glass. Alternatively, one can consider the photostability of ICG when the incident intensity is adjusted to result in the same signal intensities on silver and glass. In this case (Figure 21, bottom) photobleaching is slower on the silver surfaces. The fact that the photobleaching is not accelerated for ICG or silver indicates that the increased intensities on silver *are not due* to an increased rate of excitation.

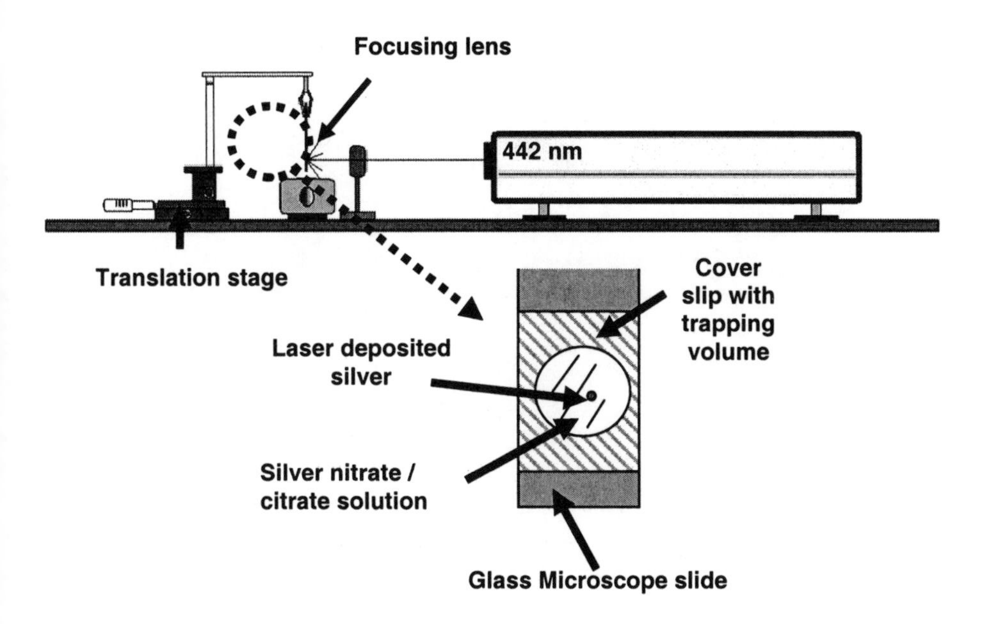

Figure 19. Experimental setup for light-induced deposition of silver on APS-coated glass microscope slides. Adapted from reference 25.

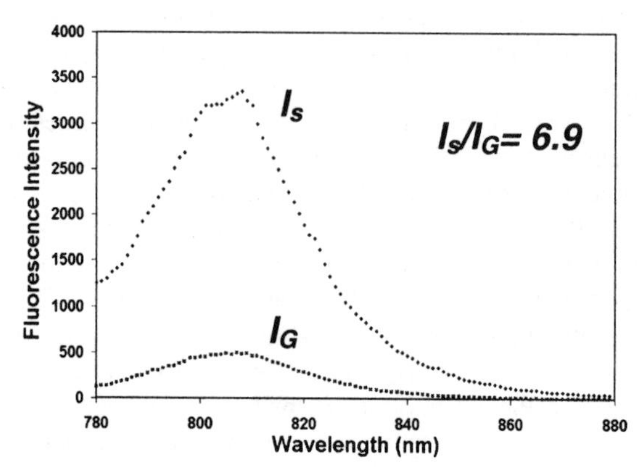

Figure 20. Fluorescence spectra of ICG-HSA on glass and on light-deposited silver. I_S-intensity on silver, I_G-intensity on glass. Adapted from reference 25.

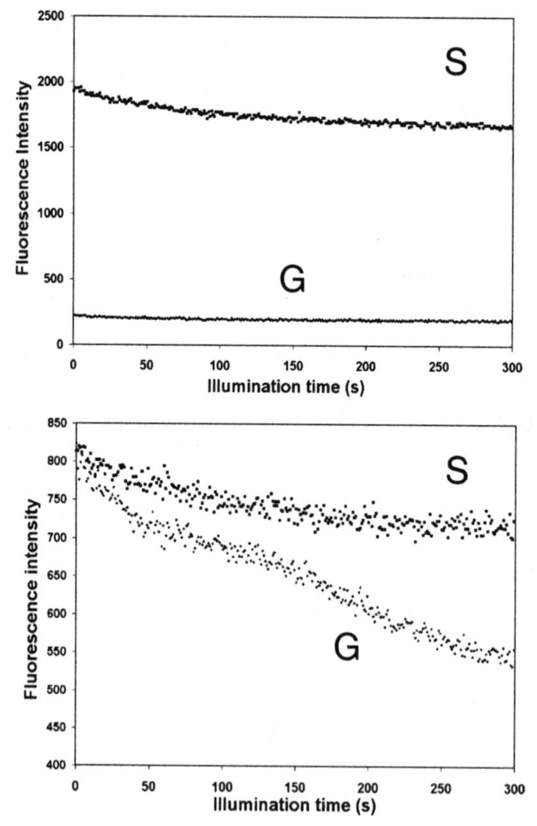

Figure 21. Photostability of ICG–HSA on (G) glass, and (S) laser deposited silver, measured with the same excitation power at 760 nm (Top), and with the laser power at 760 nm adjusted for the same initial fluorescence intensity (Bottom). Laser-deposited samples were made by focusing 442 nm laser light onto APS coated glass slides immersed in a $AgNO_3$ citrate solution for 15 min. The OD of the sample was ~0.3.

7.0. LIGHT-DEPOSITION OF SILVER FROM ELECTRICALLY PRODUCED COLLOIDS

We investigated the use of other possible methods and substrates for producing localized silver surfaces for the applications of metal-enhanced fluorescence. One of our methods of silver deposition was to pass a controlled current between two electrodes in pure water (Figure 22, Top). The two silver electrodes were mounted in a quartz cuvette containing deionized (Millipore) water. The silver electrodes had dimensions of 9 x 35 x 0.1 mm separated by a distance of 10 mm. For the production of silver colloids, a simple constant current generator circuit (60 μA) was constructed and used. After 30 min of current flow, a clear glass microscope slide was positioned within the cuvette (no chemical glass surface modifications) and was illuminated (HeCd, 442 nm). We observed silver deposition on the glass microscope slide, the amount depending on the illumination time. Simultaneous electrolysis and 442-nm laser illumination resulted in the deposition of metallic silver in the targeted illuminated region, 5-mm-focused spot size (Figure 22, Top). Absorption spectrum of the deposited silver is shown in Figure 22 (Bottom). A single absorption band is present on glass indicating that the silver particles are somewhat spherical. We examined ICG-HSA when coated on glass (G) or silver particles (S). The emission intensity was increased about 18-fold on the silver particles (Figure 23).

We examined the photostability of ICG-HSA when near silver particles on glass. We found a dramatic increase in the photostability near the silver particles (Figure 24). This very encouraging result indicates that a much higher signal can be obtained from each fluorophore prior to photodestruction and that more photons can be obtained per fluorophore before the ICG on silver eventually degrades. Our photostability data presented here are very encouraging and suggest the use of metal-enhanced fluorescence in fluorescence surface assays and lab-on-a-chip-type technologies, which are inherently prone to fluorophore instability and inadequate fluorescence signal intensity.

8.0. ELECTROPLATING OF SILVER ON SUBSTRATES

This method is similar to the one employed in the previous section, except that the silver cathode electrode was replaced with an ITO-coated glass electrode (Figure 25-Top). The current was again 60 μA. After a short period of time, silver readily deposited on the ITO surface (no laser illumination), the extent of which was again dependent on the exposure time. Absorption spectra of the deposited silver are shown in Figure 25 (Middle). Two maxima were found on ITO, which eventually formed one large broad band. This suggests that the particles are elongated and display both transverse and longitudinal resonances. Enhanced fluorescence emission from ICG-HSA was also found for silver particles on ITO (Figure 25, Bottom), but the enhancement was typically less and there appeared to be a small blue shift on silvered ITO.

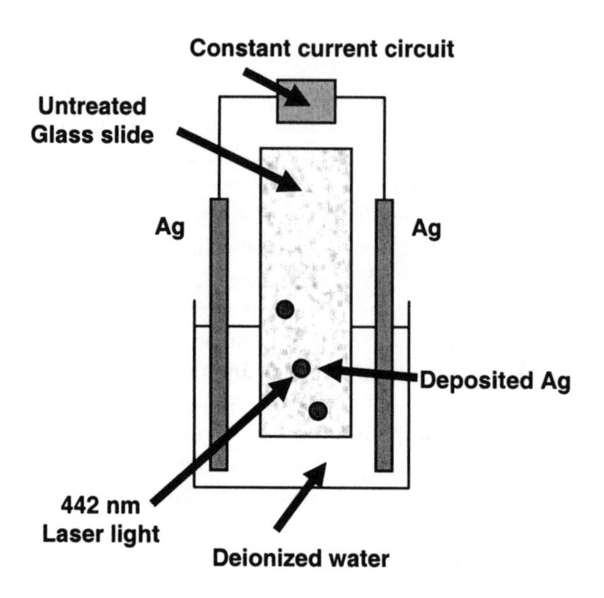

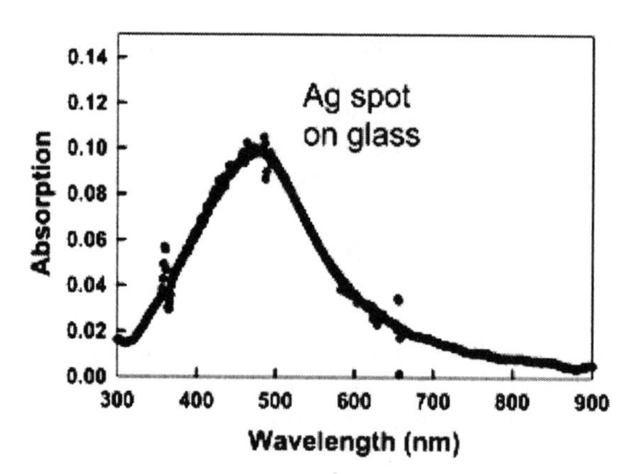

Figure 22. Light-deposited silver produced electrochemically. Constant current circuit (Top), absorbance spectrum of silver spot on glass (Bottom). Adapted from reference 26.

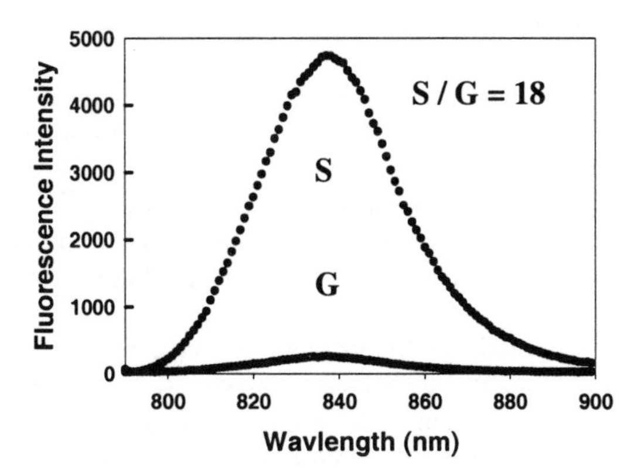

Figure 23. Fluorescence spectra of ICG-HSA on glass (G) and on light-deposited silver (S).

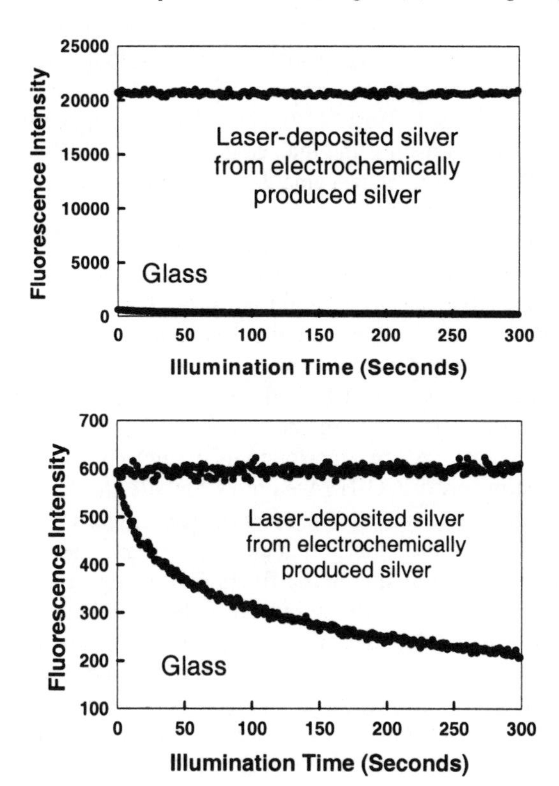

Figure 24. Photostability of ICG-HSA on glass and laser-deposited silver produced via electrolysis, measured using the same excitation power at 760 nm (Top) and with power *adjusted* to give the same initial fluorescence intensities (Bottom). In all the measurements, vertically polarized excitation was used, while the fluorescence emission was observed at the magic angle, that is, 54.7°. Adapted from reference 26.

9.0. ROUGHENED SILVER ELECTRODES

In many applications of fluorescence, it might be advantageous to have localized silver deposition for spatially selective analysis, such as lab-on-a-chip technologies. Localized silver colloid formation has been accomplished with reagents in laminar flow,[27] by nanolithography,[26-29] and by electroplating insulators. [30] Given the extensive use of roughened silver electrodes for surface-enhanced Raman scattering (SERS),[31,32] we investigated their use for spatially selective MEF due to the high surface areas of fractal-like structures.

Figure 26 shows the schematics of the method for preparing roughened electrodes. In a typical preparation, commercially available silver electrodes are placed in deionized water 10 mm apart. A constant current of 60 μA was supplied across the two electrodes for 10 minutes by a constant current generator. Figure 27 shows the time dependent growth of the silver nanostructures on the silver cathode. In comparison, the anode was relatively unperturbed.

Binding the ICG-HSA to both the silver anode and cathode after electrolysis was accomplished by soaking the electrodes in a solution of ICG-HSA overnight, followed by rinsing with water to remove the unbound material. As a control sample, an unused silver electrode was also coated with ICG-HSA. A roughened silver cathode was also dipped in 10^{-4} M NaCl for 1 hour before washing and then coated with ICG-HSA, so as to place our findings in context with the huge enhancements in Raman signals, typically obtained after chloride dipping the electrodes.[33-34]

Three electrodes were coated with ICG-HSA and studied; the roughened cathode, the anode, and an unroughened electrode. Essentially no emission was seen from ICG-HSA on an unroughened, bright silver surface (Ag in Figure 28 (Top), the control). However, a dramatically larger signal was observed on the roughened cathode and a somewhat smaller signal was observed on the anode. In all our experiments we typically found that the roughened silver cathode was ~ 20–100 fold more fluorescent than the unroughened control Ag electrode. In comparison the Ag anode was 5–50 times more fluorescent than the Ag control. When we increased the time-for-roughening to over 1 h, the intensities of both electrodes after coating with ICG-HSA were essentially the same, but still 50-fold more fluorescent than the unroughened Ag control. The emission spectra on the two electrodes probably had the same emission maximum, where the slight shift seen in Figure 28 (Bottom) is thought to be due to the filters used to reject the scattered light. It should be noted that the amount of material coated on both surfaces was approximately the same, and the effect was not due to an increased surface area and therefore increased protein coverage on the roughened surface. The dramatic and favorable increase in fluorescence intensities shown in Figure 28 could have several explanations. Two possibilities include an increased rate of excitation, due to the enhanced incident field around the metal, or increased amounts of protein bound to the fractal structure. For both eventualities the fluorescence lifetimes are expected to remain the same. Examining the intensity decay of ICG-HSA for the cathode we found that the lifetime was dramatically shortened to < 10 ps, (data not shown), in fact, so short that it was difficult to determine the absolute values with a system time-resolution of ~ 50 ps fwhm. However, a decreased lifetime with increased fluorescence intensity strongly supports an increase in the radiative decay rate, Γ_m.

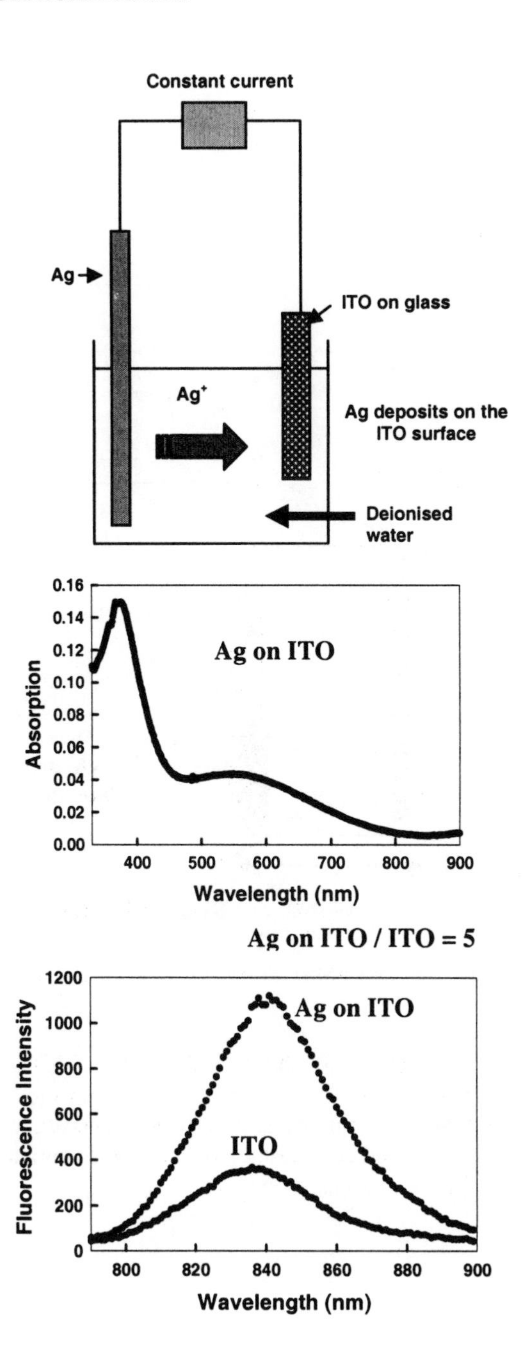

Figure 25. Electroplating of silver on substrates (Top). Absorption spectrum of electroplated silver on substrates (Middle), fluorescence emission spectra of ICG-HSA on ITO and silver deposited on ITO (Bottom). Adapted from 26.

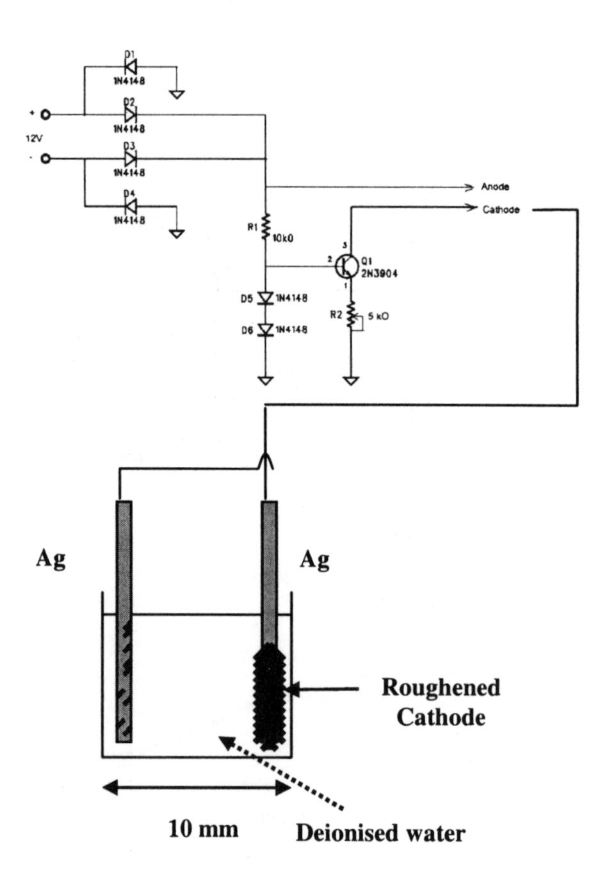

Figure 26. Experimental setup for the production of roughened silver electrodes. Adapted from reference 35.

Interestingly, Figure 1 (Top-Right) shows and image of ICG-HSA coated silver electrodes revealing spatially-localized fluorescence *hot-spots*, which we believe are due to the areas demonstrating superior fluorescence enhancements. Indeed, this is likely to occur with most silver systems studied, and thus our measurements to date are typically spatially averaged over large excitation spot sizes.

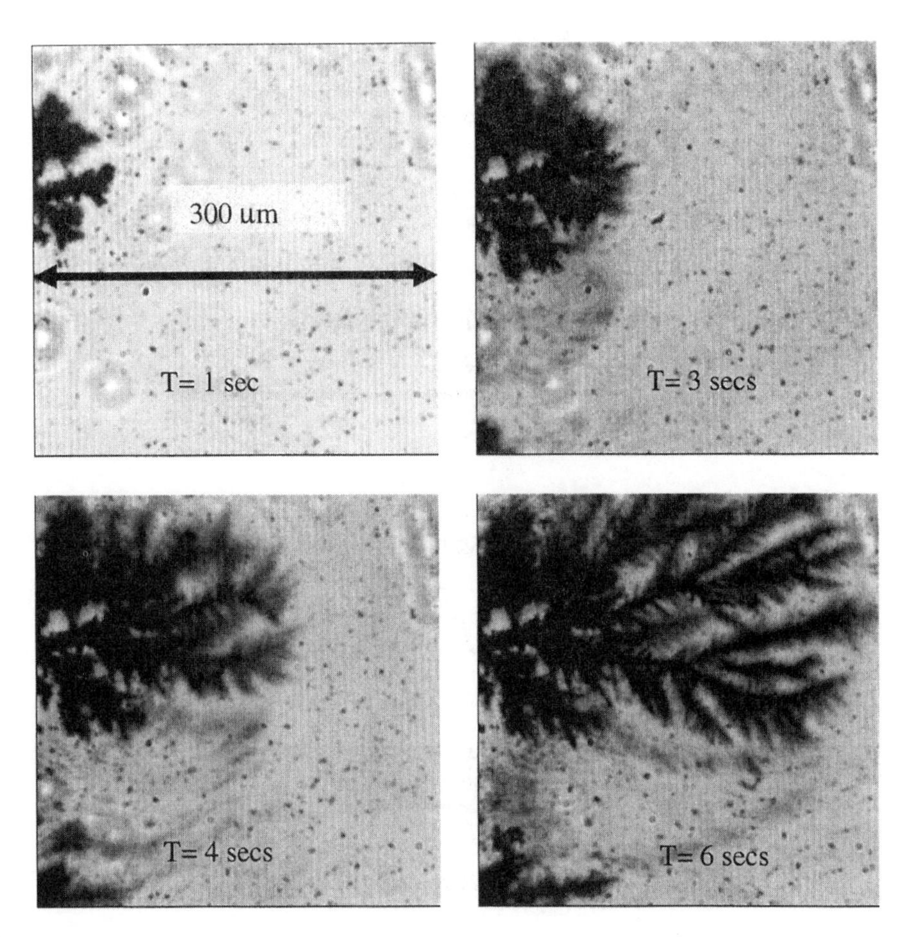

Figure 27. Fractal-like silver growth on the silver cathode as a function of time, visualized using transmitted light. This structure was characteristic of the whole electrode.

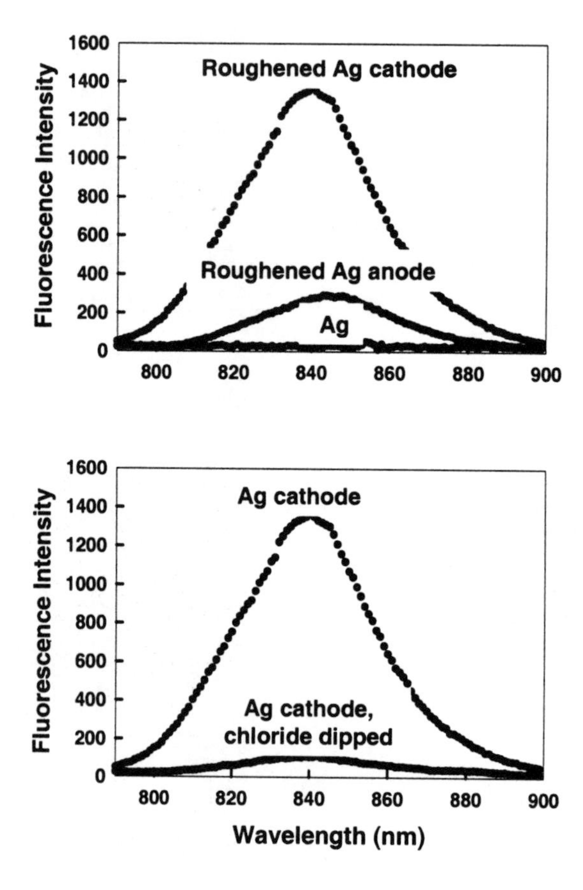

Figure 28. Fluorescence emission spectra of ICG-HSA on roughened silver electrodes (Top) and silver cathode (Bottom). Adapted from reference 35.

10. SILVER FRACTAL-LIKE STRUCTURES ON GLASS

Fractal-like silver structures were also generated on glass using two silver electrodes held between two glass microscope slides, Figure 29 (Top). The electrodes were 10 x 35 x 0.1 mm, with about 20 mm between the two electrodes. Deionized water was placed between the slides. A direct current of 10 μA was passed between the electrodes for about 10 min, during which the voltage started near 5 V and decreased to 2 V. During the current flow, fractal silver structures grew on the cathode and then on the glass near the cathode (Figure 29-Bottom), thus producing silver nanostructures on glass, compared to those on the silver electrodes as described in the previous section. Similarly to the silver electrodes, the structures grew rapidly but appeared to twist as they grew. These structures are similar to those reported recently during electroplating of insulators.[30] In addition we found that dipping the slides in 0.001 mg/dl $SnCl_2$ for 30 min, *before electrolysis*, resulted in structures that were firmly bound to the glass during working. Without the $SnCl_2$, similar structures were formed but were partially removed during washing. Following passage of the current, the silver structures on glass were soaked in 10 μM FITC-HSA overnight at 4°C, which is thought to result in a monolayer of surface bound HSA.

For the FITC-HSA-coated fractal-silver surfaces on glass, we were able to measure a fluorescence image very similar to that of the fractal silver surface alone (bright field image) using the same apparatus (Figures 30). Interestingly, regions of high and low fluorescence intensity were observed (Figure 30-Bottom). This result is roughly consistent with recent SERS data, which showed the presence of intense signals that appeared to be located between clusters of particles.[33-34]

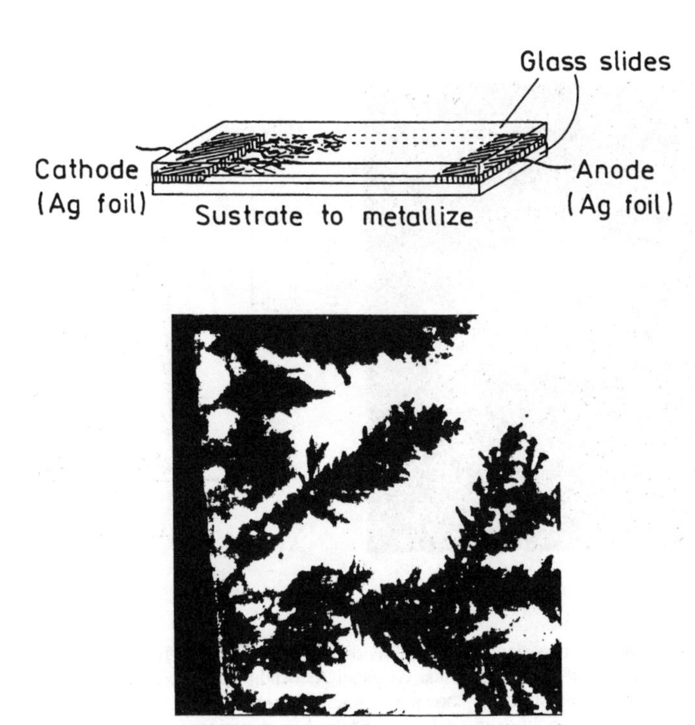

Figure 29. Configuration for creation of fractal-like silver surfaces on glass (Top) and bright-field image of fractal-like silver surfaces on glass (Bottom).

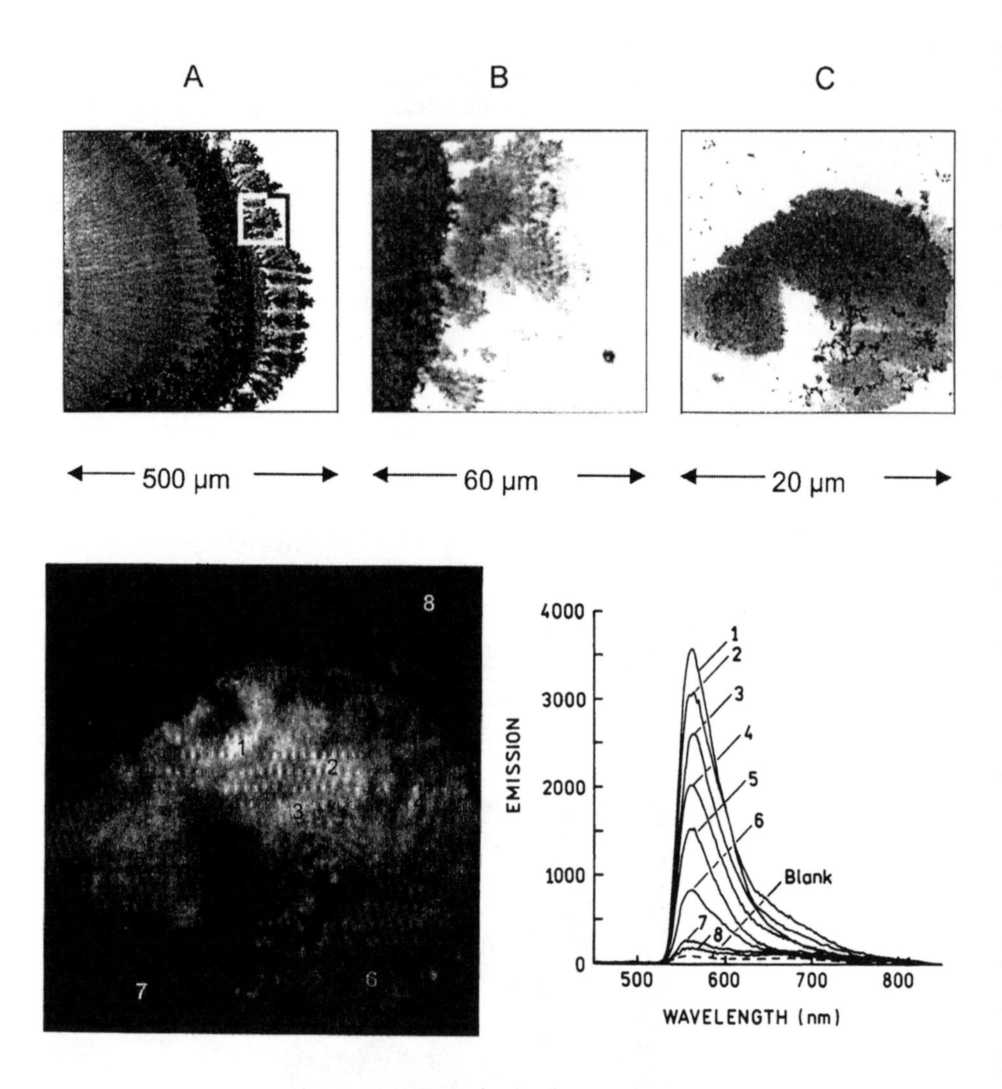

Figure 30. Silver nanostructures deposited on glass during electroplating (Top-Panel A). Panels B and C are consecutive magnification of the marked area on panel A. Bright-field image. Fluorescence image of FITC-HSA deposited on the silver structure shown in Panel C (Bottom-Left), and the emission spectra of the numbered areas shown on the right (Bottom-Right). Adapted from reference 35.

Emission spectra were collected from eight selected regions of varying brightness. In all cases, the emission spectra appeared to be that of fluorescein (Figure 30, Bottom-Right), where the blue edge of the fluorescein emission is cut off by the emission filter. As a control the silver structures were coated with unlabeled HSA. The resulting signal was substantially lower than any of the silvered areas and lower than regions of the unsilvered glass treated with FITC-HSA. Similarly to the roughened silvered electrodes, we investigated the nature of the enhanced fluorescence intensities observed in Figure 30 (Bottom-Right). If the radiative decay rate is increased, then the lifetime should decrease. We measured the frequency-domain intensity decays of FITC-HSA bound to unsilvered glass and fractal silver on glass (data not shown). The amplitude-weighted lifetime of FITC-HSA bound to glass is ~80 ps, which is in agreement with previous measurements of self-quenched fluorescein on HSA.[10] On fractal silver, the amplitude weighted lifetime is dramatically reduced to about ~ 3 ps. We carefully considered whether this decrease was due to the detection of scattered light. The background signal from unlabeled HSA on fractal silver was less than 1%. The emission filter combination of a 540-nm interference filter and a solution of CrO_4^{2-} / $Cr2O_7^{2-}$ were selected for low emission from the filter when exposed to scattered light from the sample. However, we do believe a small part of the increased intensity is due to the release of fluorescein self-quenching because of the silver surfaces, but this is estimated to be very small in comparison to the > 500-fold increases shown in Figure 30 (Bottom-Right).

We also studied the photostability of FITC on the fractal silver surface, silver island films, and uncoated quartz. Although the relative photobleaching is higher on fractal silver, the increased rate of photobleaching is less than the increase in intensity (Figure 31). From the areas under these curves we estimate ≈ 16-fold and ≈ 160-fold more photons can be detected from the FITC-HSA on SiFs or fractal silver, respectively, relative to quartz, before photobleaching.

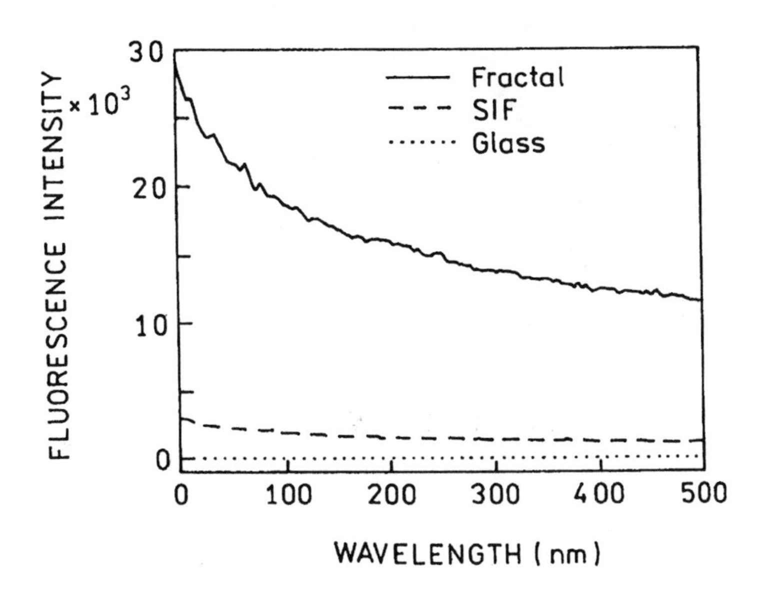

Figure 31. Photostability of FITC-HSA deposited surfaces. The samples were excited at 514 nm.

11. ROUGHENED PLATINUM ELECTRODES FOR MEF

In an effort to identify other metals for MEF we also examined metallic platinum. The surface of a strip of platinum (in essence an electrode) is roughened by oscillating voltages, analogous to the procedure used for surface enhanced Raman scattering (SERS).[31,32] In this regard, a standard 10 mm cuvette is filled with 1.0 M H_2SO_4 and 2 platinum foil (0.1 mm thick, 99.99%, Aldrich) electrodes are installed. The platinum electrodes are roughened by an electrical oxidation and reduction process, which involved potentio-dynamically cycling the voltage between the electrodes for 20 ms in the range +1.2 V to -0.4 V, for 1 min. The electrode is then washed with deionized water for 3 min and then the whole electrode is coated with ICG-HSA. For the roughening procedure, only half the Pt electrode is immersed in 1.0 M H_2SO_4, allowing the unroughened portion to act as a control sample when coated with ICG-HSA.

Figure 32 shows the emission spectra of ICG-HSA on a smooth and roughened platinum surface. We found no significant effect, about 2-fold lower intensity on the roughened surface as compared to the smooth surface. Since it is likely that the roughened surface binds more ICG-HSA than the smooth surface, the extent of quenching is probably even higher. These results suggest platinum will not be useful for metal-enhanced fluorescence, at least for long wavelength fluorophores.

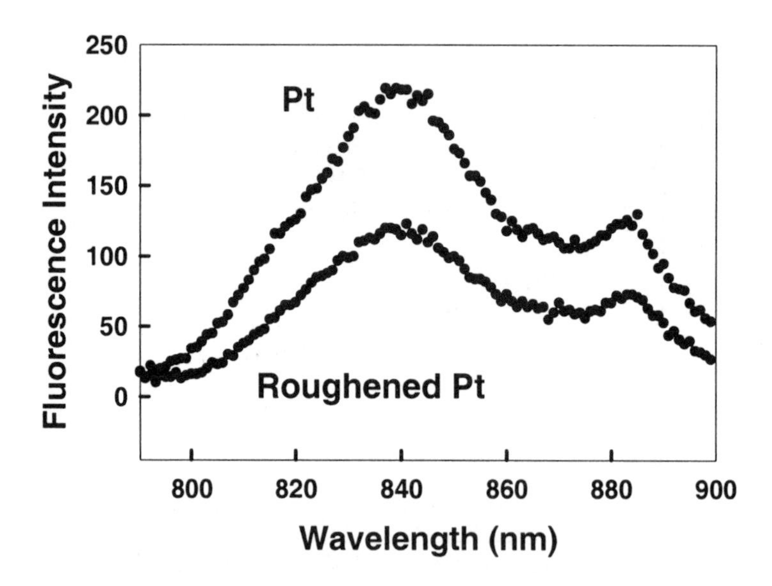

Figure 32. ICG emission on both roughened and unroughened platinum. Adapted from reference 36.

12. APPLICATIONS OF NOBEL METAL SURFACES TO DRUG DISCOVERY

Metal-enhanced fluorescence appears to be most suitable to the fluorescence assays used in drug discovery and DNA analysis. As we have shown, silver can be readily deposited on glass or polymer substrates by a variety of methods. Silver colloids are easily prepared and can be attached to surfaces functionalized with amine or sulfhydryl groups. We can imagine the bottom of multi-well plates or DNA arrays being coated with silver particles. A variety of new assay formats are possible. Assays could be based on the lightning rod effect, i.e., E_m modifications. The biochemical affinity interactions could bring the fluorophore close to the metal surface, for localized excitation, eliminating the washing steps. Another approach could be to use low quantum yield fluorophores, and the increased quantum yield of fluorophores brought in close proximity with the metal (Figure 33, Top). These effects might be coupled with another remarkable property of metal-fluorophore interactions. If the metal is close to a semi-transparent metallic surface the emission can couple into the metal and become directional rather than isotropic (Figure 33-Bottom).

Metal-enhanced fluorescence is not limited to single metallic particles. Theory predicted that the lightening rod effect is much stronger between two metallic spheres than for isolated spheres. If the biochemical affinity reaction brings particles into closer proximity then excitation fields may be substantially increased in the spaces between particles (Figure 34-Top). Multi-photon excitation is also known to be increased near metallic surfaces, and may be even more efficient between metal particles. Also, proximity to the particles can result in long range energy transfer (Figure 34-Bottom), which would allow selective detection of macromolecule complexes.

In this review Chapter, we have summarized the preparation techniques; those which were developed in our laboratories, for surfaces coated with noble-metal nanostructures for applications in metal-enhanced fluorescence. We have shown the favorable effects of enhanced fluorescence emission intensity, (quantum yield), reduced lifetime (increased photostability) for fluorophores in close proximity to appropriately sized metallic silver nanostructures. Our findings have indeed revealed that >> 500-fold enhancement in emission intensity can be realized, which is likely to offer multifarious applications to medical diagnostics and the clinical sciences. Metal-fluorophore interactions are relatively unexplored phenomenon, but we believe will receive much academic and industrial interest in the future.

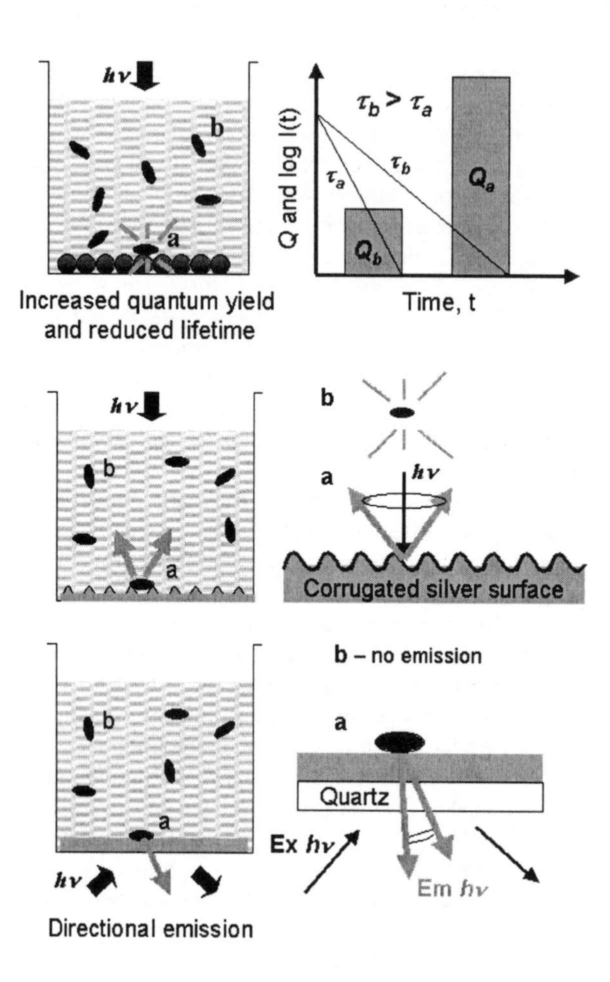

Figure 33. Potential uses of metal-enhanced fluorescence in drug discovery based on local increases in quantum yield (top) or directional emission (bottom). Adapted from reference 68.

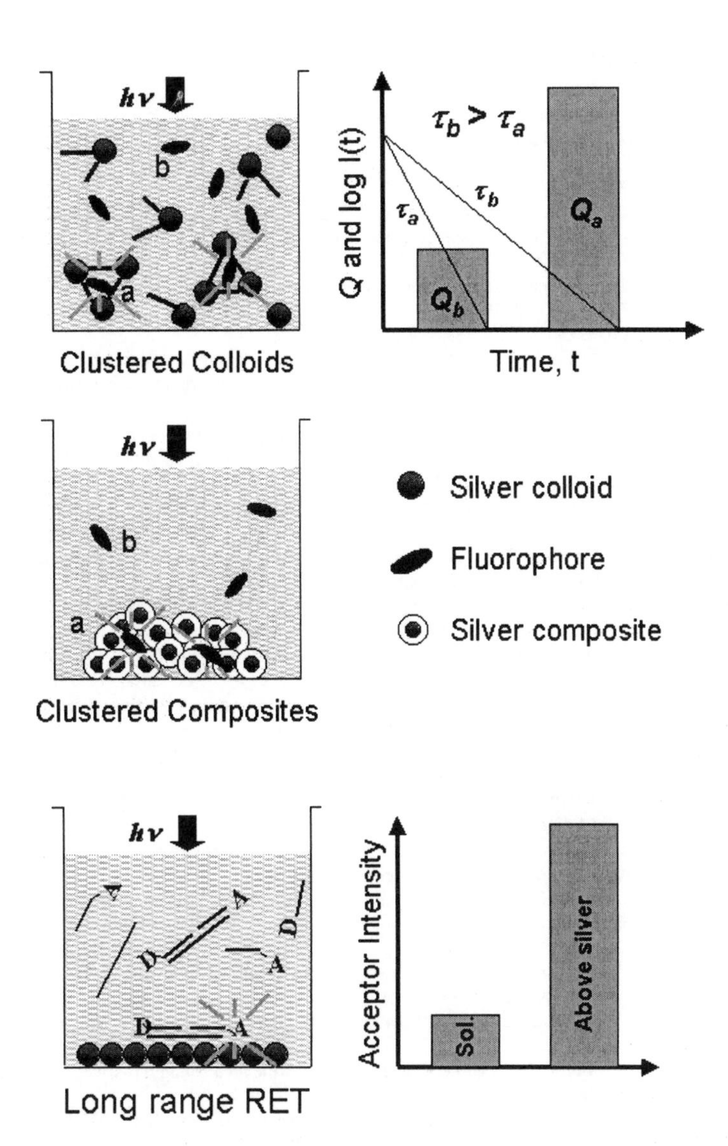

Figure 34. Potential uses of metal-enhanced fluorescence based on colloid clustering multi-photon excitation (top) or long range RET (bottom). Adapted from reference 37.

13. ACKNOWLEDGMENTS

This work was supported by the NIH, National Center for Research Resources, RR-01889. Financial support to J. R. Lakowicz and I. Gryczynski from the University of Maryland Biotechnology Institute is also gratefully acknowledged.

14. REFERENCES

1. K. H. Drexhage, Interaction of light with monomolecular dye lasers. *In Progress in Optics,* edited by Wolfe, E. (North-Holland, Amsterdam, 1974), pp. 161–232.
2. R. R. Chance, A. Prock, and R. Silbey, Molecular fluorescence and energy transfer near interfaces. *Adv. Chem. Phys.* **37,** 1–65 (1978).
3 D. A. Weitz, S. Garoff, J. I. Gersten, and A. Nitzan, The enhancement of Raman scattering, resonance Raman scattering, and fluorescence from molecules absorbed on a rough silver surface. *J. Chem. Phys.* **78**(9), 5324–5338 (1983)
4. J. Gersten, A. Nitzan, Spectroscopic properties of molecules interacting with small dielectric particles, *J. Chem. Phys.* **75**, 1139–1152 (1981).
5. P. C. Das, H. Metiu, *J. Chem. Phys.* **89**, 4680–4689 (1985)
6. C. D. Geddes and J. R. Lakowicz, Metal-enhanced fluorescence, *J. Fluoresc.* **12**(2),121–129 (2002)
7. J. R. Lakowicz, Radiative decay engineering: Biophysical and biomedical applications, *Appl. Biochem.* **298**, 1–24 (2001).
8. J. R. Lakowicz, Y. Shen, S. D'Auria, J. Malicka, J. Fang, Z. Grcyzynski and I. Gryczynski, Radiative decay engineering 2. Effects of silver island films on fluorescence intensity, lifetimes, and resonance energy transfer, *Anal.Biochem.* **301**, 261–77 (2002).
9. J. Malicka, I. Gryczynski, C. D. Geddes, J. R. Lakowicz, Metal-enhanced emission from indocyanince green: a new approach to in vivo imaging, *J Biomedical Optics*, **8(3)**,472-478 (2003).
10. J. R. Lakowicz, J. Malicka, S. D'Auria and I. Gryczynski, Release of the self-quenching of fluorescence near silver metallic surfaces, *Anal. Biochem.* **320**, 13–20 (2003).
11. J. R. Lakowicz, J. Malicka and I. Gryczynski, I., Silver particles enhance emission of fluorescent DNA oligomers, *BioTechniques.* **34**, 62–68 (2001).
12. Th. Forster, Intermolecular energy migration and fluorescence *Ann Phys* **2** 55–75 (1948) (Transl. Knox R S, Department of Physics and Astronomy, University of Rochester, Rochester, NY 14627)
13. K. Sokolov, G. Chumanov, T. M. Cotton, Enhancement of molecular fluorescence near the surface of colloidal metal films. *Anal. Chem.* **70**, 3898-3905 (1998).
14 J. Turkevich, P. C.Stevenson, J. Hillier, A study of the nucleation and growth processes in the synthesis of colloidal gold. *J. Discuss. Faraday Soc.* **11**, 55-75 (1951).
15. A. Henglein, A., Giersig, Formation of colloidal silver nanoparticles: capping action of citrate. *J. Phys. Chem. B* **103**, 9533-9539 (1999).
16 L. Rivas, S. Sanchez-Cortes, J. V. Garcia-Romez, G. Morcillo, Growth of silver colloidal particles obtained by citrate reduction to increase the Raman enhancement factor. *Langmuir* **17**, 574-577 (2001).
17. C. D. Geddes, H. Cao I. Gryczynski, Z. Gryczynski, J. Fang, J. R. Lakowicz . Metal-enhanced fluorescence due to silver colloids on a planar surface: potential applications of Indocyanine green to in vivo imaging. J Phys Chem A; **107**, 3443-3449 (2003).
18. K. Aslan, J. R. Lakowicz, and C. D. Geddes, Deposition of silver nanorods on to glass substrates and applications in metal-enhanced fluorescence, J. Phys. Chem. B (submitted).
19. G. C. Schatz, R. P. van Duyne. In Handbook of Vibrational Spectroscopy; J. M. Chalmers, P. R. Griffiths, Eds; Wiley: New York, 2002.
20. S. Chen, D. L. Carroll. Synthesis and characterization of truncated triangular silver nanoplates. Nano Letters **2(9)**, 1003-1007 (2003).
    rhodamine 6G molecules on large Ag nanocrystals. *J. Am. Chem. Soc.* **121,** 9932–9939 (1999).
21. W. C. Bell and M. L. Myrick, J. Colloid Interface Sci. **242,** 300 (2001).
22. G. Rodriguez-Gattorno, D. Diaz, L. Rendon, and G. O. Hernandez- Segura, J. Phys. Chem. B **106,** 2482 (2002).
23. J. P. Abid, A. W. Wark, P. F. Breve, and H. H. Girault, Chem. Commun. **7,** 792 (2002).
24. I. Pastoriza-Santos, C. Serra-Rodriguez, and L. M. Liz-Marzan, J. Collloid Interface Sci. **221,** 236 (2002).
25. C. D. Geddes, A. Parfenov, and J. R. Lakowicz, Photodeposition of silver can result in metal-enhanced fluorescence, *Applied Spectroscopy* **57**(5), 526-531 (2003).
26. C. D. Geddes, A. Parfenov, D. Roll, J. Fang, J. R. Lakowicz, Electrochemical and laser deposition of silver for use in metal-enhanced fluorescence, *Langmuir*, 19(15), 6236-6241 (2003).
27. R. Keir, E. Igata, M. Arundell, W. E. Smith, D. Graham, C. McHugh, and J. M. Cooper, SERRS: In situ substrate formation and improved detection using microfluidics. *Anal. Chem.* **74**(7), 1503–1508 (2002).

28. C. L. Haynes, A. D. McFarland, M. T. Smith, J. C. Hulteen, and R. P. Van Duyne, Angle-resolved nanosphere lithography:Manipulation of nanoparticle size, shape, and interparticle spacing. *J. Phys. Chem. B* **106,** 1898–1902 (2002).
29. F. Hua, T. Cui, and Y. Lvov, Lithographic approach to pattern self-assembled nanoparticle multilayers. *Langmuir* **18**, 6712–6715 (2002).
30. V. Fleury, W. A. Watters, L. Allam, and T. Devers, Rapid electroplating of insulators, *Nature* **416,** 716–719 (2002).
31. M. Fleischmann, P. J. Hendra, and A. J. McQuillan, Raman spectra of pyridine adsorbed at a silver electrode. *Chem. Phys. Letts.* **26**(2), 163–166 (1974)
32. E. Roth, G. A. Hope, D. P. Schweinsberg, W. Kiefer, and P. M. Fredericks, Simple technique for measuring surface-enhanced fourier transform Raman spectra of organic compounds.*Appl. Spec.* **47**(11), 1794–1800 (1993)
33. A. M. Michaels, J. Jiang, and L. Brus, Ag nanocrystal junctions as the site for surface-enhanced Raman scattering of single rhodamine 6G molecules. *J. Phys.Chem. B.* **104,** 11965–11971 (2000).
34. A. M. Michaels, M. Nirmal, and L. E. Brus Surface enhanced Raman spectroscopy of individual rhodamine 6G molecules on large Ag nanocrystals. *J. Am. Chem. Soc.* **121,** 9932–9939 (1999).
35. A. Parfenov, I. Gryczynski, J. Malicka, C. D. Geddes, J. R. Lakowicz, Enhanced fluorescence from fluorophores on fractal silver surfaces, *J Phys Chem B*, 107(34), 8829–8833 (2003).
36. C. D. Geddes, A. Parfenov, D. Roll, J. Uddin, J. R. Lakowicz. Fluorescence Spectral Properties of Indocyanine Green on a Roughened Platinum Electrode: Metal-Enhanced Fluorescence. J. Fluores. 13(6), 453-457 (2003).
37. C. D. Geddes, I. Gryczynski, J. Malicka, Z. Gryczynski, J. R. Lakowicz, Metal-Enhanced Fluorescence: Potential Applicatiosn in HTS, *Comb Chem and High Throughput Scr* , 6:109-117 (2003).
38. C. D. Geddes, A. Parfenov, D. Roll, I. Gryczynski, J. Malicka and J. R. Lakowicz, Silver fractal-like structures for metal-enhanced fluorescence: Enhanced fluorescence intensities and increased probe photostabilities, *J. Fluoresc.* **13**(3), 267–276 (2003)

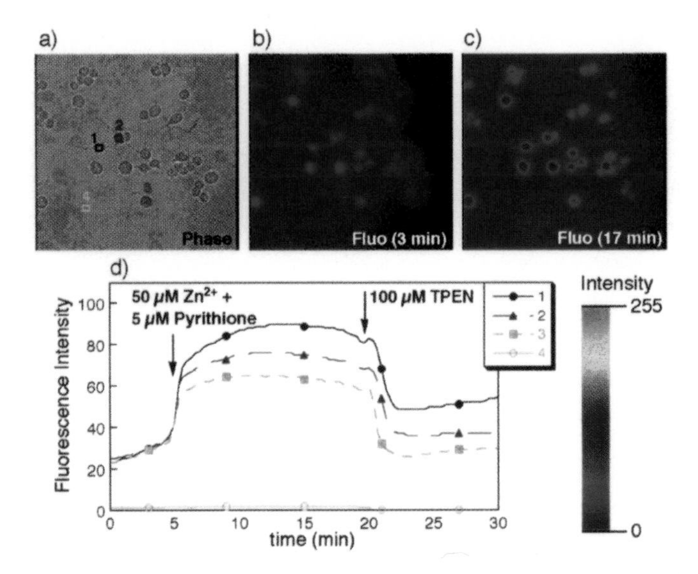

Chapter 4, Figure 12, Page 66. a) Bright-field and b) c) fluorescence images of RAW 264.7 loaded with ZnAF-2 DA. The cells were incubated with 10 μM ZnAF-2 DA for 0.5 hours at 37°C. Then the cells were washed with PBS and the fluorescence excited at 470-490 nm was measured at 20 s intervals. At 5 min, 5 μM pyrithione and 50 μM $ZnSO_4$ were added to the medium, and 100 μM TPEN was added at 20 min. Fluorescence images are shown in pseudo-color and correspond to the fluorescence intensity data in d), which shows the average intensities of the corresponding areas (1 – 3: intracellular region, 4: extracellular region).

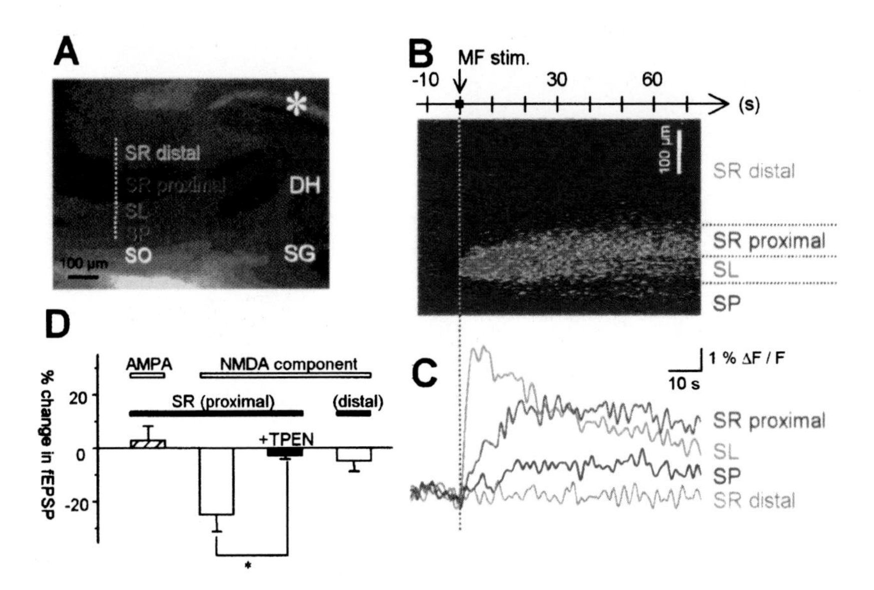

Chapter 4, Figure 13, Page 66. Extracellular Zn^{2+} release, diffusion and heterosynaptic inhibition of NMDA receptors in stratum radiatum after MF activation. A, An image of the dentate-CA3 area of a hippocampal slice perfused with ZnAF-2. The confocal ZnAF-2 signal is shown as a green-colored scale, superimposed on a transmitted beam image. Bipolar electrodes (•) were placed in the stratum granulosum (SG) to stimulate the MFs. The dotted line marks the transect of illumination during line-scan imaging. B, Line-scan image of ZnAF-2 taken at the points indicated in A. The temporal resolution was 1 s per line. "Hotter" colors correspond to increased $[Zn^{2+}]_o$ on an arbitrary pseudo-color scale. C, Data extracted from the image in B, along the time axis. Each point in time is the average %ΔF/F value across the spatial axis of the region separated by the horizontal dotted lines in B, i.e., the stratum radiatum far from stratum lucidum (SR distal, brown), the stratum radiatum near stratum lucidum (SR proximal, green), stratum lucidum (SL, red) and stratum pyramidale (SP, blue). The MFs were tetanized at 100 Hz for 2 s (MF stim.) at the time indicated by the vertical dotted line. D, Summary data for the effect of MF stimulation on AMPA and NMDA responses in proximal and distal stratum radiatum (SR). The ordinate indicates the average change in fEPSPs in associational/commissural fiber-CA3 pyramidal cell synapses 15 s after MF stimulation (100 Hz for 2 s). $^{*}p < 0.01$; Student's *t*-test. Data are means ± SEM of . 5-7 slices.

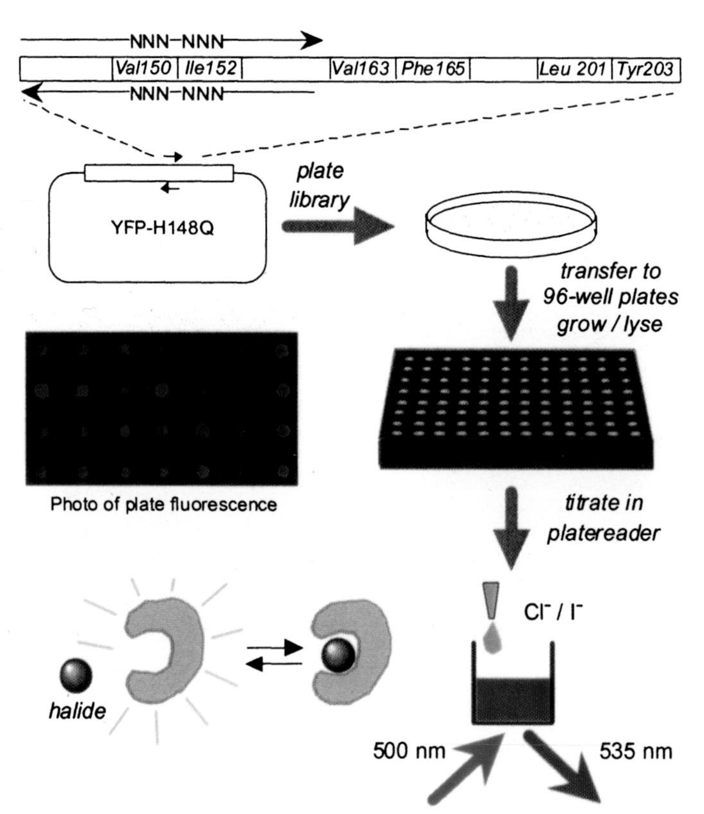

Chapter 6, Figure 2, Page 90. Strategy for generating and screening of YFP mutational libraries. Indicated pairs of amino acids were randomly mutated using appropriate primers. Transformed bacterial colonies were transferred to 96–well microplates for growth, replication to agar plates, lysis and screening. Left, middle: Photograph of a section of a replicate agar plate showing bacteria with differing amounts of fluorescence. Adapted from ref. 34.

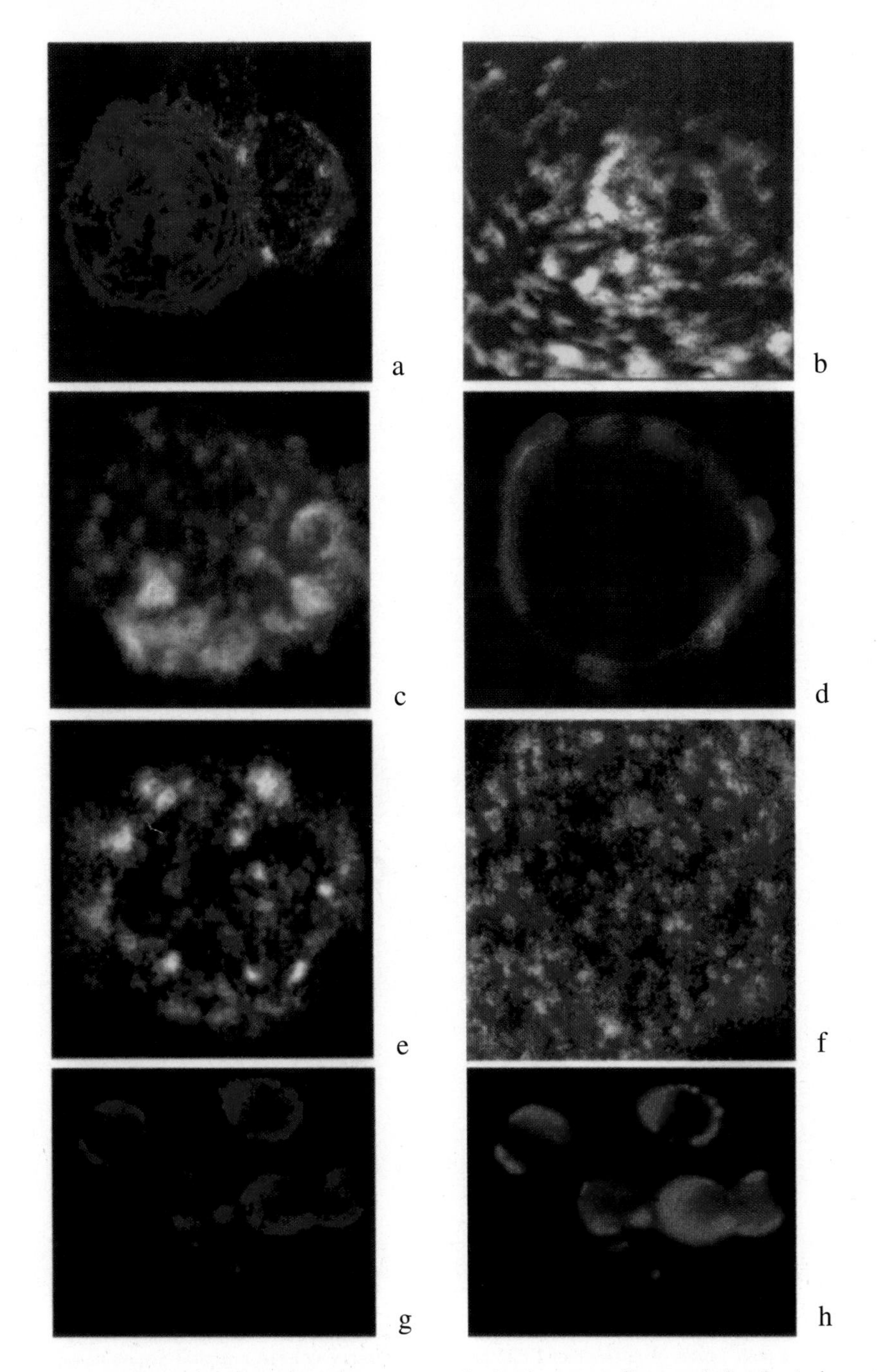

Chapter 7, Figure 2, Page 106. Fluorescence confocal microscopy allows analysis of submicrometer molecular colocalization of membrane constituents in live cells. (**a.**) Immunological synapse between human B cells (antigen presenting cell; green) and cytotoxic T lymphocytes (red). (**b.**) SNOM image of a human B cell surface: intact MHC-I molecules (labeled with SF-X-W6/32 mAb, green) and β2m-free MHC heavy chains (labeled with Cy5-HC10 mAb; red) are shown in a 15x15μm representative area. There is a substantial overlap (yellow color) between the two labels (cross-correlation coefficient: 0.771) suggesting high degree of molecular co-clustering of the two forms of MHC-I. (**c.**) Colocalization of GM1 gangiosides (labeled with Alexa488CTX; green) and MHC-II (I-E^d) molecules (labeled with TM-rhodamine-Ab; red) in murine B cells (3D reconstruction, overlap in yellow color). (**d.**) Central optical slice of a cell labeled as in (**c.**), but after cholesterol depletion of the PM. Note the remarkable segregation of the labeled species upon raft-disruption. (**e.**) IL-2R alpha subunit and MHC class I double labeled with red and green fluorophores for confocal microscopy. The overlaping clusters of these proteins result in yellow spots. (**f.**) IL-2R alpha (green) and the transfrerrin receptor (red) exhibit clusters of distinctly different size that hardly overlap. (**g.**) GPI-linked CD48 proteins on a T leukemia cell are tagged with specific monoclonal Alexa488-conjugated Fab and crosslinked with GAMIG to induce capping. (**h.**) IL-2R alpha subunit labeled with Cy3-conjugated anti-Tac Fab on the same cell show identical localization with CD48 indicating co-capping of the two raft-resident proteins.

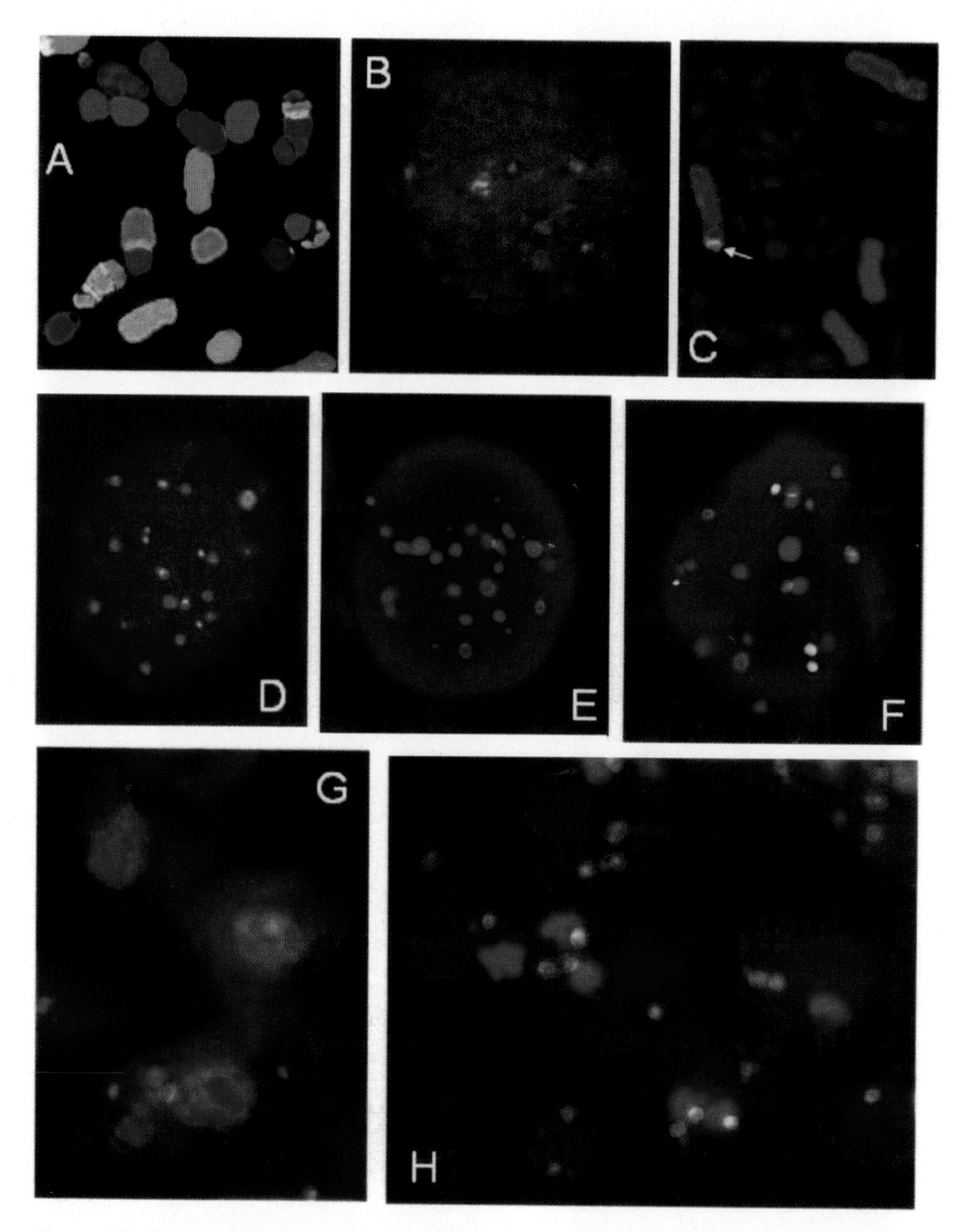

Chapter 11, Figure 4, Page 236. Color composite images of metaphase chromosome spreads and interphase nuclei hybridized with panels of multi-color FISH probes. A: Pseudo-colored image of a metaphase spread hybridized with the SpectraVysion™ M-FISH probe set. Each color represents a different label combination, and therefore a different chromosome. B: Image of a single blastomere (XY, -13,+21) hybridized at Reproductive Genetics Institute, Chicago, IL, with the MultiVysion™ PGT probe set. C: Image of metaphase chromosomes from a patient with an abnormal chromosome 4 hybridized with SO-WCP® 8 and SGr WCP® 4. D: Image of a bladder cell isolated from the urine of a patient monitored for recurrence, and hybridized with the 4-color UroVysion™ probe set. E: Image of a cell from a bronchial washing of a patient with lung cancer hybridized with SR-LSI®EGFR, SG-LSI® 5p15, SGo-LSI® MYC, and SA-CEP® 1. F: Image of a cell from ductal lavage fluid taken from a patient with breast cancer and hybridized with the Vysis Breast Cancer Aneusomy Probe Set. G: Image of cells in a formaldehyde-fixed paraffin-embedded breast tumor specimen hybridized with SO-LSI® TOP2A, SGr-LSI® HER2, and SA-CEP® 17. H: Image of cells in a formaldehyde-fixed paraffin-embedded larynx tumor specimen hybridized with SO-LSI® EGFR, and SGr-CEP® 7. All probe sets are from Vysis/Abbott Laboratories. Metaphase chromosomes and interphase nuclei in parts C through H are stained blue with DAPI counterstain.

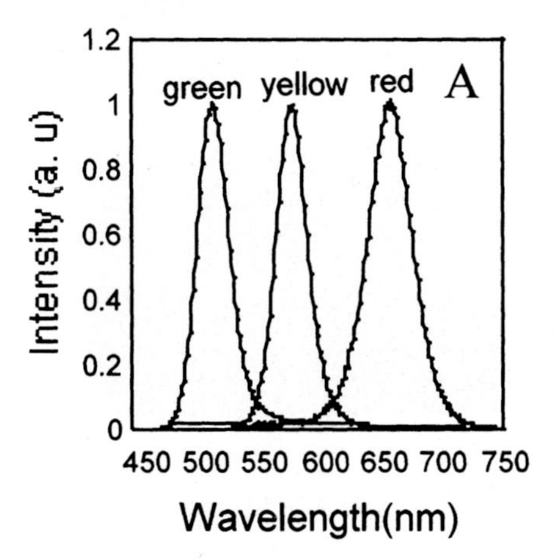

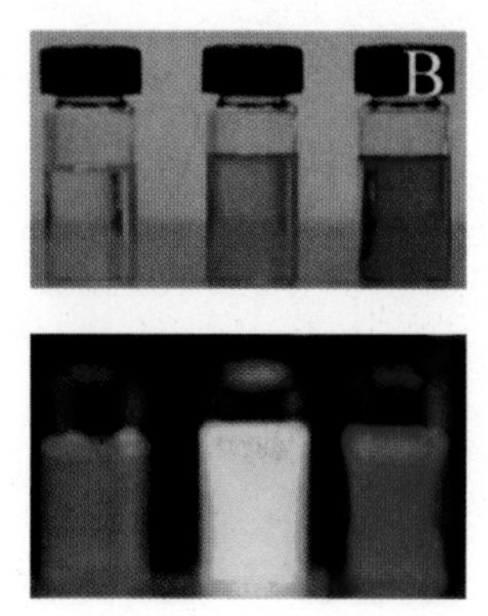

Chapter 12, Figure 1, Page 249. (A) Emission spectra of CdSe-ZnS QDs of 3 (green), 4 (yellow) and 6 (red) nm in diameter. (B) Color photos of QDs solution illuminated with visible (up) and UV (down) light.

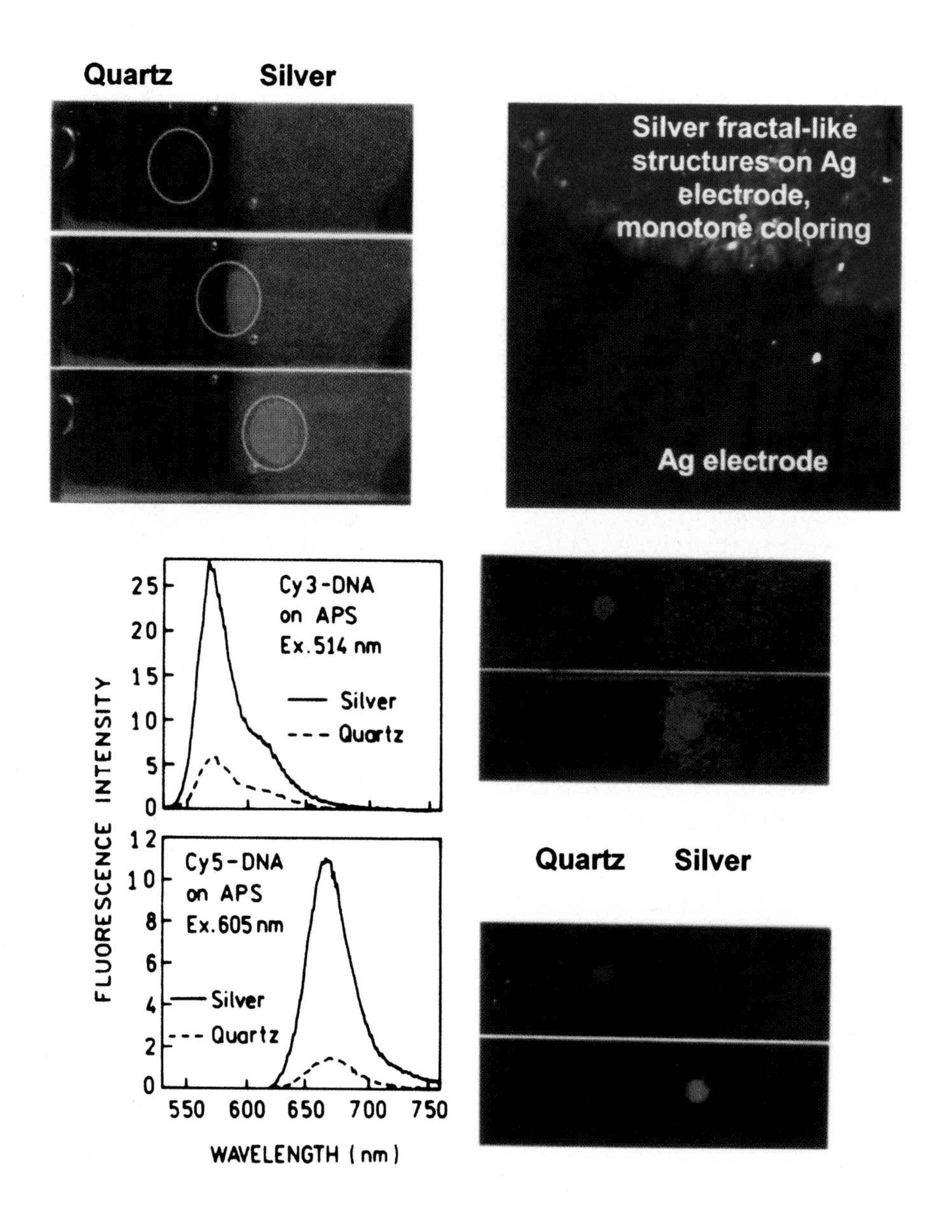

Chapter 16, Figure 1, Page 366. Photograph of fluorescein-labeled HSA (molar ratio of fluorescein/HSA =7) on quartz (Top-Left) and on SiFs (Top-Left) as observed with 430-nm excitation and a 480-nm long-pass filter. The excitation was progressively moved from the quartz side to the silver, Top to Bottom, respectively (Reference 10). Silver fractal-like structures grown on silver electrodes (Top-Right) (Reference 38). Emission spectra of Cy3-DNA (Bottom-Left) and Cy5-DNA (Bottom-Left) between quartz plates with and without SiFs. Photographs of corresponding fluorophores (Bottom-Right) (Reference 11).

INDEX